Edible Coatings and Films to Improve Food Quality

SECOND EDITION

Edible Coatings and Films to Improve Food Quality

SECOND EDITION

EDITED BY
Elizabeth A. Baldwin
Robert Hagenmaier • Jinhe Bai

CRC Press
Taylor & Francis Group
Boca Raton London New York

CRC Press is an imprint of the
Taylor & Francis Group, an **informa** business

Cover photo credits, from top to bottom, left to right: Peggy Greb, USDA, Agricultural Research Service; USDA, Anita Daniels, Agricultural Research Service; Shutterstock Images LLC; Tracy Griffith, NewGem Foods; Guiwen Cheng, JBC FoodTech; Shutterstock Images LLC.

CRC Press
Taylor & Francis Group
6000 Broken Sound Parkway NW, Suite 300
Boca Raton, FL 33487-2742

© 2012 by Taylor & Francis Group, LLC
CRC Press is an imprint of Taylor & Francis Group, an Informa business

No claim to original U.S. Government works

Printed in the United States of America on acid-free paper
Version Date: 20110517

International Standard Book Number: 978-1-4200-5962-5 (Hardback)

This book contains information obtained from authentic and highly regarded sources. Reasonable efforts have been made to publish reliable data and information, but the author and publisher cannot assume responsibility for the validity of all materials or the consequences of their use. The authors and publishers have attempted to trace the copyright holders of all material reproduced in this publication and apologize to copyright holders if permission to publish in this form has not been obtained. If any copyright material has not been acknowledged please write and let us know so we may rectify in any future reprint.

Except as permitted under U.S. Copyright Law, no part of this book may be reprinted, reproduced, transmitted, or utilized in any form by any electronic, mechanical, or other means, now known or hereafter invented, including photocopying, microfilming, and recording, or in any information storage or retrieval system, without written permission from the publishers.

For permission to photocopy or use material electronically from this work, please access www.copyright.com (http://www.copyright.com/) or contact the Copyright Clearance Center, Inc. (CCC), 222 Rosewood Drive, Danvers, MA 01923, 978-750-8400. CCC is a not-for-profit organization that provides licenses and registration for a variety of users. For organizations that have been granted a photocopy license by the CCC, a separate system of payment has been arranged.

Trademark Notice: Product or corporate names may be trademarks or registered trademarks, and are used only for identification and explanation without intent to infringe.

Library of Congress Cataloging-in-Publication Data

Edible coatings and films to improve food quality / editors, Elizabeth A. Baldwin, Robert Hagenmaier, Jinhe Bai. -- 2nd ed.
    p. cm.
Includes bibliographical references and index.
ISBN 978-1-4200-5962-5 (hardback)
1. Edible coatings. 2. Food--Quality. I. Baldwin, Elizabeth A. II. Hagenmaier, Robert D. III. Bai, Jinhe.

TP451.E3E35 2012
664--dc22   2011006378

Visit the Taylor & Francis Web site at
http://www.taylorandfrancis.com

and the CRC Press Web site at
http://www.crcpress.com

Contents

Preface		vii
The Editors		ix
The Contributors		xi
1	**Introduction** Elizabeth A. Baldwin and Robert D. Hagenmaier	1
2	**Protein-based films and coatings** Maria B. Pérez-Gago	13
3	**Edible coatings from lipids, waxes, and resins** David J. Hall	79
4	**Polysaccharide coatings** Robert Soliva-Fortuny, María Alejandra Rojas-Graü, and Olga Martín-Belloso	103
5	**Gas-exchange properties of edible films and coatings** Robert D. Hagenmaier	137
6	**Role of edible film and coating additives** Roberto de Jesús Avena-Bustillos and Tara H. McHugh	157
7	**Coatings for fresh fruits and vegetables** Jinhe Bai and Anne Plotto	185
8	**Coatings for minimally processed fruits and vegetables** Sharon Dea, Christian Ghidelli, Maria B. Pérez-Gago, Anne Plotto	243
9	**Applications of edible films and coatings to processed foods** Tara H. McHugh and Roberto de Jesús Avena-Bustillos	291

10	**Application of commercial coatings**	**319**
	Yanyun Zhao	
11	**Encapsulation of flavors, nutraceuticals, and antibacterials**	**333**
	Stéphane Desobry and Frédéric Debeaufort	
12	**Overview of pharmaceutical coatings**	**373**
	Anthony Palmieri III	
13	**Regulatory aspects of coatings**	**383**
	Guiwen A. Cheng and Elizabeth A. Baldwin	
	Index	417

Preface

Many interesting and exciting developments have occurred since the publication of the first edition in 1994. These developments result from the ever-increasing quality of coatings research and make it timely to prepare this updated and completely revised second edition, designed for the benefit of all involved in buying, selling, regulating, developing, or using coatings to improve the quality and safety of foods.

This book is a study of the coatings, films, wraps, and surface treatments used for foods. Specifically, it covers coating ingredients and additives (Chapters 2, 3, 4, and 6), their permeability properties (Chapter 5), coatings for specific applications (Chapters 7, 8, 9, and 12), the technology of coatings (Chapters 10 and 11), and regulatory aspects (Chapter 13). But, of course, the world of coatings could not be so neatly divided, so some overlap between chapters is inevitable and desirable. For example, in the discussion of a coating made with a certain ingredient, the author might include information on additives used or allowed, how the coating was applied, what foods it is best used for, and so on. Therefore, in researching any topic, it is best to cross-search other chapters.

The editors and authors hope that this book will serve as a useful reference for all aspects of the coating field. We expect that the information within will lead to further developments in this field for food and pharmaceutical products that reduce plastic waste, improve applications, lead to greater efficacy, make regulatory decisions easier in a global climate, and ultimately, economically maintain quality of food and pharmaceutical products.

The Editors

Elizabeth A. Baldwin is currently research leader and research horticulturist of the U.S. Department of Agriculture, Agricultural Research Service (USDA-ARS), Citrus and Subtropical Products Laboratory in Winter Haven, Florida. Her research interests include postharvest physiology and overall quality of fresh, fresh-cut, and processed fruits and vegetables, with an emphasis on the use of edible coatings and flavor quality of citrus, tomatoes, and tropical/subtropical products. She received a BA in anthropology from Hunter College, City University of New York; a BS in plant and soil science from Middle Tennessee State University, and a MS and PhD in horticulture from the University of Florida.

Robert D. Hagenmaier worked until retirement as a research chemist for USDA-ARS, Citrus and Subtropical Products Laboratory at Winter Haven, Florida. He holds a PhD in physical chemistry from Purdue University. His research interests focused first on coconut food products and later on how the quality of fresh fruit depends on permeability properties of coatings.

Jinhe Bai is a food technologist at USDA-ARS, Citrus and Subtropical Products Laboratory at Winter Haven, Florida. He received a BS from Shanxi Agriculture University, China; MS from Northwest Agriculture University, China; and a PhD from Osaka Prefecture University, Japan, on the effects of modified atmosphere (MA) packaging on volatile production of fruits. His current research interests are focused on the development of controlled atmosphere (CA) storage, MA packaging and edible coating technologies, and the discovery of how internal and environmental factors influence metabolism and further impact flavor and nutritional quality of fruits and vegetables.

The Contributors

Roberto de Jesus Avena-Bustillos
Department of Biological and
 Agricultural Engineering
University of California, Davis
Davis, California
roberto.avena@ars.usda.gov

Jinhe Bai
Citrus and Subtropical Products
 Laboratory
U.S. Department of Agriculture,
 Agricultural Research Service
 (USDA-ARS)
Winter Haven, Florida
jinhe.bai@ars.usda.gov

Elizabeth A. Baldwin
Citrus and Subtropical Products
 Laboratory
U.S. Department of Agriculture,
 Agricultural Research Service
 (USDA-ARS)
Winter Haven, Florida
liz.baldwin@ars.usda.gov

Guiwen A. Cheng
JBC FoodTech
Lakeland, Florida
alvin.cheng@jbtc.com

Sharon Dea
Citrus and Subtropical Products
 Laboratory
U.S. Department of Agriculture,
 Agricultural Research Service
 (USDA-ARS)
Winter Haven, Florida
sharon.dea@ars.usda.gov

Frédéric Debeaufort
Université de Bourgogne–EMMA
 EA 581
Institut Universitaire de Technologie
Dijon, France
frederic.debeaufort@u-bourgogne.fr

Stéphane Desobry
Laboratoire d'Ingenierie des
 Biomolecules
Nancy-Université-INPL-ENSAIA
Vandoeuvre, France
stephane.desobry@ensaia.inpl-nancy.fr

Christian Ghidelli
Department of Postharvest
Instituto Valenciano de Investigaciones
 Agrarias–Fundación Agroalimed
Moncada (Valencia), Spain
christian.ghidelli@gmail.com

Robert D. Hagenmaier
Citrus and Subtropical Products Laboratory, retired
U.S. Department of Agriculture, Agricultural Research Service (USDA-ARS)
Winter Haven, Florida
rhagenmaier@verizon.net

David J. Hall
HDH Agri-Products
Tavares, Florida
djhall@hdhagri.com

Olga Martín-Belloso
Department of Food Technology
University of Lleida
Lleida, Spain
omartin@tecal.udl.cat

Tara H. McHugh
Processed Foods Research
Western Regional Research Center
U.S. Department of Agriculture, Agricultural Research Service (USDA-ARS)
Albany, California
tara.mchugh@ars.usda.gov

Anthony Palmieri III
Department of Pharmaceutics
University of Florida
Gainesville, Florida
palmieri@cop.ufl.edu

Maria B. Pérez-Gago
Department of Postharvest
Instituto Valenciano de Investigaciones Agrarias–Fundación Agroalimed
Moncada (Valencia), Spain
perez_mbe@gva.es

Anne Plotto
Citrus and Subtropical Products Laboratory
U.S. Department of Agriculture, Agricultural Research Service (USDA-ARS)
Winter Haven, Florida
anne.plotto@ars.usda.gov

María Alejandra Rojas-Graü
Department of Food Technology
University of Lleida
Lleida, Spain
margrau@tecal.udl.cat

Robert Soliva-Fortuny
Department of Food Technology
University of Lleida
Lleida, Spain
rsoliva@tecal.udl.cat

Yanyun Zhao
Department of Food Science and Technology
Oregon State University
Corvallis, Oregon
yanyun.zhao@oregonstate.edu

1 Introduction

Elizabeth A. Baldwin and Robert D. Hagenmaier

Contents

1.1	Definition of an edible coating	1
1.2	History of edible coatings	2
1.3	Uses for edible coatings	2
1.4	Components of edible coatings	4
1.5	Important properties of edible coatings	6
1.6	Determination of coating properties	7
1.7	Why read this book	7
References		9

1.1 Definition of an edible coating

It seems appropriate to start with a definition of *edible coatings*. Probably all authors of chapters in this book would agree that edible coatings are substances applied to the exterior of food so that the final product is fit for consumption. However, agreement on a specific definition may be difficult to reach. To some, coatings that are edible are those that are legal and safe to use on food products. To others, edibility might also require that the coated food be acceptable to consumers. Finally, others might say that edibility of coatings implies that they have nutritional value.

Furthermore, regulations differ in different countries. For example, fresh fruit processed in the European Union (EU), cannot be coated with morpholine-containing fruit coatings, although such coatings are routinely used in the United States (Hagenmaier, 2004). Morpholine has limited approval, because like other amines, it can react to form carcinogens (Hagenmaier, 2004; Kielhorn and Rosner, 1996). Attempts to make morpholine-free coatings in the United States are, in fact, underway (Hagenmaier, 2004; Hagenmaier and Baker, 1997). Nevertheless, the definition of edibility would, therefore, differ in the United States versus the EU.

Finally, consider the word *coatings*, which is usually taken to mean that the added substance forms an exterior layer on the object coated. However, the surface of some foods is chemically treated during processing with a substance that evaporates before consumption or soaks into the food, leaving no layer on the outer surface. Are these coatings? Finally, what is the difference between a *film*, a *coating*, and a *wrap* (Duan et al., 2007; Ravishankar et al., 2009)? Then on a microscale, there are substances used to encapsulate flavors, nutrients, and drugs (Chen et al., 2006b; Kaushik and Roos, 2007; Manojlovic et al., 2008; Porzio, 2004; Reineccius, 1994, 2009), partly through new techniques that move from micro- to nanotechnology (Chen et al., 2006a). Fortunately for the reader, the editors make no attempt to resolve these issues, leaving the chapter authors to interpret *edible coatings* as they wish.

1.2 History of edible coatings

Edible coatings have been used for centuries to protect foods and prevent moisture loss. The first recorded use was in China in the twelfth century on citrus (Hardenburg, 1967) and later in England using lard or fats, called *larding* (Contrereas-Medellin and Labuza, 1981) to prolong shelf life of meat products. Since the early to mid twentieth century, coatings have been used to prevent water loss and add shine to fruits and vegetables (Baldwin, 1994), as casings using collagen or collagen-like material for sausages (Becker, 1938, 1939), and as some sort of sugary coating on confectionaries, including chocolate (Biquet and Labuza, 1988). For example, with candies, shellac or protein coatings allow candies to "melt in your mouth and not in your hand," preventing the hand from becoming soiled by coloring matter used for identification and appearance or by the underlying chocolate (Dangaren et al., 2006; Hicks, 1961). Finally, gelatin has also long been used to coat meat products (Antoniewski et al., 2007).

1.3 Uses for edible coatings

The earliest use of edible coatings and the most use today is for fruits and vegetables, to reduce moisture loss and subsequent softening and shriveling (Woods, 1990) due to loss of turgor, and also to improve appearance by imparting shine (Bai et al., 2002, 2003a, 2003b; Baldwin, 1994). Fruits and vegetables undergo metabolic activities such as respiration, where the cells use oxygen and produce carbon dioxide; fruit also produce ethylene, which acts as a hormone to accelerate ripening and senescence; and fresh produce undergo transpiration, which is loss of water vapor from fruit intercellular spaces to the environment (Wills et al., 1981). For climacteric fruit (fruit that ripen after harvest), shelf life can be extended by storage in a controlled or modified atmosphere of relatively high carbon dioxide and low oxygen, which slows ripening reactions such as respiration and ethylene production (Kader, 1986). Atmosphere modification can be achieved by specialized storage chambers, packaging, or by edible coatings. Depending on their formulations, coatings applied to fruits or vegetables can have the same effect on fruit interior gas concentration (creating an internal atmosphere of elevated carbon dioxide and reduced oxygen content) as controlled atmosphere chambers or modified atmosphere packaging have on the outer fruit environmental atmosphere, as is shown with apples in Figure 1.1. However, if oxygen levels fall too low, the anaerobic pathway of fruit respiration can be initiated (Wills et al., 1981) and ethanol is produced (Figure 1.1), which can cause off-flavor (Bai et al., 2002; Baldwin, 1994). Fresh produce is often stored in chambers, containers, or packaging that provide high relative humidity to reduce water loss. Hydrophobic coatings can also prevent water loss (Baldwin et al., 1997). To complicate matters, the relative humidity of the storage environment also tends to alter coating permeability to gases, usually making hydrophilic coatings more gas permeable (Hagenmaier and Shaw, 1990, 1992).

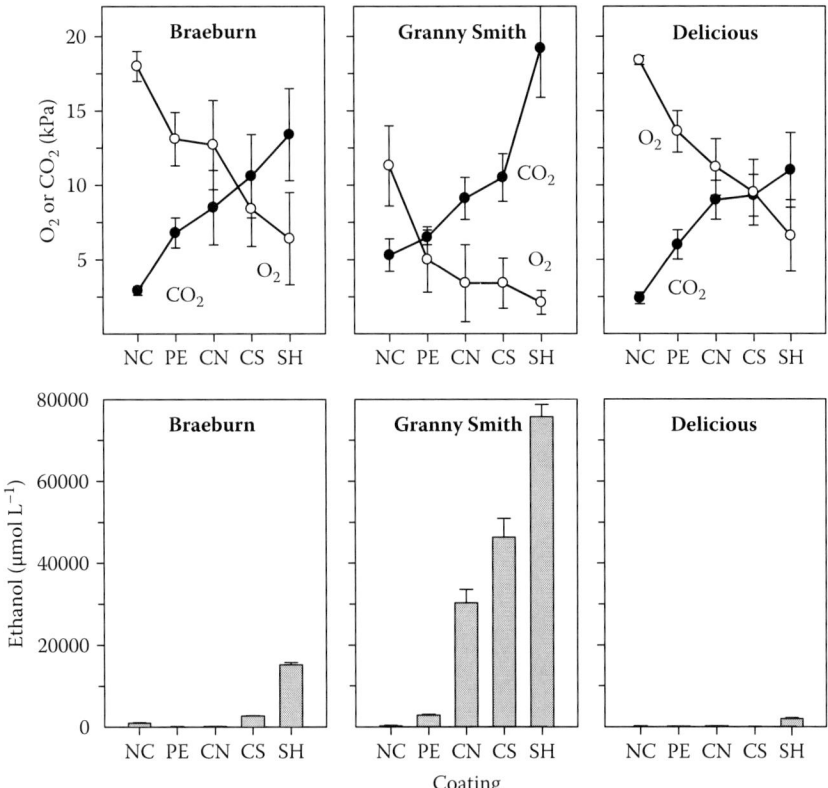

Figure 1.1 Internal gases (top) and ethanol concentrations (bottom) of 'Braeburn', 'Granny Smith', and 'Delicious' apples after 5 months controlled or room atmosphere storage, then coated and held at 20°C for 2 weeks. (NC: noncoating; PE: polyethylene; CN: candelilla; CS: carnauba-shellac; SH: shellac.)

Natural cuticle waxes on the surface of fruits and vegetables help to protect against excessive transpiration and thus water loss (Woods, 1990). In cleaning fresh citrus fruit, this natural barrier is disturbed or washed away and needs to be replaced, hence the development of citrus coatings, which was one of the first uses of coatings on fruit (Baldwin, 1994). Another simple example of water loss prevention is the enrobement of strawberries with a chocolate covering or frosting on a cake. In both cases, the chocolate or the frosting keeps the coated product from losing moisture in addition to adding flavor and sweetness. Chocolate, sugar, or other carbohydrate coatings on nuts help to prevent rancidity by providing a barrier to oxygen (Baker et al., 1994; Baldwin and Wood, 2006; Cosler, 1957).

For processed products, edible coatings have been used to impart sweet flavor and improve texture of cereals (Baker et al., 1994; Tribelhorn, 1991). Mineral oil has been used to reduce stickiness and clumping and reduce moisture loss from dried fruit like raisins as well as to improve their appearance (Kochhar and Rossell, 1982). Edible coatings have been used with heterogeneous foods, separating components of a food with differing water activity (Guilbert, 1986), such as dry cereal and raisins

(Lowe et al., 1962; Watters and Brekke, 1959), ice cream and ice cream cones (Rico-Peña and Torres, 1990), or dried fruit pieces in cake mix (Shea, 1970). Coatings have been used to coat meats and seafood (Baker et al., 1994; Kester and Fennema, 1986; Ustunol, 2009) as well as "lightly processed" or fresh-cut fruits and vegetables (Baldwin et al., 1995; Rojas-Graü et al., 2009).

Coatings can also be used as carriers of functional ingredients, such as fungicides for fruit coatings (Baldwin, 1994; Brown, 1984; Miller et al., 1988). Other antimicrobials, such as preservatives, antibrowning agents, antioxidants, and firming agents have all been added to coatings to improve the coated product microbial stability, appearance, and texture (Baldwin et al., 1996; Cuppet, 1994; Martín-Belloso and Rojas-Graü, 2009; Torres et al., 1985). Often, essential oils, acids, or natural plant extracts that have antimicrobial activity are incorporated into coatings (Almajano et al., 2007; du Plooy et al., 2009; Maizura et al., 2007; Ponce et al., 2008). Sometimes the coating can be antimicrobial as has been reported for chitosan coatings (Baldwin, 1994; Coma et al., 2002; No et al., 2007). This has been shown for coated meat (Juneja et al., 2006), cheese (Duan et al., 2007), and fruit (El Ghaouth et al., 1991; Narciso et al., 2007; Park et al., 2005; Romanazzi et al., 2002). Chitin is the second most abundant polysaccharide after cellulose and is widely distributed as the supporting material of crustacean (*Postharvest News and Information*, 1991). Finally, coatings can be used for flavor or nutrient encapsulation (Chen et al., 2006a, 2006b; Reineccius, 1994, 2009) by spray drying or extrusion processes (Manojlovic et al., 2008).

1.4 Components of edible coatings

A glance at the Table of Contents makes it obvious that much of this book is devoted to coating ingredients. To avoid repetition, this topic will be addressed in a very general manner here.

Polymers are the main ingredient of many edible coatings. This should not be surprising, because polymers are also the main ingredient of nonedible coatings, for example, ordinary house paint. For edible coatings, polymers such as protein, starch, and gums are used. In addition, many edible polymers are nontoxic, simple derivatives of cellulose, one of the most abundant natural polymers in nature, being a component of plant structure (Baldwin, 1999; Kester and Fennema, 1986). Suitability of edible polymers for use in edible coatings is often studied by measuring tensile properties accomplished by drying the edible coating or film on glass or another flat surface. Such studies have also helped to show how plasticizers and other minor ingredients and particle size or microstructure affect the integrity of the coating structure and function (Anker et al., 2000; McHugh and Krochta, 1994a, 1994b).

Polymers, however, are not a necessary ingredient of coatings. Many edible coatings are made with waxes or resins, with optional addition of polymers to form either bilayer or composite coatings (Baldwin, 1994). It seems interesting that coatings made from polymeric edible films are generally designed to be flexible and tough, whereas successful wax coatings are generally not even self-supporting when formed into thin layers or mixed with aqueous solvents forming emulsions (Baldwin et al., 1997). In fact, wax-based coatings have been so successful, that in the fresh fruit industry the word *wax* is regularly and erroneously used to mean any fruit coating, whether or not the coating actually contains any wax-like substance.

Basically, coating film formers consist of polysaccharides, proteins, lipids, or resins, alone or, more often, in combination. Polysaccharide coatings are hydrophilic and intermediate among coatings materials in gas exchange properties but are poor barriers to moisture. These include cellulose derivatives, starch derivatives, chitosan, pectin, carrageenans, alginate, and gums (Nisperos-Carriedo, 1994). Proteins are similar in properties, being also hydrophilic, and include corn zein, wheat gluten, peanut, soy, collagen, gelatin, egg, whey, and casein (Gennadios et al., 1994; Krochta, 2002). Lipids and waxes tend to be more permeable to gasses but present a better barrier to water vapor and include carnauba, candelilla, and rice bran waxes; beeswax; petroleum-based waxes such as paraffin and mineral oils; vegetable oils (corn, soy, and palm); acetoglycerides; and oleic acid (Baldwin et al., 1997; Kester and Fennema, 1986, 1989). Resins are the least permeable to gases and intermediate in resistance to water vapor and include shellac, wood rosin, and coumarone indene resin (Bai et al., 2003a; Hernandez, 1994).

A digression here seems needed because coatings for fruit are the largest usage of edible coatings, and natural and chemically derived waxes and resins are very important ingredients. In discussions of ingredients used for fruit coatings, it is not always clear which meaning of *wax* applies—namely, fruit coating or chemical substance. For example, a fruit coating made with shellac as the major ingredient might be called shellac wax. As it happens, the type of shellac used to make such coatings is known as *dewaxed* shellac, which is made by removal from unrefined shellac of a waxy substance called *shellac wax*. We return now to the discussion of polymers and waxes. Some coatings contain polymers but no waxes. Others contain waxes but no polymers. For fruit coatings, however, many contain both.

Besides polymers and wax, other ingredients are ubiquitous in food coatings. Fatty acids and other surfactants are often used to emulsify waxes and lower surface tension to improve spread. Plasticizers, which are small molecules such as glycerol, propylene glycol, or polyethylene glycol, are used to control viscosity of the liquid formulation, add flexibility and tensile strength, and control surface tension. However, these ingredients can affect coating performance adversely by altering gas permeability or affecting characteristics like shine or gloss, as is shown when adding propylene glycol to a zein coating (Figure 1.2). Ammonia and morpholine are used to solubilize and disperse fatty acids, waxes, and polymers. Alcohol is used as a solvent and to decrease microbial growth during storage of the liquid coating and also to hasten drying after application to the product. Finally, the bulk of a coating is made up of a solvent, often water or aqueous alcohol (Bai et al., 2002, 2003; Baldwin et al., 1997; Hagenmaier, 2004). Other minor ingredients besides surfactants and plasticizers include proteins, antifoam agents, as well as useful additives such as antibrowning or antimicrobial agents (Baldwin et al., 1996, 1997).

Coatings are difficult to formulate; therefore, the compositions of those used in the food industry are proprietary information. Food labeling laws and sales literature sometimes make it possible to identify the major ingredients, but minor ingredients can have a big effect on coating properties and are hard to identify in commercial

6   Edible coatings and films to improve food quality

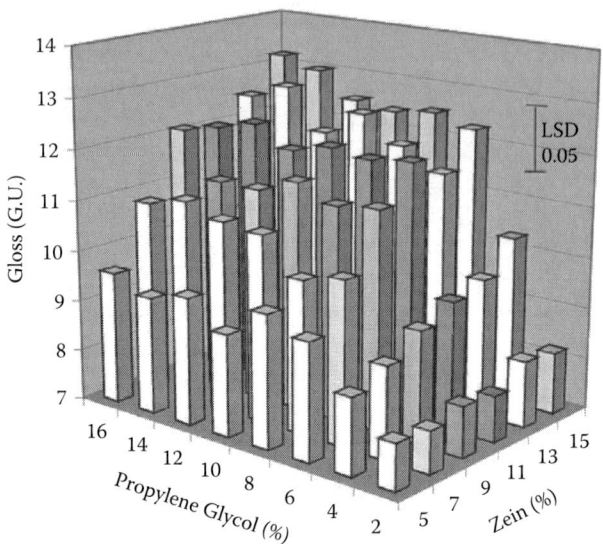

Figure 1.2 Effect of zein and propylene glycol contents on gloss of 'Gala' apples as measured using a gloss meter. Noncoated control was 7.3 GU ($n=100$). (Reprinted from Bai, J., Alleyne, V., Hagenmaier, R., Mattheis, J., and Baldwin, E.A., *Postharvest Biol. Technol.* 28, 259, 2003. With permission.)

coating products. Ingredients, however, must be food grade (generally recognized as safe, GRAS) or approved for food contact (FDA 21 CFR). In some cases, materials are allowed to coat fruits or vegetables where the peel is removed prior to consumption that are not otherwise allowed on produce where the peel is likely to be consumed (e.g., oranges versus apples) (Baldwin, 1994). Another example is the use of waxes to coat cheese wheels, which are approved for food contact but are not eaten (Baranowski, 1990; Johnson and Peterson, 1974).

1.5 Important properties of edible coatings

A coating must meet many demands for legality, safety, and performance (Table 1.2). During storage, it must not ferment, coagulate, separate, develop off-flavors, or otherwise spoil. Successful application requires that it spread evenly, dry quickly, not foam, and be easy to remove from equipment. Once applied, it should not crack, discolor, or peel during handling and storage, including when in contact with condensate. It should not be tacky or adhere to packaging, not react adversely with the food, and not impair sensory quality. Coatings should allow passage or restrict gas exchange of oxygen, carbon dioxide, aroma volatiles, and water vapor as required. In the case of fruit, for example, a coating should permit enough gas exchange to prevent the fruit from going anaerobic, but provide enough restriction to retard ripening and senescence, while retarding loss of water vapor to prevent shriveling. For fatty products such as nuts and meat, the coating should be a barrier to oxygen, to prevent rancidity, while for cereals, the coating should be a barrier to moisture to prevent sogginess

(Baker et al., 1994). Often, coatings are used to improve appearance, and most commonly in this regard are expected to impart gloss to a product, and sometimes color, that must last through shipping, handling, and marketing. Examples of this are coatings that result in shiny apples (Figure 1.2) and M&M candies.

1.6 Determination of coating properties

Several properties of edible coatings are important for their performance, depending on their function. It is useful to know mechanical properties, such as thickness using a micrometer, tensile strength, elongation, and elasticity, which can be measured by stretching a film to its breaking point, puncturing, or measuring deformation using an Instron or similar texture measuring equipment (Perez-Gago and Krochta, 2000; Park et al., 1994). Another important parameter to measure, as has been discussed, is permeability to gases. This can be measured by using Fick's first law of diffusion and Henry's law of solubility to determine steady-state permeability of a permeate through a nonporous barrier (assuming no imperfections). The term *permeance* is often used when the film is heterogeneous (nonhomogeneous barrier) and thickness may not be known (Donhowe and Fennema, 1994). Permeability to gases can be measured by applying the permeate to one side of the film at elevated pressure, or establishing a concentration gradient of permeate across a film, while maintaining equal pressure on both sides of the film (Donhowe and Fennema, 1994). These methods are often used to measure permeability to oxygen or other gases. Permeability to water vapor is normally measured using a 100% to 0% relative humidity gradient. All this can be done at various temperatures as well (Donhowe and Fennema, 1994) by gravimetric method (McHugh and Krochta, 1994a, 1994b; McHugh et al., 1993). Finally, there is coating appearance, usually the degree of gloss, which can be measured by a gloss meter (Figure 1.2), and coloration or discoloration, as the case may be, can be measured with a chromometer (Baldwin et al., 1996).

1.7 Why read this book

The best reasons for reading this book are summarized in Table 1.1 and Table 1.2. Table 1.1 shows major benefits that can be achieved with edible coatings. Table 1.2 is, however, the more important of the two, because often those benefits are not easily achieved. This book was written to help achieve the benefits and avoid the undesirable if unintended consequences of coating formulation. For example, the discussion of food additive regulations should help processors and consumers to decide on safe formulations. Understanding the structure and function relationships of polymers, waxes, and resins and the chemical effects of other additives will help in the design of better coatings. Ingredient costs are likely to be better controlled if the processor has knowledge of alternate ingredients for achieving the same purpose. Further, inadequate or excessive restriction of ingredient exchange by the coating can be avoided with knowledge of gas exchange properties of coatings. Finally, food quality concerns are better dealt with by knowledge of the physical and organoleptic properties of different edible coating ingredients.

Table 1.1 Benefits of edible coatings

Benefit	Foods to be coated	Main components of some coatings
Inhibit transfer of moisture and oxygen	Nuts	Zein
	Raisins	Whey protein
	Frozen chicken	Alginate
	Fish fillets	Chitosan
	Eggs	Protein and vegetable oil
Reduce transfer of moisture	Cake	Sugar and butter
	Crackers with cheese	Vegetable oil
	Sausage	Collagen
	Fresh fruit	Shellac, carnauba wax
	Breaded poultry	Gellan gum
Reduce mold growth	Cheese	Agar and galactomannan
	Smoked fish	Whey protein
	Baked goods	Carrageenan with fungicides
Reduce frying fat uptake	Potato	Methyl cellulose
	Doughnut	
Improve appearance	Fresh fruit	Shellac
	Chocolate candy	Zein
Reduce adhesion to cooking surface	Lasagna	Sauce with cheese or meat
	Poultry	Cellulose and protein

Table 1.2 Areas of concern with edible coatings

Chemical safety	As with all food ingredients and additives, safety is a fundamental requirement. However, much is still unknown about the safety of all food additives, including coating ingredients.
Cost	Ingredients and method of application.
Barrier properties	Ideal coatings form an acceptable barrier for gas exchange between food and atmosphere, or between two phases of the same food item, neither too restrictive nor too permeable.
Food quality	Coatings tend to change appearance, flavor, and mouthfeel, and efforts are needed to achieve changes that are good, not harmful.
Nutritive value	Some coatings are so thick that they change the nutritional value (e.g., frosting increases the caloric value of cake).
Environment	Volatile organic compounds (usually alcohol) are sometimes released when edible coatings dry.

The authors of this book have dealt with these and other concerns in their presentation of the science and practice of using edible coatings. We hope you benefit from their efforts.

References

Almajano, M.P., Carbó, R., Delgado, M.E., and Gordon, M.H. 2007. Effect of pH on the antimicrobial activity and oxidative stability of oil-in-water emulsions containing caffeic acid. *J. Food Sci.* 72:C258–C263.

Anker, M., Stading, M., and Hermansson, A.M. 2000. Relationship between the microstructure and mechanical and barrier properties of whey protein films. *J. Agric. Food Chem.* 48:3806–3816.

Anonymous. *Postharvest News and Information*. 1991. General news: waxed fruit banned in Norway. *Postharvest News and Information* 1:435.

Antoniewski, M.N., Barringer, S.A., Knipe, C.L., and Zerby, H.N. 2007. Effect of gelatin coating on the shelf life of fresh meat. *J. Food Sci.* 72:E382–E387.

Bai, J., Hagenmaier, R.D., and Baldwin, E.A. 2003a. Coating selection for 'Delicious' and other apples. *Postharv. Biol. Technol.* 28:259–268.

Bai, J., Alleyne, V., Hagenmaier, R.D., Mattheis, J.P., and Baldwin, E.A. 2003b. Formulation of zein coatings for apples (*Malus domestica* Borkh). *Postharvest Biol. Technol.* 28:259–268.

Bai, J., Baldwin, E.A., and Hagenmaier, R.H. 2002. Alternatives to shellac coatings provide comparable gloss, internal gas modification, and quality for 'Delicious' apple fruit. *HortScience.* 37:559–563.

Baker, R.A., Baldwin, E.A., and Nisperos-Carriedo, M.O. 1994. Edible coatings and films for processed foods. In: *Edible Coatings and Films to Improve Food Quality.* J.M. Krochta, E.A. Baldwin, and M. Nisperos-Carriedo (Eds.), Technomic, Lancaster, PA, pp. 89–104.

Baldwin, E.A., and Wood, B. 2006. Use of edible coatings to preserve pecans (*Carya illinoinensis*) at room temperature. *HortScience.* 41:188–192.

Baldwin, E.A. 1999. Surface treatment and edible coating in food preservation. In: *Handbook of Food Preservation.* M.S. Rahman (Ed.), Marcel Dekker, New York, pp. 577–609.

Baldwin, E.A., Nisperos, M.O., Hagenmaier, R.D., and Baker, R.A. 1997. Use of lipids in coatings for food products. *Food Technol.* 51(6):56–64.

Baldwin, E.A., Nisperos, M.O., Chen, X., and Hagenmaier, R.D. 1996. Improving storage life of cut apple and potato with edible coating. *Postharv. Biol. Technol.* 9:151–163.

Baldwin, E.A., Nisperos-Carriedo, M.O., and Baker, R.A. 1995. Use of edible coatings to preserve quality of lightly (and slightly) processed products. *Crit. Rev. Food Sci. Nut.* 35:509–524.

Baldwin, E.A. 1994. Edible coatings for fresh fruits and vegetables: past, present and future. In: *Edible Coatings and Films to Improve Food Quality.* J.M. Krochta, E.A. Baldwin, and M. Nisperos-Carriedo (Eds.), Technomic, Lancaster, PA, pp. 25–64.

Baranowski, E.S. 1990. Miscellaneous food additives. In: *Food Additives.* A.L. Branen, P.M. Davidson, and S. Salminen (Eds.), Marcel Dekker, New York, pp. 511–578.

Becker, O.W. 1939. U.S. Patent 2,161,908. Method of and apparatus for making artificial sausage casings.

Becker, O.W. 1938. U.S. Patent 2,115,607. Sausage casing.

Biquet, B., and Labuza, T.P. 1988. Evaluation of the moisture permeability characteristics of chocolate films as an edible moisture barrier. *J. Food. Sci.* 53:989–998.

Brown, G.E. 1984. Efficacy of citrus postharvest fungicides applied in water or resin solution water wax. *Plant Dis.* 68:415–418.

Chen, H., Weiss, J., and Shahidi, F. 2006a. Nanotechnology in nutraceuticals and functional foods. *Food Technol.* 60(3):30–32, 34–36.

Chen, L., Romondetto, G.E., and Subirade, M. 2006b. Food protein-based materials as nutraceutical delivery systems. *Trends in Food Sci. Technol.* 17:272–283.

Coma, V., Martial-Gros, A., Garreau, S., Copinet, A., Salin, F., and Deschamps, A. 2002. Edible antimicrobial films based on chitosan matrix. *J. Food Sci.* 67: 1162–1169.

Contreras-Medellin, R., and Labuza, T.P. 1981. Prediction of moisture protection requirements for foods. *Cereal Food World* 26:335–340.

Cosler, H.B. 1957. Method of producing zien-coated confectionery. U.S. Patent 2,791,509.

Cuppett, S.L. 1994. Edible coatings as carriers of food additives, fungicides and natural antagonists. In: *Edible Coatings and Films to Improve Food Quality.* J.M. Krochta, E.A. Baldwin, and M. Nisperos-Carriedo (Eds.), Technomic, Lancaster, PA, pp. 121–136.

Dangaran, K.L., Nantz, J.R., and Krochta, J.M. 2006. Whey protein-sucrose coating gloss and integrity stabilization by crystallization inhibitors. *J. Food Sci.* 71:E152–E157.

Donhowe, G., and Fennema, O. 1994. Edible films and coatings: characteristics, formation, definitions and testing methods. In: *Edible Coatings and Films to Improve Food Quality.* J.M. Krochta, E.A. Baldwin, and M. Nisperos-Carriedo (Eds.), Technomic, Lancaster, PA, pp. 1–24.

Du Plooy, W., Regnier, T., and Combrink, S. 2009. Essential oil amended coatings as alternatives to synthetic fungicides in citrus postharvest management. *Postharvest Biol. Technol.* 53:117–122.

Duan, J., Park, S.-I., Daechel, M.A., and Zhao, Y. 2007. Antimicrobial chitosan-lysozyme (CL) fims and coatings for enhancing microbial safety of mozzarella cheese. *J. Food Sci.* 72:M355–M362.

El Ghaouth, A., Ponnompalam, R., Castaigne, F., and Boulet, M. 1991. Chitosan coating effect on storability and quality of fresh strawberries. *J. Food Sci.* 56:1618–1631.

FDA, 21 CFR, U.S. Food and Drug Administration, Code of Federal Regulations Title 21. 2010. http://www.accessdata.fda.gov/scripts/cdrh/cfdocs/cfcfr/cfrsearch.cfm.

Gennadios, A., McHugh, T.H., Weller, C.L., and Krochta, J.M. 1994. Edible coatings and films based on proteins. In: *Edible Coatings and Films to Improve Food Quality.* J.M. Krochta, E.A. Baldwin, and M. Nisperos-Carriedo (Eds.), Technomic, Lancaster, PA, pp. 201–277.

Guilbert, S., 1986. Technology and application of edible protective films. In: *Food Packaging and Preservation; Theory and Practice.* M. Mathlouthi (Ed.), Elsevier Applied Science, London, UK, pp. 371–394.

Hagenmaier, R.D. 2004. Fruit coatings containing ammonia instead of morpholine. *Proc. Fla. State Hort. Soc.* 117:396–402.

Hagenmaier, R.D., and Baker, R.A. 1997. Edible coatings form morpholine-free wax microemulsions. *J. Agric. Food Chem.* 45:349–352.

Hagenmaier, R.D., and Shaw, P.E. 1992. Gas permeability of fruit coating waxes. *J. Amer. Soc. Hort. Sci.* 117:105–109.

Hagenmaier, R.D., and Shaw, P.E. 1990. Moisture permeability of edible films made with fatty acid and (hydroxypropyl) methylcellulose. *J. Agric. Food Chem.* 38:1799–1803.

Hardenburg, R.E. 1967. Wax and related coatings for horticultural products. A Bibliography. *Agr. Res. Bull.* 51-15, U.S. Department of Agriculture, Washington, DC.

Hernandez, E. 1994. Edible coatings from lipids and resins. In: *Edible Coatings and Films to Improve Food Quality.* J.M. Krochta, E.A. Baldwin, and M. Nisperos-Carriedo (Eds.), Technomic, Lancaster, PA, pp. 279–303.

Hicks, E. 1961. *Shellac, Its Origin and Applications.* Chemical, New York.
Johnson, A.H., and Peterson, M.S. (Eds.) 1974. *Encyclopedia of Food Technology.* AVI, Westport, CT, pp. 178–183.
Juneja, V.K., Thippareddi, H., Bari, L., Inatsu, Y., Kawamoto, S., and Friedman, M. 2006. Chitosan protects cooked ground beef and turkey against *Clostridium perfringens* spoors during chilling. *J. Food Sci.* 71:M236–M240.
Kader, A.A., Morris, L.L., and Maxie, E.C. 1986. Biochemical and physiological basis for effects of controlled and modified atmospheres on fruits and vegetables. *Food Technol.* 40:99–104.
Kaushik, V., and Roos, Y.H. 2007. Limonene encapsulation in freeze-drying of gum Arabic-sucrose-gelatin systems. *LWT.* 40:1381–1391.
Kester, J.J., and Fennema, O.R. 1989. Resistance of lipid films to water vapor transmission. *J. Amer. Oil Chem. Soc.* 66:1139–1146.
Kester, J.J., and Fennema, O.R. 1986. Edible films and coatings: a review. *Food Technol.* 40(12): 47–59.
Kielhorn, J., and Rosner, G. 1996. Environmental health criteria 179: Morpholine. International Programme on Chemical Safety, World Health Organization, Geneva, p. 65.
Kochhar, S.P., and Rossell, J.B. 1982. A vegetable oiling agent for dried fruits. *J. Food Technol.* 17:661–668.
Krochta, J.M. 2002. Proteins as raw materials for films and coatings: definitions, current status and opportunities. In: *Protein-Based Films and Coatings.* A. Gennadios (Ed.), CRC Press, Boca Raton, FL, pp. 1–41.
Lowe, E., Durkee, E.L., and Hamilton, W.E. 1962. U.S. Patent 3,046,143. Process for coating food products.
Maizura, M., Fazilah, A., Norziah, M.H., and Karim, A.A. 2007. Antibacterial activity and mechanical properties of partially hydrolyzed sago starch-alginate edible film containing lemongrass oil. *J. Food Sci.* 72: C324–C330.
Manojlovic, V., Rajic, N., Djonlagic, J., Obradovic, B., Nedovic, V., and Bugarski, B. 2008. Application of electrostatic extrusion—flavor encapsulation and controlled release. *Sensors.* 8:1488–1496.
Martín-Belloso, M., and Rojas-Graü. 2009. Delivery of flavor and active ingredients using edible films and coatings. In: *Edible Films and Coatings for Food Applications.* M.E. Embuscado and K.C. Huber (Eds.), Springer, New York, pp. 295–313.
McHugh, T.H., and Krochta, J.M. 1994a. Water vapor permeability properties of edible whey protein-lipid emulsion films. *J. Amer. Oil Chem. Soc.* 71:307–312.
McHugh, T.H., and Krochta, J.M. 1994b. Dispersed phase particle size effects on water vapor permeability of whey protein-beeswax edible emulsion films. *J. Food Process. Preserv.* 18:173–188.
McHugh, T.H., Avena-Busteillos, R., and Rochta, J.M. 1993. Hydrophilic edible films: modified procedure for water vapor permeability and explanation of thickness effect. *J. Food Sci.* 58:899–903.
Miller, W.R., Chun, D., Risse, L.A., Hatton, T.T., and Hinsch, T. 1988. Influence of selected fungicide treatments to control the development of decay in waxed or film wrapped Florida grapefruit. *Proc. Sixth Int. Cit. Cong.* 3:1471–1477.
Narciso, J., Baldwin, E.A., Plotto, A., and Ference, C. 2007. Preharvest peroxyacetic acid sprays slow down decay and extend shelf-life of strawberries. *HortScience.* 42:617–621.
Nisperos-Carriedo, M.O. 1994. Edible coatings and films based on polysaccharides. In: *Edible Coatings and Films to Improve Food Quality.* J.M. Krochta, E.A. Baldwin, and M. Nisperos-Carriedo (Eds.), Technomic, Lancaster, PA, pp. 305–335.

No, H.K., Meyers, S.P., Prinyawiwatkul, W., and Xu, Z. 2007. Applications of chitosan for improvement of quality and shelf life of foods: a review. *J. Food Sci.,* R87–R100.

Park, J.W., Testin, R.F., Park, H.J., Vergano, P.J., and Weller, C.L. 1994. Fatty acid concentration effect on tensile strength, elongation, and water vapor permeability of laminated edible films. *Food Sci.* 59:916–919.

Park, S.I., Stan, S.D., Daeschel, M.A., and Zhao, Y. 2005. Antifungal coatings on fresh strawberries (Fragariaxananassa) to control mold growth during cold storage. *J. Food Sci.* 70:M202–M207.

Perez-Gago, M.B., and Krochta, J.M. 2000. Drying temperature effect on water vapor permeability and mechanical properties of whey protein-lipid emulsion films. *J. Agric. Food Chem.* 48:2687–2692.

Ponce, A.G., Roura, S.I., Del Valle, C.E., and Moreira, M.R. 2008. Antimicrobial and antioxidant activities of edible coatings enriched with natural plant extracts: *in vitro* and *in vivo* studies. *Postharvest Biol. Technol.* 49:294–300.

Porzio, M. 2004. Flavor encapsulation: a convergence of science and art. *Food Technol.* 58(7):40–47.

Ravishankar, S., Shu, L., Olsen, C.W., McHugh, T.H., and Friedman, M. 2009. Edible apple film wraps containing plant antimicrobials inactivate foodborne pathogens on meat and poultry products. *J. Food Sci.* 74:M440–M445.

Reineccius, G.A. 2009. Edible films and coatings for flavor encapsulation. In: *Edible Films and Coatings for Food Applications.* M.E. Embuscado and K.C. Huber (Eds.), Springer, New York, pp. 269–294.

Reineccius, G.A. 1994. Flavor encapsulation. In: *Edible Coatings and Films to Improve Food Quality.* J.M. Krochta, E.A. Baldwin, and M. Nisperos-Carriedo (Eds.), Technomic, Lancaster, PA, pp. 105–120.

Rico-Peña, D.C., and Torres, J.A. 1990. Edible methylcellulose-based films as moisture-impermeable barriers in sundae ice cream cones. *J. Food Sci.* 55:1468–1469.

Rojas-Graü, M.A., Soliva-Fortuny, R., and Martín-Belloso, O. 2009. Use of edible coatings for fresh-cut fruits and vegetables. In: *Advances in Fresh-Cut Fruits and Vegetables Processing.* O. Martín-Bellos and R. Soliva-Fortuny (Eds.), CRC Press, Boca Raton, FL, pp. 285–231.

Romanazzi, G., Nigro, F., Ippolito, A., Di Venere, D., and Salerno, M. 2002. Effects of pre- and postharvest chitosan treatments to control storage grey mold of table grapes. *J. Food Sci.* 67:1862–1867.

Shea, R.A. 1970, June 23. U.S. Patent 3,516,836. Fruit containing baking mixes.

Torres, J.A., Motoki, M., and Karel, M. 1985. Microbial stabilization of intermediate moisture food surfaces I. Control of surface preservative concentration. *J. Food Proc. Preserv.* 9:75–92.

Tribelhorn, R.E. 1991. Breakfast cereals. In: *Handbook of Cereal Science and Technology.* K.J. Lorenz and K. Kulp (Eds.), Marcel Dekker, New York, pp. 741–761.

Ustunol, Z. 2009. Edible films and coatings for meat and poultry. In: *Edible Films and Coatings for Food Applications.* M.E. Embuscado and K.C. Huber (Eds.), Springer, New York, pp. 245–268.

Watters, G.G., and Brekke, J.E. 1959. U.S. Patent 2,909,435. Coating of raisins and other foods.

Wills, R.H., Lee, T.H., Graham, D., McGlasson, W.B., and Hall, E.G. 1981. *Postharvest, an Introduction to the Physiology and Handling of Fruit and Vegetables.* AVI, Westport, CT.

Woods, J.L. 1990. Moisture loss from fruits and vegetables. *Postharvest News and Information.* 1:195–199.

2 Protein-based films and coatings

Maria B. Pérez-Gago

Contents

2.1	Introduction	14
2.2	Zein films and proteins	15
	2.2.1 Film formation	15
	2.2.2 Functional properties of zein protein films	16
	2.2.3 Zein coatings	24
2.3	Soy film and coatings	24
	2.3.1 Formation of soy protein films	25
	2.3.2 Functional properties of soy proteins	25
	2.3.3 Soy protein coatings	29
2.4	Wheat gluten proteins	29
	2.4.1 Formation of wheat gluten protein films	30
	2.4.2 Functional properties of wheat gluten protein films	30
	2.4.3 Wheat gluten protein coatings	33
2.5	Whey proteins	34
	2.5.1 Formation of whey protein films	34
	2.5.2 Functional properties of whey protein films	35
	2.5.3 Whey protein coatings	40
2.6	Casein films and coatings	40
	2.6.1 Formation of casein films	41
	2.6.2 Functional properties of casein films	41
	2.6.3 Casein protein coatings	44
2.7	Cottonseed proteins	44
	2.7.1 Formation of cottonseed protein films	44
	2.7.2 Functional properties of cottonseed protein films	45
2.8	Collagen and gelatin proteins	46
	2.8.1 Formation of collagen and gelatin protein films	46
	2.8.2 Functional properties of collagen films	47
	2.8.3 Functional properties of gelatin films	47
	2.8.4 Collagen and gelatin coatings	48
2.9	Myofibrillar and sarcoplasmic proteins	49
	2.9.1 Formation of myofibriller protein films	49
	2.9.2 Functional properties of myofibriller and sarcoplasmatic protein films	50
	2.9.3 Myofibriller protein coatings	51

2.10   Egg white proteins   51
    2.10.1   Formation of egg white protein films   52
    2.10.2   Functional properties of egg white protein films   52
    2.10.3   Egg white protein coatings   53
2.11   Keratin films   53
    2.11.1   Formation of keratin protein films   54
    2.11.2   Functional properties of keratin protein films   54
    2.11.3   Keratin coatings   54
2.12   Protein films of limited availability   54
    2.12.1   Peanut protein   54
    2.12.2   Rice protein   55
    2.12.3   Pea protein   56
    2.12.4   Pistachio protein   56
    2.12.5   Lupin protein   57
    2.12.6   Grain sorghum protein   57
    2.12.7   Winged bean protein   57
    2.12.8   Cucumber pickle brine protein   58
2.13   Conclusions   58
References   58

2.1 Introduction

Functional properties of proteins include the ability to form films and coatings. Proteins are polymers with specific amino acid sequences and molecular structure. Depending on the sequential order of the amino acids, the protein will assume different structures along the polymer chain which will determine the secondary, tertiary, and quaternary structures. The secondary, tertiary, and quaternary structures of proteins can be easily modified to optimize the protein configuration, protein interactions, and resulting film properties. Films and coatings based on proteins are edible or biodegradable, depending on formulation, formation method, and modification treatments (Krochta, 2002). As long as food-grade proteins and additives are used and only protein changes due to heating, pH modification, salt addition, enzymatic modification, and water removal occur, the resulting film or coating is edible (Krochta and De Mulder-Johnston, 1997).

Water vapor permeability (WVP) and oxygen permeability (OP) are the barrier properties that usually determine the ability of the edible film to protect the food product from the environment. Mechanical properties are also of interest to assess the ability of the film and coating to maintain the mechanical integrity and protect the food. Both barrier and mechanical properties depend on the film composition and structure. The development of protein films involves disulfide bonds, hydrophobic interactions, and hydrogen bonds. Factors affecting these interactions modify the barrier and mechanical properties of the films. This way, (1) interaction between proteins and small molecules, including water, plasticizers, lipids, and other additives dispersed in the matrix; (2) application of physical treatments, such as heat treatments and irradiations; and (3) application of enzymatic treatments, contribute to the barrier and mechanical properties of the films (Park et al., 2002a).

Protein film-forming materials are derived from different animal or plant sources. Much research has been done to forming various protein films and quantifying the protein film properties. This chapter provides an overview of edible protein films and coatings by compiling the most recent work found in the literature on the subject. The information presents the film formation ability of a number of proteins, the associated functional properties of the films, and some coating applications described in the literature. Plant-origin proteins of wide and limited availability derived from corn, soybean, wheat, cottonseed, peanut, rice, pea, pistachio grain sorghum, and animal-origin proteins like casein, whey proteins, collagen, gelatin, myofibrilar proteins, and egg albumen proteins will be considered in this chapter.

2.2 Zein films and proteins

Zein comprises a group of alcohol-soluble proteins (prolamins) found in corn endosperm. Zein occurs as aggregates linked by disulfide bonds in whole corn. Those bonds may be cleaved by reducing agents during extraction or wet-milling operations (Reiners et al., 1973). Based on solubility, molecular size, and charge differences, zein consists of three protein fractions: α-zein (75% to 85% of total zein), β-zein (10% to 15% of total zein), and γ-zein (5% to 10% of the total zein) (Esen, 1987). These fractions present a high content of the nonpolar amino acids leucine, alanine, and proline, which make zein water insoluble. In addition, the high level of glutamine is believed to promote hydrogen bonding between zein chains that also contribute to its water insolubility (Reiners et al., 1973).

Commercial zein is essentially a by-product of the corn wet-milling industry. In the process, corn gluten is extracted with 80% to 90% aqueous isopropyl alcohol containing 0.25% sodium hydroxide at 60% to 70°C, and the extract is chilled to precipitate the zein. Additional extractions and precipitations increase zein purity, with the possibility to obtain ultrapure zein lattices (Boundy et al., 1967; Cook and Shulman, 1994).

2.2.1 Film formation

Zein edible films are generally cast from alcohol solutions. Generally, zein is dissolved in warm (65 to 85°C) aqueous ethanol, acetone, or isopropanol with added plastizicers. The solution is cooled to 40 to 50°C, allowing bubbling to cease prior to casting, and is then poured over the casting plate (Gennadios et al., 1993a). Formation of films is believed to involve development of hydrophobic, hydrogen, and limited disulfide bonds, due to the low content of cystine, between zein protein chains in the film matrix (Gennadios et al., 1994). Near-infrared Fourier transform Raman spectroscopy showed that the secondary structure of zein protein remains unaltered during film formation. Raman results indicated that hydrophobic interaction played an important role in the formation of zein films, and disulfide bonding might be responsible for the structural stability of zein protein molecule during film formation (Hsu et al., 2005). Films prepared with aqueous ethanol disintegrate when they come into contact with water, whereas films produced by using aqueous acetone have good water resistance (Yamada et al., 1995).

Zein films may also be drawn from wet, freshly precipitated resins or formed by extrusion of dry resin pallets (Padua and Wang, 2002). Laboratory preparation of drawn films involves (1) a plastification step of a zein solution with long-chain fatty acids in a warm aqueous ethanol solution; (2) a resin formation step in which cold water is added to the solution to precipitate the plasticized zein; and (3) a kneading step where the precipitate is collected and kneaded into a cohesive and elastic moldable mass.

2.2.2 Functional properties of zein protein films

Films formed upon solvent evaporation are tough, glossy, scuff resistant, and grease resistant (Pomes, 1971). In general, zein films are relatively good water barriers compared to other protein-based edible films but are much poorer than low density polyethylene (LDPE) and ethylene-vinyl alcohol copolymer (EVOH) (Table 2.1) (Smith, 1986; Krochta, 1997a). Films are extremely brittle and therefore require plasticizers to increase flexibility. Incorporation of plasticizers and increase of the relative humidity (RH) that surrounds the film increases the WVP of zein films (Lai and Padua, 1998; Weller et al., 1998). Normal plasticizers used in zein films include glycerol (Gly), polyethylene glycol (PEG), polypropylene glycol (PPG), and fatty acids (oleic, stearic, lauric acid, etc.). Park et al. (1994a) observed Gly migration from the bulk of the film matrix to the surface, and this was attributed to a weak interaction with the protein molecules that caused excess Gly to migrate through the film matrix. Mixtures of Gly and PEG or PPG have shown lower migration rates of the plasticizers through the protein matrix than Gly alone (Park et al., 1994a; Park and Chinnan, 1995). Unsaturated fatty acids are also effective plasticizers, even at temperatures below freezing or in low relative humidity environments (Park et al., 1994b). In addition, incorporation of hydrophobic plasticizers yields films with lower water absorption rates (Santosa and Padua, 1999; Lawton, 2004) which translates into lower WVP. For example, a combination of PEG and lauric acid (LA) reduced the increase in film WVP compared to the use of PEG alone due to the hydrophobic character of fatty acids (Tillekeratne and Easteal, 2000; Paramawati et al., 2001). However, Wang and Padua (2004, 2006) observed a temperature effect on WVP of zein films plasticized with oleic acid (OA). At 4°C, WVP of film decreased with OA concentration while it increased at 25°C (Table 2.1). This response was attributed to OA phase changes due to temperature. At 4°C, OA is a crystalline solid that limits water diffusion through the film; whereas at 25°C, OA is a liquid that increases the system free volume, allowing water diffusion.

The moisture barrier can be improved by the addition of lipid materials, the use of cross-linking agents, or preparation of bilayer films (i.e., pouring, spraying, or rolling a second film directly on top of a previously prepared one). Application of a carnauba wax layer decreased WVP by one order of magnitude due to the high moisture barrier of waxes (Weller et al., 1998). Cross-linking agents such as polymeric dialdehyde starch (PDS), 1,2-epoxy-3-chloropropane (ECP), 1-[3-dimethylaminopropyl]-3-ethyl-carbodiimide hydrochloride (EDC), and N-hydroxysuccinimide (NHS) improved moisture barrier compared to unplasticized zein films (Yamada et al., 1995; Parris and Coffin, 1997; Parris et al., 2001; Kim et al., 2004). Multilayer films of different

Table 2.1 Water vapor permeability of protein-based edible films compared to packaging films

Film	Film solution/treatment	Film test conditions[a]	Permeability (g mm m^{-2} d^{-1} kPa^{-1})	Reference
Zein, unplasticized	80% EtOH	30°C, 0/100% RH	14.7	Parris and Coffin (1997)
Zein:Gly (3.9:1)	EtOh	26°C, 50/100% RH	35.2	Aydt et al. (1991)
Zein-30% Gly:PEG (1:3)	80% EtOH	30°C, 0/100% RH	25.4	Parris and Coffin (1997)
Zein:OA 2:8	95% EtOH	25°C, 51/0% RH	6.7	Wang and Padua (2006)
Zein:OA 2:8	95% EtOH	4°C, 69/0% RH	0.7	Wang and Padua (2006)
SPI:Gly (1.7:1)	60°C, 10 min, pH 6	25°C, 50/100% HR	262.0	Brandenburg et al. (1993)
SPI:Gly (1:1)	95°C, 45 min, pH 6	25°C, 55/77.5% RH	91.2	Park et al. (2002a)
SPI:Gly (1:1)	95°C, 45 min, pH 8	25°C, 55/77.5% RH	48.0	Park et al. (2002a)
WG:Gly (2.4:1)	100°C, 52% alk EtOH	26°C, 50/100% HR	108.0	Aydt et al. (1991)
WG:Gly (4.3:1)	50°C, 32% EtOH, pH 4	25°C, 100/0% RH	11.8	Micard et al. (2000)
WG:BW:Gly (5:1.5:1)	40°C, 45% EtOH, pH 4	30°C, 100/0% RH	3.0	Gontard et al. (1994)
WPI:Gly (1.6:1)	90°C, 30 min	25°C, 0/65% RH	119.8	McHugh et al. (1994)
WPI:Sor (1.6:1)	90°C, 30 min	25°C, 0/79% RH	62.0	McHugh et al. (1994)
WPI:BW:Gly (15:4:1)	90°C, 30 min	25°C, 0/92% RH	34.1	Shellhammer and Krochta (1997)
WPI:BW:Gly (15:24:1)	90°C, 30 min	25°C, 0/98% RH	7.7	Shellhammer and Krochta (1997)
WPI:BW:Gly (3:9:1)	90°C, 30 min, dried 80°C	25°C, 0/99% RH	2.0	Pérez-Gago and Krochta (2000)
SC	23°C	25°C, 0/81% RH	36.7	Avena-Bustillos and Krochta (1993)
SC		23°C, 50/100% RH	44.0	Schou et al. (2005)
SC:Gly (3.1:1)		23°C, 50/100% RH	72.0	Schou et al. (2005)
SC:LA (4:1)		25°C, 0/92% RH	9.6	Ho (1992)
Cottonseed flower		20°C, 0/100% RH	12.5	Marquié (1996)
MP:Gly (3:1)	pH 3	25°C, 0/100% RH	6.1	Cuq et al. (1995)

(Continued)

Table 2.1 Water vapor permeability of protein-based edible films compared to packaging films (Continued)

Film	Film solution/treatment	Film test conditions[a]	Permeability ($g\ mm\ m^{-2}\ d^{-1}\ kPa^{-1}$)	Reference
EW:400PEG (2.5:1)	45°C, 20 min, pH 11	25°C, 50/72% RH	199.2	Handa et al. (1999b)
LDPE		38°C, 90/0% RH	0.08	Smith (1986)
EVOH (68% VOH)		38°C, 90/0% RH	0.25	Foster (1986)
Cellophane		38°C, 90/0% RH	7.3	Taylor (1986)

Notes: SPI, soy protein isolate; WG, wheat gluten; WPI, whey protein isolate; SC, sodium caseinate; MP, myofibrillar protein; EW, egg white; Gly, glycerol; PEG, polyethylene glycol; Sor, sorbitol; OA, oleic acid; LA, lauric acid; BW, beeswax; EtOH, ethanol; LDPE, low density polyethylene; EVOH, ethylene-vinyl alcohol copolymer; VOH, vinyl alcohol.

[a] Relative humidity (RH) values are those on the top and bottom sides of the film (top/bottom).

film-forming polymer materials also reduce WVP. Bilayer films have been produced by combining zein films with methylcellulose (MC) films (Park et al., 1994b, 1996), ethylcellulose films (Romero-Bastida et al., 2004), soy protein and wheat gluten films (Foulk and Bunn, 2001; Pol et al., 2002), and high-amylose corn starch (HACS) films (Ryu et al., 2002). Effectiveness of a multilayer film depended on the preparation technique and the presence of plasticizers. Foulk and Bunn (2001) investigated the possibility of forming bilayer films by three different techniques: (1) liquid protein spread by pouring, spraying, or rolling on top of a dry film; (2) heat pressing dry films; and (3) solvent laminating dry films. These authors observed that pouring or spaying zein on soy films caused film surface to have an oil feel due to excessive plasticizer at the surface; whereas, zein could be rolled onto wheat and soy film without problems. Heat pressing and solvent laminating dry films formed zein-soy protein and zein-wheat gluten films with some wrinkles as if the materials were not properly fed through the calendar rolls. However, WVP of zein-soy protein and zein-wheat gluten films was not affected by the bilayer formation technique. The type and content of plasticizer also affect the moisture barrier of bilayer films. WVP of MC/zein-fatty acid films decreased as chain length and fatty acid concentration increased (Park et al., 1994b, 1996). Zein/ethylcellulose (EC) films plasticized with stearic acid (SA) presented lower water sorption than films plasticized with PEG (Romero-Bastida et al., 2004).

Zein films have also been prepared in ethanol or acetone aqueous solution without addition of plasticizers and using various controlled drying conditions that ranged from 30 to 45°C and 5 to 90% RH (Yoshino et al., 2002). The lowest WVP was obtained for films formed in acetone solution and dried at 45°C and 5% RH. Differences in the WVP were also found between the sides of the zein films (i.e., the air side of the zein film had a higher WVP than the basal side of the zein film when the films were exposed to high humidity during testing). This result suggested a rearrangement of the hydrophobic–hydrophilic parts of the zein molecules upon drying that affected the surface microstructure of pure zein films as observed by atomic force microscopy (Yoshino et al., 2000), showing a relationship between the WVP and the contact angle of the zein film.

Oxygen and carbon dioxide permeability values of zein edible films are low at low RH (Gennadios and Weller, 1990; Aydt et al., 1991; Gennadios et al., 1993a; Park and Chinnan, 1995; Yoshino et al., 2002). OP is lower than polysaccharide-based films (Greener and Fennema, 1989). Compared to other protein films, OP of zein films appeared to be one order of magnitude greater than for collagen at 0% RH and higher than for wheat gluten films (Table 2.2) (Gennadios et al., 1993a; Krochta, 1997a). OP is also significantly lower than common plastic films such as LDPE, propylene, plyestyrene, and polyvinyl chloride (Billing, 1989). Zein films have selective gas permeability depending on the drying conditions, with drying RH more determinant than drying temperature. Zein films dried at 35°C and 90% RH have higher OP than carbon dioxide permeability. Whereas, reducing RH to 5% RH decreased film OP 10 times, with values similar to carbon dioxide permeability. This was attributed to a change in protein network due to drying rate that reduced polymer mobility to a greater extent when the drying rate was increased (Yoshino et al., 2002). OP also increased with increased RH during test conditions and with the addition of plasticizer (Park and Chinnan, 1995; Yamada et al., 1995). OP of soy protein laminated

Table 2.2 Oxygen permeability of protein-based edible films compared to packaging films

Film	Film solution/treatment	Film test conditions	Permeability ($g\ mm\ m^{-2}\ d^{-1}\ kPa^{-1}$)	Reference
Zein:Gly (3.9:1)	Ethanol	38°C, 0% RH	67.4	Gennadios et al. (1993a)
SPI:Gly (1.7:1)	60°C, 10 min, pH 6	25°C, 0% HR	4.5	Brandenburg et al. (1993)
SPI:Gly (1.7:1)	60°C, 10 min, pH 12	25°C, 0% HR	1.6	Brandenburg et al. (1993)
WG:Gly (2.5:1)	75°C, 15 min, 95% EtOH	25°C, 0% HR	3.9	Gennadios et al. (1993e)
WPI:Gly (2.3:1)	90°C, 30 min	23°C, 50% RH	59.0	Pérez-Gago and Krochta (2001b)
WPI:Gly (2.3:1)	Native	23°C, 50% RH	78.0	Pérez-Gago and Krochta (2001b)
WPI:Sor (2.3:1)	90°C, 30 min	23°C, 50% RH	4.3	McHugh and Krochta (1994b)
CO_2C:Gly (3:1)		23°C, 50% RH	144.0	Tomasula et al. (2003)
CaC:Gly (3:1)		23°C, 50% RH	86.0	Tomasula et al. (2003)
Collagen		RT, 0% RH	<0.04–0.5	Lieberman and Guilbert (1973)
Collagen		RT, 63% RH	23.3	Lieberman and Guilbert (1973)
MP:Gly (3:1)	pH 3	25°C, 0% RH	1.9	Gontard et al. (1996)
EW:Gly (2.9:1)	80°C, 20 min, pH 10.5, TGase	25°C, 50% RH	70.0	Lim et al. (1998)
LDPE		23°C, 50% RH	1870	Salame (1986)
Cellophane		23°C, 0% RH	0.7	Salame (1986)
Cellophane		23°C, 50% RH	16	Taylor (1986)
Cellophane		23°C, 95% RH	252	Taylor (1986)
EVOH (70% VOH)		23°C, 0% RH	0.1	Salame (1986)
EVOH (70% VOH)		23°C, 95% RH	12	Salame (1986)
Polyester		23°C, 50% RH	15.6	Halon (1992)

Notes: SPI, soy protein isolate; WG, wheat gluten; WPI, whey protein isolate; CO_2C, carbon dioxide precipitated casein; CaC, calcium caseinate; MP, myofibrillar protein; EW, egg white; Gly, glycerol; Sor, sorbitol; EtOH, ethanol; TGase, transglutaminase; LDPE, low density polyethylene; EVOH, ethylene-vinyl alcohol copolymer; VOH, vinyl alcohol.

with either single or double coats of corn-zein remained unaltered as compared with the base soy film (Pol et al., 2002).

Mechanical properties of edible films are generally evaluated by three parameters: tensile strength (TS), elongation at break (E), and Young's modulus (YM). Mechanical properties depend on the plasticizer type and amount, and the temperature and RH of the ambient. TS of zein films is similar to that of wheat gluten films (Table 2.3) and two- to threefold lower than that of methycellulose (MC) and hydroxypropyl cellulose (HPC) films (Gennadios et al., 1993b). Gennadios et al. (1993b) showed that the TS of zein films decreased linearly with increased RH and quadratically with temperature increase. The type of plasticizer used also determined the extent of the changes seen in the tensile properties of films stored at different RH values. In general, films' TS and YM increases as RH increases. Zein films containing dibutyl tartrate, triethylene glycol, levulinic acid, or PEG 300 showed an increase in E with increasing RH. But films containing Gly, OA, or no plasticizer did not show any increase in E as RH increased (Lawton, 2004). The moisture absorption behavior of films plasticized with PEG 400 and PEG 1000 is similar at low RH but significantly different at higher RH. Incorporation of up to about 30% PEG substantially enhanced the TS, with PEG 1000 being more effective than PEG 400 (Tillekeratne and Easteal, 2000). Mechanical properties have also been improved by combining different plasticizers. Zein films plasticized with Gly/PPG (1:3 weight ratio) had elongation values almost 50 times greater than those of Gly-plasticized films (Parris and Coffin, 1997). The use of PEG-LA composite plasticizers increased film flexibility as the portion of PEG was increased (Paramawati et al., 2001). Films with mixed Gly and sorbitol (Sor) as plasticizers showed better mechanical properties than pure films (Muthuselvi and Dhathathreyan, 2006). The presence of endogenous free fatty acids in zein isolated from dry-milled, bulk yellow dent, and ground hybrid corn formed films with 40% to 200% higher E than films without free fatty acids (Parris et al., 2002).

The addition of cross-linking agents such as formaldehyde, glutaraldehyde, or citric acid increased TS up to twofold (Table 2.3) (Yang et al., 1996; Parris and Coffin, 1997). Lamination of zein films with flax oil or tung oil also increased TS and E of the films with respect to uncoated films (Wang and Padua, 2005).

Films formed from zein resin (zein-oleate mass) are translucent, flexible, ductile, and heat sealable. Resin zein films can be drawn by hand or formed by extrusion. It was observed that stiffness and E were affected by processing conditions, and quality differences are believed due to film structure. Under the microscope, zein-oleate films show a fiber network structure with pinholes and structural gaps (Padua and Wang, 2002). Differential scanning calorimetry (DSC) showed that there was no phase separation between protein and oleic acid upon heating, which suggested a strong interaction between these components (Lai and Padua, 1997; Wang et al., 2003). Small-angle X-ray scattering (SAXS) showed a strong periodicity across the zein-oleate film plane. A structural model was proposed consisting of a layered structure of zein planes altering with double layers of oleic acid (Lai et al., 1999). Surface plasmon resonance (SPR) was used to further investigate interactions between zein and oleic acid. It was suggested that the structure development involves hydrophilic

Table 2.3 Mechanical properties of protein-based edible films compared to packaging films

Film	Film solution/treatment	Film test conditions	TS (MPa)	YM (MPa)	E (%)	Reference
Zein, unplasticized	80% EtOH	RT, 52% RH	10.9	551	3.4	Parris and Coffin (1997)
CZ:Gly+PPG (3:1)	80% EtOH	RT, 52% RH	5.1	135	117.8	Parris and Coffin (1997)
CZ:Gly (6.7:1)	80% EtOH	RT, 52% RH	5.2	499	4.4	Parris and Coffin (1997)
CZ:Gly (6.7:1)	80% EtOH-glutaraldehyde	RT, 52% RH	19.8	412	3.0	Parris and Coffin (1997)
SPI:Gly (1:1)	70°C, 20 min, pH 10	23°C, 50% HR	3.7	—	124.2	Gennadios et al. (1998a)
SPI:Gly (1:1)	70°C, 20 min, pH 10, UV 103.7 J/m^2	23°C, 50% HR	6.1	—	85.8	Gennadios et al. (1998a)
SPI:Sor (1.7:1)	70°C, 20 min, pH 8	25°C, 50%RH	4.2	—	101.8	Tang et al. (2005)
SPI:Sor (1.7:1)	70°C, 20 min, pH 8, TG 4 U × g^{-1}	25°C, 50%RH	4.5	—	27.3	Tang et al. (2005)
WG:Gly (4.3:1)	50°C, 32% EtOH, pH 4	20°C, 60% RH	1.7	13	501.0	Micard et al. (2000)
WG:Gly (4.3:1)	50°C, 32% EtOH, pH 4, formaldehyde	20°C, 60% RH	8.1	70	90.0	Micard et al. (2000)
WG:Gly (2:1)	Press heating, 80°C	20°C, 60% RH	0.3	—	468.0	Cuq et al. (2000)
WG:Gly (2:1)	Press heating, 135°C	20°C, 60% RH	2.0	—	236.0	Cuq et al. (2000)
WPI:Gly (2.3:1)	90°C, 30 min	23°C, 50% RH	13.9	475	30.8	McHugh and Krochta (1994b)
WPI:Sor (2.3:1)	90°C, 30 min	23°C, 50% RH	14.0	1040	1.6	McHugh and Krochta (1994b)
WPI:Gly (2.3:1)	70°C, 5 min	23°C, 50% RH	3.4	156	7.0	Pérez-Gago and Krochta (2001b)
WPI:Gly (2.3:1)	90°C, 20 min	23°C, 50% RH	13.2	472	15.6	Pérez-Gago and Krochta (2001b)
SC		23°C, 50% RH	37.0	2100	2.5	Schou et al. (2005)
SC:Gly (3:1)		23°C, 50% RH	7.0	—	66.0	Tomasula et al. (2003)

CottonS:Gly (4:1)	Cross-linked formaldehyde	RT, 56% RH	4.0	—	90.0	Marquié and Guilbert (2002)
Collagen casings		23°C, 50% RH	11.2	3	35.4	Simelane and Ustunol (2005)
MP:Gly (3:1)	pH 3	20°C, 59% RH	17	—	23	Cuq et al. (1995)
MP:Gly (2.9:1)	40°C, 30 min, pH 2.7	20°C, 50% RH	3–5	2–5	50–100	García and Sobral (2005)
EW:400PEG (2.5:1)	45°C, 20 min, pH 11		5.0	—	62.3	Handa et al. (1999b)
LDPE			9–17	—	500	Briston (1986)
PVC			45–55	—	120	Briston (1986)
OPP			165	—	50–75	Briston (1986)
PET			175	—	70–100	Briston (1986)
Cellophane			48–110	—	15–25	Briston (1986)

Notes: CZ, zein; SPI, soy protein isolate; WG, wheat gluten; WPI, whey protein isolate; CottonS, cottonseed protein; MP, myofibrillar protein; EW, egg white; Gly, glycerol; PEG, polyethylene glycol; Sor, sorbitol; EtOH, ethanol; LDPE, low density polyethylene; PVC, polyvinyl chloride; OPP, oriented polypropylene; PET, polyethylene terephthalate.

adsorption of fatty acids onto the zein surface followed by hydrophobic associations leading to a layered structure organization that promotes plastification compared to the casting ethanol solution method (Wang et al., 2003; Wang et al., 2004). Recent studies show the effect of the zein/oleic acid film-forming process on film structure, establishing the molecular structure of films formed by solution casting, by the stretching of moldable resins, and by blown film extrusion (Wang et al., 2005).

To ameliorate the contamination problem of food surface, food additives with antimicrobial activity are incorporated into the edible films and coatings. Antiomicrobial zein edible films have been prepared by incorporating benzoic acid, nisin, lysozyme, EDTA (Padgett et al., 2000; Hoffman et al., 2001; Janes et al., 2002; Dawson et al., 2003; Hsu et al., 2005; Mecitoglu et al., 2006). Physical entrapment or hydrophobic interaction was crucial to the incorporation of benzoic acid into the protein matrix, and FT-Raman spectroscopy showed no modification of the secondary structure of the zein matrix (Hsu et al., 2005).

2.2.3 Zein coatings

Zein is one of the few proteins, such as collagen and gelatin, employed as an edible coating for foods and pharmaceuticals. Zein coatings are used for nuts, candies, confectionary, and other foods (Andres, 1984), and for controlled ingredients release (Gennadios and Weller, 1990; Yoshimaru et al., 2000). Recent works found in the literature show the effect of zein coatings on quality of apples, mango, and shell eggs (Wong et al., 1996; Bai et al., 2002, 2003; Hoa et al., 2002). Zein coatings are also able to reduce oil uptake of deep-fat frying of cowpea paste and of starchy products like mashed potato balls (Mallikarjunan et al., 1997; Huse et al., 1998). Antimicrobial zein edible coatings containing nisin, calcium proprionate, or sorbic acid have been studied to control *Listeria monocytogenes* of ready-to-eat chicken and cooked sweet corn (Janes et al., 2002; Carlin et al., 2001). Microencapsulation with zein protein coatings of L-lysine, for improving the balance of amino acids in ruminants, and fish oil have increased digestion and absorption of the bioactive compounds (Yoshimaru et al., 2000; Xu et al., 2000). The low oxygen permeability and grease-resistant properties of zein films have led to the development of recyclable coated paper (Parris et al., 1998, 2000; Trezza et al., 1998).

2.3 Soy film and coatings

Most of the protein in soybeans can be classified as globulin. Soy protein is classified into 2S, 7S, 11S, and 15S fractions according to the relative sedimentation rates, with 7S and 11S fractions principal components (Gennadios et al., 1994). Soy protein is high in asparagines and glutamine residues. Both 7S and 11S fractions are tightly folded due to intramolecular disulfide bonds (Kinsella, 1979).

Soy protein used in the food industry is classified based on protein content. Defatted soy flour (DSF) contains 50% to 59% protein and is obtained by grinding defatted soy flakes. Soy protein concentrate (SPC) contains 65% to 72% protein and is obtained by aqueous liquid extraction or acid leaching process. Soy protein isolate

(SPI) contains more than 90% protein and is obtained by aqueous or mild alkali extraction followed by isolectric precipitation (Park et al., 2002a).

2.3.1 Formation of soy protein films

Soy protein films are usually prepared by drying thin layers of cast film-forming solutions (Park et al., 2002a). Other film-forming techniques that have been studied include extrusion, spinning of SPI in a coagulating buffer, thermal compaction of mixtures of SPI and glycerol, compression molding, and uniaxial drawing of SPI (Ghorpade and Hanna, 1996; Naga et al., 1996; Rampon et al., 1999; Cunningham et al., 2000; Ogale et al., 2000; Wu and Zhang, 2001; Kurose et al., 2006). Soy protein films can be also produced from soymilk upon heating to near the boiling point or by using twin screw extrusion (Wu and Bates, 1972; Sian and Ishak, 1990; Gennadios and Weller, 1991; Hwang et al., 2003). The dried film formed from the soymilk is a soy protein-lipid film, because typical soymilk composition contains around 28% lipids (dry basis) (Wu and Bates, 1972).

Film formation from aqueous SPI dispersions is believed to proceed through protein polymerization and solvent evaporation. Soy protein polymerization is promoted by heating at temperatures above 60°C and alkaline conditions below pH 10.5 (Kelley and Pressey, 1966). Formation of films is believed to involve development of hydrophobic, disulfide, and hydrogen bonds between protein polymer chains. Heating of the film-forming solution is very important to disrupt the protein structure, cleave native disulfide bonds, and expose sulfhydryl groups and hydrophobic groups, and then to form new bonds between protein chains during film drying (Gennadios et al., 1994).

As with other protein films, soy films require the use of plasticizers to improve flexibility and prevent cracking. Among the different plasticizers incorporated into edible films, Gly and Sor are the most commonly used plasticizers for soy protein–based films (Gennadios et al., 1994).

2.3.2 Functional properties of soy proteins

SPI films are transparent and flexible when plasticized but are poor moisture barriers. Compared to LDPE films, SPI films have greater WVP values. At comparable plasticizer content and test conditions, SPI films appear to be similar moisture barriers to wheat gluten films prepared from alkaline solution and somewhat poorer moisture barriers than corn zein films (Table 2.1) (Krochta, 1997a). Soy protein films are potent oxygen barriers at low RH (Ghorpade et al., 1995a). At similar plasticizer levels and test conditions, soy protein films have similar OP than wheat gluten but one order of magnitude lower than zein films (Table 2.2) (Krochta, 1997a). Mechanical properties of soy protein films are moderate compared to plastic films such as polyethylene, polypropylene, and polyvinylidene chloride (Gennadios et al., 1994; Krochta and De Mulder-Johnston, 1997). Compared to other protein films, SPI films present similar TS as zein and wheat gluten films, but lower E, indicating that SPI films are less tough (Table 2.3) (Krochta, 1997a).

Protein concentration of the film-forming solution is an important factor that contributes to film properties. Films prepared with 5%, 8%, or 10% (w/w) of DSF, SPC, and SPI presented different TS and E (Park et al., 2002a). As the protein content of the product increases (from DSF to SPI) and the concentration of the soy product increases (from 5% to 10%), the TS and E of the formed film increases.

Heat and pH denaturation of soy protein helps in the film-formation process, promoting the formation of disulfide bonds within the structure of the dried films (Gennadios et al., 1993c). Soy protein films can be formed at both alkaline and acidic conditions. Cast SPI films formed at alkaline conditions (pH 8 to 10) had greater TS, greater E, and lower WVP than films formed at acidic conditions (pH 1 to 3) (Gennadios et al., 1993c). SPI films prepared at pH 6 had higher WVP and OP and lower TS and E than films prepared at pH 8, 10, or 12 (Brandenburg et al., 1993; Park et al., 2002a). This was attributed to partial inhibition of the thiol-disulfide interchange reactions and thiol oxidation reactions at the acidic pH value of 6. A recent study shows that protein films obtained from SPI solutions at different pH values (2, 8, and 11) maintained the same type of interactions, but the intensity of each type of interaction (predicted from solubility tests in buffers with different chemical action) depended on the pH of the initial solution. Sodium dodecyl sulfate-polyacrylamide gel (SDS-PAGE) patterns indicated that films were mainly formed by β-conglycinin and glycinin, which aggregated in different forms during film formation, depending on the pH of the initial solutions (Mauri and Anon, 2006).

Heat treatments on dried films also modified properties of soy protein films (Gennadios et al., 1996a; Rangavajhyala et al., 1997; Rhim et al., 2000; Kim et al., 2002a). Heat curing of protein-based films promotes cross-linking between intra- and intermolecules, such as lysine and cystine, and polar groups of the protein chains, resulting in films with increased TS and reduced E, as well as enhanced film hydrophobicity (Cheftel et al., 1985; Park et al., 2002a). For example, SPI films subjected to heat curing (90°C for 24 hours) increased film TS from 8.2 to 14.7 MPa, decreased E from 30% to 6%, and reduced moisture content of the film (Rhim et al., 2000). Heat treatment under vacuum reduced the WVP of SPI films faster than heat curing at atmospheric pressure. High TS values and low E values were also reached within short heating times under vacuum. However, vacuum treatment increased the size and number of cavities in cured films as evidenced by scanning electron micrographs (Kim et al., 2002a).

Covalent cross-linking between aromatic amino acids can be promoted by ultraviolet (UV) irradiation. This effect has been observed in UV-irradiated SPI films, where SDS-PAGE patterns for UV-treated samples revealed bands of aggregates, increasing in intensity with UV dosage, which were absent in the control sample. This translates on SPI films with increased TS and decreased E (Table 2.3) (Gennadios et al., 1998a) and lower protein solubility than control films (Rhim et al., 2000). However, WVP was not affected by UV treatment (Gennadios et al., 1998a). The effects of ionizing radiation, such as γ-radiation, on film properties have also been studied on SPI films. γ-radiations up to 30 kGy did not affect film mechanical properties (Ghorpade et al., 1995b), whereas 50 kGy decreased WVP by 13% and increased TS by two times (Lee et al., 2005). γ-Radiation combined with thermal treatments have also been applied to form cross-linked soy and whey protein-based films, with improvement of film mechanical properties (Sabato et al., 2001).

Plasticizer is incorporated into SPI films to induce flexibility. Gly is the most widely used plasticizer for SPI films because of its small size and hydrophilic nature, which make it compatible with SPI films (Cho and Rhee, 2002; Choi et al., 2003; Wan et al., 2005). Increased Gly or Sor content, alone or in combination, increased flexibility of SPI films. Plasticizer and RH exert a synergic effect on mechanical and barrier properties. RH effects on mechanical properties of SPI films were varied with plasticizers and their concentration. Films of lower Gly content were more sensitive to RH variation as compared to the higher Gly samples, whereas Sor concentration affected the RH region where a sharp decrease in TS value occurred (Cho and Rhee, 2002). Choi et al. (2003) determined the sorption isotherms of SPI films as affected by Gly content and a_w, observing an increase in the equilibrium moisture isotherm as a_w increased and values were significantly different among Gly-plasticized films at $a_w > 0.33$. These results showed a correlation with the molecular dynamics of the SPI films determined by nuclear magnetic resonance (NMR). Combination of less hygroscopic plasticizers, such as propylene glycol (PG), PEG, Sor, or Sucrose (Suc) with Gly improved SPI film moisture barrier but reduced film flexibility (Wan et al., 2005). An addition of as little as 25% of a less hygroscopic plasticizer in the mixture induced significant reduction in WVP. However, at least 50% of the mixture needs to be Gly to show significant improvement in TS. In this work, 50:50 Gly:Sor was the recommended combination because of its comparatively low WVP value and relatively high flexibility and strength, whereas the Gly:PEG plasticizer mixture showed incompatibility with surface migration of PEG from the film matrix.

The use of enzymes that promote cross-linking between protein chains has also been studied in SPI films. Horseradish peroxidase enzyme did not affect WVP of SPI films, but it increased TS and protein solubility and decreased E, suggesting some protein degradation by enzyme addition (Stuchell and Krochta, 1994). Transglutaminase (TG) is another enzyme able to catalyze cross-linking reactions between proteins. TG treatment increases TS; decreases E, moisture content, total soluble matter, and transparency; and enhances surface hydrophobicity of SPI films (Table 2.3) (Tang et al., 2005). TG has also been used to combine SPI or protein fractions (11S) with other proteins or polysaccharides, such as casein, whey protein, and pectin, improving mechanical and barrier properties of the films and decreasing film solubility (Motoki et al., 1987a; Yildirim and Hettiarachchy, 1998a; Mariniello et al., 2003; Di Pierro et al., 2005).

From sulfur-containing proteins, cysteine groups can undergo polymerization upon heat denaturation to form a continuous covalent network upon cooling. Therefore, incorporation of cysteine to SPI solutions enhances cross-linking of the protein polymer chains, increasing SPI film TS and decreasing E (Liu et al., 2006). For example, addition of 1% cysteine (w/w) to SPI film-forming solutions at pH 7 increased the number of disulfide bonds by 576.6%, which translated in an increase in TS (Were et al., 1999).

Other cross-linking agents that have been studied in SPI films include ferulic acid, calcium salts, glucono-δ-lactone (GDL), propyleneglycol alginate (PGA), and low molecular weight aldehydes. Ferulic acid can react with amino and thiol groups in proteins, and additionally with tyrosine and with itself to form diferulic acid, which acts as a bridge between protein molecules (Ou et al., 2005). Calcium bridges maximize interactions between negatively charged molecules, increasing protein network and stability. GDL promotes aggregation of the unfolded protein by increasing hydrophobicity and insolubilization (Park et al., 2001). PGA interacts with the amino group of lysine, reducing WVP and solubility of SPI films (Rhim et al., 1999a). Low molecular weight aldehydes (e.g., formaldehyde, glutaraldehyde, dialdehyde starch) react with primary amino groups and sulfhydryl groups in proteins, forming intra- and intermolecular cross-links, resulting in films with higher TS, lower solubility, and WVP (Ghorpade et al., 1995a; Park et al., 2000a; Rhim and Weller, 2000; Rhim et al., 2000).

Addition of the anionic surfactant sodium dodecyl sulfate (SDS) to film-forming solutions prior to casting greatly modifies the properties of SPI films. SDS reduced film TS by as much as 43% for films with 40% SDS, and increased E. Films with SDS had smaller moisture content and larger solubility than control SPI films. Films containing 10% or more SDS had lower WVP values than control SPI films by as much as 50%. Changes in tensile, solubility, and water vapor barrier properties of SPI films due to the addition of SDS were largely attributed to disruption of hydrophobic associations among neighboring protein molecules as the nonpolar portions of the SDS molecules attached onto hydrophobic amino acid residues within the film structure (Rhim et al., 2002). The disruption by SDS of the low-energy intermolecular bonds that maintain the protein conformation is also confirmed by a decrease in thermal stability and the glass transition temperature of the films formed by SPI-SDS (Schmidt et al., 2005).

Fractionation of SPI by ultrafiltration to obtain SPI fractions of different molecular weights (100 and 300 kDa) did not modify WVP of the films. However, TS and %E at break of films increased with molecular weight of soy protein formed films. In addition, protein solubilities of fractionated films were in the range of 3.5 to 4.6 g/100 g of dry film, whereas in nonfractionated films, solubility was 11.9 g of proteins/100 g of dry SPI film (Seung and Rhee, 2004).

The water barrier of hydrophilic films, like SPI, can be improved through combination with lipid materials forming an edible composite film. Fatty acids (i.e., lauric, myristic, palmitic, or oleic acid) were incorporated into SPI film-forming solutions at 10% to 30% w/w of SPI, observing a reduction in WVP of SPI-fatty-acid-based films. Among these fatty acids, OA showed a plasticizer effect on the films, increasing E of the films (Gennadios et al., 1998b). Increasing the concentration of sorghum wax paste in SPI films decreased WVP, E, and total soluble matter values of SPI-sorghum-wax-paste composite films. TS values were lower than the control upon addition of 5% and 10% wax paste; however, TS values increased at 15% and 20% wax concentrations (Kim et al., 2002b, 2003).

Soy protein–based films have also been prepared from water extract of soybeans and proteinaceous fibrous material (PFSP) produced from soybean fermented by *Bacillus natto* (Cao and Chang, 2002; Park and Bae, 2006). Soy extract films were relatively weak but very elastic compared to SPI films, and pH 10 produced films with the highest TS (Cao and Chang, 2002). To produce films from PFSP, pH values of 7 to 9 and heat treatment of 70°C to 90°C were needed. Alkaline pH and temperature caused a decrease in TS and WVP of the films. The combination treatment that provided the best combination of barrier and mechanical properties was the PFSP film prepared at pH 7 with addition of 1% cysteine. These films were good oxygen barriers (Park and Bae, 2006).

2.3.3 Soy protein coatings

Soy proteins have been tested as potential controlled-release carriers for insect sex pheromones (Atterholt et al., 1998) and microencapsulating agents of flavors and medications (Petersen et al., 1999). Grease-resistant paper has been produced by coating paper with SPI (Park et al., 2000b), and improved coated board properties can be obtained by using soy protein as cobinders in coatings for paper and paperboard (Rhim et al., 2006). Soy protein and gellan gum films, alone or in combination, are able to reduce fat intake in deep-fried foods (Albert and Mittal, 2002; Rayner et al., 2000). Soy protein films have reduced lipid oxidation and retained moisture of pink salmon fillets during frozen storage and precooked beef patties (Sathivel, 2005; Wu et al., 2000).

SPI films carrying antimicrobial compounds (nisin and natural extracts) have shown inhibitory effects against *Listeria monocytogenes* on ready-to-eat meat products (Theivendran et al., 2006). A combination of SPI films containing natural antimicrobials, in combination with γ-irradiation, was effective in reducing aerobic microorganisms and *Pseudomona putida* of precooked shrimp with at least 12 days extension of shelf life (Ouattara et al., 2001). Eswaranandam et al. (2006), after showing the effectiveness of SPI-organic acid films controlling *L. monocytogenes*, *Escherichia coli* O157:H7, and *Salmonella gaminara*, studied the effects of these coatings on the sensory attributes of whole apples and fresh-cut cantaloupes, showing that incorporation of the organic acids did not affect sensory properties. However, incorporation of thyme oil and trans-cinnamaldehyde reduced the acceptability scored for taste and odor of precooked shrimp (Ouattara et al., 2001).

2.4 Wheat gluten proteins

Wheat gluten (WG) is the cohesive and elastic mass that is leftover after starch is washed away from wheat flour dough. Vital WG consists mainly of wheat storage protein (70% to 80%, dry matter basis) with traces of starch and nonstarch polysaccharides (10% to 14%), lipids (6% to 8%), and minerals (0.8% to 1.4%) (Guilbert et al., 2002).

There are four wheat protein classes based on solubility in different solvents: albumins, globulins, gliadins, and glutenins. The albumins and globulins (15% to 22% of total protein) are removed with starch granules during gluten processing due to their solubility in water and salt, respectively. Whereas the gliadins or prolamine, which are alcohol soluble, and the glutenins, which are soluble (or at least dispersible) in dilute acid or alkali solutions, are being collected into gluten. Both proteins amount to up to 85% (w/w) of total wheat flour protein and are evenly distributed into gluten (Wrigley and Bietz, 1988). Gliadin is characterized by its high content of nonpolar amino acids proline and glutamine. The terminal amide group of glutamine promotes extensive hydrogen bonding among polypeptide chains. Gliadin globular structure is mainly due to hydrophobic interactions between nonpolar amino acids and development of some intramolecular disulfide bonds (Kasarda et al., 1989; Lásztity, 1986). Glutenin is believed to be one of the largest natural polymeric molecules with estimated molecular weight of over 10^7. It has a similar amino acid composition to gliadin, but with a slightly lower content of hydrophobic amino acids (Wrigley and Bietz, 1988). It is believed that glutenin results from cross-linking of polypeptide subunits by end-to-tail disulfide bonds (Kasarda, 1999). The gluten complex is believed to be a protein network held in place by extensive covalent and noncovalent bonding. Extended random-coiled glutenin polypeptides provide the frame of the structure, and smaller globular gliadin polypeptides are packed into the network (Bietz and Wall, 1980).

2.4.1 Formation of wheat gluten protein films

Film formation from WG solutions or dispersions has been extensively studied (Gennadios and Weller, 1990; Gennadios et al., 1993d, 1993e, 1993f; Park and Chinnan, 1995; Guilbert et al., 2002). WG protein films are usually prepared by drying thin layers of cast film-forming solutions or by extrusion (Guilbert et al., 2002). WG films have also been produced by collecting the surface skin formed during heating of WG solutions to temperatures near boiling (Watanable and Okamoto, 1976). Gliadins and glutenin films can also be prepared after extraction of both proteins from WG (Hernandez-Muñoz et al., 2003).

The casting method requires a complex solvent system with basic or acidic conditions in the presence of alcohol and disulfide bond reducing agents. Generally, changing the pH of the medium disrupts hydrogen and ionic interactions, while ethanol disrupts hydrophobic interactions. Intermolecular (for glutenins) and intramolecular (for gliadins and glutenins) covalent SS bonds are cleaved and reduced to SH groups when dispersing WG in an alkaline environment (Okamoto, 1978). Reducing agents, such as sodium sulfite, cysteine, or mecaptoethanol, should be used in acidic environments. During drying, solvents evaporate, allowing active sites to become free for bond formation (Guilbert et al., 2002). The extensive intermolecular interactions in WG result in quite brittle films that require the addition of plasticizers (Krochta, 1997a). The most common plasticizers used in WG films are Gly and ethanolamine; other plasticizers that have been studied are Sor, manitol, diglycerol, PPG, triethylene glycol, maltitol, polyvinyl alcohol, and PEG.

The extrusion method involves a thermal treatment of plasticized WG followed by a thermoforming step. Plastification of WG reduces glass transition temperature and enables processing at temperatures below those that lead to protein decomposition, which means that film formation can be performed by using techniques that are conventionally used with synthetic polymers (e.g., extrusion, injection, and molding) (Redl et al., 1999; Gallstedt et al., 2004; Mangavel et al., 2004).

2.4.2 Functional properties of wheat gluten protein films

The barrier and mechanical properties of WG films depend on processing conditions, the addition of plasticizers, lipids, and other cross-linking agents, and external conditions as temperature and RH.

WG films show low TS but high E compared to other films (Table 2.3) (Gennadios et al., 1993c). Gly and water act as plasticizers of WG films, reducing TS and increasing E (Gontard et al., 1993; Tanada-Palmu et al., 2000). However, at very low water content and low RH (below 15 g/100 g dry matter and 60%, respectively), hydration has a positive effect on film mechanical resistance, probably due to formation of supplementary hydrogen bonds among protein chains (Gontard et al., 1993). Irissin-Mangata et al. (2001) selected diethanolamine and triethanolamine among several low and high molecular weight polyols and amines tested as potential plasticizers of wheat gluten film. In comparison with Gly, their use did not significantly affect the solubility in water, the opacity of the film (both increased very slightly), and the water vapor barrier properties. However, they significantly increase extensibility and elasticity of the films. At 58% RH and 20 g/100 g dry matter of di- or triethanolamine, the E was five times higher than film plasticized with Gly. At 98% RH, water appeared to compete with the amines to act as the plasticizer.

Considering the importance of hydrogen bonding between polymer chains on mechanical properties, Mangavel et al. (2003) studied the effect of various hydrophilic plasticizers (differing in their chain length) on the mechanical properties of cast gluten films, taking into account their molar amount in the film as well as their ability to share hydrogen bonds with the protein network. Two series of organic molecules were chosen for this purpose: (1) the ethyleneglycol/diethyleneglycol/triethyleneglycol/tetraethyleneglycol series, which consists of the repeat of the ethylene oxide $-CH_2-CH_2-O$ group, and (2) the ethyleneglycol/1,3-propanediol/1,4-butanediol series (addition of one methyl group $-CH_2-$ at each increment). In both cases, the results were compared with Gly. When plasticizer efficiencies were considered on a molar basis, there appeared to be a clear influence of their molecular weights on the mechanical properties of cast films below a molar ratio specific to each molecule. A finer analysis revealed that these differences of efficiency were partly related to the ability of the plasticizers to supply hydrogen bonds to the proteins. The effect of Gly was somewhat different from that of the other plasticizers examined, because the threshold value occurred at a higher plasticizer amount but led to lower stress and higher strain for films; this was attributed to the specific chemical structure of this molecule, such as the number and position of hydroxide groups in the molecule. In terms of applications, Gly was able to provide cast films with a wider range of variations of E than the other plasticizers studied.

Increasing the processing temperature of pressure-molded WG films plasticized with Gly from 80 to 135°C induces an increase in TS and a decrease in E of the films (Table 2.3) (Cuq et al., 2000). Thermoplastic processed WG films plasticized with Gly behave like rubber, and the effect of temperature and water content can be correlated with changes on glass transition temperature (Gontard and Ring, 1996). Increasing curing temperature and exposure time of solvent-cast WG films also increased TS and decreased E of the films (Ali et al., 1997; Micard et al., 2000). Heat curing induces protein insolubilization, probably due to an increase in covalent binding between glutenins, leading to changes in the mechanical properties (Micard et al., 2000). For example, heat curing at 140°C increased TS and E up to 300% and 43%, respectively, of the value obtained when the films were heated at 80°C.

In addition, results showed an interaction between temperature and exposure time, because the 140°C, 1.5 minutes treatment gave values comparable to those obtained after a 110°C, 15 minutes treatment (Micard and Guilbert, 2000). Drying temperature and RH also affect mechanical properties of cast WG-Gly films, with the effect of drying temperature being more important than the effect of RH. At 35% RH, TS increased when drying temperature increased. However, at 70% RH, TS decreased when temperature increased (Dawson et al., 2003).

Common cross-linking agents (such as formaldehyde, calcium chloride, SDS, transglutaminase, 1-ethyl-3-3-dimethylaminopropyl carbodiimide, and N-hydroxysuccinimide) and radiation treatments (such as UV and γ-radiation) have been shown to improve the mechanical properties of WG films as a consequence of an increase in covalent bonds between polymer chains (Table 2.3) (Rhim et al., 1999b; Larre et al., 2000; Micard and Guilbert, 2000; Micard et al., 2000; Domenek et al., 2004; Tropini et al., 2004).

Gennadios et al. (1993g) studied the effect of pH on mechanical properties of WG film. WG films formed within pH 2 to 4 and 9 to 13, and film formation was inhibited by poor protein dispersion around the isolectric pH region (pH 7.6). Gontard et al. (1992) and Herald et al. (1995) showed that WG films made at low pH from an ethanol solution were stronger than films obtained from alkaline conditions. Whereas Kayserilioglu et al. (2001) observed that films prepared at pH 11 were stronger than films prepared at pH 4 and 6. The solubility of the WG films in aqueous and 1.5% SDS solution decreased with increasing pH of films, and tricine sodium dodecyl sulfate-polyacrylamide gel (SDS-PAGE) patterns of wheat gluten and film showed the formation of high molecular weight fractions, indicating the formation of irreversible intermolecular cross-linking upon film formation at pH 11.

The WVP of WG films is equivalent to that of other protein- or polysaccharide-based films but is relatively high compared to synthetic films (Table 2.1). At comparable plasticizer content and test conditions, WG films prepared from alkaline solutions appear to be somewhat poorer moisture barriers than corn zein (Krochta, 1997a). WVP of WG films is greatly modified by plasticizer type and content and RH (Gontard et al., 1993; Roy et al., 2000). For example, at 30°C the WVP is tenfold lower at 50% RH than at 93% RH, and WVP increased linearly with increasing Gly concentration (Gontard et al., 1993). The use of hydrophobic plasticizers, such as fatty acids, improves the barrier properties of pressure-mold WG films compared to films plasticized with Gly (Pommet et al., 2003). WVP of solvent-cast WG films is less affected by heat curing, UV and γ-irradiation, pH, and application of formaldehyde vapor treatments, although the effect might depend on treatment conditions and doses (Ali et al., 1997; Rhim et al., 1999b; Micard et al., 2000; Kayserilioglu et al., 2001).

WVP of WG films can be optimized by incorporating lipid materials to obtain an emulsion film (if the lipid is dispersed within the WG protein matrix) or a bilayer film (if the lipid is deposited as a layer onto the surface of preformed WG film). Gontard et al. (1995) reduced WVP of WG films laminated with beeswax (BW) by 200-fold compared to an uncoated control GW film, whereas the WVP of emulsion films containing beeswax (BW) was reduced lower than threefold

(Gontard et al., 1994, 1995). However, in BW-WG bilayer films, the lipid layer became easily detached from the WG-based film. Lipid adhesion was improved by the incorporation of diacetyl tartaric ester of monglyceride in the WG:Gly base film, which also helped reduce WVP.

Combination of WG with other proteins or polysaccharides, such as cellulose, soy protein, corn zein, and starch, has also been prepared as edible films with different barrier and physical properties (Fringant et al., 1998; Were et al., 1999; Foulk and Bunn, 2001; Fakhouri et al., 2004).

The gas (O_2, CO_2, and ethylene) barrier properties of WG-based films are low at low RH conditions, and values are close to those for EVOH and much lower than those of LDPE (Table 2.2) (Aydt et al., 1991; Gennadios et al., 1993f). Gas barrier properties are highly dependent on RH and temperature. This effect is more pronounced for hydrophilic gases (CO_2) than for hydrophobic gases (O_2) (Gontard et al., 1996; Mujica-Paz and Gontard, 1997). Thus, the CO_2/O_2 selectivity coefficient can rise from 3 to more than 50 when RH increases from 0 to 100% and temperature increases from 5 to 45°C (Guilbert et al., 2002).

Recent works present the mechanical and gas barrier properties of gliadins and glutenins-rich film. Glutenin-rich films presented higher TS values and lower E and WVP values than gliadin-rich films. Gliadin films disintegrate when immersed in water (Hernandez-Muñoz et al., 2003). The mechanical and gas barrier properties of these films are improved by cross-linking agents, such as cysteine and aldehydes (Hernandez-Muñoz et al., 2004a, 2004b, 2004c, 2004d). Both films became stronger and less extensible with lower WVP after heat curing treatment, showing a sharp change in physical and properties in the range of 55 to 75°C (Hernandez-Muñoz et al., 2004e). Mechanical and WVP properties of glutenin-rich films plasticized with Gly change dramatically over time, whereas the properties of films plasticized with triethanolamine and Sor remained stable during storage (Hernandez-Muñoz et al., 2004f).

Nisin has been incorporated into WG films to form antimicrobial films. The filmforming method affected the retention of biologically active nisin (nisaplin) and release of activity at different temperatures. Cast WG films retained 15.8% of the original activity after film formation, whereas heat-pressed WG films retained 7.4%. Similarly, cast films presented higher nisin activity than heat-pressed films (Dawson et al., 2003). Temperature dependence of nisin diffusion in all films followed the Arrhenius model (Teerakarn et al., 2002). Sorbate diffusion in WG films also follows the Arrhenius model, and diffusivity was decreased when a lipid was incorporated into the WG film (Redl et al., 1996).

2.4.3 Wheat gluten protein coatings

WG coatings have reduced weight loss of cherry tomatoes and sharon fruits compared to uncoated fruits (Tanada-Palmu et al., 2000). A bilayer film of WG and BW significantly lowered weight loss from coated cheese cubes compared to a single-layer coating of WG (Tanada-Palmu et al., 2000). WG coatings also reduced moisture loss and lipid oxidation of precooked beef patties during refrigeration (Wu et al., 2000).

2.5 Whey proteins

Whey proteins remain in the milk serum after cheese or casein manufacture (DeWit, 1989; Morr and Ha, 1993). Whey protein, which represents approximately 20% of total milk proteins (Brunner, 1977), is a mixture of proteins with diverse functional properties. The main five proteins are α-lactalbumins, β-lactoglobulins, bovine serum albumin, immunoglobulins, and proteose-peptones. β-Lactoglobulins comprise approximately 57% of the protein in whey (Dybing and Smith, 1991). Crystallization analysis has shown that β-lactoglobulins exist in a globular form, with stabilizing hydrophobic and SH groups located in the interior. α-Lactalbumin is the second most abundant whey protein, accounting for about 19% of the total whey protein (Dybing and Smith, 1991). It is a globular protein that contains four S–S bonds. This molecule has a low content of organized secondary structure. However, bound calcium and the four S–S bridges maintain the globular structure of α-lactalbumin and stabilize it against denaturation (Kinsella and Whitehead, 1989). Bovine serum albumin is a large globular protein and represents approximately 7% of the proteins in whey (Dybing and Smith, 1991). It contains 17 S–S bonds and one free thiol group, which makes it highly structured (Morr and Ha, 1993). The immunoglobulins and proteose-peptone fractions represent 13% and 4% of the protein in whey, respectively.

Depending on the industrial processes used for whey protein recovery, whey protein products can be classified according to their protein content as whey protein concentrate (WPC) and whey protein isolate (WPI), which contain 20% to 80% and >90% protein, respectively.

2.5.1 Formation of whey protein films

Formation of whey protein films has mainly involved heat denaturation of whey proteins in aqueous solutions. Heating modifies the three-dimensional structure of the protein, exposing internal SH and hydrophobic groups (Shimada and Cheftel, 1998), which promote intermolecular S–S and hydrophobic bonding upon drying (McHugh and Krochta, 1994a). McHugh et al. (1994) produced WPI films by heat-treating 8% to 12% (w/w) solutions of whey proteins at temperatures between 75°C and 100°C. Whey protein concentrations greater than 12% (w/w) gelled upon heating. Optimal conditions for production of these films were heating of 10% whey protein solutions at 90°C for 30 minutes. These conditions formed films with a consistent structure, and X-ray diffraction indicated that whey proteins were irreversibly denatured (Lent et al., 1998). Lower protein concentration values (down to 5% WPI) have also been shown to form good whey protein-based films whenever the amount of solids deposited per casting surface unit was kept constant (Pérez-Gago et al., 1999). Pérez-Gago et al. (1999) also observed that native whey protein (i.e., whey protein that has not undergone heat denaturation) had good film-formation properties in a pH range from 4 to 8. Because native whey proteins maintain their globular structure with most of the hydrophobic and SH groups buried in the interior of the molecule, intermolecular forces that promote cohesion in native protein films rely mainly on hydrogen bonding and hydrophobic bonds (Pérez-Gago and Krochta, 2002). The intermolecular disulfide bonds of heat denatured whey protein films imparted insolubility to the resulting films; whereas, native whey protein formed soluble films.

Banerjee and Chen (1995) tested various denaturing temperature and pH conditions to optimize the film-forming conditions for WPC. They reported that heating WPC solutions at 75°C for 30 minutes after pH adjustment to 6.6 using $2M$ NaOH produced solutions that ensured uniform film production.

Compression-molded WPI films can be also be formed using 30% to 50% moisture content or Gly content at 104°C to 160°C for 2 minutes (Sothornvit et al., 2003). Water-plasticized films are quite brittle due to loss of moisture during the compression molding process. Addition of Gly to the WPI for compression molding results in quite flexible films. The degree of film flexibility depends on the amount of plasticizer. Molding temperature affects the degree of film solubility, allowing one to obtain soluble or insoluble films depending on film application.

2.5.2 Functional properties of whey protein films

Whey protein films are transparent, bland, and flexible, and have excellent oxygen, aroma, and oil barrier characteristics. On the other hand, these films are poor moisture barriers due to their hydrophilic character. Compared to synthetic materials, plasticized whey protein films have WVP nearly one order of magnitude greater than cellophane and nearly four orders of magnitude greater than LDPE under comparable test conditions and with appropriate level of plasticizer. Even at low RH and low amount of plasticizer, WVP of WPI films is three orders of magnitude greater than that of LDPE. Compared to other protein films, WPI films appear to be poorer moisture barriers than wheat gluten, soy protein, casein, and zein films (Table 2.1). Incorporation of lipid materials improves film moisture barrier properties by increasing hydrophobicity. WPI-lipid composite films can be accomplished by laminating the protein film with a lipid layer, forming a bilayer film, or it can be added to film-forming solutions to form emulsion composite films. In general, bilayer composite films tend to delaminate because of the high surface energy existing between the two components (Shellhammer and Krochta, 1997). Nevertheless, Anker et al. (2002) prepared a WPI-acetylated monoglyceride bilayer film and compared its moisture barrier property to an emulsion film. Lipid lamination decreased the WVP 70 times compared with the WPI film, whereas the WVP of the emulsion film was half the value of the WPI film. Among the lipids that have been added to whey protein films to form composite emulsion films are acetylated monoglycerides, waxes, fatty alcohols, and fatty acids. Increasing the lipid content of films decreased WVP, but the rate at which it decreased depended on the lipid type. Beeswax and fatty acids were more effective at reducing WVP of WPI-based emulsion films than fatty acid alcohols. Increasing the chain length of both fatty acids and fatty alcohols also reduced the WVP of WPI emulsion films, which correlates with the magnitude of lipid polarity (McHugh and Krochta, 1994a; Yoshida and Antunes, 2004). However, increasing soy oil concentration did not affect WVP of WPI composite films (Shaw et al., 2002a). Shellhammer and Krochta (1997) observed that the moisture barrier of WPI-based emulsion films correlated with the viscoelastic properties of the lipids and not with the WVP of the lipid materials. Emulsion films prepared with candelilla wax

(CanW) and carnauba wax (CarW), the materials with the lower WVPs, gave emulsion films with the highest WVPs. In contrast, emulsion films prepared with BW and a hard anhydrous milkfat fraction (HAMFF), the materials with the highest WVPs, gave emulsion films with the lower WVPs. In addition, films could be made with up to 70% to 80% BW and HAMFF, but with greater than 40% CanW or CarW films did not form. A sharp drop in WVP was observed at 40% to 50% BW and HAMFF content, and further increase in lipid content did not affect WVP. These films had up to 90% lower WVPs than films with no lipid content. These results suggested that the more viscoelastic lipids (BW and HAMFF) may have yielded more easily to the internal forces related to shrinkage of the drying protein structure. In addition, above 40% to 50% content, the more easily deformable BW and HAMFF may form an interconnecting lipid network internal to the film that causes the large drop in film WVP (Krochta, 1997b).

McHugh and Krochta (1994a) observed that, given a constant volume fraction of BW, decreasing the lipid particle size significantly correlated with a linear decrease in film WVP of WPI-based emulsion films. Pérez-Gago and Krochta (2001a) showed that the effect of lipid particle size on film WVP and mechanical properties was influenced by lipid content and film orientation during WVP measurements. As lipid content increased, a decrease in lipid particle size reduced the WVP of the WPI-BW emulsion films, probably due to an increase in protein immobilization at the lipid–protein interface as lipid content became more important in the film. This effect seemed to be supported by the increase in film TS. For those films that showed lipid phase separation during drying, WVP was not affected by lipid particle size when the enriched lipid phase was facing the high RH side. This was attributed to the greater barrier that the water vapor experienced when the enriched lipid phase was exposed to the high RH side during the WVP measurement, making any effect due to lipid particle size overwhelmed by this factor.

Because lipid distribution in the matrix has been shown to affect the moisture barrier properties of emulsion films, Pérez-Gago and Krochta (1999) examined the role of emulsion stability, as affected by pH, on the final WVP of WPI-BW emulsion films. Phase separation occurred in emulsion films except at pH 5 ($\approx$pI). In general, the results showed no significant differences among WVPs of films prepared at pH different from the pI. However, at pH close to the pI of the WPI (4 and 5), the WVP was significantly higher. At these pH values, emulsion viscosity increased significantly as a consequence of protein–protein aggregation at the pI, lowering lipid mobility and, thus, phase separation.

Lipid distribution was also modified by changes of the film drying temperature. Pérez-Gago and Krochta (2000) observed a reduction in film WVP as drying temperature increased from 40°C to 80°C. A large drop was observed in WVP of WPI-BW emulsion films at 20% BW content, reaching values as low as those found by Shellhammer and Krochta (1997), even though the plasticizer level was five times larger and lipid content was two times lower than that used by these authors (Table 2.1).

As with other proteins, WPI film formation requires the addition of plasticizers to overcome brittleness. Plasticizer and RH have the most significant effects on WVP and mechanical properties of WPI films. In the case of whey protein edible films,

Gly, Sor, and PEG have been commonly used as plasticizers. McHugh et al. (1994) studied the effects of various RH and plasticizers, including Gly, Sor, and PEG of different chain lengths, on WVP of WPI films. The results showed an exponential-type relationship between WVP and RH. The effect of RH on WVP was more pronounced for Gly-plasticized films compared to Sor-plasticized films. Sor provided the highest flexibility increase per unit increase in WVP among the studied plasticizers at comparable concentrations and RH conditions. Increases in film flexibility with increasing plasticizer content are accompanied by decreases in glass transition temperature and increases in equilibrium moisture contents of films, suggesting that plasticizers may function by altering the moisture content of films (Galietta et al., 1998; Kim and Ustunol, 2001; Shaw et al., 2002a). Shaw et al. (2002b) attributed the relative effectiveness of different plasticizers (Gly, Sor, and Xylitol) in imparting flexibility to WPI films to their different hygroscopicities. Anker et al. (2000) correlated film microstructure and the mechanical and barrier properties of WPI films. These authors observed that the microstructure of WPI-plasticized films was dependent on the concentration of WPI, the plasticizer type, and the pH. When WPI concentration increased, a more aggregated structure was formed with a denser protein network and larger pores. This resulted in increased WVP and decreased OP. When Gly was used as a plasticizer instead of Sor, the microstructure was different, and the moisture content and WVP approximately doubled. The pores tend to be more and smaller when Sor was used, compared to the fewer and larger pores seen when Gly was used. When the pH increased from 7 to 9, a denser protein structure was formed, the strain at break increased, and the OP decreased.

Sothornvit and Krochta (2001) studied the effects of six plasticizers over a range of concentrations on the mechanical properties of β-lactoglobuline films. Results indicate that plasticizer efficiency generally decreased in the order Gly, PEG 200, PEG 400, Sor, and Suc, on the bases of mole plasticizer-oxygen-atom/mole β-lactoglobuline and mass plasticizer/mass β-lactoglobuline, whereas the order was reversed when the basis was changed to mole plasticizer/mole β-lactoglobuline. The latter results clearly reflect the effect of plasticizer number of O atoms. These results were consistent with the glass transition temperature of the β-lactoglobuline-plasticized films (Sothornvit et al., 2002).

Sucrose-plasticized WPI films have been determined to be highly glossy compared to WPI films plasticized with PPG, Gly, or PEG (Lee et al., 2002a). As sucrose concentration increased, WPI films became tougher and more durable, having moderate TS and larger elongation values (Dangaran and Krochta, 2003). However, sucrose-plasticized films became brittle and hazy during storage, probably due to changes caused by the transition of sucrose from the amorphous state to crystalline state. Sucrose crystallization in WPI films has deleterious effects on the desirable properties of low OP, high gloss, and good durability. However, the transition of amorphous to crystalline sucrose can be slowed by cross-linking the WP matrix through thermal denaturation and by the addition of an inhibitor, such as raffinose and modified starch (Dangaran et al., 2006).

Talens and Krochta (2005) studied the possible plasticizing effect of BW (viscoelastic wax) and CarWax (elastic wax) on mechanical and moisture barrier properties of WPI, observing that the incorporation of beeswax produced a plasticizing effect in WPI:glycerol films, whereas CarWax produced an antiplasticizing effect. The moisture barrier properties of WPI:Gly films benefit from the addition of BW, by both increase of the hydrophobic character and decrease of the amount of hydrophilic plasticizer required in the film.

Fairley et al. (1996) concluded that the nature and extent of cross-linking had no effect on WVP of WPI films, as these authors studied the effect of blocking free thiol groups with N-ethylmaleimide (NEM) and reducing intermolecular S–S bonds with cysteine. However, blocking free SH groups with NEM had an unexpected effect on mechanical properties and film solubility. As NEM increased, film E decreased, whereas film solubility decreased. The authors suggested that possible prevention of sulfhydryl/disulfide interchange enabled the protein to adopt conformations that resulted in increased hydrogen bonding, increased hydrophobicity, increased disulfide–disulfide interchange, or increased sulfhydryl–sulfhydryl reaction.

Properties of native and heat-denatured whey protein films also confirm that S–S bonds, whether intra- or intermolecular, play a very small role in determining the moisture barrier properties of WPI-based films. However, these films possess different OP, solubility, and mechanical properties (Table 2.2 and Table 2.3) (Pérez-Gago and Krochta, 2001b). OP of native WPI films is significantly higher than OP of heat-denatured films, but of the same order of magnitude. The lower OP values for heat-denatured films may be related to their more linear (unfolded) structure, leading to higher cohesive energy density and lower free volume among polymer chains. The unfolded structure of heat-denatured whey proteins and the covalent S–S bonding during drying leads to film insolubility in water and produces films that are stronger and can withstand higher deformations (Pérez-Gago et al., 1999). The degree of protein denaturation and unfolding as heating time and temperature increase affects the degree and nature of protein–protein cross-linking and, as a consequence, the solubility and mechanical properties of the films (Pérez-Gago and Krochta, 2001b). Similar results are obtained in WPI films formed from different proportions of heated and unheated whey protein solutions. As the proportion of heated WPI solution increases E, TS and YM increase, and film solubility decreases, whereas WVP remains unaltered (Guckian et al., 2006).

Sothornvit and Krochta (2000a) studied the effects of whey protein hydrolysis on film mechanical properties at several Gly-plasticizer levels. Both 5.5% and 10% degree of hydrolysis WPI had significant effects on film tensile properties compared to unhydrolyzed WPI. Hydrolyzed WPI required less Gly to achieve the same mechanical properties compared to those of unhydrolyzed WPI.

Chemical agents and physical treatments have also been used for covalent cross-linking of whey proteins. Transglutaminase has been used to produce films by polymerization of whey proteins (Mahmoud and Savello, 1992, 1993; Yildirim and Hettiarachchy, 1998a). Mahmoud and Savello (1992) reported that transglutaminase-cross-linked films in the presence of Ca^{+2} did not affect water vapor transmission rate and moisture content of β-lactoglobulin, α-lactalbumin, and WPI films. Yildirim and Hettiarachchy (1998a) reported lower solubility of transglutaminase-cross-linked films and TS values twofold greater than those of the control. However, the WVP of transglutaminase cross-linked films was lower than that for control films. The presence of transglutaminase in chitosan-whey protein films also induced an enhancement in film mechanical resistance and a reduction in their deformability, and the barrier efficiency toward oxygen and carbon dioxide was found to be markedly improved in the cross-linked films, which also showed a lower WVP (Di Pierro et al., 2006). Other chemical agents such as glutaraldehyde, formaldehyde, dialdehyde starch, and carbonyldiimidazole enhanced the TS and insolubility behavior of WPI cross-linked films, whereas E was not affected. Chemical cross-linking with these agents increased WVP and decreased OP of the films (Galietta et al., 1998; Ustunol and Mert, 2004). The addition of a low concentration of $CaCl_2$ in the film formulation did not significantly affect the functional properties of the films (Galietta et al., 1998).

Contrary to observations with other protein films, treatment of the WPI film-forming solution with 324 J/cm^2 UV light had no effect on film solubility, WVP, and OP (Ustunol and Mert, 2004). Similarly, γ-irradiation had no effect on film solubility of WPI and WPC films (Vachon et al., 2000), whereas insolubility and mechanical and WVP properties of calcium caseinate–whey protein composite films were improved by γ-irradiation treatments. Calcium caseinate and WPI produced a synergistic effect. The strongest combined effect was obtained for caseinate:WPI (75:25) with a reduction of the WVP by half (Lacroix et al., 2002, Ouattara et al., 2002). This effect can also be modified by adding polysaccharides to the formulations (Ciesla et al., 2006a, 2006b, 2006c).

Whey protein films are excellent oxygen barriers at low to moderate RH. The OPs of whey protein films are lower than those of high-density polyethylene (HDPE) and low-density polyethylene (LDPE) and are comparable to those of ethylene vinyl alcohol films (EVOH) under similar RH conditions and proper selection of plasticizer (Table 2.2). RH also has an exponential effect on the OP of WPI films (McHugh and Krochta, 1994b). WPI films had lower OP when plasticized with Sor than with Gly (McHugh and Krochta, 1994b). Maté and Krochta (1996) observed an increase in OP of WPI and β-lactoglobulin films when Gly content and RH increased. However, despite the different molecular structures in WPI films compared to β- lactoglobulin films and the relative differences in hydrogen bonds, hydrophobic interactions, and S–S bonds, no differences between WPI and β-lactoglobulin OP were detected. Sothornvit and Krochta (2000b) studied the effect of a wide range of plasticizers with different composition, shape, and size. Sucrose-plasticized β-Lg films gave the best oxygen barrier. Gly- and PG-plasticized films had similar OP values, and both had higher OP than Sor-plasticized films. PEG 200- and PEG 400-plasticized films were the poorest oxygen barriers. Hydrolysis of WPI did not influence film OP (Sothornvit and Krochta, 2000b). The use of hydrolyzed WPI allowed the achievement of desired film flexibility with less glycerol and with a smaller increase in OP.

Numerous works have been published of WPI films containing antimicrobials. Studies have been done with polypeptic antimicrobials, such as lactoferrin, lactoferrin hydrolysate, and lactoperoxidase systems (LPOS) (Min and Krochta, 2005; Min et al., 2005a, 2005b), and plant extracts oregano, rosemary, and garlic essential oils (Seydim and Sarikus, 2006). The effectiveness of these antimicrobials depended on the target microorganism, the growing media, and concentration. The effect of LPOS incorporation at concentrations between 0.15 and 0.25 g LPOS/g film (dry basis) on OP and mechanical properties resulted in a plasticizing effect on EM and TS, but in an anti-plasticizer effect on the E and OP (Min et al., 2005b). However, at LPOS concentrations below 0.06 g/g, film (dry basis) did not affect mechanical and OP of the WPI films (Min and Krochta, 2005; Min et al., 2005a, 2005b).

2.5.3 Whey protein coatings

Whey protein coatings have been studied as potential controlled released carriers for insect pheromones (Atterholt et al., 1998). Coating applications on coated peanuts have reduced rancidity, improving gloss of the samples (Maté and Krochta, 1998; Lee et al., 2002b; Dangaran et al., 2006; Lin and Krochta, 2006). WPI coatings have shown to reduce enzymatic browning of fresh-cut apples and potato (Le Tien et al., 2001; Pérez-Gago et al., 2005, 2006) and to improve postharvest quality of fresh whole apples if appropriate thickness is applied to avoid anaerobic respiration (Cisneros-Zevallos and Krochta, 2003). Heat-sealed WPI films containing p-aminobenzoic acid forming casings were effective controlling the growth of *Listeria monocytogenes* on hot dogs (Cagri et al., 2003) and *Listeria monocytogenes*, *Escherichia coli* O157:H7, and *Salmonella enterica* on bologna and summer sausage (Cagri et al., 2002).

2.6 Casein films and coatings

Casein is the major protein in milk. It is a unique protein, because it is only synthesized in the mammary gland and is found nowhere else in nature (Fox and McSweeney, 1998). Chemical and physical properties of casein have been well characterized (Brunner, 1981; Fox and McSweeney, 1998). Four principal components, α_{s1}-, α_{s2}-, β-, and κ-caseins are identified. Caseins are phosphoproteins containing about 0.85% phosphorus, which is sterified to serine forming phosphate centers that bind calcium cations strongly through hydrophobic interactions (Fox and McSweeney, 1998). The amino acid composition in casein is characterized by low levels of cysteine, so they cannot form extensive covalent inter- or intramolecular disulfide bonds to form water-insoluble films. Caseins have little secondary or tertiary structure, probably due to the high content of proline, which disrupts α-helices and β-sheets (Fox and McSweeney, 1998). The casein molecules have an open, flexible, and mobile conformation that is often referred as "random-coil" structure (Chen, 2002).

Commercially available caseins are acid casein, rennet caseins, caseiantes, and carbon dioxide precipitated casein. Caseins precipitate when skim milk is acidified to the casein isoelectric point at pH of approximately 4.6 (Dalgleish, 1989). Under acidic condition, calcium phosphate solubilizes and individual casein molecules associate to form insoluble acid casein. The acid casein can be converted to soluble caseinates through neutralization with an alkali, obtaining sodium, calcium, mangnesium, potassium, and ammonium caseinates (Chen, 2002). Rennet casein is obtained by protein coagulation with a proteolytic enzyme (rennin) that destabilizes the casein micelles and promotes coagulation of casein in the presence of calcium cations. Rennet casein is insoluble in water or alkali, but it can be dissolved by adding calcium-sequestering agents such as polyphosphates (Muller, 1982). Casein obtained by injecting high-pressure carbon dioxide into skim milk at elevated temperature has a high calcium content, equivalent to that of calcium caseinate, and contains more phosphorus than acid casein, which is an indication that the integrity of the casein micelles is preserved (Tomasula et al., 1995, 1997).

2.6.1 Formation of casein films

Caseins form films from aqueous solutions without further treatment due to their random-coil nature and their ability to hydrogen bond. Extensive hydrogen and electrostatic bonds and hydrophobic interactions facilitate the formation of intermolecular interactions to form casein edible films (McHugh and Krochta, 1994c). Casein films are highly soluble in water (Vachon et al., 2000). Film solubility can be decreased with buffer treatments at the isoelectric point of the films (Krochta et al., 1990; Chen, 2002), with cross-linking of the protein using irradiation (Vachon et al., 2000), and with the transglutaminase enzyme (Nielsen, 1995). Casein films can also be prepared by using the wet spinning process at pH 9 (Frinault et al., 1997). Kozempel and Tomasula (2004) developed a continuous process to make casein films from carbon dioxide precipitated casein. In the film-forming process, the caseinate–Gly solution is wetted and spread on polyethylene or Mylar belts using a Meier rod, and the film is easily removed after drying.

2.6.2 Functional properties of casein films

Casein- and caseinate-based films are transparent and flexible but have poor moisture barrier compared to synthetic films. At comparable test conditions, caseinate films appear to be similar moisture barriers to WG and SPI films and somewhat poorer moisture barrier than CZ films (Table 2.1) (Krochta, 1997). Different casein products may result in films of different WVPs. Different works have observed that WVP ascended in the following order: magnesium caseinate < calcium caseinate < micellar casein < sodium caseinate < rennet casein (Ho, 1992; Banerjee and Chen, 1995; Chen, 2002). The greater moisture barrier of calcium and magnesium caseinate films than sodium caseinate films has been attributed to cross-linking effect of the divalent cations. Treating sodium caseinate films with calcium chloride at pH 9.6 induced protein cross-linking and reduced film WVP by 42% (Avena-Bustillos and Krochta, 1993). Casein cross-linking can also be induced by γ-irradiation, UV-irradiation, heat treatment, and the use of transglutaminase. Lacroix et al. (1999) reported a reduction of 65% in WVP of γ-irradiated calcium caseinate films. Isothermal calorimetry study of calcium caseinate films cross-linked by heating and γ-irradiation showed that combination of both treatments enhanced the generation of cross-links (Letendre et al., 2002). However, the use of transglutaminase and UV-irradiation did not affect WVP of casein and sodium casein films, respectively, even though total soluble matter was reduced (Rhim et al., 1999b; Oh et al., 2004).

Some attempts have been made to enhance protein interaction within the protein film matrix by pH adjustment toward the pI of the protein (i.e., pH 4.6). This way, sodium acetate buffer rendered sodium caseinate films insoluble but had little effect on film WVP. However, treatment of sodium caseinate-acetylated monoglyceride emulsion films with the same buffer, followed by soaking in water for 2 minutes reduced WVP by 50% (Krochta et al., 1990). Differences between both films were attributed to the hygroscopic nature of Gly in the film that was probably lost in the soaking process of the former sodium caseinate film, whereas in the emulsion film the effect was overwhelmed by the hydrophobic character of acetylated monoglyceride. Avena-Bustillo and Krochta (1993) also reported that sodium caseinate films treated in various buffer solutions at the pI reduced WVP of the films.

Increasing moisture content and plasticizer content increased WVP of casein films (Siew et al., 1999; Schou et al., 2005). The increase in WVP of sodium caseinate films plasticized with Gly increased at Gly:protein ratio above 0.24 (Schou et al., 2005). Siew et al. (1999) found that PEG-plasticized films presented higher WVP than Gly-plasticized films. This was correlated to the higher hydration rate of PEG and to the higher percentage of random coil structure of the PEG-plasticized films. Chick and Ustunol (1998) showed that Sor-plasticized films were more effective moisture barriers than Gly-plasticized films, due to the lower hygroscopic character of Sor.

Moisture barrier properties of casein films are improved by incorporation of hydrophobic materials. Krochta et al. (1990) reported that incorporation of paraffin, CarW, BW, or acetylated monoglycerides reduced the water vapor transmission rate by 40% compared to control films. Among these hydrophobic components, acetylated monoglycerides were considered better film additives because they have lower melting temperatures and are easier to produce emulsions. Avena-Bustillos and Krochta (1993) concluded that BW was more effective than SA and acetylated monoglyceride in improving moisture barrier of sodium caseinate films, which was attributed to the greater hydrophobicity and crystalline structure of BW. This caseinate-BW composite film had WVP 90% lower than pure sodium caseinate films. Adding acetylated monoglyceride to various caseinate films reduced WVP, obtaining a greater reduction with sodium caseinate films (Banerjee and Chen, 1995; Chen, 2002). Lactic acid casein films improved their moisture barrier when CarW or CandW was incorporated into the system, with CandW more effectively reducing WVP than CarW. Scanning electron microscopy showed a partial phase separation of wax and protein (Chick and Hernandez, 2002).

The origin of the casein used to form films also influences the moisture barrier of the film. Chick and Ustunol (1998) reported that lactic acid casein films had slightly lower WVP than rennet precipitated casein films. Carbon dioxide precipitated casein formed films that were less water soluble (7% water soluble) than calcium caseinate films (100% water soluble) and with 20% lower WVP values (Tomasula et al., 1998).

Mechanical properties of casein films show a large variation in the literature (Chen, 2002). Differences in film preparation techniques, moisture equilibrium, testing temperatures, and RH control, in addition to changes in film formulation are the reason for this large variation. Tomasula et al. (1998) reported that a minimal Gly content of 20% was necessary to prepare casein films suitable for mechanical properties (Table 2.3). Calcium caseinate films, which have significantly high TS and low E, are characterized as harder and more brittle than sodium or potassium caseinate films (Banerjee and Chen, 1995; Chen, 2002). This was attributed to the cross-linking effect of the divalent calcium ion. Siew et al. (1999) reported that plasticizers were effective within a range (i.e., too little plasticizer resulted in brittle films and too much plasticizer rendered the films too tacky to yield reliable mechanical data). Within the effective range of plasticizers, the TS of sodium caseinate films decreased with increased plasticizer content. When comparing plasticizer type, these authors observed a crossover phenomenon. Gly caused more pronounced changes to the film TS compared to PEG. At low plasticizer content, films had higher TS with Gly than with PEG, whereas at high plasticizer content films they had lower TS with Gly than with PEG. This behavior can be attributed to the differences in plasticizer size and hydrogen bonding interactions. Schou et al. (2005) observed that maximum load, TS, and YM of sodium caseinate films decreased with increasing Gly content, whereas E only increased at Gly:caseinate ratios of 0.24 and 0.32.

Chick and Ustunol (1998) reported that rennet casein films had higher TS (by 27% to 98% at various plasticizer levels) than lactic acid casein films. This was attributed to the higher ash content of the rennet casein, which probably provided additional cross-linking between divalent cations and negatively charged groups along the protein chains. In this study, Sor-plasticized films were stronger than Gly-plasticized films. However, films plasticized with Gly were more extensible.

Carbon dioxide precipitated casein films were more than 50% stronger than calcium caseinate films at low Gly content (20% w/w), but the differences diminished when Gly content increased. These films were also stiffer than and approximately as elastic as caseinate films (Tomasula et al., 1998). The stronger and stiffer nature of carbon dioxide precipitated casein films was attributed to the fact that the integrity of the casein micelles is preserved in the CO_2-casein, whereas in acid casein most of the micellar calcium phosphate dissolved. In a later study, no significant differences were found in TS between calcium caseinate and carbon dioxide casein films (Tomasula et al., 2003). In this work, films were cast from 4% (w/w) protein solutions as compared to 6% (w/w) protein solutions from the previous work, so the effects of the micellar calcium and phosphate linkages appear not to contribute to film strength.

Blending casein with other structural materials alters the mechanical properties of the resulting films. Casein-starch films decreased in TS and increased in E and YM (Arvanitoyannis et al., 1996; Jagannath et al., 2003). Casein films incorporating zein hydrolysate and also cross-linked by transglutaminase improved film flexibility, reducing the level of plasticizer required to maintain film flexilbility without sacrificing their WVP (Oh et al., 2004). Radiation treatments, such as UV, did not affect mechanical properties of sodium caseinate films (Rhim et al., 1999b). However, γ-irradiation treatments improved significantly the puncture strength of calcium caseinate (Lacroix et al., 2002).

Permeability of casein films to oxygen has been less studied. Buonocore et al. (2005) used a mathematical model to describe the OP of casein films. OP values were similar to those of chitosan and zein films at intermediate RH but increased significantly at high RH. Tomasula et al. (2003) determined the OP of CO_2-precipitated casein, calcium caseinate, and acylated casein films. At similar test conditions, Gly-plasticized CO_2-precipitated casein films had higher OP than calcium caseinate and acylated casein films, which can be attributed to the higher hydrophobic character due to its greater concentration in calcium and phosphorus. OPs for the calcium caseinate and acylated casein-Gly films are within the range of OP reported for hydrolyzed WPI films. OP for the films of this study are less than those reported for the LDPE and high density polyethylene (HDPE) films but greater than OP reported for the polyvinylidene chloride (PVDC) or EVOH films (Table 2.2).

2.6.3 Casein protein coatings

Casein edible coatings have been applied to whole and fresh-cut fruits and vegetables, such as carrots, zucchini, whole fruit apples, celery sticks, and kinnow citrus (Avena-Bustillos et al., 1993, 1994a, 1994b, 1997; Alam and Paul, 2001). The coatings have reduced white blush, respiration rate, and moisture loss of whole and peeled carrots (Avena-Bustillos et al., 1993, 1994a). Optimized calcium caseinate-acetylated monoglyceride formulations have reduced water loss of zucchini and celery sticks (Avena-Bustillos et al., 1994b, 1997). Shelf life of kinnow fruits was extended by 20 days when fruits were coated with casein formulations and stored in high density polyethylene bags (Alam and Paul, 2001). Transglutaminase cross-linked casein films have been studied as enzyme immobilizing systems, maintaining the enzymatic activity on repeated usage (Motoki et al., 1987b). A more recent work shows the potential of casein protein–based thermoplastics and composites as alternative biodegradable polymers for biomedical applications (Vaz et al., 2003).

2.7 Cottonseed proteins

Cottonseed proteins make up 30% to 40% (w/w) of the cottonseed kernel. Other important cottonseed components include lipids, soluble carbohydrates, cellulose, minerals, phytates, and polyphenolic pigments. The protein components are mainly globulines (60%) and albumins (30%) with lesser proportion of prolamins (8.6%) and gluteins (0.5%) (Saroso, 1989).

Globulines include two protein fractions, the 11S and 7S types called gossypin and congossypin, respectively (Cho et al., 1992). They are insoluble in water at pH 6.8 and can be readily extracted by the salt-in process. Albumin proteins are water souble at pH 6.8 (Martinez et al., 1970).

Cottonseed proteins have a high content of ionizable amino acids (aspartic and glutamic acids, argidine, histidine, and lysine) and low content of sulfur amino acids. These proteins are more soluble at basic pH than at acidic pH, and present the isoelectric pH at around 5. These proteins are denatured by thermal treatment above 80°C, which leads to the loss of solubility (Marquié and Guilbert, 2002).

2.7.1 Formation of cottonseed protein films

Initially, cottonseed film formation involved soaking kernels in hot water to prepare an "oilseed milk." Films were successively formed on the surface of the heated (90 ± 5°C) liquid (Wu and Bates, 1973). Films obtained by this technique present poor mechanical properties, which makes them of limited use as edible coatings.

Cottonseed protein–based films are obtained directly from cottonseed flours using a casting process (Marquié et al., 1995). The process involves solubilizing cottonseed flour proteins under appropriate conditions (i.e., solvent, pH, temperature, addition of salts, plasticizers, and dissociating agents) to minimize protein–protein interactions. The dispersion is then centrifuged to remove insoluble substances, and the supernatant is homogenized. Plasticizers are added to the solutions to obtain films with good mechanical properties.

The conditions required to obtain films from cottonseed-flour solutions are difficult to determine due to the complexity of the raw material, which contains proteins, lipids, ash, cellulose, and carbohydrates. Numerous works have established as optimal values to obtain cottonseed films: pH 8 to 12, temperature 20 to 60°C, solid/solvent ratio 10% to 50% w/v, plasticizer content 10% to 50% w/w (dry basis), and use of dispersive agents (Marquié and Guilbert, 2002). The chemical composition of cottonseed film differs depending on the cottonseed flour type. Films made from glandless and glanded flour had a similar protein–lipid ratio and only differed in the level of gossypol. Films obtained from defatted glandless flour were richer in protein than those obtained from nondefatted flour, and they contained less than 2% lipids (Marquié et al., 1995).

2.7.2 Functional properties of cottonseed protein films

Films from cottonseed flour are highly hydrophilic due to their protein content (Marquié and Guilbert, 2002). Measured WVP values of cottonseed flour (glanded, glandless, and glandless delipidated) films were about 0.52 g mm/m^2 h kPa at 20°C and 0/100% RH gradient (Marquié, 1996). These values are similar to those reported in the literature for other protein-based films (Table 2.1).

Cottonseed films are weaker than other protein and synthetic films. These films are also excessively brittle, and they require the presence of plasticizers. Mechanical properties of cottonseed films depend on the flour source (i.e., glandless, glanded, and glandless delipidated), chemical cross-linking agents, temperature, RH, and Gly level (Marquié, 1996). Marquié et al. (1995) established the optimum of Gly between 10% and 30%. Below 10% Gly (w/w, dry basis) cottonseed films are too brittle, and above 30% (w/w, dry basis) the films became sticky. From 10% to 30% Gly, film puncture strength decreased rapidly.

Cottonseed films from nondefatted flour (i.e., containing lipids) were weaker than those from delipidated flour, showing lower puncture strength and deformation values (Marquié et al., 1995). These can be explained by the protein content difference between both types of flour. Differences in flour composition also influenced film solubility. The higher level in gossypol of glanded flour formed films with lower solubility than glandless flour (delipidated or not). In protein films, gossypol strengthens the protein network, entrapping lipids inside the protein structure and thus reducing film solubility.

Puncture strength of cottonseed flour films varied as a function of film moisture content or water activity. Films from glandless flour (delipidated or not) with low moisture content (2% to 3% w/w, dry basis) were very brittle. Then, puncture strength increased sharply for film moisture contents within 3% to 7% (w/w, dry basis). Above 7% moisture content, puncture strength decreased. The increase in puncture strength at low moisture content may be interpreted as an antiplasticizing effect of water (Marquié, 1996).

Mechanical properties of cottonseed films are improved by cross-linking agents, such as formaldehyde, glyosal, or glutaraldehyde (Table 2.3) (Marquié et al., 1995). Among these cross-linking agents, formaldehyde produced more resistant films than glyosal, or glutaraldehyde. The percentage of lysine that reacts with these compounds was closely correlated with the puncture strength of the films (Marquié et al., 1997). Almost the entire reactive lysine content of cottonseed proteins reacts with glyosal and glutaraldehyde, but only 50% of the amino acid is involved in cross-linking by formaldehyde. Even though mechanical properties of cottonseed films are improved by cross-linking agents, these films were five- to tenfold weaker in terms of tensile strength than other protein and synthetic films (Marquié and Guilbert, 2002). However, cottonseed flour films that contained carded cotton fibers improved significantly their mechanical properties, reaching tensile strength values close to HDPE films (Marquié et al., 1996).

2.8 Collagen and gelatin proteins

Collagen is a protein constituent of skin, tendon, and connective tissues. It is a fibrous protein produced by self-assembly of collagen molecules in the extracellular matrix, and it provides tensile strength to animal tissue (Trotter et al., 2000). Collagen fibers can be dissolved in dilute acid or alkali and in neutral solutions. Two major components are identified: α (molecular weight 100,000 Da) and β (molecular weight 200,000 Da) (Harrington, 1966; Piez et al., 1968). Collagen is rich in glycine, hydroxyproline, proline, and hydroxylysine. Its higher content in acidic, basic, and hydroxylated amino acid residues than in lipophilic residues makes it responsible for the hydrophilic character of this protein. Therefore, it swells in polar liquids with high solubility parameters.

Gelatin is a protein resulting from partial hydrolysis of collagen. The hydrolysis involves alkali or acid treatment followed by or accompanied by heating in the presence of water (Eastoe and Leach, 1977). The acid process yields to Type A gelatin with an isoelctric point between 7 and 9, whereas the alkaline process yields to Type B gelatin with an isoelectric point between 4.6 and 5.2 (Rose, 1987). The molecular weight of gelatin covers a wide range, from 3000 to 200,000 Da, depending on the raw material employed during gelatin production (Young, 1967).

2.8.1 Formation of collagen and gelatin protein films

Collagen films are commercially produced as casings. *In vivo,* collagen exits as fibers embedded in mucopolysacharides and other proteins (Balian and Bowes, 1977). Production of industrial collagen casings can be accomplished using a dry or wet process with some similarity, including (a) alkaline treatment to dehair and remove collagen from carbohydrates and other proteins; (b) acid swelling and homogenization to form a ≈4.5% moisture gel (wet process) or ≈10% moisture gel dough (dry process); (c) extrusion into a tube; and (d) neutralization of the extruded tube, washing the tube of salts, treating the tube with plasticizer and cross-linkers, and drying to 10% to 14% moisture, with the order depending on whether the wet or dry process is used (Krochta, 1997a).

Stand-alone gelatin films can be formed by hot or cold casting of aqueous film-forming solutions containing plasticizers, such as Gly and Sor (Bradbury and Martin, 1952; Guilbert, 1986).

2.8.2 Functional properties of collagen films

No data on the WVP of collagen film have been reported in the scientific literature. Leiberman and Gilbert (1973) studied the effects of moisture, plasticizer content, denaturing, and cross-linking on the permeability of collagen films to gases (Table 2.2). Collagen films have excellent oxygen barrier at 0% RH, superior to LDPE and comparable to the best synthetic oxygen barrier films EVOH and PVDC (Krochta, 1997a). Increasing RH and the level of aliphatic diol plasticizers increased OP rapidly. Interestingly, addition of Sor lowered permeability, which was attributed to interaction with the collagen chain-active side in such a way as to restrict segmental mobility. Heat-denatured collagen also reduced OP of collagen films, probably due to a decrease in the free volume in the more random structure of the denatured protein. Contrary to expectations, cross-linking with formaldehyde or chromium ions increased gas permeability. This was attributed to an increase in moisture uptake of cross-linked films due to greater spacing between collagen chains.

Collagen films have good mechanical properties. TS of collagen casings depends on whether longitudinal or transverse sections are tested. The range of TS values for both the wet and dry process is similar to that for LDPE, whereas E is relatively low, ranging between that of cellophane and oriented polypropylene (OPP) (Krochta, 1997a).

Hood (1987) determined tensile properties of collagen casings prepared by the wet and dry process and containing Gly or Sor, respectively. Casings prepared by the dry process had higher TS and E than casings prepared by the wet process. However, it is difficult to draw a valid conclusion because of the complexity of interactions among ingredients (e.g., cellulose, fat, and carboxy methylcellulose) present in the casings.

Simelane and Ustunol (2005) put collagen films through a meat-processing scheme typical of Polish sausage manufacture to study the effect of meat processing on mechanical properties, showing no change on the properties during the multistage cooking process.

Differences in mechanical properties were found between collagen films from different fish species. Films formed from collagen of New Zealand hoki and New Zealand ling had greater E, TS, and elasticity in comparison with similar films from Irish fish species (O'Sullivan et al., 2006).

2.8.3 Functional properties of gelatin films

Gelatin edible films can be good gas barriers but poor water barriers. Gelatin-plasticized films are clear, flexible, and strong. The main parameters affecting film-forming properties of gelatin are raw material source, extraction method, molecular weight, film preparation method, and degree of hydration or presence of plasticizer (Arvanitoyannis, 2002).

Solutions of 5% (w/v) gelatin dried at 20°C (cold casting) formed films with higher TS than solutions dried at 60°C (hot casting) (Bradbury and Martin, 1952). The difference in TS was attributed to the higher degree of crystallization in films cast at lower temperature (Bradbury and Martin, 1952) and configuration of gelatin molecule depending on drying temperature (Robinson, 1953).

Guilbert (1986) found that cross-linking/denaturing gelatin films with calcium ions had no effect on WVP. However, treatment with lactic or tannic acid improved moisture barrier, but films became less flexible and less transparent. Segmental orientation induced by hot and cold and rapid drawing and cross-linking have been used to improve mechanical properties of gelatin films. However, at draw ratios above 4, film modulus and stress at break decreased due to extensive changes in the network affecting contacts or joins between crystallites and microfibrils, resulting in chain scission or disassociation (Fakirov et al., 1996, 1997). Taylor et al. (2002) treated Gly-plasticized gelatin films with transglutaminase. The amount of Gly added significantly affected TS of the films. Increasing the concentration of cross-linking agent gave products with higher TS that were less soluble in water and had improved water absorption properties.

Blends consisting of gelatin–polysaccharide (soluble starch, modified starch, or chitosan), and plasticizer (water, sugars, or polyols), combining low and high temperature processes, have been prepared by Arvanitoyannis et al. (1997, 1998a, 1998b). The film that produced the highest TS (130 Mpa) was made from a combination of gelatine/chitosan/sorbitol and water using a low casting process. The lowest TS (34.4 Mpa) film was produced from a combination of gelatine/soluble starch/sorbitol and water using a high casting process; however, this film had the highest E at 44.7%. Oxygen and carbon dioxide permeability of these films changes considerably below and above the glass transition temperature. At temperatures above the Tg, molecules have to create their own interstitial spaces by separating interchain polymer contacts, and the permeant diffuses through the polymer matrix along cylindrical voids created by the synchronized rotation of polymer chain segments. On the contrary, at temperatures below Tg, gas molecules can diffuse through existing interstitial space, requiring lower activation energy (Arvanitoyannis et al., 1998a).

Recent work shows the ability of edible films made from tuna-fish gelatin to act as carrier of antioxidant extracts of two different murta ecotype leaves (Gómez-Guillen et al., 2007a). Microscopy studies denoted a certain degree of interference of polyphenolic compounds in the arrangement of gelatin molecules. The addition of the murta extracts led to transparent films with increased protection against UV light as well as antioxidant capacity. Mechanical and WVP properties of the films depended on the extract type.

2.8.4 Collagen and gelatin coatings

Collagen is used to make the most commercially successful edible protein films. Collagen casings are used for sausage production and other meat products.

Collagen films are normally eaten with the meat product after removal of the netting (Krochta, 1997a). Other applications that have been studied include the effect of collagen film overwrap on exudation and lipid oxidation in both refrigerated and frozen/thawed beef round steak (Farouk et al., 1990).

Polysaccharide–gelatin interactions have been applied for microencapsulation (gelatin–gellan), drug release (gelatin–dextran), and tissue adhesion (gelatin–polycarboxylic acid) (Chilves and Morris, 1987; Kurisawa and Yui, 1998; Otani et al., 1998). Edible coatings based on native starches and gelatin have been tested in the conservation and sensory acceptance of grapes. Among the different native starches studied, the most effective at extending the shelf life of grapes were sorghum- and rice-gelatin coating, with a 10 day increase (Fakhouri et al., 2007). Gelatin coatings containing antimicrobial agents have also been tested in food products. Antimicrobial gelatine-chitosan-based edible coatings containing oregano or rosemary extracts in combination with high-pressure processing prevented lipid oxidation and inhibited microbial growth of cold-smoke sardine (Gómez-Estaca et al., 2007b). Antimicrobial gelatin coatings containing benzoic acid were also effective at reducing microbial loads of tilapia fillets without affecting the sensory attributes (Ou et al., 2002). The combination of gum acacia, gelatin, and calcium chloride formed edible coatings that were effective for the prevention of after-cooking darkening of water-blanched potatoes (Mazza and Qi, 1991). However, a comparative evaluation of different edible coatings to reduce fat uptake in deep-fried cereal products showed that gelatin was not suitable as single material coating for that purpose (Albert and Mittal, 2002).

2.9 Myofibrillar and sarcoplasmic proteins

Myofibrillar proteins are found in meats (e.g., beef, sheep, pork, and poultry) and fish. These proteins represent the main component of muscles and are mainly composed of myosin and actin, regardless of whether the source of protein is mammalian or fish. Myosin is a fibrous protein that associates two subunits known as heavy chains (200 kDa) and two pairs of light chains (16 to 20 kDa). Actin exits in globular form (G-actin) and in a fibrous form (F-actin) formed by linear polymerization of G-actin. The association of myosin and actin forms contractible proteins named actomyosin (Cuq, 2002).

Sarcoplasmic proteins also form part of fish muscle tissue. This fraction constitutes 25% to 30% of the total protein content. These proteins are mainly composed of myoalbumin, globulin, and enzymes. They are soluble in neutral salt solutions of low ionic strength (<0.15 M).

2.9.1 Formation of myofibriller protein films

Myofibrillar proteins can only be used for film-forming applications after purification and concentration from meat or fish (Cuq et al., 1995). This involves successive washing treatments to remove undesirable compounds. The process is known as surimi (or fish and meat mince) and it results in semipurified protein fractions containing high concentrations of myofibrillar proteins. The addition of cryoprotective agents (e.g., sucrose and sorbitol) to the resulting surimi product permits long-term storage at –20 to –30°C without changes in protein functional properties (Cuq, 2002).

To prepare films or coatings, surimi can be used directly after thawing (Cuq et al., 1995), or it can be lyophilized (Monterrey-Quintero, 1998) or air dried (Cuq et al., 1997a) and ground into powder. Two methods have been used to prepare myofibrillar protein films: (1) the solvent process and (2) the dry process. The solvent process requires dissolution of myofibrillar proteins by adjusting the pH of the film-forming solution with acetic acid, lactic acid, or ammonium hydroxide. Optimal conditions for preparing film-forming solutions based on myofibrillar proteins were determined at pH 3 and 2 g protein/100 g solution by Cuq et al. (1995), and at pH 2.5 and 1.25 g protein/100 g solution by Monterrey-Quintero (1998). Plasticizers had to be incorporated into the film-forming solution to improved film mechanical properties. In the dry process, myofibrillar protein-based powders, hydrated at various levels by water addition, are heated at temperatures above the Tg value. This produces a soft rubbery material that can be shaped into specific forms such as films (thermomolding process). Cuq et al. (1997a) produced glassy translucent materials with moisture content of 2.2% and temperatures less than 200°C, which avoid thermal degradation of the protein.

Recently, it was observed that sarcoplasmatic proteins obtained from fish were capable of forming films (Paschoalick et al., 2003). Proteins are extracted in a similar way to the method described by Monterrey-Quintero and Sobral (2000) for myofibrillar proteins. The proteins were obtained by fine grinding the fish muscle, followed by separating the connective tissue and freeze-drying after liquid nitrogen freezing. The films were prepared by the casting technique, with Gly being added as plasticizer and the film-forming solution being adjusted to pH 2.7 with acetic acid (Paschoalick et al., 2003).

2.9.2 Functional properties of myofibriller and sarcoplasmatic protein films

Myofibrillar proteins form films that are able to maintain their integrity after being immersed in water for 24 hours (Cuq et al., 1997b; Monterrey-Quintero and Sobral, 2000). The high density of molecular interactions and the presence of intermolecular covalent bonds are responsible for the partial insolubility of these films. Solubility of myofibrillar films is strongly influenced by plasticizer type and amount (Cuq et al., 1998; Monterrey-Quintero and Sobral, 2000).

Myofibrillar and sarcoplasmatic protein films present good oxygen barrier and mechanical properties (Table 2.2 and Table 2.3). OP of fish myofibrillar protein-based films was lower by three orders of magnitude than that of LDPE (Gontard et al., 1996). Myofibrillar protein films present higher TS than other protein-based films, such as WG, SPI, and corn zein films. However, these films are less elastic than other biopolymers. Compared to synthetic films, myofibrillar protein films have similar TS than LDPE, while polyester is 10 times stronger (Lacroix and Cooksey, 2005). Mechanical properties of both myofibrillar and sarcoplasmatic protein films are greatly influenced by protein content, plasticizer amount, and moisture content (Paschoalick et al., 2003; Garcia and Sobral, 2005; Sobral et al., 2005). For example, the addition of Gyl, Sor, or Suc decreased film TS and YM; however, it increased E (Cuq et al., 1997c). Strain at failure increased 100-fold when the RH increased from 11% to 95%, and 12-fold when the Gly content increased (Cuq, 2002). Sobral et al. (2005) observed that films prepared with 2 g proteins/100 g film-forming solution were more resistant to break than films prepared with 1 g protein/100 g film-forming solution. Garcia and Sobral (2005) observed that thermal denaturation of the film-forming solutions also influenced mechanical properties of the films (Table 2.3). Solutions heated at 65°C for 30 minutes formed films that were more resistant and more rigid than the films treated at 40°C for 30 minutes. This behavior was more pronounced in the case of films with higher protein content (2 g protein/100 g film-forming solution). Shiku et al. (2003) studied the effect of pH on the preparation of myofibrillar protein films. Myofibrillar protein-based films were formed within pH 2 to 3 and 7 to 12, whereas films were not formed between pH 4 and 6 because of the poor protein dispersion around the isoelectric point. TS of the films was higher when prepared at the acidic (pH 2, 3) and alkaline (pH 11, 12) conditions, whereas E was almost constant irrespective of pH.

The moisture barrier of myofibrillar and sarcoplasmatic protein films is similar to that of other hydrocolloid-based films (Cuq et al., 1995; Paschoalick et al., 2003).

The WVP of these films is greatly influenced by plasticizer type and amount, as well as by moisture content (Cuq et al., 1997c; Paschoalick et al., 2003). An increase of the heat denaturation temperature of film-forming solution caused an increase in the WVP of myofibrillar and sarcoplasmatic protein films obtained from Thai Tilapia fish (Paschoalick et al., 2003). However, the pH of the film-forming solution had no effect on WVP of fish myofibrillar protein films (Shiku et al., 2003). Sobral (2000) observed that TS, WVP, color, and opacity of myofibrillar protein films increased linearly as thickness increased.

Properties such as color and opacity have also been studied in myofibrillar and sarcoplasmic protein films. In general, color and opacity of both protein films increased as protein content increased (Garcia and Sobral, 2005) and Gly content decreased (Paschoalick et al., 2003). Films prepared at low or high pH (pH 2, 3, 11, or 12) possessed a transparency close to synthetic film, which can be correlated with a more stable protein network at those pH values (Shiku et al., 2003).

2.9.3 Myofibriller protein coatings

Application of myofibrillar protein-based edible coatings was tested on sliced fish pieces (Cuq, 1996). The coatings slightly modified the texture profiles of the fish pieces compared to the uncoated control, mainly by increasing mechanical resistance. Generally, mechanical resistance of the coated meat increased as coating thickness increased. The tested edible coatings seemed to be too resistant for such applications and would have to be weakened (Cuq, 1996).

2.10 Egg white proteins

Egg white is a mixture of eight globular proteins: ovalbumin, ovotransferrin, ovomucoid, ovomucin, lysozyme, G2 globulin, G3 globulin, and avidin. Ovalbumin, ovotransferrin, and ovomucoid constitute 54%, 12%, and 11% of the total protein weight, respectively. Ovalbumin is the only fraction that contains free sulfhydryl groups. Ovotransferrin, ovomucoid, and lysozyme contain disulfide bonds (Mine, 1995).

2.10.1 Formation of egg white protein films

Formation of egg white protein films involves denaturation of egg white protein in aqueous solution, followed by the casting process. Gennadios et al. (1996b) prepared films from 9% egg white protein aqueous solutions under alkaline conditions with the addition of different plasticizers. The egg white solutions were denatured at 45°C for 20 minutes. These authors reported that a pH range of 10.5 to 12 was necessary to obtain smooth and homogeneous egg white films, and pH above 12 should be avoided as the film-forming solution becomes too viscous. A similar pH range (10.5 to 11.5) was suggested by Okamoto (1978) and Handa et al. (1999a).

The mechanism of film formation is hypothesized to involve inter- and intramolecular S–S bonds. At alkaline conditions, the S–S bonds are reduced to SH groups, facilitating dispersion of the protein. Heat denaturation of the protein unfolds the protein chains, exposing more SH and hydrophobic groups. During the drying process, the SH groups form inter- and intramolecular S–S covalent bonds. These covalent bonds, together with hydrogen bonding and hydrophobic and electrostatic interactions form three-dimensional networks resulting in the final film (Gennadios et al., 1996b).

2.10.2 Functional properties of egg white protein films

Egg white protein films are clear and transparent and have properties similar to other proteins (Gennadios et al., 1996b). To improve mechanical and barrier properties, different cross-linking treatments have been studied. Chemical cross-linking agents, such as transglutaminase and dialdehyde starch, have been added to egg white protein solutions upon heat denaturation to induce protein cross-linking (Gennadios et al., 1998c; Lim et al., 1998; Di Pierro et al., 2007). The resulting films increased TS properties, reduced film solubility, and reduced the degree of swelling. SDS-PAGE patterns showed polymerization of protein molecules (Lim et al., 1998). Handa et al. (1999a) found a clear correlation between concentration of SH groups and film TS, E, and film total soluble matter. Thermal and alkaline denaturation of the protein solution increased the concentration of SH groups in the egg white protein solutions. Films cast from heat denatured solutions had higher TS than those formed from native protein, but TS was not modified by pH. Conversely, films prepared from heated solutions had higher E than unheated solutions only at the most alkaline test conditions (pH 11.5). UV radiation has also been studied as a possible physical cross-linking treatment for egg albumin films. Films were treated with 253.7 nm UV radiation for 24 hours. Egg albumin films had slightly but significantly lower total soluble matter contents than untreated films (Rhim et al., 1999b). These authors suggested that the UV-induced cross-linking in these films involve aromatic amino acids, such as phenylalanine and tyrosine, present in egg white proteins.

Varying the amount of added plasticizers can form egg white films with a spectrum of mechanical and oxygen barrier properties (Lim et al., 2002). Gennadios et al. (1996b) showed that regardless of the added plasticizer (Gly, Sor, or PEG 400), TS decreased and E increased with increasing plasticizer level. Moreover, PEG 400 gave films with higher TS and E than films obtained from Gly or Sor. To obtain similar mechanical properties, higher amounts of Sor than Gly are required, which correlates with the smaller particle size of Gly than Sor.

Moisture conditions also affect the film mechanical properties (Lim et al., 1998). Increasing Gly content and RH increased TS of transglutaminase cross-linked egg white films. The effect of RH on TS depended on the plasticizer level. At high plasticizer content, TS values were higher and more sensitive to RH variations. At low plasticizer levels, films became weaker and less sensitive to RH variations (Lim et al., 1998).

The limited moisture barrier of egg white protein films can be improved reducing the hydrophilicity of the protein matrix. WVP of egg white films decreased from 6.21 to 5.62 g·mm/m$_2$·h·kPa by adding egg yolk solids (50% w/w of total egg solids), which contain egg yolk lipids (32% to 36% on the wet basis in egg yolk) (Gennadios et al., 1996b). However, incorporation of two milkfat fractions to egg white protein films did not affect film WVP (Handa et al., 1999b). This was attributed to the high content of short- and medium-chain (C_4–C_{10}) fatty acids in the milkfat, because moisture barrier ability of lipids decreases with decreasing hydrocarbon chain length.

Lim et al. (1998) studied the OP of transglutaminase cross-linked egg white films. Under low RH conditions, OP of the films was low but increased strongly as RH increased. Films with reduced glycerol content were better oxygen barriers but were more sensitive to RH variation. The permeability values eventually became similar for the two Gly contents studied as RH approached saturation.

2.10.3 Egg white protein coatings

Egg albumen has been studied for its potential to retain moisture inside raisins in cereal/raisin mixtures (Bolin, 1976) and inside meat products (Reutimann et al., 1996). Combinations of egg white, whey protein, and pregelatinized starch have also been prepared to separate food components of different moisture content, such as pastry, bread, pizza bases, and so forth, from toppings, fillings, garnishes, sauces, and so on (Berberat and Wissgott, 1993).

2.11 Keratin films

Keratins are proteins found in hair, wool, feathers, nail, horns, and other epithelial coverings. In wool, keratins are the main constituents of long cortical cells and weigh 30% to 60% of the total fiber (Yamauchi and Yamauchi, 2002). In feathers, keratin accounts for about 90% (Nissen, 1995). Keratins are characterized by their high content of cysteine and half-cystine residues (Arai et al., 1996). Keratin fibrils may thus be considered as polymers, which are formed by intra- and intermolecular oxidative S–S bonds of cysteine residues between the monomeric keratins.

2.11.1 Formation of keratin protein films

Stable aqueous solutions of reduced keratins can readily be prepared from wool with a mixture of urea, 2-mercaptoethanol, and water at 40 to 60°C. This extraction method cleaves hydrogen bonds and covalent S–S bonds between proteins, causing keratin fibrils to dissociate and be reduced simultaneously into monomeric reduced keratins. SDS can be added to improve the extraction and also to help stabilize the aqueous protein solution. The aqueous solution of reduced keratin can be cast to form clear keratin films that are insoluble in water and in most organic solvents (Yamauchi and Yamauchi, 2002).

2.11.2 Functional properties of keratin protein films

Keratin protein films are insoluble in organic solvents. These films swelled by 30% to 60% lengthwise in water and shrank to 90% in boiling water. However, they dissolved upon heating in an aqueous solution of 2-mercaptoethanol due to reductive cleavage of S–S bridges (Yamauchi and Yamauchi, 2002). As a consequence of the high percentage of covalent S–S bonds, Gly-plasticized keratin films are mechanically stronger than similarly prepared collagen films (Yamauchi and Yamauchi, 2002). Nakagaki and Yonese (1971) determined that keratin films presented lower apparent permeability values to glucose, urea, sodium chloride, and proteins such as bovine serum albumin than collagen films cross-linked by glutaraldehyde prepared under similar conditions. Generally, apparent permeability through keratin films decreased with increasing molecular weight of the permeating species (Yamauchi and Yamauchi, 2002).

2.11.3 Keratin coatings

Keratin proteins have been used as microencapsulating agents. Keratin vesicles have shown superior encapsulation properties for hydrophobic compounds than similarly prepared vesicles from egg yolk albumin, BSA, soy protein, and myoglobin proteins (Yamauchi and Khoda, 1997). Various hydrophobic materials including glycerides and dyes have been encapsulated in yields greater than 95% within keratin microcapsules. However, microencapsulation of antioxidants such as vitamins C and E in keratins was not successful (Suslick and Grinstaff, 1990).

2.12 Protein films of limited availability

2.12.1 Peanut protein

Peanut protein concentrates (PPCs) and isolates (PPIs) are commercially produced from defatted peanut flour. PPC has about 70% protein content, whereas PPI is 90% to 95% protein. Fractionated peanut proteins are classified as albumins, arachin, and conarachin, with the two latter fractions composed mainly of complex, high molecular weight globulins (Natarajan, 1980; Prakash and Rao, 1986).

Peanut protein films can be obtained by two different methods. First, peanut protein-lipid films can be formed on the surface of heated peanut milk (Wu and Bates, 1973; Aboagye and Stanley, 1985). These films present lower tensile strength than films prepared from soymilk under similar conditions (Wu and Bates, 1973). The second method involves casting the PPC or PPI solutions. Jangchud and Chinnan (1999a, 1999b) prepared PPC films by casting PPC solutions, previously adjusted to different pH values (6, 7.5, or 9) and dried at 70, 80, or 90°C. Gly, Sor, PEG, or PPG were added as plasticizers at different concentrations. Gly was determined to be the best performing plasticizer. Films formed at a higher pH (7.5 or 9) and higher drying temperature (80 or 90°C) were less moist and sticky at the surface than films formed at pH 6 and a drying temperature of 70°C. TS and E increased but WVP and OP decreased as drying temperature increased. At pH 9 and 90°C, the film had the lowest WVP and OP and the highest TS. It was hypothesized that the reduction of WVP and the increase in mechanical properties with increased drying temperature were due to increased protein denaturation and greater cross-linking, resulting in a tight and compact protein film structure (Jangchud and Chinnan, 1999a, 1999b). Patrick et al. (2001) prepared peanut protein films with and without natural fat of the peanuts. Films contained Sor as plasticizer at three different levels (37.5%, 44.5%, and 50%). As Sor content increased, higher TS and higher YM were achieved for the films without fat. However, films with fat exhibited maximum strength and modulus at 44.5% Sor content. Jiang et al. (2006) observed a decrease in the water content, total soluble matter, and transparency of transglutaminase-treated PPI films, with no effect on film mechanical properties.

2.12.2 Rice protein

Rice bran and polish contain 12% to 15.6% and 11.8% to 13% protein, respectively (Houston, 1972). Rice bran protein concentrates are prepared from commercially available unstabilized or heat-stabilized rice bran by alkaline extraction and isoelectric precipitation (Gnanasambandam and Hettiarachchy, 1995). Alternatively, rice flour or kernels may be treated with enzymes to partially remove the starch component (Chen and Chang, 1984). The majority of proteins in rice bran are albumins and globulins, which are water or salt soluble (Park et al., 2002b).

Films from rice bran protein solutions (70% protein, dry basis) have been prepared using Gly as a plasticizer (2% w/v), adjusting the pH to either 9.5 or 3, heated to 80°C, poured onto polyethylene plates, and dried at 60°C (Gnanasambandam et al., 1997). No substantial differences were observed with regard to WVP between films prepared at pH 3 and 9.5. OP was affected by pH, with films prepared at pH 3 showing lower values than films prepared at pH 9.5. Reduced OP was related to the lower solubility of rice bran protein at pH 3, which may result in a tighter film structure. TS was greater for rice bran protein films prepared at pH 9.5 than for those prepared at pH 3 (Gnanasambandam et al., 1997).

Shih (1996) prepared films from mixtures of rice protein concentrate and pullulan. Film TS and WVP were a function of rice protein concentrate content. The protein–pullulan mixture with up to 50% protein concentrate had higher TS (up to 30 Mpa) and lower WVP than films with high content (>50%) of rice protein. Film TS and water vapor resistance were improved by the addition of small amounts of propylene glycol alginate under alkaline condition. Oils were also incorporated into the film for improved water vapor resistance.

2.12.3 Pea protein

Pea proteins are separated from starch and fiber by multistep solubilization at pH 2.5 to 3, followed by centrifugation (Nickel, 1981). These protein isolates have a mean crude protein content of 85.3% and an ash content of 4.1% to 5%. They show a lower fat absorption than SPI, suggesting the presence of more numerous hydrophilic than hydrophobic groups on the surface of protein molecules (Naczk et al., 1986). Pea protein consists primarily of globulins (>80% of total protein) and a small fraction of albumins. Pea protein possesses good nutritional quality (i.e., protein efficiency ratio and essential amino acid content) and has potential as a dietary protein fortifier (Buffo and Han, 2005), although a relatively poor functionality has been reported with respect to protein solubility, emulsion stability, and gelation characteristics (Sumner et al., 1981; Jackman and Yada, 1989).

Owing to their globular structure, pea proteins need unfolding under alkaline conditions prior to film formation. An industrial pea protein concentrate (70.6% protein) has been utilized to cast films plasticized with various polyols (e.g., mono-, di-, tri-, and tetraethylene glycol, Gly and propane diol) and dried at 60°C. Gly-plasticized pea-protein films are sticky, while propane diol-plasticized films become brittle rapidly. Ethylene glycol gives films of greater TS and E, and lower WVP, than di-, tri-, or tetraethylene glycol. Overall, pea-protein isolate films show lower TS and E, and greater WVP (by one or two orders of magnitude), than LPDE. Cross-linking with formaldehyde substantially reduces the water solubility and increases the TS of these films, an expected effect due to the high lysine content of pea protein (Gueguen et al., 1995, 1998).

Films made from denatured pea-protein isolate (ca. 85% protein) show physical and mechanical properties similar to those of soy protein and whey protein films, and possess the strength and elasticity to resist handling. Increasing the concentration of Gly as a plasticizer in the film decreased the TS and YM but increased the E and WVP. Film solubility was not affected by the amount of plasticizer (Choi and Han, 2001). Heat treatment of pea protein isolate solutions for 5 minutes at 90°C to induce denaturation increases the TS and E and decreases WVP and the YM. Fourier transform infrared spectroscopy showed more water in the nonheated than in the heat-denatured ones, while electrophoresis suggests that intermolecular disulfide bonds are created during heat denaturation, which most feasibly results in the observed greater film integrity with respect to the nonheated control films (Choi and Han, 2002).

2.12.4 Pistachio protein

Protein content of pistachio kernel ranges between 35% and 40%. A protein isolate (about 95% protein) can be prepared by extraction from ground kernels, followed by a bleaching process to change the color from brownish to white using hydrogen peroxide, and a coagulation step to obtain the isolate (Ayranci and Dalgiç, 1992; Ayranci and Çetin, 1995). The amino acid composition of the obtained protein isolate shows a high content of hydrophilic amino acids glutamic and aspartamic acid (Ayranci and Dalgiç, 1992).

Ayranci and Dalgiç (1992) combined HPMC and pistachio protein isolate to cast films plasticized with PEG-400. Film WVP increased as protein content increased due to the high hydrophilic amino acid content. In addition, films swell by absorbing water, showing a greater thickness with respect to the control HPMC film.

2.12.5 Lupin protein

Lupin or lupine are trivial names for plants of the genus *Lupinus* belonging to the Leguminosae family. Lupin seeds are rich in protein, reaching protein levels as high as 42% for *L. luteus* (Kiosseoglou et al., 1999). Lupin protein concentrate (73% protein) has been prepared by alkali solubilization method (Sathe et al., 1982). Lupin protein isolates have been prepared by extraction at pH 8 to 9, acid coagulation, ultrafiltration/diafiltration, isoelectric precipitation, or dialysis (Ruiz and Jove, 1976; Duranti et al., 1983; Kiosseoglou et al., 1999).

Lupin proteins are either albumins (13%) or globulins (87%) (Duranti et al., 1983). Amino acid composition varies with lupin species, but all species are deficient in methionine, cystine, and tryptophan (Gross et al., 1989).

Lupin protein films have been prepared on the surface of heated lupin seed milk (Karara, 1989; Chango et al., 1995). Karara (1989) obtained an optimum protein extractability of 94% for film formation after soaking lupin seeds at 22°C for 5 hours. Chango et al. (1995) investigated the optimum pH value for lupin protein film formation at slightly alkaline conditions (pH 8.5).

2.12.6 Grain sorghum protein

Grain sorghum ranks fifth among cereals produced in the world (Dendy, 1995). The main protein fraction in the sorghum kernel is prolamine, known as kafirin. Kafirin is similar to corn zein in molecular weight, structure, and amino acid composition (Shull et al., 1991). Kafirin is more difficult to solubilize than zein due to the more hydrophobic character of kafirin and more disulfide cross-linked (El Nour et al., 1998; Duodu et al., 2002). According to its prolamin nature, kafirin can be readily extracted with alcohol or a reducing agent such as β-mercaptoethanol.

Films prepared from ethanol-extracted kafirin and plasticized with Gly and PEG presented similar TS, E, and WVP than those of films prepared from commercial corn zein. However, kafirin films were darker than zein films, probably due to a higher content in pigments (Buffo et al., 1997). Gao et al. (2005) studied the effect of the extracting method on film preparation and related the protein secondary structure to the film-forming properties. The inclusion of sodium hydroxide in the aqueous ethanol extractant with a reducing agent significantly increased the extraction yield of kafirin. The best films were obtained with kafirin containing a large proportion of nativelike α-helical structures with little intermolecular β-sheet content.

2.12.7 Winged bean protein

Winged bean protein is similar to soybean in terms of protein and fat content. Winged seed oil and protein range between 7.2% to 21.5% and 20.7% to 45.9%, respectively, depending on the plant origin and cultivar (Hildebrand et al., 1981). Of the total proteins in winged been seed, globulins and albumins account for 29% to 33% and 15% to 22%, respectively. Compared to peanut and soybean protein, winged bean protein is richer in lysine but has a low content of methionine (Kantha and Erdman, 1984).

Sian and Ishak (1990) prepared protein films on the surface of heated winged bean milk. Film formation occurred within the pH ranges of 2 to 3 and 6.3 to 11. At high pH values (7.5, 9, or 10), films had higher protein, carbohydrate, and ash contents and lower fat content than films formed at pH 6.7 or 2.5.

2.12.8 Cucumber pickle brine protein

A proteinaceous fibrous material can be recovered from the surface of brine during cucumber fermentation. The mean proximate composition of the freeze-dried proteinaceous material was reported as 72.5% protein, 3.2% ash, 0.7% lipid, 20.2% total carbohydrate, and 3% moisture (Hettiarachchy et al., 1998).

Yildirim and Hettiarachchy (1998b) prepared films from aqueous solutions plasticized with Gly. Heating the solution at 70 or 90°C for 30 minutes prior to casting gave clearer and more uniform films than unheated solutions. Adjustment of pH to 11 produced films highly soluble in water, whereas solutions heated at 70°C and adjusted to pH 7 or 9 yield films with greater TS and WVP and lower solubility than solutions heated at 90°C at pH 7 or 9.

2.13 Conclusions

Hydrophilic polymers that can associate through hydrogen or ionic bonding, like proteins, are able to form edible films and coatings with good oxygen and aroma barrier at low relative humidity, similar to the best synthetic polymer films. However, due to their hydrophilic nature, they present a poor moisture barrier, which is several orders of magnitude greater than the same synthetic films. Protein mechanical properties are also inferior to those of synthetic films but are sufficient for most applications.

Much protein film research has been conducted to improve protein mechanical and moisture barriers. Research has also been conducted to study the ability to extrude protein films as an alternative to solvent casting. Recent research has focused on studying the effectiveness of adding antimicrobials, and to a lesser extent, other food additives, to protein films. Investigated protein coating applications demonstrate protein potential on extending the shelf life of fruits and vegetables, dry fruits and nuts, meat products, and bakery and confectionary products. Much research is needed to demonstrate protein coating effectiveness at reducing moisture change, oxidation, migration of aroma and oil, and improving film integrity in order to help protein coatings reach their full potential.

References

Aboagye, Y., and D.W. Stanley. 1985. Texturization of peanut proteins by surface film formation 1. Influence of process parameters on film forming properties. *Can Inst Food Sci Technol J* 18:12–20.

Alam, Md.S., and S. Paul. 2001. Efficacy of casein coating on storage behavior of kinnow. *J Food Sci Technol* 38:235–238.
Albert, S., and G.S. Mittal. 2002. Comparative evaluation of edible coatings to reduce fat uptake in a deep-fried cereal product. *Food Res Int* 35:445–458.
Ali, Y., V.M. Ghorpade, and M.A. Hanna. 1997. Properties of thermally-treated wheat gluten films. *Ind Crop Prod* 6:177–184.
Andres, C. 1984. Natural edible coating has excellent moisture and grease barrier properties. *Food Process* 45:48–49.
Anker, M., M. Stading, and A.M. Hermansson. 2000. Relationship between the microstructure and the mechanical and barrier properties of whey protein films. *J Agric Food Chem* 48:3806–3816.
Anker, M., J. Berntsen, A.M. Hermansson, and M. Stading. 2002. Improved water vapor barrier of whey protein films by addition of an acetylated monoglyceride. *Innov Food Sci Emerg* 3:81–92.
Arai, K., S. Naito, V.B. Nagasawa, and M. Hirano. 1996. Crosslinking structure of keratin. VI. Number, type, and location of disulfide crosslinkages in low-sulfur protein of wool fiber and their relation to permanent set. *J Appl Polym Sci* 60:169–179.
Arvanitoyannis, I. 2002. Formation and properties of collagen and gelatin films and coatings. In *Protein-based films and coatings*, ed. A. Gennadios, 275–304. Boca Raton, FL: CRC Press.
Arvanitoyannis, I., E. Psomiadou, and A. Nakayama. 1996. Edible films made from sodium caseinate, starches, sugars or glycerol. Part 1. *Carbohydr Polym* 31:179–192.
Arvanitoyannis, I., E. Psomiadou, A. Nakayama, S. Aiba, and N. Yamamoto. 1997. Edible films made from gelatin, soluble starch and polyols: Part 3. *Food Chem* 60:593–604.
Arvanitoyannis, I., A. Nakayama, and S.I. Aiba. 1998a. Chitosan and gelatin based edible films: State diagram, mechanical and permeation properties. *Carbohydr Polym* 37:371–382.
Arvanitoyannis, I., A. Nakayama, and S.I. Aiba. 1998b. Edible films made from hydroxypropyl starch and gelatin and plasticised by polyols and water. *Carbohydr Polym* 36:105–119.
Atterholt, C.A., M.J. Delwiche, R.E. Rice, and J.M. Krochta. 1998. Study of biopolymers and paraffin as potential controlled-release carriers for insect pheromones. *J Agric Food Chem* 46:4429–4434.
Avena-Bustillos, R.J., and J.M. Krochta. 1993. Water vapor permeability of caseinate-based edible films as affected by pH, calcium crosslinking and lipid content. *J Food Sci* 58:904–907.
Avena-Bustillos, R.J., L.A. Cisneros-Zevallos, J.M. Krochta, and M.E. Saltveit. 1993. Optimization of edible coatings on minimally processed carrots using response surface methodology. *Trans ASAE* 36:801–805.
Avena-Bustillos, R.J., L.A. Cisneros-Zevallos, J.M. Krochta, and M.E. Saltveit. 1994a. Application of casein-lipid edible film emulsions to reduce white blush on minimally processed carrots. *Postharvest Biol Technol* 4:319–329.
Avena-Bustillos, R.J., J.M. Krochta, M.E. Saltveit, R.J. Rojas-Villegas, and J.A. Sauceda-Pérez. 1994b. Optimization of edible coating formulations on zucchini to reduce water loss. *J Food Engr* 21:197–214.
Avena-Bustillos, R.J., J.M. Krochta, and M.E. Saltveit. 1997. Water vapor resistance of red delicious apples and celery sticks coated with edible caseinate-acetylated monoglyceride films. *J Food Sci* 62:351–354.
Aydt, T.P., C.L. Weller, and R.F. Testin. 1991. Mechanical and barrier properties of edible corn and wheat protein films. *Trans ASAE* 34:207–211.
Ayranci, E., and A.C. Dalgiç. 1992. Preparation of protein isolates from *Pistacia terebinthus* L. and examination of some functional properties. *Lebensm Wiss Technol* 25:442–444.

Ayranci, E., and E. Çetin. 1995. The effect of protein isolate of *Pistacia terebinthus* L. on moisture transfer properties of cellulose-based edible films. *Lebensm Wiss Technol* 28:241–244.

Bai, J., E.A. Baldwin, and R.H. Hagenmaier. 2002. Alternatives to shellac coatings provide comparable gloss, internal gas modification, and quality for 'Delicious' apple fruit. *HortSci* 37:559–563.

Bai, J., V. Alleyne, R.D. Hagenmaier, J.P. Mattheis, and E.A. Baldwin. 2003. Formulation of zein coatings for apples (*Malus domestica* Borkh). *Postharvest Biol Technol* 28:259–268.

Balian, G., and J.H. Bowes. 1977. The structure and properties of collagen. In *The science and technology of gelatin*, ed. A.G. Ward, and A. Courts, 1–30. New York: Academic Press.

Banerjee, R., and H. Chen. 1995. Functional properties of edible films using whey protein concentrates. *J Dairy Sci* 78:1673–1683.

Berberat, A., and U. Wissgott. 1993. US Patent No. 5,248,512.

Bietz, J.A., and J.S. Wall. 1980. Identity of high molecular weight gliadin and ethanol-soluble glutenin subunits of wheat: Relation to gluten structure. *Cereal Chem* 57:415–421.

Billing, O. 1989. *Flexible packaging*. Sweden: Akerlund & Rausing.

Bolin, H.R. 1976. Texture and crystallization control in raisins. *J Food Sci* 41:1316–1319.

Boundy, J.A., J.E. Turner, J.S. Wall, and R.J. Dimler. 1967. Influence of commercial processing on composition and properties of corn zein. *Cereal Chem* 44:281.

Bradbury, E., and C. Martin. 1952. The effect of the temperature of preservation on the mechanical properties and structure of gelatin films. *Proc Roy Soc A* 214:183–192.

Brandenburg, A.H., C.L. Weller, and R.F. Testing. 1993. Edible films and coatings from soy protein. *J Food Sci* 5:1086–1089.

Briston, J.H. 1986. Films, plastic. In *The Wiley encyclopedia of packaging technology*, ed. M. Bakker, 329–335. New York: John Wiley & Sons.

Brunner, J.R. 1977. Milk proteins. In *Food proteins*, ed. J.R. Whitaker, and S.R. Tannenbaum, 175–208, Westport, CT: AVI.

Brunner, J.R. 1981. Cow milk proteins: Twenty-five years of progess. *J Dairy Sci* 64:1038–1054.

Buffo, R.A., and J.H. Han. 2005. Edible films and coatings from plant origin proteins. In *Innovations in food packaging*, ed. J.H. Han, 277–300. London: Elsevier Academic Press.

Buffo, R., C.W. Curtis, and A. Gennadios. 1997. Films from Laboratory-Extracted Sorghum Kafirin. *Cereal Chem* 74:473–475.

Buonocore, G.G., A. Conte, and M.A. Del Nobile. 2005. Use of a mathematical model to describe the barrier properties of edible films. *J Food Sci* 70:E142–E147.

Cagri, A., Z. Ustunol, and E.T. Ryser. 2002. Inhibition of three pathogens on bologna and summer sausage using antimicrobial edible films. *J Food Sci* 67:2317–2324.

Cagri, A., Z. Ustunol, W. Osburn, and E.T. Ryser. 2003. Inhibition of *Listeria monocytogenes* on hot dogs using antimicrobial whey protein-based edible casings. *J Food Sci* 68:291–299.

Cao, Y.M., and K.C. Chang. 2002. Edible films prepared from water extract of soybeans. *J Food Sci* 67:1449–1454.

Carlin, F., N. Gontard, N. Reich, and C. Nguyen-The. 2001. Utilization of zein coating and sorbic acid to reduce *Listeria monocytogenes* growth on cooked sweet corn. *J Food Sci* 66:1385–1389.

Chango, A., C. Villaume, H.M. Bau, J.P. Nicolas, and L. Méjean. 1995. Fractionation by thermal coagulation of lupin proteins: Physicochemical characteristics. *Food Res Intern* 28:91–99.

Cheftel, J.C., J.L. Cuq, and D. Lorient. 1985. Amino acids, peptides, and proteins. In *Food Chemistry*, ed. O.R. Fennema, 245–369. New York: Marcel Dekker.

Chen, H. 2002. Formation and properties of casein films and coatings. In *Protein-based films and coatings*, ed. A. Gennadios, 181–211. Boca Raton, FL: CRC Press.

Chen, W.P., and Y.C. Chang. 1984. Production of high-fructose rice syrup and high-protein rice flour from broken rice. *J Sci Food Agric* 35:1128–1135.

Chick, J., and Z. Ustunol. 1998. Mechanical and barrier properties of lactic acid and rennet precipitated casein-based edible films. *J Food Sci* 63:1024–1027.

Chick, J., and R.J. Hernandez. 2002. Physical, thermal, and barrier characterization of casein-wax-based edible films. *J Food Sci* 67:1073–1079.

Chilves, G.R., and V.J. Morris. 1987. Coacervation of gelatin-gellan gum mixtures and their use in microencapsulation. *Carbohydr Polym* 7:111–120.

Cho, S.Y., and C. Rhee. 2002. Sorption characteristics of soy protein films and their relation to mechanical properties. *Lebensm Wiss Technol* 35:151–157.

Cho, S.G., P.S.W. Park, E.T. Adams, and K.C. Rhee. 1992. A simple rapid and simultaneous preparation of glandless cottonseed 7S and 11S protein fractions and characterization of some physicochemical properties. *J Food Qual* 15:357–367.

Choi, W.S., and J.H. Han. 2001. Physical and mechanical properties of pea-protein-based edible films. *J Food Sci* 66:319–322.

Choi, W.S., and J.H. Han. 2002. Film-forming mechanism and heat denaturation effects on the physical and chemical properties of pea-protein-isolate edible films. *J Food Sci* 67:1399–1406.

Choi, S.G., K.M. Kim, M.A. Hanna, C.L. Weller, and W.L. Kerr. 2003. Molecular dynamics of soy-protein isolate films plasticized by water and glycerol. *J Food Sci* 68:2516–2522.

Ciesla, K., S. Salmieri, and M. Lacroix. 2006a. γ-Irradiation influence on the structure and properties of calcium caseinate-whey protein isolate based films. Part 1. Radiation effect on the structure of protein gels and films. *J Agric Food Chem* 54:6374–6384.

Ciesla, K., S. Salmieri, and M. Lacroix. 2006b. γ-Irradiation influence on the structure and properties of calcium caseinate-whey protein isolate based films. Part 2. Influence of polysaccharide addition and radiation treatment on the structure and functional properties of the films. *J Agric Food Chem* 54:8899–8908.

Ciesla, K., S. Salmieri, and M. Lacroix. 2006c. Modification of the properties of milk protein films by gamma radiation and polysaccharide addition. *J Sci Food Agric* 86:908–914.

Cisneros-Zevallos, L., and J.M. Krochta. 2003. Whey protein coatings for fresh fruits and relative humidity effects. *J Food Sci* 68:176–181.

Cook, R., and M.L. Shulman. 1994. Aqueous ultrapure zein lattices as functional ingredients and coatings. Corn Utilization Conference V, Bedford, MA. 1–4.

Cunningham, P., A.A. Ogale, P.L. Dawson, and J.C. Acton. 2000. Tensile properties of soy protein isolate films produced by a thermal compaction technique. *J Food Sci* 65:668–671.

Cuq, B. 1996. Formation and characterization of biomaterials from myofibrillar proteins. Ph.D. Dissertation. University of Montpellier, France, pp. 211.

Cuq, B. 2002. Formation and properties of fish myofibrillar protein films and coatings. In *Protein-based films and coatings*, ed. A. Gennadios, 213–232. Boca Raton, FL: CRC Press.

Cuq, B., C. Aymard, J.L. Cuq, and S. Guilbert. 1995. Edible packaging films based on fish myofibrillar proteins: Formation and functional properties. *J Food Sci* 60:1369–1374.

Cuq, B., N. Gontard, and S. Guilbert. 1997a. Thermoplastic properties of fish myofibrillar proteins: Application to the biopackaging fabrication. *Polymer* 38:4071–4078.

Cuq, B., N. Gontard, J.L. Cuq, and S. Guilbert. 1997b. Selected functional properties of fish myofibrillar protein-based films as affected by hydrophilic plasticizers. *J Agric Food Chem* 45:622–626.

Cuq, B., N. Gontard, N. Aymard, and S. Guilbert. 1997c. Relative humidity and temperature effects on mechanical and water vapor barrier properties of myofibrillar protein-based films. *Polym Gels Netw* 5:1–15.

Cuq, B., N. Gontard, J.L. Cuq, and S. Guilbert. 1998. Packaging films based on myofibrillar proteins: Fabrication, properties and applications. *Nahrung/Food* 42:260–263.

Cuq, B., F. Boutrot, A. Redl, and V. Lullien-Pellerin. 2000. Study of the temperature effect on the formation of wheat gluten network: Influence on mechanical properties and protein solubility. *J Agric Food Chem* 48:2954–2959.

Dalgleish, D.G. 1989. Milk proteins—Chemistry and physics. In *Food proteins*, ed. J.E. Kinsella, and W.G. Soucie, 155–178. Champagne: American Oil Chemists Society.

Dangaran, K.L., and J.M. Krochta. 2003. Aqueous whey protein coatings for panned products. *Manuf Confect* 83:61–65.

Dangaran, K.L., J. Renner-Nantz, and J.M. Krochta. 2006. Whey protein-sucrose coating gloss and integrity stabilization by crystallization inhibitors. *J Food Sci* 71:E152–E157.

Dawson, P.L., D.E. Hirt, J.R. Rieck, J.C. Acton, and A. Sotthibandhu. 2003. Nisin release from films is affected by both protein type and film-forming method. *Food Res Int* 36:959–968.

Dendy, D.A.V. 1995. Sorghum and millets: Production and importance. In *Sorghum and millets: Chemistry and technology*, ed. D.A.V. Dendy, 11–26. St. Paul: American Association of Cereal Chemists.

DeWit, J.N. 1989. Functional properties of whey proteins. In *Developments in dairy chemistry*, ed. P.F. Fox, 285–321. New York: Elsevier Applied Science.

Di Pierro, P., L. Mariniello, C.V.L. Giosafatto, P. Masi, and R. Porta. 2005. Solubility and permeability properties of edible pectin-soy flour films obtained in the absence or presence of transglutaminase. *Food Biotechnol* 19:37–49.

Di Pierro, P., B. Chico, R. Villalonga, L. Mariniello, A.E. Damiao, P. Masi, and R. Porta. 2006. Chitosan-whey protein edible films produced in the absence or presence of transglutaminase: Analysis of their mechanical and barrier properties. *Biomacromolecules* 7:744–749.

Di Pierro, P., B. Chico, R. Villalonga, L. Mariniello, P. Masi, and R. Porta. 2007. Transglutaminase-catalyzed preparation of chitosan-ovalbumin films. *Enzyme Microb Tech* 40:437–441.

Domenek, S., L. Brendel, M.H. Morel, and S. Guilbert. 2004. Swelling behavior and structural characteristics of wheat gluten polypeptide films. *Biomacromolecules* 5:1002–1008.

Duodu, K.G., A. Nunes, I. Delgadillo, M.L. Parker, E.N.C. Mills, P.S. Belton, and J.R.N. Taylor. 2002. Effect of grain structure and cooking on sorghum and maize *in vitro* protein digestibility. *J Cereal Sci* 35:161–175.

Duranti, M., A. Riccardi, P. Restan, and P. Cerletti. 1983. Isolation and purification of proteins from lupine seed and corn germ. In *Progress in food engineering*, eds. C. Cantarelli, and C. Peri, 561–562. Küsnacht: Foster-Verlag AG.

Dybing, S.T., and D.E. Smith. 1991. Relation of chemistry and processing procedures to whey protein functionality: A review. *Cultured Dairy Prod J* 26:4–12.

Eastoe, J.E., and A.A. Leach. 1977. Chemical constitution of gelatin. In *The science and technology of gelatin*, ed. A.G. Ward, and A. Courts, 73–107. New York: Academic Press.

El Nour, I.N.A., A.D.F. Peruffo, and A. Curioni, 1998. Characterisation of sorghum kafirins in relations to their cross-linking behaviour. *J Cereal Sci* 28:197–207.

Esen, A. 1987. A proposed nomenclature for the alcohol-soluble proteins zeins of maize (*Zea mays* L.). *J Cereal Sci* 5:117–128.

Eswaranandam, S., N.S. Hettiarachchy, and J.F. Meullenet. 2006. Effect of malic and lactic acid incorporated soy protein coatings on the sensory attributes of whole apple and fresh-cut cantaloupe. *J Food Sci* 71:S307–S313.

Fairley, P., F.J. Monahan, J.B. German, and J.M. Krochta. 1996. Mechanical properties and water vapor permeability of edible films from whey protein isolate and n-ethylmaleimide or cysteine *J Agric Food Chem* 44:3789–3792.

Fakhouri, F.M., P.S. Tanada-Palmu, and C.R.F. Grosso. 2004. Characterization of composite biofilms of wheat gluten and cellulose acetate phthalate. *Braz J Chem Eng* 21:261–264.

Fakhouri, F.M., L.C.B. Fontes, P.V.D.M. Goncalves, C.R. Milanez, C.J. Steel, and F.P. Collares-Queiroz. 2007. Films and edible coatings based on native starches and gelatin in the conservation and sensory acceptance of Crimson grapes. *Ciencia e Tecnologia de Alimentos* 27:369–375.

Fakirov, S., Z. Sarac, T. Anbar, B. Boz, I. Bahar, M. Evstatiev, A.A. Apostolov, J.E. Mark, and A. Kloczkowski. 1996. Mechanical properties and transition temperatures of cross-linked oriented gelatin. 1. Static and dynamic mechanical properties of cross-linked gelatin. *Colloid Polym Sci* 274:334–341.

Fakirov, S., Z. Sarac, T. Anbar, B. Boz, I. Bahar, M. Evstatiev, A.A. Apostolov, J.E. Mark, and A. Kloczkowski. 1997. Mechanical properties and transition temperatures of crosslinked oriented gelatin. 2. Effect of orientation and water content on transition temperatures. *Colloid Polym Sci* 275:307–314.

Farouk, M.M., J.F. Price, and A.M. Salih. 1990. Effect of an edible collagen film overwrap on exudation and lipid oxidation in beef round steak. *J Food Sci* 55:1510, 1563.

Foster, R. 1986. Ethylene-vinyl alcohol copolymers (EVOH). In *The Wiley encyclopedia of packaging technology*, ed. M. Bakker, 270–275. New York: John Wiley & Sons.

Foulk, J.A., and J.M. Bunn. 2001. Physical and barrier properties of developed bilayer protein films. *Appl Eng Agric* 17:635–641.

Fox, P.F., and P.L.H. McSweeney. 1998. *Dairy chemistry and biochemistry*. London: Blackie Academic & Professional.

Frinault, A., D.J. Gallant, B. Bouchet, and J.P. Dumont. 1997. Preparation of casein films by a modified wet spinning process. *J Food Sci* 62:744–747.

Fringant, C., M. Rinaudo, N. Gontard, S. Guilbert, and H. Derradji. 1998. A biogradable starch based coating to waterproof hydrophilic materials. *Starch-Starke* 50:292–296.

Galietta, G., L. Di Gioia, S. Guilbert, and B. Cuq. 1998. Mechanical and thermomechanical properties of films based on whey proteins as affected by plasticizer and crosslinking agents. *J Dairy Sci* 81:3123–3130.

Gallstedt, M., A. Mattozzi, E. Johansson, and M.S. Hedenqvist. 2004. Transport and tensile properties of compression-molded wheat gluten films. *Biomacromolecules* 5:2020–2028.

Gao, C., J. Taylor, N. Wellner, Y.B. Byaruhanga, M.L. Parker, E.N.C. Mills, and P.S. Belton. 2005. Effect of preparation conditions on protein secondary structure and biofilm formation of kafirin. *J Agric Food Chem* 53:306–312.

Garcia, F.T., and P.J.D.A. Sobral. 2005. Effect of the thermal treatment of the filmogenic solution on the mechanical properties, color and opacity of films based on muscle proteins of two varieties of Tilapia. *Lebensm Wiss Technol* 38:289–296.

Gennadios, A., and C.L. Weller. 1990. Edible films and coatings from wheat and corn proteins. *Food Technol* 44:63–69.

Gennadios, A., and C.L. Weller. 1991. Edible films and coatings from soymilk and soy protein. *Cereal Foods World* 36:1004–1009.

Gennadios, A., C.L. Weller, and R.F. Testin. 1993a. Temperature effect on oxygen permeability of edible protein-based films. *J Food Sci* 58:212–214, 219.

Gennadios, A., H.J. Park, and C.L. Weller. 1993b. Relative humidity and temperature effects on tensile strength of edible protein and cellulose ether films. *Trans ASAE* 36:1867–1872.

Gennadios, A., A.H. Bradenburg, C.L. Séller, and R.F. Testin. 1993c. Effect of pH on properties of wheat gluten and soy protein isolate films. *J Agric Food Chem* 41:1835–1839.

Gennadios, A., C.L. Weller, and R.F. Testin. 1993d. Property modification of edible wheat gluten-based films. *Trans ASAE* 36:465–470.

Gennadios, A., C.L. Weller, and R.F. Testin. 1993e. Modification of physical and barrier properties of edible wheat gluten-based films. *Cereal Chem* 70:426–429.

Gennadios, A., C.L. Weller, and R.F. Testin. 1993f. Temperature effect on oxygen permeability of edible protein-based films. *J Food Sci* 58: 212–214, 219.

Gennadios, A., A.H. Brandenburg, C.L. Weller, and R.F. Testin. 1993g. Effect of pH on properties of wheat gluten and soy protein isolate films. *J Agric Food Chem* 41:1835–1839.

Gennadios, A., T.H. McHugh, C.L. Weller, and J.M. Krochta. 1994. Edible coatings and films based on proteins. In *Edible coatings and films to improve food quality*, ed. J.M. Krochta, E.A. Baldwin, and M. Nisperos-Carriedo, 201–277. Lancaster, PA: Technomic.

Gennadios, A., V.M. Ghorpade, C.L. Séller, and M.A. Hanna. 1996a. Heat curing of soy protein films. *Trans ASAE* 39:575–579.

Gennadios, A., C.L. Weller, M.A. Hanna, and G.W. Froning. 1996b. Mechanical and barrier properties of egg albumin films. *J Food Sci* 61:585–589.

Gennadios, A., J.W. Rhim, A. Handa, C.L. Séller, and M.A. Hanna. 1998a. Ultraviolet radiation affects physical and molecular properties of soy protein films. *J Food Sci* 63:225–228.

Gennadios, A., C. Cezeirat, C.L. Weller, and M.A. Hanna. 1998b. Emulsified soy protein-lipid films. In *Paradigm for successful utilization of renewable resources*, ed. D.J. Sessa, and J.L. Willett, 213–226. Champaign: AOCS Press.

Gennadios, A., A.Handa, G.W. Froning, C.L. Weller, and M.A. Hanna. 1998c. Physical properties of egg white-dialdehyde starch films. *J Agric Food Chem* 46:1297–1302.

Ghorpade, V.M., H. Li, A. Gennadios, and M.A. Hanna. 1995a. Chemical modified soy protein films. *Trans ASAE* 38:1805–1808.

Ghorpade, V.M., H. Li, A. Gennadios, M.A. Hanna, and C.L. Weller. 1995b. Soy protein isolate/poly(ethylene oxide) films. *Cereal Chem* 72:559–563.

Ghorpade, V.M., and M.A. Hanna. 1996. Mechanical properties of soy protein-polyethylene ribbon and film extrudates. *Trans ASAE* 39:611–615.

Gnanasambandam, R., and N.S. Hettiarachchy. 1995. Protein concentrates from unstabilized and stabilized rice bran: Preparation and properties. *J Food Sci* 60:1066–1069, 1074.

Gnanasambandam, R., N.S. Hettiarachchy, and M. Coleman. 1997. Mechanical and barrier properties of rice bran films. *J Food Sci* 62:395–398.

Gómez-Guillen, M.C., M. Ihl, V. Bifani, A. Silva, and P. Montero. 2007a. Edible films made from tuna-fish gelatin with antioxidant extracts of two different murta ecotypes leaves (*Ugni molinae* Turcz). *Food Hydrocol* 21:1133–1143.

Gómez-Estaca, J., P. Montero, B. Gimenez, and M.C. Gomez-Guillen. 2007b. Effect of functional edible films and high pressure processing on microbial and oxidative spoilage in cold-smoked sardine (*Sardina pilchardus*). *Food Chem* 105:511–520.

Gontard, N., S. Guilbert, and J.L. Cuq. 1992. Edible wheat gluten films: Influence of the main process variables on film properties using response surface methodology. *J Food Sci* 57:190–195, 199.

Gontard, N., S. Guilbert, and J.L. Cuq. 1993. Water and glycerol as plasticizers affect mechanical and water vapor barrier properties of an edible wheat gluten film. *J Food Sci* 58:206–211.

Gontard, N., C. Duchez, J.L. Cuq, and S. Guilbert. 1994. Edible composite films of wheat gluten and lipids: Water vapor permeability and other physical properties. *Int J Food Sci Technol* 29:39–50.

Gontard, N., S. Marchesseau, J.L. Cuq, and S. Guilbert. 1995. Water vapor permeability of edible bilayer films of wheat gluten and lipids. *Int J Food Sci Technol* 30:49–56.

Gontard, N., and S. Ring. 1996. Edible wheat gluten film: Influence of water content on glass transition temperature. *J Agric Food Chem* 44:3474–3478.

Gontard N., R. Thibault, B. Cuq, and S. Guilbert. 1996. Influence of relative humidity and film composition on oxygen and carbon dioxide permeabilities of edible films. *J Agric Food Chem* 44:1064–1069.

Greener, I.K., and O. Fennema. 1989. Barrier properties and surface characteristics of edible, bilayer films. *J Food Sci* 54:1393–1399.

Gross, R., F. Koch, I. Malaga, A.F. Miranda, H. Schoeneberger, and L.C. Trugo. 1989. Chemical composition and protein quality of some local Andean food sources. *Food Chem* 34:25–34

Guckian, S., C. Dwyer, M. O'Sullivan, E.D. O'Riordan, and F.J. Monahan. 2006. Properties of and mechanisms of protein interactions in films formed from different proportions of heated and unheated whey protein solutions. *Eur Food Res Technol* 223:91–95.

Gueguen, J., G. Viroben, and J. Barbot. 1995. Preparation and characterization of films from pea protein isolates. In *Proceedings of second European conference on grain legumes*, 358–359. Copenhagen, Denmark.

Gueguen, J., G. Viroben, P. Noireaux, and M. Subirabe. 1998. Influence of plasticizers and treatments on the properties of films from pea proteins. *Industr Crops Products* 7:149–157.

Guilbert, S. 1986. Technology and application of edible protective films. In *Food packaging and preservation: theory and practice*, ed. M. Mathlouthi, 371–394. Essex: Elsevier Applied Science.

Guilbert, S., N. Gontard, M.H. Morel, P. Charlier, V. Micard, and A. Redl. 2002. Formation and properties of wheat glute films and coatings. In *Protein-based films and coatings*, ed. A. Gennadios, 43–67. Boca Raton, FL: CRC Press.

Handa, A., A. Gennadios, G.W. Froning, N. Kuroda, and M.A. Hanna. 1999a. Tensile, solubility, and electrophoretic properties of egg white films as affected by surface sulfhydryl groups. *J Food Sci* 64:82–85.

Handa, A., A. Gennadios, M.A. Hanna, C.L. Weller, and N. Kuroda. 1999b. Physical and molecular properties of egg-white lipid films. *J Food Sci* 64:860–864.

Harrington, W.F. 1966. Collagen. In *Encyclopedia of polymer science and technology*, ed. H.F. Mark, N.G. Gaylord, and N.M. Bikales, 1–16. New York: Interscience

Herald, T.J., R. Gnanasambandam, B.H. Mcguire, and K.A. Hachmeister. 1995. Degradable wheat gluten films: Preparation, properties and applications. *J Food Sci* 60:1147–1150.

Hernandez-Muñoz, P., A. Kanavouras, P.K.W. Ng, and R. Gavara. 2003. Development and characterization of biodegradable films made from wheat gluten protein fractions. *J Agric Food Chem* 51:7647–7654.

Hernandez-Muñoz, P., A. Kanavouras, R. Villalobos, and A. Chiralt. 2004a. Characterization of biodegradable films obtained from cysteine-mediated polymerized gliadins. *J Agric Food Chem* 52:7897–7904.

Hernandez-Muñoz, P., J.M. Lagaron, A. Lopez-Rubio, and R. Gavara. 2004b. Gliadins polymerized with cysteine: Effects on the physical and water barrier properties of derived films. *Biomacromolecules* 5:1503–1510.

Hernandez-Muñoz, P., R. Villalobos, and A. Chiralt. 2004c. Effect of cross-linking using aldehydes on properties of glutenin-rich films. *Food Hydrocolloid* 18:403–411.

Hernandez-Muñoz, P., A. Lopez-Rubio, J.M. Lagaron, and R. Gavara. 2004d. Formaldehyde cross-linking of gliadin films: Effects on mechanical and water barrier properties. *Biomacromolecules* 5:415–421.

Hernandez-Muñoz, P., R. Villalobos, and A. Chiralt. 2004e. Effect of thermal treatments on functional properties of edible films made from wheat gluten fractions. *Food Hydrocolloid* 8:647–654.

Hernandez-Muñoz, P., A. Lopez-Rubio, V. Del-Valle, E. Almenar, and R. Gavara. 2004f. Mechanical and water barrier properties of glutenin films influenced by storage time. *J Agric Food Chem* 52:79–83.

Hettirachchy, N.S., M. Yildirim, and R.W. Buescher. 1998. Composition and functional properties of an unusual proteinaceous fibrous material from cucumber fermentation. *J Agric Food Chem* 46:825–828.

Hildebrand, D.F., C. Chaven, T. Hymowitz, H.H. Bryan, and A.A. Duncan. 1981. Protein and oil content of winged bean seeds as measured by near-infrared light reflectance. *Agronomy J* 73:623–625.

Ho, B. 1992. Water vapor permeability and structural characteristics of casein films and casein-lipid emulsion films. M.S. Thesis, University of California, Davis.

Hoa, T.T., M.N. Ducamp, M. Lebrun, and E.A. Baldwin. 2002. Effect of different coating treatments on the quality of mango fruit. *J Food Quality* 25:471–486.

Hoffman, K.L., I.Y. Han, and P.L. Dawson. 2001. Antimicrobial effects of corn zein films impregnated with nisin, lauric acid, and EDTA. *J Food Protect* 64:885–889.

Hood, L.L. 1987. Collagen in sausage casting. In *Advances in meat research*, ed. A.M. Pearson, T.R. Dutson, and A.J. Bailey, 109–129. New York: Van Nostrand Reinhold.

Houston, D.F. 1972. Rice bran and polish. In *Rice chemistry and technology*, ed. D.F. Houston, 272–300. St. Paul, MN: American Association of Cereal Chemists.

Hsu, B.L., Y.M. Weng, Y.H. Liao, and W. Chen. 2005. Structural investigation of edible zein films/coatings and directly determining their thickness by FT-Raman spectroscopy. *J Agric Food Chem* 53:5089–5095.

Huse, H.L., P. Mallikarjunan, M.S. Chinnan, Y.C. Hung, and R.D. Phillips. 1998. Edible coatings for reducing oil uptake in production of akara (deep-fat frying of cowpea paste). *J Food Process Preserv* 22:155–165.

Hwang, S.L., S.J. Liu, W.L. Chen, C.Y. Wang, and S.W. Chou. 2003. Experimental development of processing technology for yuba-like films using twin screw extrusion. *Int Polym Proc* 18:343–348.

Irissin-Mangata, J., G. Bauduin, B. Boutevin, and N. Gontard. 2001. New plasticizers for wheat gluten films. *Eur Polym J* 37:1533–1541.

Jackman, R.L., and R.Y. Yada. 1989. Functional properties of whey-pea protein composite blends in a model system. *J Food Sci* 54:1287–1292.

Jagannath, J.H., C. Nanjappa, D.K. Das Gupta, and A.S. Bawa. 2003. Mechanical and barrier properties of edible starch-protein-based films. *J Appl Polym Sci* 88:64–71.

Janes, M.E., S. Kooshesh, and M.G. Johnson. 2002. Control of *Listeria monocytogenes* on the surface of refrigerated, ready-to-eat chicken coated with edible zein film coatings containing nisin and/or calcium propionate. *J Food Sci* 67:2754–2757.

Jangchud, A., and M.S. Chinnan. 1999a. Peanut protein film as affected by drying temperature and pH of film forming solution. *J Food Sci* 64:153–157.

Jangchud, A., and M.S. Chinnan. 1999b. Properties of peanut protein film: Sorption isotherm and plasticizer effect. *Lebensm Wiss Technol* 32:89–94.

Jiang, Y., Q.B. Wen, C.H. Tang, and X.Q. Yang. 2006. Effect of transglutaminase on properties of cast films from food proteins. *J S China University Technol* 34:110–115.

Kantha, S.S., and J.R. Erdman Jr. 1984. The winged bean as an oil and protein source: A review. *JAOCS* 61:515–525.

Karara, H.A. 1989. Formation of protein-lipid film from lupinseed. *Fat Sci Technol* 91:412–416.

Kasarda, D.D. 1989. Glutenin structure in relation to wheat quality. In *Wheat is unique: Structure, composition, processing, end-use properties, and products*, ed. Y. Pomeranz, 277–302. St. Paul, MN: American Association of Cereal Chemists.

Kasarda, D.D. 1999. Glutenin polymers: The *in vitro* to *in vivo* transition. *Cereal Food World* 44:566–572.

Kayserilioglu, B.S., W.M. Stevels, W.J. Mulder, and N. Akkas. 2001. Mechanical and biochemical characterisation of wheat gluten films as a function of pH and co-solvent. *Starch-Starke* 53:381–386.

Kelley, J.J., and R. Pressey. 1966. Studies with soybean protein and fiber formation. *Cereal Chem* 43:195.

Kim, S.J., and Z. Ustunol. 2001. Solubility and moisture sorption isotherms of whey-protein-based edible films as influenced by lipid and plasticizer incorporation. *J Agric Food Chem* 49:4388–4391.

Kim, K.M., C.L. Weller, M.A. Hanna, and A. Gennadios. 2002a. Heat curing of soy protein films at selected temperatures and pressures. *Lebensm Wiss Technol* 35:140–145.

Kim, K.M., K.T. Hwang, C.L. Weller, and M.A. Hanna. 2002b. Preparation and characterization of soy protein isolate films modified with sorghum wax. *JAOCS* 79:615–619.

Kim, K.M., D.B. Marx, C.L. Weller, and M.A. Hanna. 2003. Influence of sorghum wax, glycerin, and sorbitol on physical properties of soy protein isolate films. *JAOCS* 80:71–76.

Kim, S., D.J. Sessa, and J.W. Lawton. 2004. Characterization of zein modified with a mild cross-linking agent. *Industrial Crops Products* 20:291–300.

Kinsella, J.E. 1979. Functional properties of soy protein. *JAOCS* 56:242.

Kinsella, J.E., and D.M. Whitehead. 1989. Proteins in whey: Chemical, physical and functional properties. *Adv Food Nutr Res* 33:343–438.

Kiosseoglou, A., G. Doxastakis, S. Alevisopoulos, and S. Kasapis. 1999. Physical characterization of thermally induced networks of lupin protein isolates prepared by isoelectric precipitation and dialysis. *Int J Food Sci Technol* 34:253–263.

Kozempel, M., and P.M. Tomasula. 2004. Development of a continuous process to make casein films. *J Agric Food Chem* 52:1190–1195.

Krochta, J.M. 2002. Proteins as raw materials for films and coatings: Definitions, current status, and opportunities. In *Protein-based films and coatings*, ed. A. Gennadios, 1–41. Boca Raton, FL: CRC Press.

Krochta, J.M. 1997a. Edible protein films and coatings. In *Food proteins and their applications in foods*, ed. S. Damodaran, and A. Paaraf, 529–549. New York: Marcel Dekker.

Krochta, J.M. 1997b. Edible composite moisture-barrier films. In *Packaging yearbook: 1997*, ed. B. Blakistone, 38–51. Washington: National Food Processors Association.

Krochta, J.M., A.E. Pavlath, and N. Goodman. 1990. Edible films from casein-lipid emulsions for lightly-processed fruits and vegetables. In *Engineering and food, Vol 2, Preservation processes and related techniques*, ed. W.E.L. Spiess, and H. Schubert, 329–340. London: Elsevier Applied Science.

Krochta, J.M., and C. De Mulder-Johnston. 1997. Edible and biodegradable polymer films: Challenges and opportunities. *Food Technol* 51:61–74.

Kurisawa, M., and N. Yui. 1998. Gelatin/dextran intelligent hydrogels for drug delivery: Dual-stimuli-responsive degradation in relation to miscibility in interpretating polymer networks. *Macromol Chem Phys* 199:1547–1554.

Kurose, T., K. Urman, J.U. Otaigbe, R.Y. Lochhead, and S.F. Thames. 2006. Effect of uniaxial drawing of soy protein isolate film on mechanical properties. In *Annual Technical Conference—ANTEC, Conference Proceedings*, 1489–1493.

Lacroix, M., and K. Cooksey. 2005. Edible films and coatings from animal-origin proteins. In *Innovations in food packaging*, ed., J.H. Han, 301–317. London: Elsevier Academic Press.

Lacroix, M., M. Ressouany, B. Ouattara, H. Yu, and C. Vachon. 1999. Radiation cross-linked protein films and their physicochemical properties. In *Proceedings of the fifth conference of food engineering,* 602–607, Dallas, TX.

Lacroix, M., T.C. Le, B. Ouattara, H. Yu, M. Letendre, S.F. Sabato, M.A. Mateescu, and G. Patterson. 2002. Use of γ-irradiation to produce films from whey, casein and soya proteins: Structure and functional characteristics. *Radiat Phys Chem* 63:827–832.

Lai, H.M., and G.W. Padua. 1997. Properties and microstructure of plasticized zein films. *Cereal Chem* 74:771–775.

Lai, H.M., and G.W. Padua. 1998. Water vapor barrier properties of zein films plasticized with oleic acid. *Cereal Chem* 75:194–199.

Lai, H.M., P.H. Geil, and G.W. Padua. 1999. X-ray diffraction characterization of the structure of zein-oleic acid films. *J Appl Polym Sci* 71:1267–1281.

Larre, C., C. Desserme, J. Barbot, and J. Gueguen. 2000. Properties of deamidated gluten films enzymatically cross-linked. *J Agric Food Chem* 48:5444–5449.

Lásztity, R. 1986. Recent results in the investigation of the structure of the gluten complex. *Die Nahrung* 30:235–244.

Lawton, J.W. 2004. Plasticizers for zein: Their effect on tensile properties and water absorption of zein films. *Cereal Chem* 81:1–5.

Lee, M., S. Lee, and K.B. Song. 2005. Effect of γ-irradiation on the physicochemical properties of soy protein isolate films. *Radiat Phys Chem* 72:35–40.

Lee, S.Y., K.L. Dangaran, and J.M. Krochta. 2002a. Gloss stability of whey-protein/plasticizer coating formulation on chocolate surface. *J Food Sci* 67:1121–1125.

Lee, S.Y., K.L. Dangaran, J.X. Guinard, and J.M. Krochta. 2002b. Consumer acceptance of whey-protein-coated as compared with shellac-coated chocolate. *J Food Sci* 67:2764–2769.

Lent, L.E., L.S. Vanasupa, and P.S. Tong. 1998. Whey protein edible film structures determined by atomic force microscope *J Food Sci* 63:824–827.

Letendre, M., G. D'Aprano, G. Delmas-Patterson, and M. Lacroix. 2002. Isothermal calorimetry study of calcium caseinate and whey protein isolate edible films cross-linked by heating and γ-irradiation. *J Agric Food Chem* 50:6053–6057.

Le Tien, C., C. Vachon, M.A. Mateescu, and M. Lacroix. 2001. Milk protein coatings prevent oxidative browning of apples and potatoes. *J Food Sci* 66:512–516.

Lieberman, E.R., and S.G. Gilbert. 1973. Gas permation of collagen films as affected by crosslinkage, moisture and plasticizer content. *J Polym Sci Symp* 41:33–43.

Lim, L.T., Y. Mine, and M.A. Tung. 1998. Transglutaminase cross-linked egg white protein films: tensile properties and oxygen permeability. *J Agric Food Chem* 46:4022–4029.

Lim, L.T., Y. Mine, I.J. Britt, and M.A. Tung. 2002. Formation and properties of egg white films and coatings. In *Protein-based films and coatings,* ed. A. Gennadios, 233–252. Boca Raton, FL: CRC Press.

Lin, S.Y., and J.M. Krochta. 2006. Whey protein coating efficiency on mechanically roughened hydrophobic peanut surfaces. *J Food Sci* 71:E270–E275.

Liu, G., L. Li, B. Li, Q. Lu, L. Chen, J. Chen, and S. Guo. 2006. Mechanical properties and sorption characteristics of stearic acid-cysteine-soy protein isolate blend films. *Trans Chinese Soc Agric Eng* 22:153–158.

Mahmoud, R., and P.A. Savello. 1992. Mechanical properties of and water vapor transferability through whey protein films. *J Dairy Sci* 75:942–946.

Mahmoud, R., and P.A. Savello. 1993. Solubility and hydrolyzability of films produced by transglutaminase catalytic crosslinking of whey protein. *J Dairy Sci* 76:29–35.

Mallikarjunan, P., M.S. Chinnan, V.M. Balasubramaniam, and R.D. Phillips. 1997. Edible coatings for deep-fat frying of starch products. *Lebensm Wiss Technol* 30:709–714.

Mangavel, C., J. Barbot, J. Gueguen, and Y. Popineau. 2003. Molecular determinants of the influence of hydrophilic plasticizers on the mechanical properties of cast wheat gluten films. *J Agric Food Chem* 51:1447–1452.

Mangavel, C., N. Rossignol, A. Perronnet, J. Barbot, Y. Popineau, and J. Gueguen. 2004. Properties and microstructure of thermo-pressed wheat gluten films: A comparison with cast films. *Biomacromolecules* 5:1596–1601.

Mariniello, L., P. Di Pierro, C. Esposito, A. Sorrentino, P. Masi, and R. Porta. 2003. Preparation and mechanical properties of edible pectin-soy flour films obtained in the absence or presence of transglutaminase. *J Biotechnol* 102:191–198.

Marquié, C. 1996. Formation and characterization of biodegradable films obtained from cottonseed flour. Ph.D. Dissertation. University of Montpellier, France.

Marquié, C., and S. Guilbert. 2002. Formation and properties of cottonseed protein films and coatings. In *Protein-based films and coatings*, ed. A. Gennadios, 139–158. Boca Raton, FL: CRC Press.

Marquié, C., C. Aymard, J.L. Cuq, and S. Guilbert. 1995. Biodegradable packaging made from cottonseed flour: Formation and improvement by chemical treatments with gossypol, formaldehyde, and glutaraldehyde. *J Agric Food Chem* 43:2762–2767.

Marquié, C., E. Héquet, S. Guilbert, A.M. Tessier, and V. Vialettes. 1996. Biodegradable material made from cottonseed flour. In *Proceedings of the 23rd International Cotton Conference*, 145–156, Faserinstitut, Germany.

Marquié, C., A.M. Tessier, C. Aymard, and S. Guilbert. 1997. HPLC determination of the reactive lysine content of cottonseed protein films to monitor the extent of cross-linking by formaldehyde, glutaraldehyde, and glyosal. *J Agric Food Chem* 45:922–926.

Martinez, W.H., L.C. Berardi, and L.A. Goldblatt. 1970. Cottonseed protein products. *J Agric Food Chem* 18:961–968.

Maté, J.I., and J.M. Krochta. 1996. Comparison of oxygen and water vapor permeabilities of whey protein isolate and β-lactoglobulin edible films. *J Agric Food Chem* 44:3001–3004.

Maté, J.I., and J.M. Krochta. 1998. Oxygen uptake model for uncoated and coated peanuts. *J Food Eng* 35:299–312.

Mauri, A.N., and M.C. Anon. 2006. Effect of solution pH on solubility and some structural properties of soybean protein isolate films. *J Sci Food Agric* 86:1064–1072.

Mazza, G., and H. Qi. 1991. Control of after-cooking darkening in potatoes with edible film-forming products and calcium chloride. *J Agric Food Chem* 39:2163–2166.

McHugh, T.H., and J.M. Krochta. 1994a. Water vapor permeability properties of edible whey protein-lipid emulsion films. *JAOCS* 71:307–312.

McHugh, T.H., and J.M. Krochta. 1994b. Sorbitol- vs glycerol-plasticized whey protein edible films: Integrated oxygen permeability and tensile property evaluation. *J Agric Food Chem* 52:841–845.

McHugh, T.H., and J.M. Krochta. 1994c. Milk-protein-based edible films and coatings. *Food Technol* 48:97–103.

McHugh, T.H., J.F. Aujard, and J.M. Krochta. 1994. Plasticized whey protein edible films: Water vapor permeability properties. *J Food Sci* 59:416–423.

Mecitoglu, C., A. Yemenicioglu, A. Arslanoglu, Z.S. Elmaci, F. Korel, and A.E. Cetin. 2006. Incorporation of partially purified hen egg white lysozyme into zein films for antimicrobial food packaging. *Food Res Int* 39:12–21.

Micard, V., and S. Guilbert. 2000. Thermal behavior of native and hydrophobized wheat gluten, gliandin and glutenin-rich fractions by modulated DSC. *Int J Biol Macromol* 27:229–236.

Micard, V., R. Belamri, M.H. Morel, and S. Guilbert. 2000. Properties of chemically and physically treated wheat gluten films. *J Agric Food Chem* 48:2948–2953.

Min, S., and J.M. Krochta. 2005. Inhibition of *Penicillium commune* by edible whey protein films incorporating lactoferrin, lactoferrin hydrolysate, and lactoperoxidase systems. *J Food Sci* 70:M87–M94.

Min, S., L.J. Harris, J.H. Han, and J.M. Krochta. 2005a. *Listeria monocytogenes* inhibition by whey protein films and coatings incorporating lysozyme. *J Food Protect* 68:2317–2325.

Min, S., L.J. Harris, J.H. Han, and J.M. Krochta. 2005b. *Listeria monocytogenes* inhibition by whey protein films and coatings incorporating the lactoperoxidase system. *J Food Sci* 70:M317–M324.

Mine, Y. 1995. Recent adavances in the understanding of egg white protein functionality. *Trends Food Sci Tech* 6:225–232.

Monterrey-Quintero, E.S. 1998. Physico-chemical characterization of myofibrillar proteins and film formation. Ms. Dissertation in Zootechnology, Sao Paolo, Brazil.

Monterrey-Quintero, E.S., and P.J.D.A. Sobral. 2000. Extraction and properties of Nile Tilapia myofibrillar proteins for edible films. *Pesqui Agropecu Bras* 35:179–189.

Morr, C.V., and E.Y.W. Ha. 1993. Whey protein concentrates and isolates: Processing and functional properties. *Crit Rev Food Sci Nutr* 33:431–476.

Motoki, M., N. Nio, and K. Takinami. 1987a. Functional properties of heterologous polymer prepared by transglutaminase between milk casein and soybean globulin. *Agric Biol Chem* 51:237–239.

Motoki, M., H. Aso, K. Seguro, and N. Nio. 1987b. Immobilization of enzymes in protein films prepared using transglutaminase. *Agric Biol Chem* 51:997–1002.

Mujica-Paz, H., and N. Gontard. 1997. Oxygen and carbon dioxide permeability of wheat gluten film: Effect of relative humidity and temperature. *J Agric Food Chem* 45:4101–4105.

Muller, L.L. 1982. Manufacture of casein, caseinates and coprecipitates. In *Developments in dairy chemistry—1 Proteins*, ed. P.F. Fox, 315–337. London: Applied Science.

Muthuselvi, L., and A. Dhathathreyan. 2006. Contact angle hysteresis of liquid drops as means to measure adhesive energy of zein on solid substrates. *J Phys* 66:563–574.

Naczk, M., L.J. Rubin, and F. Shahidi. 1986. Functional properties and phylate content of pea protein preparations. *J Food Sci* 51:1245–1247.

Naga, M., S. Kirihara, Y. Tokugawa, F. Tsuda, and M. Hirotsuka. 1996. U.S. Patent No. 5,569,482.

Nakagaki, S., and M. Yonese. 1971. Permeability and its temperature dependency on gelatin, collagen membranes and bovine eye lens capsule. *J Pharm Soc Jpn* 91:1211–1216.

Natarajan, K.R. 1980. Peanut protein ingredients: Preparation, properties, and food uses. In *Advances in food research*, Vol 26, ed. C.O. Chichester, E.M. Mrak, and G.F. Stewart, 215–273. New York: Academic Press.

Nickel, G.B. 1981. Canadian Patent No. 1,104,871.

Nielsen, P.M. 1995. Reactions and potential industrial applications of transglutaminase: Review of literature and patents. *Food Biotechnol* 9:119–156.

Nissen, H. 1995. Hydrolyzing primer. *Render* 24:10–13.

Ogale, A.A., P. Cunningham, P.L. Dawson, and J.C. Acton. 2000. Viscoelastic, thermal, and microstructural characterization of soy protein isolate films. *J Food Sci* 65:672–679.

Oh, J.H., B. Wang, P.D. Field, and H.A. Aglan. 2004. Characteristics of edible films made from dairy proteins and zein hydrolysate cross-linked with transglutaminase. *Int J Food Sci Technol* 39:287–294.

Okamoto, S. 1978. Factors affecting protein film formation. *Cereal Food World* 23:256–262.

O'Sullivan, A., N.B. Shaw, S.C. Murphy, J.W. Van De Vis, H. Van Pelt-Heerschap, and J.P. Kerry. 2006. Extraction of collagen from fish skins and its use in the manufacture of biopolymer films. *J Aqua Food Pro Tech* 15:21–32.

Otani, Y., Y. Tabata, and Y. Ikada. 1998. Preparation of rapid curable hydrogels from gelatin and poly(carboxylic acid) and their adhesion to skin. *Macromol Symp* 130:169–177.

Ou, C.Y., S.F. Tsay, C.H. Lai, and Y.M. Weng. 2002. Using gelatin-based antimicrobial edible coating to prolong shelf-life of tilapia fillets. *J Food Qual* 25:213–222.

Ou, S., Y. Wang, S. Tang, C. Huang, and M.G. Jackson. 2005. Role of ferulic acid in preparing edible films from soy protein isolate. *J Food Eng* 70:205–210.

Ouattara, B., S.F. Sabato, and M. Lacroix. 2001. Combined effect of antimicrobial coating and gamma irradiation on shelf life extension of pre-cooked shrimp (*Penaeus* spp.). *Int J Food Microbiol* 68:1–9.

Ouattara, B., L.T. Canh, C. Vachon, M.A. Mateescu, and M. Lacroix. 2002. Use of γ-irradiation cross-linking to improve the water vapor permeability and the chemical stability of milk protein films. *Radiat Phys Chem* 63:821–825.

Padgett, T., I.Y. Han, and P.L. Dawson. 2000. Effect of lauric acid addition on the antimicrobial efficacy and water permeability of corn zein films containing nisin. *J Food Process Preserv* 24:423–432.

Padua, G.W., and Q. Wang. 2002. Formation and properties of corn zein films and coatings. In *Protein-based films and coatings*, ed. A. Gennadios, 43–67. Boca Raton, FL: CRC Press.

Paramawati, R., T. Yoshino, and S. Isobe. 2001. Properties of plasticized-zein film as affected by plasticizer treatments. *Food Sci Technol Res* 7:191–194.

Park, H.J., and M.S. Chinnan. 1995. Gas and water barrier properties of edible films from protein and cellulosic materials. *J Food Engr* 25:497–507.

Park, H.J., J.M. Bunn, C.L. Weller, P.J. Vergano, and R.F. Testin. 1994a. Water vapor permeability and mechanical properties of grain-protein-based films as affected by mixtures of polyethylene glycol and glycerin plasticizers. *Trans ASAE* 37:1281–1285.

Park, J.W., R.F. Testin, H.J. Park, P.J. Vergano, and C.L. Weller. 1994b. Fatty acid concentration effect on tensile strength, elongation, and water vapor permeability of laminated edible films. *J Food Sci* 59:916–919.

Park, J.W., R.F. Testin, P.J. Vergano, H.J. Park, and C.L. Weller. 1996. Fatty acid distribution and its effect on oxygen permeability in laminated edible films. *J Food Sci* 61:401–406.

Park, S.K., and D.H. Bae. 2006. Film-forming properties of proteinaceous fibrous material produced from soybean fermented by *Bacillus natto*. *J Microbiol Biotechnol* 16:1053–1059.

Park, S.K., D.H. Bae, and K.C. Rhee. 2000a. Soy protein biopolymers cross-linked with glutaraldehyde. *JAOCS* 77:879–883.

Park, H.J., S.H. Kim, S.T. Lim, D.H. Shin, S.Y. Choi, and K.T. Hwang. 2000b. Grease resistance and mechanical properties of isolated soy protein-coated paper. *JAOCS* 77:269–273.

Park, S.K., C.O. Rhee, D.H. Bae, and N.S. Hettiarachchy. 2001. Mechanical properties and water-vapor permeability of soy-protein films affected by calcium salts and glucono-δ-lactone. *J Agric Food Chem* 49:2308–2312.

Park, S.K., S.N. Hettiarachchy, Z.Y. Ju, and A. Gennadios. 2002a. Formation and properties of soy protein films and coatings. In *Protein-based films and coatings*, ed. A. Gennadios, 123–137. Boca Raton, FL: CRC Press.

Park, H.J., J.W. Rhim, C.L. Weller, A. Gennadios, and M.A. Hanna. 2002b. Films and coatings from proteins of limited availability. In *Protein-based films and coatings*, ed. A. Gennadios, 305–327. Boca Raton, FL: CRC Press.

Parris, N., and D.R. Coffin. 1997. Composition factors affecting the water vapor permeability and tensile properties of hydrophilic zein films. *J Agric Food Chem* 45:1596–1599.

Parris, N., P.J. Vergano, L.C. Dickey, P.H. Cooke, and J.C. Craig. 1998. Enzymatic hydrolysis of zein-wax-coated paper. *J Agric Food Chem* 46:4056–4059.

Parris, N., L.C. Dickey, J.L. Wiles, R.A. Moreau, and P.H. Cooke. 2000. Enzymatic hydrolysis, grease permeation, and water barrier properties of zein isolate coated paper. *J Agric Food Chem* 48:890–894.

Parris, N., L.C. Dickey, P.M. Tomasula, D.R. Coffin, and P.J. Vergano. 2001. Films and coatings from commodity agroproteins. *ACS Symposium Series* 786:118–131.

Parris, N., L.C. Dickey, M.J. Powell, D.R. Coffin, R.A. Moreau, and J.C. Craig. 2002. Effect of endogenous triacylglycerol hydrolysates on the mechanical properties of zein films from ground corn. *J Agric Food Chem* 50:3306–3308.

Paschoalick, T.M., F.T. Garcia, P.J.A. Sobral, and A.M.Q.B. Habitante. 2003. Characterization of some functional properties of edible films based on muscle proteins of Nile Tilapia. *Food Hydrocolloid* 17:419–427.

Patrick, N., Y.X. Gan, and H.A. Aglan. 2001. Development and characterization of edible peanut protein films. *Life Support Biosphere Sci: Int J Earth Space* 8:15–22.

Pérez-Gago, M.B., and J.M. Krochta. 1999. Water vapor permeability of whey protein emulsion films as affected by pH. *J Food Sci* 64:695–698.

Pérez-Gago, M.B., and J.M. Krochta. 2000. Drying temperature effect on water vapor permeability mechanical properties of whey protein-lipid emulsion films. *J Agric Food Chem* 48:2687–2692.

Pérez-Gago, M.B., and J.M. Krochta. 2001a. Lipid particle size effect on water vapor permeability and mechanical properties of whey protein/beeswax emulsion films. *J Agric Food Chem* 49:996–1002.

Pérez-Gago, M.B., and J.M. Krochta. 2001b. Denaturation time and temperature effects on solubility, tensile properties, and oxygen permeability of whey protein edible films. *J Food Sci* 66:705–710.

Pérez-Gago, M.B., and J.M. Krochta. 2002. Formation and properties of whey protein films and coatings. In *Protein-based films and coatings*, ed. A. Gennadios, 159–180. Boca Raton, FL: CRC Press.

Pérez-Gago, M.B., P. Nadaud, and J.M. Krochta. 1999. Water vapor permeability, solubility and tensile properties of heat-denatured versus native whey protein films. *J Food Sci* 64:1034–1037.

Pérez-Gago, M.B., M. Serra, M. Alonso, M. Mateos, and M.A. Del Rio. 2005. Effect of whey protein- and hydroxypropyl methylcellulose-based edible composite coatings on color change of fresh-cut apples. *Postharvest Biol Tec* 36:77–85.

Pérez-Gago, M.B., M. Serra, and M.A. Rio. 2006. Color change of fresh-cut apples coated with whey protein concentrate-based edible coatings. *Postharvest Biol Tec* 39:84–92.

Petersen, K., V.P. Nielsen, G. Bertelsen, M. Lauther, M.B. Olsen, N.H. Nilsson, and G. Mortensen. 1999. Potential of biobased materials for food packaging. *Trends Food Sci Tech* 10:52–68.

Piez, K.A., P. Bornstein, and A.H. Kang. 1968. The chemistry and biochemistry of interchain crosslinks in collagen. In *Symposium on fibrous proteins*, ed. W.G. Grewther, 205–211. New York: Plenum Press.

Pol, H., P. Dawson, J. Acton, and A. Ogale. 2002. Soy protein isolate/corn-zein laminated films: Transport and mechanical properties. *J Food Sci* 67:212–217.

Pomes, A.F. 1971. Zein. In *Encyclopedia of polymer science and technology: Plastics, resins, rubbers, fibers*, Vol. 15, ed. H.F. Mark, N.G. Gaylord, and N.M. Bikales, 125–132. New York: Interscience.

Pommet, M., A. Redl, M.H. Morel, and S. Guilbert. 2003. Study of wheat gluten plasticization with fatty acids. *Polymer* 44:115–122.

Prakash, V., and M.S.N. Rao. 1986. Physicochemical properties of oilseed proteins. *CRC Crit Rev Biochem* 20:265–363.

Rampon, V., P. Robert, N. Nicolas, and E. Dufour. 1999. Protein extructure and network orientation in edible films prepared by spinning process. *J Food Sci* 64:313–316.
Rangavajhyla, N., V.M. Ghorpade, and M.A. Hanna. 1997. Solubility and molecular properties of heat-cured soy protein films. *J Agric Food Chem* 45:4204–4208.
Rayner, M., V. Ciolfi, B. Maves, P. Stedman, and G.S. Mittal. 2000. Development and application of soy-protein films to reduce fat intake in deep-fried foods. *J Sci Food Agric* 80:777–782.
Redl, A., N. Gontard, and S. Guilbert. 1996. Determination of sorbic acid diffusivity in edible wheat gluten and lipid based films. *J Food Sci* 61:116–120.
Redl, A., M.H. Morel, J. Bonicel, B. Vergnes, and S. Guilbert. 1999. Extrusion of wheat gluten plasticized with glycerol: Influence of process conditions on flow behavior, rheological properties and molecular size distribution. *Cereal Chem* 76:361–370.
Reiners, R.A., J.S. Wall, and G.E. Inglett. 1973. Corn proteins: Potential for their industrial use. In *Industrial uses of cereals*, ed. Y. Pomeranz, 285–302. St. Paul, MN: American Association of Cereal Chemists.
Reutimann, E.J., D.V. Valdehra, and E.R. Wedral. 1996. U.S. Patent No. 5,567,453.
Rhim, J.W., Y. Wu, C.L. Weller, and M. Schnepf. 1999a. Physical characteristics of emulsified soy protein-fatty acid composite films. *Sciences des Aliments* 19:57–71.
Rhim, J.W., A. Gennadios, D. Fu, C.L. Weller, and M.A. Hanna. 1999b. Properties of ultraviolet irradiated protein films. *Lebensm Wiss Technol* 32:129–133.
Rhim, J.W., and C.L. Weller. 2000. Properties of formaldehyde adsorbed soy protein isolate films. *ASAE Annual Intenational Meeting, Technical Papers: Engineering Solutions for a New Century* 2:1413–1423.
Rhim, J.W., A. Gennadios, A. Handa, C.L. Weller, and M.A. Hanna. 2000. Solubility, tensile, and color properties of modified soy protein isolate films. *J Agric Food Chem* 48:4937–4941.
Rhim, J.W., A. Gennadios, C.L. Weller, and M.A. Hanna. 2002. Sodium dodecyl sulfate treatment improves properties of cast films from soy protein isolate. *Ind Crop Prod* 15:199–205.
Rhim, J.W., J.H. Lee, and S.I. Hong. 2006. Water resistance and mechanical properties of biopolymer (alginate and soy protein) coated paperboards. *Lebensm Wiss Technol* 39:806–813.
Robinson, C. 1953. The hot and cold forms of gelatin. In *Nature and structure of collagen*, ed. J.T. Randell, 96–105. New York: Academic Press.
Romero-Bastida, C.A., E. Flores-Huicochea, M.O. Martin-Polo, G. Velazquez, and J.A. Torres. 2004. Compositional and moisture content effects on the biodegradability of zein/ethylcellulose films. *J Agric Food Chem* 52:2230–2235.
Rose, P.I. 1987. Gelatin. In *Encyclopedia of polymer science and engineering*, Vol 7, ed. H.F. Mark, N.M. Bikales, C.G. Overberger, and G. Mengues, 488–513. New York: John Wiley & Sons.
Roy, S., A. Gennadios, C.L. Weller, and R.F. Testin. 2000. Water vapor transport parameters of a cast wheat gluten film. *Ind Crop Prod* 11:43–50.
Ruiz, L.P., and E.L. Hove. 1976. Conditions affecting production of a protein isolate from lupin seed kernels. *J Sci Food Agric* 27:667–674.
Ryu, S.Y., J.W. Rhim, H.J. Roh, and S.S. Kim. 2002. Preparation and physical properties of zein-coated high-amylose corn starch film. *Lebensm Wiss Technol* 35:680–686.
Sabato, S.F., B. Ouattara, H. Yu, G. D'Aprano, G. Le Tien, M.A. Mateescu, and M. Lacroix. 2001. Mechanical and barrier-properties of cross-linked soy and whey protein based films. *J Agric Food Chem* 49:1397–1403.
Salame, M. 1986. Barrier polymers. In *The Wiley encyclopedia of packaging technology*, ed. M. Bakker, 48–54. New York: John Wiley & Sons.

Santosa, F.X.B., and G.W. Padua. 1999. Tensile properties and water absorption of zein films plasticized with oleic and linoleic acids. *J Agric Food Chem* 47:2070–2074.

Saroso, B. 1989. Chemical properties of protein in cottonseed kernels. *Industr Crop Res J* 1:60–65.

Sathe, S.K., S.S. Deshpande, and D.K. Salunkhe. 1982. Functional properties of lupin seed (*Lupinus mutabilis*) protein and protein concentrates. *J Food Sci* 47:491–497, 502.

Sathivel S. 2005. Chitosan and protein coatings affect yield, moisture loss, and lipid oxidation of pink salmon (*Oncorhynchus gorbuscha*) fillets during frozen storage. *J Food Sci* 70:E455–E459.

Schmidt, V., C. Giacomelli, and V. Soldi. 2005. Thermal stability of films formed by soy protein isolate-sodium dodecyl sulfate. *Polym Degrad Stabil* 87:25–31.

Schou, M., A. Longares, C. Montesinos-Herrero, F.J. Monahan, D. O'Riordan, and M. O'Sullivan. 2005. Properties of edible sodium caseinate films and their application as food wrapping. *Lebensm Wiss Technol* 38:605–610.

Seung, Y.C., and C. Rhee. 2004. Mechanical properties and water vapor permeability of edible films made from fractionated soy proteins with ultrafiltration. *Lebensm Wiss Technol* 37:833–839.

Seydim, A.C., and G. Sarikus. 2006. Antimicrobial activity of whey protein based edible films incorporated with oregano, rosemary and garlic essential oils. *Food Res Int* 39:639–644.

Shaw, N.B., F.J. Monahan, E.D. O'Riordan, and M. O'Sullivan. 2002a. Effect of soya oil and glycerol on physical properties of composite WPI films. *J Food Eng* 51:299–304.

Shaw, N.B., F.J. Monahan, E.D. O'Riordan, and M. O'Sullivan. 2002b. Physical properties of WPI films plasticized with glycerol, xylitol, or sorbitol. *J Food Sci* 67:164–167.

Shellhammer, T.H., and J.M. Krochta. 1997. Whey protein emulsion film performance as affected by lipid type and amount. *J Food Sci* 62:390–394.

Shih, F.F. 1996. Edible films from rice protein concentrate and pullulan. *Cereal Chem* 73:406–409.

Shiku, Y., P.Y. Hamaguchi, and M. Tanaka. 2003. Effect of pH on the preparation of edible films based on fish myofibrillar proteins. *Fisheries Sci* 69:1026–1032.

Shimada, K., and J.C. Cheftel. 1998. Sulfhydryl group disulfide bond interchange during heat induced gelation of whey protein isolate. *J Agric Food Chem* 37:161–168.

Shull, J.M., J.J. Watterson, and A.W. Kirleis. 1991. Propesed nomenclature for the alcohol-soluble proteins (Kafirins) of *Sorghum bicolor* (L. Moench) based on molecular weight, solubility, and structure. *J Agric Food Chem* 39:83–87.

Sian, N.K., and S. Ishak. 1990. Effect of pH on formation, proximate composition and rehydration capacity of winged bean and soybean protein-lipid film. *J Food Sci* 55:261–262.

Siew, D.C.W., C. Heilmann, A.J. Easteal, and R.P. Cooney. 1999. Solution and film properties of sodium caseinate/glycerol and sodium caseinate/polyethylene glycol edible coating systems. *J Agric Food Chem* 47:3432–3440.

Simelane, S., and Z. Ustunol. 2005. Mechanical properties of heat-cured whey protein-based edible films compared with collagen casings under sausage manufacturing conditions. *J Food Sci* 70:E131–E134.

Smith, S.A. 1986. Polyethylene, low density. In *The Wiley encyclopedia of packaging technology*, ed. M. Bakker, 514–523. New York: John Wiley & Sons.

Sobral, P.J.D.A. 2000. Thickness effects of myofibrillar protein based edible films on their functional properties. *Pesqui Agropecu Bras* 35:1251–1259.

Sobral, P.J.D.A., J.S. Dos Santos, and F.T. Garcia. 2005. Effect of protein and plasticizer concentrations in film forming solutions on physical properties of edible films based on muscle proteins of a Thai Tilapia. *J Food Eng* 70:93–100.

Sothornvit, R., and J.M. Krochta. 2000a. Oxygen permeability and mechanical properties of films from hydrolyzed whey protein. *J Agric Food Chem* 48:3913–3916.
Sothornvit, R., and J.M. Krochta. 2000b. Plasticizer effect on oxygen permeability of beta-lactoglobulin (β-Lg) films. *J Agric Food Chem* 48:6298–6302.
Sothornvit, R., and J.M. Krochta. 2001. Plasticizer effect on mechanical properties of beta-lactoglobulin (β-Lg) films. *J Food Eng* 50:149–155.
Sothornvit, R., D.S. Reid, and J.M. Krochta. 2002. Plasticizer effect on the glass transition temperature of beta-lactoglobulin (β-Lg) films. *Trans ASAE* 45:1479–1484.
Sothornvit, R., C.W. Olsen, T.H. McHugh, and J.M. Krochta. 2003. Formation conditions, water-vapor permeability, and solubility of compression-molded whey protein films. *J Food Sci* 68:1985–1989.
Stuchell, Y.M., and J.M. Krochta. 1994. Enzymatic treatments and thermal effects on edible soy protein films. *J Food Sci* 59:1332–1337.
Sumner, A.K., M.A. Nielsen, and C.G. Youngs. 1981. Production and evaluation of pea protein isolate. *J Food Sci* 46:364–366.
Suslick, K.S., and M.W. Grinstaff. 1990. Protein microcapsulation of nonaqueous liquid. *J Am Chem Soc* 112:7807–7809.
Talens, P., and J.M. Krochta. 2005. Plasticizing effects of beeswax and carnauba wax on tensile and water vapor permeability properties of whey protein films. *J Food Sci* 70:E239–E243.
Tanada-Palmu, P., H. Helen, and L. Hyvonen. 2000. Preparation, properties and applications of wheat gluten edible films. *Agr Food Sci Finland* 9:23–35.
Tang, C.H., Y. Jiang, Q.B. Wen, and X.Q. Yang. 2005. Effect of transglutaminase treatment on the properties of cast films of soy protein isolates. *J Biotechnol* 120:296–307.
Taylor, C.C. 1986. Cellophane. In *The Wiley encyclopedia of packaging technology*, ed. M. Bakker, 159–163. New York: John Wiley & Sons.
Taylor, M.M., C.K. Liu, N. Latona, W.N. Marmer, and E.M. Brown. 2002. Enzymatic modification of hydrolysis products from collagen using microbial transglutaminase. II. Preparation of films. *J ALCA* 97:225–234.
Teerakarn, A., D.E. Hirt, J.C. Acton, J.R. Rieck, and P.L. Dawson. 2002. Nisin diffusion in protein films: Effects of film type and temperature. *J Food Sci* 67:3019–3025.
Theivendran, S., N.S. Hettiarachchy, and M.G. Johnson. 2006. Inhibition of *Listeria monocytogens* by nisin combined with grape seed extract or green tea extract in soy protein film coated on turkey frankfurters. *J Food Sci* 71:M39–M44.
Tillekeratne, M., and A.J. Easteal. 2000. Modification of zein films by incorporation of poly(ethylene glycol)s. *Polym Int* 49:127–134.
Tomasula, P.M., J.C. Craig Jr., R.T. Boswell, R.D. Cook, M.J. Kurantz, and M. Maxwell. 1995. Preparation of casein using carbon dioxide. *J Dairy Sci* 78:506–514.
Tomasula, P.M., J.C. Craig Jr., and R.T. Boswell. 1997. A continuous process for casein production using high-pressure carbon dioxide. *J Food Eng* 33:405–419.
Tomasula, P.M., N. Parris, W. Yee, and D. Coffin. 1998. Properties of films made from CO_2-precipitated casein. *J Agric Food Chem* 46:4470–4474.
Tomasula, P.M., W.C. Yee, and N. Parris. 2003. Oxygen permeability of films made from CO_2-precipitated casein and modified casein. *J Agric Food Chem* 51:634–639.
Trezza, T.A., J.L. Wiles, and P.J. Vergano. 1998. Water vapor and oxygen barrier properties of corn zein coated paper. *Tappi J* 81:171–176.
Tropini, V., L.P. Lens, W.J. Mulder, and F. Silvestre. 2004. Wheat gluten films cross-linked with 1-ethyl-3-(3-dimethylaminopropyl) carbodiimide and N-hydroxysuccinimide. *Ind Crop Prod* 20:281–289.

Trotter, J.A., K.E. Kadler, and D.F. Holmes. 2000. Echinoderm collagen fibrils grow by surface-nucleation and propagation from both centers and ends. *J Mol Biol* 300:531–540.

Ustunol, Z., and B. Mert. 2004. Water solubility, mechanical, barrier, and thermal properties of cross-linked whey protein isolate-based films. *J Food Sci* 69:FEP129–FEP133.

Vachon, C., H.L. Yu, R. Yefsah, R. Alain, D. St-Gelais, and M. Lacroix. 2000. Mechanical and structural properties of milk protein edible films cross-linked by heating and γ-irradiation. *J Agric Food Chem* 48:3202–3209.

Vaz, C.M., L.A. De Graaf, R.L. Reis, and A.M. Cunha. 2003. Effect of crosslinking, thermal treatment and UV irradiation on the mechanical properties and *in vitro* degradation behavior of several natural proteins aimed to be used in the biomedical field. *J Mater Sci: Mater in Med* 14:789–796.

Wan, V.C.H., S.K. Moon, and S.Y. Lee. 2005. Water vapor permeability and mechanical properties of soy protein isolate edible films composed of different plasticizer combinations. *J Food Sci* 70:E387–E391.

Wang, Y., and G.W. Padua. 2004. Water sorption properties of extruded zein films. *J Agric Food Chem* 52:3100–3105.

Wang, Y., and G.W. Padua. 2005. Properties of zein films coated with drying oils. *J Agric Food Chem* 53:3444–3448.

Wang, Y., and G.W. Padua. 2006. Water barrier properties of zein-oleic acid films. *Cereal Chem* 83:331–334.

Wang, Y., A.M. Rakotoniriainy, and G.W. Padua. 2003. Thermal behavior of zein-based biodegradable films. *Starch* 55:25–29.

Wang, Q., P. Geil, and G. Padua. 2004. Role of hydrophilic and hydrophobic interactions in structure development of zein films. *J Polym Environ* 12:197–202.

Wang, Y., F. Lopes Filho, P. Geil, and G.W. Padua. 2005. Effects of processing on the structure of zein/oleic acid films investigated by X-ray diffraction. *Macromol Biosci* 5:1200–1208.

Watanable, K., and S. Okamoto. 1976. Formation of yuba-like films and their physical properties. *New York Industry* 18:65–77.

Weller, C.L., A. Gennadios, and R.A. Saraiva. 1998. Edible bilayer films from zein and grain sorghum wax or carnauba wax. *Lebensm Wiss Technol* 31:279–285.

Were, L., N.S. Hettiarachchy, and M. Coleman. 1999. Properties of cysteine-added soy protein-wheat gluten films. *J Food Sci* 64:514–518.

Wong, Y.C., T.J. Herald, and K.A. Hachmeister. 1996. Evaluation of mechanical and barrier properties of protein coatings on shell eggs. *Poultry Sci* 75:417–422.

Wrigley, C.W., and J.A. Bietz. 1988. Protein and amino acids. In *Wheat chemistry and technology*, Vol 1, ed. Y. Pomeranz, 159–275. St. Paul, MN: American Association of Cereal Chemists.

Wu, L.C., and R.P. Bates. 1972. Soy protein-lipid films. 1. Studies on the film formation phenomenon. *J Food Sci* 37:36–39.

Wu, L.C., and R.P. Bates. 1973. Influence of ingredients upon edible protein-lipid film characteristics. *J Food Sci* 38:783–787.

Wu, Q., and L. Zhang. 2001. Properties and structure of soy protein isolate-ethylene glycol sheets obtained by compression molding. *Ind Eng Chem Res* 40:1879–1883.

Wu, Y., J.W. Rhim, C.L. Weller, F. Hamouz, S. Cuppett, and M. Schnepf. 2000. Moisture loss and lipid oxidation for precooked beef patties stored in edible coatings and films. *J Food Sci* 65:300–304.

Xu, X., N. Hasegawa, U. Doi, H. Umekawa, T. Takahashi, K. Suzuki, and Y. Furuichi. 2000. Biological availability of docosahexaenoic acid from fish oil encapsulated in zein-coated porous starch granules in rats. *Food Sci Technol Res* 6:87–93.

Yamada, K., H. Takahashi, and A. Noguchi. 1995. Improved water resistance in edible zein films and composites for biodegradable food packaging. *Int J Food Sci Technol* 30:599–608.

Yamauchi, K., and A. Khoda. 1997. Novel proteinous microcapsules from wool keratins. *Colloid Surface B* 9:117–119.

Yamauchi, A., and K. Yamauchi. 2002. Formation and properties of wool keratin films and coatings. In *Protein-based films and coatings*, ed. A. Gennadios, 253–273. Boca Raton, FL: CRC Press.

Yang, Y., L. Wang, and S. Li. 1996. Formaldehyde-free zein fiber preparation and investigation. *J Appl Polym Sci* 59:433–441.

Yildirim, M., and N.S. Hettiarachchy. 1998a. Properties of films produced by cross-linking whey proteins and 11S globulin using transglutaminase. *J Food Sci* 63:248–252.

Yildirim, M., and N.S. Hettiarachchy. 1998b. Properties of cast films from pickle fermentation brine protein. *J Agric Food Chem* 46:4969–4972.

Yoshida, C.M.P., and A.J. Antunes. 2004. Characterization of whey protein emulsion films. *Braz J Chem Eng* 21:247–252.

Yoshimaru, T., H. Takahashi, and K. Matsumoto. 2000. Microencapsulation of L-lysine for improving the balance of amino acids in ruminants. *J Facult Agric Kyushu University* 44:359–365.

Yoshino, T., S. Isobe, and T. Maekawa. 2000. Physical evaluation of pure zein films by atomic force microscopy and thermal mechanical analysis. *JAOCS* 77:699–704.

Yoshino, T., S. Isobe, and T. Maekawa. 2002. Influence of preparation conditions on the physical properties of zein films. *JAOCS* 79:345–349.

Young, H.H. 1967. Gelatin. In *Encyclopedia of polymer science and technology: Plastics, resins, rubbers, fibers*, Vol 7, ed. H.F. Mark, N.G. Gaylord, and N.M. Bikales, 446–460. New York: Interscience.

3 Edible coatings from lipids, waxes, and resins

David J. Hall

Contents

3.1	Introduction	79
3.2	Ingredients	80
	3.2.1 Lipids	80
	3.2.2 Waxes	80
	3.2.3 Resins	84
3.3	Technology of fruit coatings	87
	3.3.1 History	87
	3.3.2 Solvent wax	87
	3.3.3 Emulsion formulations	88
	3.3.4 Resin formulations	90
3.4	Application of coatings to fruit	90
	3.4.1 Dip application	91
	3.4.2 Foam application	92
	3.4.3 Spray	92
	3.4.4 Drip	95
3.5	Design and function	96
3.6	Future developments	98
References		98

3.1 Introduction

The most commonly used edible coatings are those applied to fresh fruits for functional and cosmetic purposes, and those are given great emphasis in this chapter. Other edible coatings are used for candies, shell eggs, and some processed foods. For example, with candies, shellac or protein coatings serve to prevent the heat and moisture of the hand from melting the candy and becoming soiled by coloring matter used for identification or appearance (Dangaren et al., 2006; Hicks, 1962).

Edible coatings are also sometimes used for vegetables, especially for those that are bulky organs derived from ovaries or their associated floral parts and root crops. Coatings based on mineral oil are used for those organs derived from floral parts, such as tomatoes, cucumbers, bell peppers, eggplant, and squash. These coatings improve appearance and reduce abrasive injuries during handling and transport (Hardenburg, 1967; Hartman and Isenberg, 1956). No drying is required before or after the coating process.

The term *edible coatings* brings to mind the thought that the coating is composed of common food ingredients, for example, a gelatin coating on a meat product (Antoniewski et al., 2007). In many cases, however, edible coatings contain substances seldom associated with food, as the following discussion of coating ingredients will show.

3.2 Ingredients

3.2.1 Lipids

Lipids are usually added to food coatings in order to impart hydrophobicity and thereby reduce moisture loss (Baldwin et al., 1997). An example of a coating lipid is white mineral oil, which consists of a mixture of liquid paraffinic and naphthenic hydrocarbons, allowed for use as a food release agent and as a protective coating agent for fruits and vegetables in an amount not to exceed good manufacturing practices (Table 3.1). The principal function is to provide lubrication to the commodity as it passes through the sorting and packing processes.

3.2.2 Waxes

Waxes are the most common lipid used in coatings, so much so that the word *wax* is commonly used to denote any coating, whether or not it includes a lipid substance such as beeswax, carnauba wax, polyethylene wax, or candelilla wax. Examples of such usage are *storage wax, shipping wax, solvent wax,* and *water wax*.

Carnauba wax is an exudate of palm tree leaves from the Tree of Life (*Copernica cerifera*), found mostly in Brazil. It has the highest melting point and specific gravity of commonly found natural waxes and is added to other waxes to increase melting point, hardness, toughness, and luster (Table 3.2). Refined carnauba wax consists mostly of saturated wax acid esters with 24 to 32 hydrocarbons and saturated long-chain monofunctional alcohols (Bennett, 1975). Carnauba wax is considered a GRAS (generally recognized as safe) substance and is permitted for use in coatings for fresh fruits and vegetables, in chewing gum, in confections, and in sauces with no limitations other than good manufacturing practice (Table 3.1). There are several grades of carnauba wax available (from crude to refined), but coating companies are reluctant to indicate which grades are most used in edible coatings.

Polyethylene wax was invented to serve as an alternative to carnauba wax in the manufacture of microemulsions and today is still used for that purpose. Oxidized polyethylene is defined as the basic resin produced by the mild air oxidation of polyethylene, a petroleum by-product. It should have a minimum number average molecular weight of 1200, as determined by high-temperature vapor pressure osmometry, a maximum of 5% total oxygen by weight, and an acid value of 9 to 19 according to CFR (Table 3.1). Several grades of polyethylene wax are available to obtain desired emulsion properties. These waxes differ in molecular weight (affecting viscosity), density (affecting hardness), and softening point (Eastman Chemical Inc., 1990). This wax, although less popular of late, is used to make emulsion coatings.

Oxidized polyethylene is allowed for use as a protective coating or component of protective coatings for some fresh produce where, generally, the peel is not normally

Table 3.1 Lipids, lipid emulsifiers, resins, and rosins commonly used in food coatings with hydrophilic–lipophilic balance (HLB) and required HLB values for emulsification

Component	HLB[a,b]	Required HLB[a,b]	Uses/regulatory status[c] (21 CFR)
Acetylated monoglycerides	1.5	—	Coating agent, emulsifier (172.828)
Acetylated sucrose diester	1	—	
Ammonium lauryl sulfate	31	—	Emulsifier, scald agent (172.822)
Beeswax	3.5	9	Coating agent, candy glaze (GRAS, 184.1973)
Butyl stearate	—	11	Plasticizer (181.27)
Carnauba wax	—	15	Coating agent, confections, fruit and fruit juices (GRAS, 184.1978)
Castor oil	—	8	Coating agent, anticaking (172.876)
Chlorinated paraffin	—	12–14	Coating agent, chewing gum (172.615, 175.105, 175.250)
Cocoa butter	—	6	Coating, confections (GRAS)
Corn oil	—	8	Coating and emulsifying agent, texturizer (GRAS)
Coumarone indene resin			Coating agent (172.215, 175.300)
Ethylene glycol monostearate	2.9	—	Emulsifier, confections (73.1)
Glycerol monostearate	3.8	—	Coating agent, emulsifier (GRAS, 184.1324)
Isopropyl palmitate	12	—	
Lard	—	5	Coating agent, emulsifier (GRAS)
Lauric acid	12.8	6	Defoaming agent, lubricant (172.860)
Lauryl alcohol	9.5	14	Intermediate, flavoring agent (172.515)
Mineral oil	—	10	Coating agent, confections (172.878)
Oleic acid	1	17	Emulsifier, binder, lubricant (172.862)
Oxidized polyethylene	—	—	Coating agent, defoaming agent (172.260, 175.300, 176.200)
Palm oil	—	7	Coating agent, emulsifier, texturizer (GRAS, 184.1585)

(*Continued*)

Table 3.1 Lipids, lipid emulsifiers, resins, and rosins commonly used in food coatings with hydrophilic–lipophilic balance (HLB) and required HLB values for emulsification (Continued)

Component	HLB[a,b]	Required HLB[a,b]	Uses/regulatory status[c] (21 CFR)
Paraffin wax	—	10	Coating agent, lubricant, chewing gum (i72.6iS, 178.3800)
Potassium oleate	20	—	Emulsifier, stabilizer (172.863)
Propylene glycol monostearate	—	3.4	Emulsifier, stabilizer (172.856)
Shellac resin	—	—	Coating component (175.300)
Sodium alkyl sulfate	40	—	Emulsifier, scald agent (172.210)
Sodium oleate	18	—	Emulsifier, anticaking agent (172.863)
Sorbitan monostearate	4.7	—	Emulsifier, defoamer, stabilizer (172.515,172.842)
Soybean oil		6	Coating agent, texturizer, formulation aid (GRAS)
Soya lecithin	8	—	Emulsifier, antioxidant (GRAS, 184.1400)
Stearyl alcohol	4.9	15	Intermediate (172.864)
Triethanolamine oleate soap	12	—	
Tallow	—	6	Coating agent, texturizer (GRAS)
Wood rosin	—	—	Coating component (175.300)

[a] See Griffin, W.C. In *Kirk-Othmer Encyclopedia of Chemical Technology*, 3rd ed, vol. 8, John Wiley & Sons, New York, 1979.
[b] Shinoda, K., and H. Kunieda. In *Encyclopedia of Emulsion Technology*, vol. 1, P. Becher, ed., Marcel Dekker, New York, 1984.
[c] FDA, CFR (Code of Federal Regulations).

ingested (avocado, banana, coconut, citrus, mango, melon, pineapple, papaya, pumpkin, etc.) and certain nuts in their shells. Polyethylene emulsions can be applied at the 10% to 15% total solids level (Allied Signal, Inc., 1986; Eastman Chemical Products, 1984; Wineman, 1984). Polyethylene wax can also be combined with paraffin wax to help reduce moisture loss but at a slight sacrifice in gloss or appearance.

Paraffin wax is derived from the wax distillate fraction of crude petroleum. It is composed of hydrocarbon fractions of generic formula C_nH_{2n+2} ranging from 18 to 32 carbon units (Bennett, 1975). Synthetic paraffin wax is allowed for food use in the United States. It consists of a mixture of solid hydrocarbons resulting from the catalytic polymerization of ethylene. Both natural and synthetic paraffins are refined to

Edible coatings from lipids, waxes, and resins 83

Table 3.2 Properties of some waxes

	Melting point (°C)	Specific gravity (g/mL)	Saponification number	Refractive index	Iodine number	Hardness $(cm \times 10^{-2})$[a]
Paraffin[b]	50.0–51.1	0.880–0.915	0	1.410–1.450	0	31.3
Carnauba[b]	82.5–86.0	0.996–0.998	78–88	1.463 (60°)	7–14	4.7
Beeswax[c]	61.0–65.0	0.950–0.960	88–102	1.439–1.445 (75°)	8–11	2.0–4.0
Candelilla[b]	65.0–68.8	0.982–0.993	46–66	1.455	15–30	4.7
Polyethylene[b]	97.0–115.0	0.922	>0.1	—	—	2.0–4.0[d]

[a] At 150 g/minute/25°C.
[b] Bennett, H. *Industrial Waxes*, vol. 1, Chemical, New York, 1975.
[c] Tulloch, A.P., *Lipids*, 5, 247–258, 1970.
[d] 100 g/5 seconds/25°C.

meet U.S. Food and Drug Administration (FDA) specifications for ultraviolet absorbance (Title 21 CFR, Code of Federal Regulations, 184.1973). Synthetic paraffin wax should not have an average molecular weight lower than 500 or higher than 1200. Additional physical properties are listed in Table 3.2. Paraffin waxes are permitted for use as protective coatings for raw fruit and vegetables and for cheese. They can also be used in chewing gum base, as a defoamer, and as a component in microencapsulation of spice-flavoring substances.

Paraffin wax is frequently used to coat root crops such as rutabaga, turnip, and yucca. In some cases, the stems of sugarcane are also coated. To adhere well to the surface, vegetables to be coated with paraffin waxes must be clean. Because molten waxes are applied to vegetable surfaces well above the boiling point of water, any free water on the commodity will form *steam pockets* and will flake during subsequent handling. For this reason, the product must be completely dry before application of paraffin.

Beeswax, also known as white wax, is secreted by honeybees for comb building. The wax is harvested by centrifuging the honey from the wax combs, and then melting with hot water, steam, or solar heating. The wax is subsequently refined with diatomaceous earth and activated carbon, and is finally bleached with permanganates or bichromates. Beeswax consists mostly of monofunctional alcohols C_{24}–C_{33}, hydrocarbons C_{25}–C_{33}, and long-chain acids C_{24}–C_{34} (Tulloch, 1970). This wax is very plastic at room temperature but becomes brittle at colder temperatures. It is soluble in most other waxes and oils. Some of its properties are listed in Table 3.2. Beeswax is considered a GRAS substance and is allowed for direct use in food with some limitations (Table 3.1). Maximum allowed levels are 0.065% in chewing gum, 0.005% in confections and frostings, 0.04% in hard candy, 0.1% in soft candy, and 0.002% in all other categories.

Candelilla wax is an exudate of the candelilla plant (*Euphorbia cerifera, E. antisphylitica, Pedilanthus parvonis, P. aphyllus*), a reed-like plant that grows mostly in Mexico and southern Texas. The wax is recovered by immersing the plant in boiling water, after which the wax is skimmed off the surface, refined, and bleached. Its degree of hardness is intermediate between beeswax and carnauba. This wax sets very slowly, taking several days to reach maximum hardness. Some of its properties are listed in Table 3.2. Candelilla wax is considered a GRAS substance and is allowed for certain food uses including chewing gum and hard candy with no limitations other than good manufacturing practices (CFR, 184.1978) (Table 3.1).

3.2.3 Resins

Resins are a group of acidic substances that are usually secreted by special plant cells into long resin ducts or canals in response to injury or infection in many trees and shrubs. For the purpose of coatings, resins may be broadly defined as follows:

1. Resins may be any of numerous clear to translucent yellow or brown, solid or semisolid, viscous substances of plant origin, such as copal, rosin, and amber, used principally in lacquers, varnishes, inks, adhesives, synthetic plastics, and pharmaceuticals.

2. Resins may be any of numerous physically similar polymerized synthetics or chemically modified natural resins including thermoplastic materials such as polyvinyl, polystyrene, and polyethylene and thermosetting materials such as polyesters, epoxies, and silicones that are used with fillers, stabilizers, pigments, and other components to form plastics.
3. Resins may be shellac of insect origin, either natural or bleached with chlorine (Hicks, 1962).
4. The various protein isolates could be considered in this group as they are commonly used and behave similarly (Hall, 1980).
5. Regardless of composition, the material will be soluble in alcohol or an alkaline aqueous solution. An exception to this is coumarone indene resin that is soluble in petroleum solvents.

The U.S. Food and Drug Administration (FDA) has set standards for some of the chemically modified and synthetic resins. These are described in the FDA 21 Code of Federal Regulations (CFR). Those resins so codified are limited in their applications and may only be used on specified commodities (Table 3.3).

Shellac is the most commonly used resin in coatings for pharmaceuticals (e.g., enteric-coated aspirin), hard shell–coated chocolates, jelly beans, and so forth, and some fruits. The principal fruits coated with shellac-based coatings are citrus and apple fruits.

Shellac resin is a secretion by the insect *Laccifer lacca* and is mostly produced in central India. This resin is composed of a complex mixture of aliphatic alicyclic hydroxy acid polymers (i.e., aleuritic and shelloic acids) (Griffin, 1979). This resin is soluble in alcohols and in alkaline solutions. It is also compatible with most waxes, resulting in improved moisture barrier properties and increased gloss for coated products. Some of its properties are listed in Table 3.4. Because shellac is not a GRAS substance, it is only permitted as an indirect food additive in food coatings and adhesives (21 CFR 175.300). A petition has been submitted by its manufacturers for an upgrade to GRAS status (*Federal Register*, 54, 142).

Other natural resins permitted for use in food coatings include copal, damar, and elemi. They are used mostly in coatings for the pharmaceutical industry, but little work has been reported on their uses in food.

Wood rosins are obtained from the oleoresins of pine trees, either as an exudate or as tall oil, a by-product from the wood pulp industry. Rosin is the residue left after distillation of volatiles from the crude resin. Wood rosin is approximately 90% abietic acid ($C_{20}H_{32}O_2$) and its isomers, and 10% dehydroabietic acid ($C_{20}H_{28}O_2$). Some of its properties are listed in Table 3.4. Wood resin can be modified by hydrogenation, polymerization, isomerization, and decarboxylation to make it less susceptible to oxidation and discoloration and to improve its thermoplasticity. Drying oils, such as some vegetable oils, may be esterified with some glycol derivatives (e.g., butylene, ethylene, polyethylene, and polypropylene), to form resinous and polymeric coating components. Rosin and its derivatives are widely utilized in coatings for citrus and other fruits (Sward, 1972).

Coumarone indene resin is a petroleum or coal tar by-product. It is 100% aromatic in content and exhibits excellent resistance to alkalis, dilute acids, and moisture.

Table 3.3 FDA approved coating materials in the CFR

Coating material	Reference	Limitations
Partially hydrogenated rosin	21 CFR 172.210	Fresh citrus fruits
Pentaeryarythritol ester of malic anhydride-modified wood rosin	21 CFR 172.210	Fresh citrus fruits
Polyhydric alcohol diesters of oxidatively refined montan wax acids	21 CFR 172.210	Fresh citrus fruits
Calcium salt of partially dimerized rosin	21 CFR 172.210	Fresh citrus fruits
Coumerone-indene resin	21CFR 172.215	Fresh grapefruit, lemons, limes, oranges, tangelos, and tangerines
Oxidized polyethylene	21CFR 172.260	Fresh avocados, bananas, beets, coconuts, eggplant, garlic, grapefruit, lemons, limes, mango, muskmelons, onions, oranges, papaya, peas (in pods), pineapple, plantain, pumpkin, rutabaga, squash (acorn), sweet potatoes, tangerines, turnips, watermelon, Brazil nuts, chestnuts, filberts, hazelnuts, pecans, and walnuts (all nuts in shells)
Synthetic paraffin and succinic derivatives	21CFR 172.275	Fresh grapefruit, lemons, limes, muskmelons, oranges, sweet potatoes, and tangerines
Morpholine	21CFR 172.235	As the salt(s) of one or more of the fatty acids meeting the requirements of Section 172.860, as a component of protective coatings applied to fresh fruits and vegetables
Fatty acids	21CFR 172.860	General use

Table 3.4 Properties of some resins[b,c]

Resin	Melting point (°C)	Specific gravity (g/mL)	Saponification number	Iodine number	Acid number
Shellac	115–120	1.035–1.140	230–262	10–18	73–95
Wood rosin	82[a]	1.07–1.09	168	—	162
Wood rosin (hydrogenated)	75[a]	1.045	—	214–218[b]	160
Copal	119	1.072	—	—	139

[a] Softening point.
[b] Sward, G.G. Natural Resins. ASTM Tech. Publ., STP 500, pp. 76–91, 1972.
[c] Martin, J. In *Kirk-Othmer Encyclopedia of Chemical Technology,* Wiley Interscience, New York, 1982.

It is approved for use in citrus coatings only (Table 3.3) (Neville Chemical Co., 1988). It was widely used in *solvent wax* used for citrus.

3.3 Technology of fruit coatings

3.3.1 History

Literature reviews on use of lipids include Kester and Fennema (1986) and Guilbert (1986). Waxes such as carnauba, beeswax, paraffin, rice bran, and candelilla have been used in combination with other lipids, resins, or polysaccharides to coat fresh fruits and vegetables such as citrus and apples (Ahmad et al., 1979; Claypool, 1939; Dalal et al., 1971; Durand et al., 1984; Hitz and Haut, 1942; Lakshminarayana et al., 1974; Paredes-Lopez et al., 1974), although of these waxes, only carnauba is still commonly used today.

The practice of coating fruits and vegetables dates back to the twelfth and thirteenth centuries. In China, commodities were commonly dipped in hot waxes to encourage fermentation (Kaplan, 1986; Yehoshua, 1987). Today, coatings are used for fresh fruits and vegetables to retard moisture loss, improve appearance by imparting shine to the surface, provide a carrier for fungicides or growth regulators, and create a barrier for gas exchange between the commodity and the external atmosphere.

In the United States, early coating procedures utilized dipping, brushing, or individual wrapping in oiled papers to disperse the material over the surface of the commodity. Later, commodities were coated by spraying materials onto rollers or brushes, then allowing the tumbling action of the commodity to evenly spread the coating. Paraffin and beeswax, dispersed in an organic solvent, castor oil, and mineral oil, were used alone or in combination either in a dip, slab-wax process, hot fog, or spray method (Brooks, 1937; Hardenburg, 1967; Kaplan, 1986; Long and Leggo, 1959; Magness and Diehl, 1924; Platenius, 1939; Trout et al., 1953). Current formulations, upon demand of a dynamic fresh fruit and vegetable industry, are applied in a variety of ways and contain a variety of GRAS or ingredients approved by the FDA CFR.

3.3.2 Solvent wax

Solvent wax holds an important place in the history of citrus coatings. These formulations usually contained about 10% to 12% nonvolatile content dissolved in a mixture of aromatic and aliphatic solvents. Solvent wax citrus coatings had high shine and were easy to apply because of low viscosity and rapid drying rates. In 1980 it accounted for about 90% of the coating used on Florida citrus and more than 50% in other growing areas (Hall, 2003); however, it had serious disadvantages. First, the solvent sometimes migrated into the fruit or vegetable, affecting their flavor and wholesomeness (Hall, 1981; Kaplan, 1986). Second, solvent wax coatings formed a film almost impermeable to gases and water, making complete coverage undesirable. Third, this type of coating was not an effective carrier for postharvest fungicides. Finally, considering purchase price together with intensive equipment service, solvent waxes were the most expensive coating.

Because of their highly flammable nature, the solvent waxes required special handling.

The solvent wax solution was sprayed into an airstream sucked by fans downward through a roller conveyor carrying clean dry citrus fruit that was being rotated on a conveyor (Hall, 1981; Kaplan, 1986; Long, 1964; Sells and Porch, 1944). The coating was drawn through conveyed fruit by a partial vacuum created by venting the chamber from below. The solvent was exhausted to the atmosphere, which created significant air pollution problems, and for that reason the coatings were eventually banned.

There were other problems as well. First, a considerable amount of the coating solids were deposited in the exhaust system. As the owner and maintainer of the equipment, the solvent wax supplier took responsibility for regular cleaning of the machines. Second, there were fire and explosion hazards, normally controlled by maintaining the amount of wax in the airstream to a concentration below its flammable limit.

Before arrival at the coating bed, most if not all water must be removed from the fruit surface. Excess water only serves to disrupt the coating. Less coating is applied, and foaming may result from the interaction of the coating additives and water. If fungicides are incorporated in the coating, fungicides are diluted (Hall, 2003). Drying was usually accomplished by passing the fruit over a bed of rotating horsehair brushes, called a polisher or polisher dryer, turning at 180 to 200 rpm. These brush beds would often be 20 to 30 feet long and would typically have as many as 60 brushes. Unfortunately, if the fruit were allowed to dwell too long on these "polishers," peel injury would occur in susceptible varieties (Grierson, 1956).

One major drawback with this type of coating was that the applicator and polisher were very expensive. At the height of the solvent wax era, the applicators, and often the polishers, were owned and maintained by the solvent wax supplier.

3.3.3 Emulsion formulations

Waxes and oils are usually applied as emulsions whose composition is proprietary, but some of the science is well known. Macro- and microemulsions are both used. Macroemulsions are milky formulations with typical globule size of about 2 to 20 µm (Hernandez and Baker, 1991). Because of the large globule size, wax macroemulsions do not form a glossy finish when dried. Homogenization of macroemulsions is sometimes undertaken, not so much to change globule size as to transform wax in the globules from liquid state into solid state, to reduce flocculation, settling, and phase separation.

Microemulsion formulations of waxes, on the other hand, are quite different from macroemulsions in method of preparation, properties, and appearance, as liquids and also as dried coating. For example, homogenization or other high-energy turbulence is not needed to reduce globule size during preparation of microemulsions, despite their smaller globule size, typically about 0.5 µm for coating wax formulations (Prince, 1977). Further, the liquid formulation is clear rather than milky. As for the coating, microemulsions of hard waxes dry to form coatings with good shine, something not observed with macroemulsions. A typical microemulsion wax formulation is made up of about 17% wax, 3% fatty acid soap, with the remaining balance of water (Hagenmaier and Baker, 1994b).

A commonly used method for making wax microemulsions is the *water-to-wax method*. Briefly, mopholine is added to a mixture of food-grade oleic acid and molten wax, to which hot water is added at a controlled rate to form the microemulsion, which is promptly cooled. In the pressure emulsification method, all ingredients except the bulk of the water are heated in an agitated pressure vessel. Next, hot water is added to invert the emulsion from water-in-wax to wax-in-water and at the same time reduce solids content to 20% to 30%. The emulsion is cooled and diluted to desired concentration. Sometimes a shellac or wood rosin formulation is then added to create a more uniform and continuous coating. Wax microemulsions and resin solutions with similar pH can normally be mixed in almost any ratio, with one exception being that calcium or other multivalent cations might cause coagulation of wax globules. The detailed composition and procedures used for commercial coatings are proprietary information, but these are the methods used for small batches of emulsions for research purposes (Hagenmaier and Baker, 1994b).

The most commonly used fatty acids are mixtures made from vegetable oils with high oleic acid content, such as soy, canola, and cottonseed.

Emulsion coatings are often referred to as *storage waxes* or *shipping waxes*. The storage waxes are usually either macro- or microemulsions of various waxes and oils, designed to reduce weight loss in the commodity while it is stored to await further processing or preparation for marketing; to serve as carriers for most commonly used postharvest chemicals (fungicides, growth regulators); and to be washed off before processing or marketing. Common storage waxes are relatively simple formulations, using paraffin, mineral oil, various natural waxes, or oxidized polyethylene. The emulsifiers commonly used are alkali metal salts of fatty acids (e.g., sodium oleate) or nonionic surfactants (more common). After application, they are seldom subjected to extensive drying, if at all, as their residue is easily removed from subsequent equipment.

If the commodity is to be marketed as a raw agricultural commodity (intact), the storage wax may be removed and replaced with a *shipping wax*, as is done with lemons in California. Because the largest lemon harvest is mid-winter and the demand for fresh lemons is year round, it is necessary to store significant quantities of lemons for up to 6 months. In the packinghouse, the mature harvested fruit are washed, coated with a storage wax containing a fungicide and growth regulators, and then sorted by color. (Because citrus requires cold weather in order to lose all of the chlorophyll in its peel, some mature fruit may still be quite green.) The fruit is then placed into dark, cool storage where the chlorophyll will naturally degrade. During storage, the wax coating minimizes weight loss while allowing sufficient gas exchange for respiration (Bartholomew and Sinclair, 1951). Occasionally, market demand will require that fruit that is still green be treated with ethylene to accelerate degreening. When this is necessary, the gas permeability of the storage wax becomes even more important.

When needed for the market, the fruit are removed from storage, washed, and the shipping wax is applied. Lemon storage wax, therefore, needs to be readily removable without leaving a residue that will interfere with the shipping wax.

The shipping wax for lemons can be a wax microemulsion, a resin solution, or a mixture of the two. Shipping waxes are often called *water waxes* because wax microemulsion and resin solution waxes both contain water as the principal carrier (about 80% of weight). Resin solution waxes typically contain shellac or wood rosin, fatty acids, morpholine or ammonia, and protein (Drake et al., 1987; Hall, 1981).

3.3.4 Resin formulations

The function of resin in coating formulations is principally to impart gloss to the finished product. Accordingly, many shipping waxes are resin based. In order that it maintain its cosmetic appearance, it must be quick drying so that it does not come off or lose gloss when in contact with processing equipment. It must also be waterproof, and for that reason, the choice of emulsifiers is rather limited. At present, emulsifiers based on the fatty acid salts of either morpholine or ammonia are used for this purpose (Table 3.3). Both gas off when the coating dries, leaving a waterproof film behind. The formulations of shipping waxes are proprietary, but some information is readily available. For example, many formulations include plasticizers.

3.4 Application of coatings to fruit

All wax application systems, independent of type, should have a strategy for cleanup after each use. Failure to do so may result in pump breakdown and clogging of nozzles. Buildup of hardened coatings on brushes, rollers, and side bumpers will result in commodity injury and subsequent decay.

Brush wear should be routinely monitored. Incomplete coating coverage, or missed spots, can often be avoided by replacement of worn brushes on the brush bed and by proper coating bed length to allow adequate tumbling action over the brushes. Incomplete coverage can also be avoided by ensuring that the commodity is clean before coating it. Otherwise, soil or debris present on the coated commodity surface will accumulate on the brushes or rollers and create problems.

The coating should be dried as much as possible before packing. Some coating formulations remain "tacky" after drying, and commodities must be handled carefully to avoid the creation of missed spots on the finished product. An excellent coating job can be negated by inadequate airflow, temperature, and underfill and overfill of the drier.

The use of coatings, although increasingly under scrutiny, has become an important component of the fresh fruit and vegetable industry. The ability to control desiccation, incorporate fungicides for decay control, and control aspects of product physiology serves to lengthen the market window for commodities with a finite shelf life. Because travel to distant markets requires longer storage and holding periods, greater quality demands are made. These trends in domestic and international marketing underscore the importance of coating in the maintenance of quality in fresh fruits and vegetables.

When a high shine on the produce is desired, then the resin solution coatings are most commonly used. Commodities like citrus and apples are often coated with a high-shine coating. Shellac, rosins, and modified rosins are commonly the major components of these formulations. Because either natural or bleached shellac is an ingredient commonly found in these formulations, these waxes are often referred to as *shellac* waxes or *shellac-based* waxes.

The basic carrier for these waxes is water with a nonvolatile content between 16% and 20%. Because the resins are alkali soluble, ammonium hydroxide is often a significant component. Various proteins such as soy and casein are common along with plasticizers such as glycerin and fatty acids (Hall, 1981).

The most desirable properties of this type of coating are to have a high shine and good water resistance when dry. Also, quick drying is important so that the coating will not leave large deposits on packing equipment or be sticky when packed. The coating must also be flexible enough that it will not fracture and flake off or become dusty.

Because many commodities are shipped or stored under refrigeration, they tend to sweat when exposed to the ambient conditions of the marketplace. When that happens, some films will be lifted off the surface of the commodity (e.g., citrus, apples) and may appear as white areas that might be mistaken for pesticide residues. This undesirable condition is variously referred to as *chalking, rewetting, fracturing,* and *whiting*.

When applying the resin solution waxes, great care must be exercised not to apply too much, especially for commodities that require good gas exchange to maintain quality (Bai et al., 2002, 2003).

Regardless of the type of coating, the method of application is of critical importance. The commodity must be adequately covered to accomplish its purpose. This means that the coating must be fairly uniform. The type of formulation, of course, has considerable impact upon the method of application used.

With the exception of solvent wax and dipping applications, brushes apply most coatings. Early applicators used brushes that ran the length of the process line so that the fruit traveled along a single valley between brushes. With the development of the transverse brush applicators, better coverage was obtainable with the fruit being pushed along either mechanically or by pressure from incoming fruit. By being pushed from one brush valley to the next as it progressed through the machine, all but the most elongated varieties would tumble and get thorough coverage.

Modern brushes are usually composed of a mixture of horsehair and synthetic bristles. The horsehair wets easily and allows the brush to act as a reservoir for wax, while the synthetic bristles provide support and prevent the hair from matting. This also facilitates cleaning when needed.

3.4.1 Dip application

Platenius (1939) reported that the Florida citrus industry was the first to realize the value of coating by dip applications for fresh citrus fruit. In an industrial process, coating was accomplished by submerging the commodity into a tank of emulsion. Because of the superior shelf life and gloss of dipped rutabagas from Canada, consumers developed a preference for rutabagas showing glossy appearance; soon, rutabagas, tomatoes, and peppers were being dipped in coating emulsions in this country.

Dipping fruits and vegetables into a tub or tank of the coating material is adequate usually for small quantities of commodities. The produce is washed, dried, and then immersed in the dip tank. Time of immersion is not important, but complete wetting of the fruit or vegetable is imperative for good coverage (Long, 1964; Newhall and Grierson, 1956; Van Doren, 1944). The continuous dipping of fruits and vegetables into a largely static milieu results in unacceptable buildup of decay organisms, soil, and trash in the dip tank. Other coating methods are more desirable where considerable quantities of fruits and vegetables are to be coated. Still, dipping commodities such as rutabagas, and tropical fruit such as carambolas and pineapples, continues, especially when thicker applications of the coating are required.

Formulations of solid waxes can be applied in a variety of ways. The simplest application is the practice of dipping some tropical root vegetables, such as cassava, in molten paraffin. The earliest machine-assisted applications of coatings to fresh produce were in the citrus industry of Florida (Hall, 2003).

One of the earliest applicators of solid wax involved pressing a bar of wax against a rotating brush that ran parallel to the direction of fruit travel. The fruit would press against this brush as it was pushed along by mechanical means, gravity, or by the pressure of incoming fruit. Wax is thus transferred to the fruit and smoothed out by rotating brushes. In latter developments, a bed of brushes transverse to the direction of fruit travel became standard. The solid wax would be held in pans pressed against the underside of the first few brushes of the applicator, with subsequent brushes smoothing and polishing the coating. This type of application, which greatly increased the volume of fruit that could be treated, still has limited application.

A final development in the application of solid waxes came in the form of an applicator that would spray molten wax onto the produce in a heated chamber with rotating brushes to smooth out the wax, followed by additional brushes rotating at high speed for polishing (Brogden, 1927). Because the cooled spray residue resembled snow, the application was referred to as *snow wax*. These applicators were mechanically complicated and required laborious daily cleanup. The lack of high shine and the development of successful solvent wax systems began the decline and eventually total elimination of this type of wax.

3.4.2 Foam application

Although largely replaced by other systems, emulsions can be applied with a foam applicator. A foaming agent is added to the coating or compressed air (less than 5 psi or 35 kPa) blown into the applicator tank (Hartman and Isenberg, 1956; Long and Leggo, 1959). Even distribution of the coating is difficult to achieve because extensive tumbling action is necessary to break the foam and distribute the coating.

3.4.3 Spray

Spray application is the conventional method for applying most coatings to fruits and vegetables. The typical applicator consists of a metered spray onto the fruit as it passes over a series of rotating brushes that serve to spread the coating evenly over the fruit. The volume applied must be carefully controlled for two reasons. If too much wax is applied, the coating can interfere with respiration, and when the wax is a carrier for postharvest fungicides, there is the possibility of exceeding acceptable tolerances. A typical application for most citrus coatings is 800 kg/L, which varies with formulation, commodity, and variety.

The earliest applicators used a moving nozzle that traveled back and forth across the brush bed. An often-favored nozzle for these applicators was an oil burner nozzle that produced a hollow cone spray. Applicators had as many as four nozzles to insure complete coverage. The volume was controlled by maintaining a constant pressure on the system and using nozzles with a rated capacity. In addition, a signaling device would shut the system off when fruit flow wax interrupted or slowed. Electric eyes, sonic sensors, and mechanical switches controlled by a flag on the entry conveyor were some of the parts. The relatively high pressure used in these systems minimized nozzle plugging, but operators would often respond to increased fruit flow by increasing the pressure rather than changing to a larger nozzle, resulting in such fine atomization that the wax would drift away from the fruit and deposit on any contacted surface.

An adaptation of the traveling nozzle system used an air nozzle, especially when very viscous formulations were used. A metering pump, set to deliver a measured amount of wax, controlled the wax flow rate into a chamber behind the nozzle opening. Air was introduced there to atomize the wax into a spray onto the fruit. A problem with these dual-fluid nozzles, if not regularly cleaned, was a clogged nozzle opening from a buildup of the wax, and the air pressure would need to be increased, eventually overcoming the pressure from the metering pump, and thus effectively shutting the system down. In practice, this meant that the system needed to be regularly attended during the operating day.

Air is delivered to the nozzle head at pressures usually at or below 5 psi and the coating delivered with a metering pump at less than 40 psi. Although nozzle size is not critical, the air nozzles can be expensive. Low-pressure applicators used in the past delivered coating in excess. Often, recovery wells were utilized and excess coating was recirculated. As with dip application, dilution and contamination were of concern.

Another delivery system used irrigation drippers from a manifold above the brushes, with a metering pump to control the flow of wax. This was mechanically simple, inexpensive to construct, and easily maintained.

Most application systems have some means of controlling the application rate, such as metering pumps to adjust the rate of flow with a manual control, adjusting the pressure of single-fluid nozzles, using simple on/off solenoids controlled by flag switches, electric eyes, or programmable computer controlled systems, all with the aim of providing the optimum amount of coating to the commodity.

Later, high-pressure spray applicators, delivering coatings at 60 to 80 psi, became available. These used much less coating material and gave equal or better coverage, negating the need for recovery wells and recirculation. Nozzle size was critical because small nozzles often became plugged, and large nozzles delivered too much wax.

For citrus, apples, and pears, the coating is sprayed onto the fruit passing under a set of fixed or mobile nozzles. Fruit travels over a slowly rotating bed of brushes. The waxer brush bed is one of the most important pieces of equipment for coating purposes, because application and distribution of the coating occur largely on and across the bed. Straight-cut brushes on the brush bed are more effectively used with round or slightly elliptical commodities, whereas spiral-cut or tumble-trim brush designs are used with small, flat, irregularly shaped produce that requires more tumbling action for good coating coverage. The bed is typically composed of 12 to 14 brushes, with 4 to 6 brushes positioned after the applicator. Too many brushes after the applicator can remove the wax on the commodity surface. With all spray applications where brushes are expected to distribute the coating, all brushes are recommended to be made of a 50% mixture of horsehair and polyethylene bristles. The recommended distance between bristles on each brush is no more than 0.95 cm. Horsehair, horsehair mixtures, or very soft but durable synthetic types are superior for coating purposes because of their soft texture and high coating absorbency for recommended brush speeds of 100 rpm or less (Hall, 1981; Wardowski et al., 1987). With prolonged use, however, horsehair bristles will lose "body" and fall away from the brush, necessitating more frequent replacement. Brush wear must be checked often, and coatings must be washed from the brushes with clean water after every application.

Spray nozzles can be selected to deliver full cone, tapered, or even-edged flat, or air atomizing spray characteristics. When mounted, attention must be given to operating pressure, distance from the brush bed, and nozzle wear, because spray pattern quality and capacity will be directly affected. Full cone spray nozzles are more effectively utilized when a single mobile "wig-wag" nozzle or a single bank of nozzle heads are used to apply coatings, whereas tapered or even-edged flat spray can be chosen with multiple bank stationary nozzle mountings for overlapping or uniform coverage characteristics, respectively. Air atomizing spray nozzles are available that deliver a variety of spray patterns depending on the velocity of the atomizing air. Uniform coating coverage is easier to achieve with spray application but can be adversely affected by wind patterns under and around the nozzles. For this reason, coating equipment is often partially enclosed to minimize the effect of air currents on coating coverage.

Manifold units consisting of several stationary spray nozzles can be mounted over the brush bed. However, nozzles with lower capacity must be used because several are usually mounted on the manifold, and plugging is often a problem (Hall, 1981). Turbulence within the manifold coupled with a drop in pressure along its length result in differences in coating delivery rate across the brush bed. Nozzles of increasing capacity along the manifold and away from the pressure source could be selected to ensure uniform coating delivery. Nozzles that travel along the width of the brush bed are also used to distribute coatings to produce. Beam sliders, swing arms applicators, "lawn sprinklers," and elliptical chain applicators are known in Florida (Hall, 1981). Uniform produce coverage suffers as changes in nozzle direction occur.

Spray applicators deliver coatings over the majority of the commodity surface area. However, good, uniform coverage is not achieved when commodity flow and bed fill are not properly adjusted. Excess coating sometimes accumulates on the brushes, for example, when spray application continues after commodity flow is disrupted by accident or design. Coating can be wasted as a result. For economy and efficiency, programmable spray systems have been developed that control the delivery of coatings as a function of throughput and product type.

One relatively new programmable spray system for a high-pressure pump (suitable for high volume) delivers coatings on a percent time-on, time-off basis. Maximum benefit can be attained from this system when the pump is delivering the necessary rate of coating and fruit flow is even and steady. In addition, photosensors can be installed that "see" produce as it is introduced to the coating equipment, and solenoid valves adjust the rate of coating application from the pumps accordingly. Systems are also available that can sense fruit flow by means of a series of light steel or plastic fingers mounted above the roller bed and positioned before the spray applicator. As the produce passes through the fingers, the system senses the rate of flow, and solenoid values adjust the rate of coating application from the pump. In Israel, air nozzles have been used in concert with a metering pump to deliver the coating (Kaplan, 1986). Another sensing system not only senses the rate of commodity flow, but also compensates and adjusts for inherent differences in commodity flow rate across the width of the brush bed. An additional feature of this *programmable pulse* system is the use of a pressure pump with conventional spray nozzles, eliminating the need for air in the coating facility and expensive air nozzles. Valves that sense differences in nozzle performance are also used with these systems.

Resin solution and emulsion coatings are pressure-sprayed over rollers onto the surfaces of root and tender tropical crops and are distributed by the spray action. Brushes are not used. With some crops, such as tomatoes, cucumbers, squash, limes, and bell peppers, coatings are dripped onto a brush mounted just above the commodity. The produce becomes coated as it passes underneath and touches the overhead brush. The coating is then evenly distributed by the tumbling action of the produce over additional brushes downstream.

3.4.4 Drip

Because the oil waxes have limited application, their method of application is also fairly limited. The main application is currently to tomatoes. Cucumbers, peppers, some stone fruit, and, to a very limited extent, citrus are oil coated. In the case of citrus, limes are oil coated to reduce weight loss. The very high acid content of these fruit prevents the products of anaerobic respiration from being noticeable to the average consumer.

With tomatoes, the most common method of application is to drip the oil wax onto a long bristled brush (9 to 12 inches) that then wipes the oil onto the tomatoes as they pass below on a bed of rotating brushes. These brushes then spread the coating smoothly over the surface of the tomato. The oil is often warmed (40 to 45°C) to facilitate spreading and to minimize the thickness of the coating. The application of this type of wax is similar with cucumbers and bell peppers.

Another type of applicator, tried briefly, was the mop applicator. It used curtains similar to a string mop's head to wipe the coating onto the produce as it passed under on a roller conveyor. Coverage was obtained by using a series of "mops" positioned to contact different areas of the produce as it turned on the conveyor (Ahlburg, 1927). This process resulted in considerable cleanup labor and was never successful.

The most economical method used today to apply coatings to fruits and vegetables is the drip application method. Different sizes of emitters are available that will deliver a variety of large droplet sizes. Usually mounted on a dual-bank manifold, emitters are spaced 1 inch apart along the manifold and width of the bed. Irrigation tubing can be used to house the emitters. A metering pump is used to deliver the coating at pressures not greater than 40 psi (276 kPa). Besides the economy of drip application, the advantage of this method lies in its ability to deliver the coating either directly to the commodity surface or the brushes. Controlled drop applicators (spinners) have been used successfully to coat some produce types. Low-pressure metering pumps are used to deliver the coating to the nozzle housing, where a rotating disc effectively shatters the large drops into smaller ones that are then delivered by centrifugal force to the commodity. The rotating disc operates typically at speeds of 2000 to 5000 rpm. Lower rotating disc speeds dispense larger drop sizes. Although used successfully in Washington State to coat apples, the citrus industry was slow to utilize this application method because of difficulties in delivering a high-solids wax with incorporated fungicide. Now, controlled drop applicators are being used successfully with citrus. Controlled drop applicators can also be used with programmable controllers that, like spray systems, will vary coating flow to match fruit flow.

Depending upon the commodity and the desired effect, the application methods of emulsion waxes have many aspects in common with the resin solution waxes. If the aim of the application is a shine coating similar to that used on citrus and apples, then the application methods and equipment are essentially interchangeable. Some packers simply change the wax to suit their immediate needs. If the coating is essentially functional, to reduce weight loss or carry a postharvest fungicide, then the coating application might be a spray over brushes as in the case of cantaloupes, or a recirculating flood as in the case of lemons intended for storage.

Some commodities may require special application methods due to their shape or condition. Carambolas (star fruit) cannot travel over conventional conveyors, and they cannot tolerate more than the gentlest of handling; therefore, they must be dipped into whatever coating may be applied. Pineapples may be submerged in a tank and carried out on a conveyor. In some cases, the desire is only to coat the bottom of the pineapple where it has been cut from the plant. This serves to seal the wound and to apply a postharvest fungicide to the point that is most susceptible to decay.

Regardless of the system used, successful application requires that the equipment be maintained and adjusted as necessary. No system is maintenance free. The packinghouse management should ensure that the coating applicators receive regular attention.

3.5 Design and function

As has already been mentioned in passing, wax formulations and ingredients must be carefully chosen so that the coating fulfills its mission. The purpose of coatings may either be functional or cosmetic. The cosmetic function of a coating is obvious: maintain fruits or vegetables that appeal to the eye and sell better than ordinary looking produce (Hall, 1981). A shellac- or wax-coated shiny apple will be more appealing to the buyer than one that has a dull appearance. Coatings may also help retard color changes either by their functional nature in slowing respiration or as a carrier for substances that encourage or prevent color changes, for example, an acidified coating to retard the loss of the natural red color of fresh lychee (Rattanapnone et al., 2007).

In addition to improving appearance and controlling weight loss or shriveling, lipid coatings also serve to prevent decay, preserve internal quality, and carry postharvest fungicides (Hall and Sorenson, 2006).

Even after harvest, fresh fruits and vegetables continue to respire and carry on metabolic functions. Because they have been cut off from their source of nutrition, they must continue on their own internal resources. As they respire, they metabolize nutrients, consume oxygen, and produce carbon dioxide as well as other by-products of respiration. A coating that prevents the free exchange of gases can cause anaerobic respiration to take place which will produce by-products that often cause the development of off-flavors, making the commodity unacceptable to the consumer (Hagenmaier and Baker, 1994a; Hagenmaier and Goodner, 2002; Hall, 1981; Segall et al., 1974). Wax coatings are generally more permeable to gases than resin coatings (Bai et al., 2002; Hagenmaier and Shaw, 1991).

The control of weight loss is important in that most fresh produce is sold by weight. The packer that guarantees a specific weight to his buyer must, of necessity, overfill his containers in order to have the correct weight at arrival. A coating that will reduce the rate of weight loss will reduce the amount of overfill required. Many commodities shrivel as they lose weight and, therefore, do not look fresh in the marketplace (Khout et al., 2007). Generally, lipid-based coatings, being hydrophobic, tend to be quite effective at preventing water loss. Further, because they have higher gas permeability, they tend not to cause anaerobic respiration (Avena-Bustillos et al., 1994b; Hall, 1981).

A successful formulation for coating fresh produce depends upon several factors. It must be relatively easy to apply with traditional packline equipment requiring little to no modification. Included in that requirement is the ability to clean coating residues from the equipment at the end of the day.

It must give good coverage so that the commodity has a uniform appearance. It must be durable, holding its appearance through normal market channels. This means that the coating cannot shatter or "dust" due to vibration or friction during transport, and it must be able to withstand normal sweating when cold produce is brought out from refrigerated storage into a warm, humid room (Avena-Bustillos et al., 1994a). In some cases, the coating must also be waterproof to some extent in order to stand up to the misting used in some marketplaces.

Another requirement is that the coating needs to be compatible with any postharvest fungicides that might be used. Such incompatibility might be either physical or chemical. Some formulations of fluidoxonil fungicide have been found to be physically incompatible with some paraffin-based emulsions used in coating stone fruit, while the fungicide benomyl was chemically incompatible with many resin-based postharvest coatings (Hall, 1980).

In addition, it is important that the coating not affect the flavor of the commodity (Eswaranandam et al., 2006). For example, some impart a bitter taste.

To date, the coating formulations that have proved commercially successful have fallen into five main groups: synthetic and natural resins dissolved in a hydrocarbon solvent, solid waxes, liquid oil based compositions, water-based solutions of resinous materials, and water-based emulsions of waxy materials (Hall, 1981).

Cantaloupes have been coated with emulsions to slow ripening and reduce weight loss. Peaches are often defuzzed before packing and will shrivel badly if not coated with an emulsion to prevent water loss. Processing of citrus fruit in packinghouses involves thorough washing of the fruit to improve its appearance. A side effect of this process is that the washing is often so vigorous that it also removes natural waxes that protect citrus fruit from weight loss. Therefore, thorough washing makes it necessary to apply a coating to prevent water loss and shriveling.

With fruits like citrus and apples, a shiny appearance is desired. Therefore, a formulation that is not only functional but also will improve the fruit's appearance and gloss is desirable. Such shipping waxes usually contain mixtures of shellac, wood rosin, carnauba wax, or polyethylene wax.

3.6 Future developments

In developing a new coating for produce, the lessons of current successes and past failures must be taken into consideration. Before any new coating can be successful, it must meet some very specific criteria and must be equal to or better than what is currently being used:

> Appearance: If this is a factor for the packer, then the product must be, at least, equal to those products currently available.
> Cost: A new product must be cost competitive. If it is more expensive to use, then it must be clearly superior in other aspects to justify the added expense.
> Applicators: The method of application must either be compatible with current equipment or the end result must provide a competitive edge in the marketplace in order to justify the expense of changing.
> Maintenance: The product and its applicators must be relatively easy to maintain and clean.
> Service: The supplier must be in a position to replace the level of service currently enjoyed by the packer. If special applicators are required, they may need to be provided and maintained. The product needs to be readily available when needed. Fresh produce usually must be packed and shipped soon after harvest. The packer cannot wait for a shipment of coating to come from a distance, and most are unwilling to stockpile a large quantity.

Many new edible coating products have failed in the marketplace because of at least one of these factors. The patent literature is full of possibly excellent products and processes that have failed in the marketplace because of one or more of the above points.

Finally, any new product must take into consideration the availability of the raw materials. An examination of the materials listed in Table 3.1 will reveal that none of them are manufactured for their application as part of a coating for fresh produce. That use is only a minor fraction of the total production for other major uses. Unless a particular raw material is naturally occurring (i.e., carnauba wax), its availability is dependent upon some manufacturer having sufficient market to justify its production.

References

Ahlburg, F. 1927. Process and apparatus for coating articles. U.S. Patent 1,618,519.

Ahmad, M., Z.M. Khalid, and W.A. Farooqi. 1979. Effect of waxing and lining materials on storage life of some citrus. *Proc. Florida State Hort. Soc.*, 92:237.

Allied Signal, Inc. 1986. A-C Polyethylenes for Fruit and Vegetable Coatings. Brochure No. CTG-2, pp. 1–6.

Antoniewski, M.N., S.A. Barringer, C.L. Knipe, and H.N. Zerby. 2007. Effect of gelatin coating on the shelf life of fresh meat. *J. Food Sci.* 72(6):E382–E387.

Avena-Bustillos, R.J., L.A. Cisnero-Zevallos, J.M. Krochta, and M.E. Saltveit. 1994a. Application of casein-lipid edible film emulsions to reduce white blush on minimally processed carrots. *Postharv. Biol. Technol.* 4:319–329.

Avena-Bustillos, R.J., J.M. Krochta, M.E. Saltveit, R.J. Rojas-Villegas, and J.A. Sauceda-Pérez. 1994b. Optimization of edible coating formulations on zucchini to reduce water loss. *J. Food Engr.* 21:197–214.

Bai, J., E.A. Baldwin, and R.H. Hagenmaier. 2002. Alternatives to shellac coatings provide comparable gloss, internal gas modification, and quality for 'Delicious' apple fruit. *HortScience.* 37:559–563.

Bai, J., R.D. Hagenmaier, and E.A. Baldwin. 2003. Coating selection for 'Delicious' and other apples. *Postharv. Biol. Technol.* 28:381–390.

Baldwin, E.A., M.O. Nisperos, R.D. Hagenmaier, and R.A. Baker. 1997. Use of lipids in coatings for food products. *Food Technol.* 51:56–64.

Bartholomew, E.T., and W.B. Sinclair. 1951. *The Lemon Fruit.* Berkeley: University of California Press.

Bennett, H. 1975. *Industrial Waxes,* Vol. 1. New York: Chemical.

Brogden, E.M. 1927. Method and apparatus for treating fruit. U.S. Patent 1,641,112.

Brooks, C. 1937. Some effects of waxing tomatoes. *Proc. Amer. Soc. Hort. Sci.* 35:720.

Claypool, L.L. 1939. The waxing of deciduous fruits. *Proc. Am. Soc. Hort. Sci.* 37:443.

Dalal, V.B., W.E. Eipeson, and N.S. Singh. 1971. Wax emulsion for fresh fruits and vegetables to extend their storage life. *Indian Food Packer.* 25(5):9.

Dangaran, K.L., J.R. Nantz, and J.M. Krochta. 2006. Whey protein-sucrose coating gloss and integrity stabilization by crystallization inhibitors. *J. Food Sci.* 71(3):E152–E157.

Drake, S.R., J.K. Fellman, and J.W. Nelson. 1987. Postharvest use of sucrose polymers for extending the shelf-life of stored 'Golden Delicious' apples. *J. Food Sci.* 52:1283.

Durand, V.J., L. Orean, U. Yanko, G. Zaubennan, and N. Fuchs. 1984. Effect of waxing on moisture loss and ripening of 'Fuerte' avocado fruit. *Hort. Sci.* 19:421.

Eastman Chemical Inc. 1984. A Guide to the Use of Epolene Waxes under United States FDA Food Additive Regulations. Publication No. F-243D, pp. 1–6.

Eastman Chemical Inc. 1990. Emulsification of Epolene Waxes. Publication No. F-302, pp. 1–17.

Eswaranandam, S., N.S. Hettiarachchy, and J.-F. Meullenet. 2006. Effect of malic and lactic acid incorporated soy protein coatings in the sensory attributes of whole apple and fresh cut cantaloupe. *J. Food Sci.* 71(3):S307–S313.

Grierson, W. 1956. Reducing losses in harvesting and handling tangerines. *Proc. Fla. State Hort Soc.* 69:165–170.

Griffin, W.C. 1979. Emulsions. *Kirk-Othmer Encyclopedia of Chemical Technology,* 3rd ed, vol. 8, New York: John Wiley & Sons, pp. 913–916.

Guilbert, S. 1986. Technology and application of edible protective films. In *Food Packaging and Preservation. Theory and Practice,* M. Mathlouthi, ed., London, England: Elsevier Applied Science.

Hagenmaier, R.D., and P.E. Shaw. 1991. Permeability of shellac coatings to gases and water vapor. *J. Agric. Food Chem.* 39:825.

Hagenmaier, R.D., and R.A. Baker. 1994a. Internal gasses, ethanol content and gloss of citrus fruit coated with polyethylene wax, carnauba wax, shellac or resin at different application levels. *Proc. Fla. State Hort. Soc.* 107:261–265.

Hagenmaier, R.D., and R.A. Baker. 1994b. Wax microemulsions and emulsions as citrus coatings. *J. Agric. Food Chem.* 42:899–902.

Hagenmaier, R.D., and K. Goodner. 2002. Storage of 'Marsh' grapefruit and 'Valencia' oranges with different coatings. *Proc. Fla. State Hort Soc.* 115:303–308.

Hall, D.J. 1980. Comparative fungicidal activity of Benomyl and its breakdown product methyl 2-benzimidazolecarbamate (MBC) on citrus. *Proc. Fla. State Hort. Soc.* 93:341–344.

Hall, D.J. 1981. Innovations in citrus waxing—An overview. *Proc. Fla. State Hort. Soc.* 94:258–263.

Hall, D.J. 2003. Twentieth century developments in handling Florida's fresh citrus fruit—An overview. *Proc. Fla. State Hort. Soc.* 116:369–374.

Hall, D.J., and D. Sorenson. 2006. Washing, waxing and color-adding. In *Fresh Citrus Fruits*, 2nd ed. W.F. Wardowski, W.M. Miller, D.J. Hall, and W. Grierson, eds., Longboat Key, FL: Florida Science Source, pp. 421–450.

Hardenburg, R.E. 1967. Wax and Related Coatings for Horticultural Products—A Bibliography. USDA/ARS Publication No. 51-15.

Hernandez, E., and R.A. Baker, 1991. Candelilla wax emulsion, preparation and stability. *J. Food Sci.* 56:1382–1383.

Hartman, J., and F.M. Isenberg. 1956. Waxing Vegetables. New York Agriculture Extension Service Bulletin No. 965.

Hicks, E. 1962. *Shellac, Its Origin and Applications*. London: MacDonald and Co.

Hitz, C.W., and I.C. Haut. 1942. Effects of Waxing and Pre-Storage Treatments upon Prolonging the Edible and Storage Qualities of Apples. Bulletin, Agricultural Experiment Station, University of Maryland, No. A14, pp. 1–44.

Kaplan, H.J. 1986. Washing, waxing and color adding. In *Fresh Citrus Fruits*, W.F. Wardowdki, S. Nagy, and W. Grierson, eds., Westport, CT: AVI, p. 379.

Kester, J.J., and O.R. Fennema. 1986. Edible films and coatings: A review. *Food Tech.* 40:47.

Khout, M.P., M.A. Ritenour, and J.J. Salvatoe. 2007. BASF Freshseal® CHC helps keep packed tomatoes firmer and fresher longer. *Proc. Fla. State Hort. Soc.* 120:217–221.

Lakshminarayana, S., L. Sarmiento, and J.I. Ortiz. 1974. Extension of storage life of citrus fruits by application of candelilla wax emulsion and comparison of its efficiency with tag and flavor seal. *Proc. Fla. State Hort. Soc.* 87:325.

Long, J.K., and D. Leggo. 1959. Waxing citrus fruit. *CSIRO Food Preserv. Quart.* 19:32–37.

Long, W.G. 1964. Better handling of Florida's fresh citrus fruit. Florida Agricultural Experiment Station, Bulletin No. 681.

Magness, J.R., and H.C. Diehl. 1924. Physiological studies on apples in storage. *J. Agric. Res.* 27:1–38.

Martin, J. 1982. Shellac. In *Kirk-Othmer Encyclopedia of Chemical Technology*. New York: Wiley Interscience, pp. 737–747.

Neville Chemical Co. 1988. Production Information, Resins, Plasticizers, Nonstaining Antioxidants and Chlorinated Paraffins, Brochure No. NCCFAL88, pp. 1–15.

Newhall, W.F., and W. Grierson. 1956. Low-cost, self-polishing, fungicidal wax for citrus fruit. *Proc. Amer. Soc, Hort. Sci.* 66:146–154.

Paredes-Lopez, O., E. Camargo-Rubio, and Y. Gallardo-Navarro. 1974. Use of coatings of candelilla wax for the preservation of limes. *J. Sci. Food and Agric.* 25:1207.

Platenius, H. 1939. Wax Emulsions for Vegetables. New York Agricultural Experiment Station, Bulletin No. 723.

Prince, L.M. 1977. Formulation. In *Microemulsions. Theory and Practice*, L.M. Prince, ed., New York: Academic Press, pp. 33–49.

Rattanapnone, N., A. Plotto, and E. Baldwin. 2007. Effect of edible coatings and other surface treatments on pericarp color of Thai Lychee cultivars. *Proc. Fla. State Hort. Soc.* 120:222–227.

Segall, R.H., Alice Dow, and Paul L. Davis. 1974. Effect of waxing on decay, weight loss, and volatile pattern of cucumbers. *Proc. Fla. State Hort. Soc.* 87:250–251.

Sells, Ogden S., and Howard L. Porch. 1944. Method for treating fruit. U.S. Patent 2,342,063.

Shinoda, K., and H. Kunieda. 1983. Phase properties of emulsions. In *Encyclopedia of Emulsion Technology,* vol. 1. P. Becher, ed., New York: Marcel Dekker, pp. 337–368.

Sward, G.G. 1972. Natural Resins. ASTM Tech. Publ., STP 500, pp. 76–91.

Trout, S.A., E.G. Hall, and S.M. Sykes. 1953. Effects of skin coatings on the behavior of apples in storage. *Aust. J. Agric. Res.* 4:57–81.

Tulloch, A.P. 1970. The composition of beeswax and other waxes secreted by insects. *Lipids.* 5:247–258.

Van Doren, A. 1944. A report on the construction and operation of a grower-size apple washing machine. *Proc. Amer. Soc. Hort. Sci.* 44:183–189.

Wardowski, W.F., W.M. Miller, and W. Grierson. 1987. Packingline machinery for Florida packinghouses. Florida Cooperative Extension Bulletin No. 239.

Wineman, R.D. 1984. Water-Emulsion Fruit and Vegetable Coatings Based on Epolene Waxes. Publication No. F-257A, Eastman Chemical Products, pp. 1–4.

Yehoshua, S.B. 1987. Transpiration. In *Postharvest Physiology of Vegetables,* J. Weichmann, ed., New York: Marcel Dekker, pp. 113–170.

4 Polysaccharide coatings

Robert Soliva-Fortuny, María Alejandra Rojas-Graü, and Olga Martín-Belloso

Contents

4.1	Introduction	104
4.2	Properties of polysaccharide edible films and coatings	105
	4.2.1 Cellulose and derivatives	105
	4.2.1.1 Microcrystalline cellulose	105
	4.2.1.2 Microfibrillated cellulose	106
	4.2.1.3 Nonionic cellulose ethers	106
	4.2.1.4 Anionic cellulose ethers	107
	4.2.2 Starches and derivatives	108
	4.2.2.1 Raw starches	108
	4.2.2.2 Modified starches	109
	4.2.3 Pectins and derivatives	110
	4.2.4 Seaweed extracts	111
	4.2.4.1 Alginates	111
	4.2.4.2 Carrageenans	113
	4.2.4.3 Agar	114
	4.2.5 Exudate gums	114
	4.2.5.1 Acacia gums	114
	4.2.5.2 Gum tragacanth	115
	4.2.5.3 Gum ghatti	116
	4.2.5.4 Gum karaya	116
	4.2.6 Seed gums	117
	4.2.7 Microbial fermentation gums	118
	4.2.7.1 Xanthan gum	118
	4.2.7.2 Gellan gum and related exopolysaccharides	118
	4.2.8 Chitosan	119
4.3	Applications of polysaccharide coatings	120
	4.3.1 Fruits and vegetables	120
	4.3.2 Meats, poultry, fish, and seafood	124
	4.3.3 Nuts and cereal-based products	125
	4.3.4 Other applications	126
4.4	Future trends	127
References		127

4.1 Introduction

This chapter presents information on polysaccharides that have proven useful in the formulation of food coatings. The quantity of cellulose and starch derivatives greatly exceeds the amounts of other hydrocolloids used for food products as stabilizers, emulsifiers, thickeners, binders, and suspending or gelling agents. During the past 30 years, the range of other polysaccharides employed is increasing, and edible coatings is one of the most relevant growth areas, especially within the fields of confectionery, dairy, desserts, meat products, ready-to-eat meals, fresh and fresh-cut fruits and vegetables, and bakery.

However, polysaccharides other than starch and cellulose derivatives are gaining importance due to the development of new applications, Acacia and tragacanth gums constitute a large but fluctuating proportion of the many types of hydrocolloids used in Western countries because of their functionality as thickeners, stabilizers, or emulsifiers (Stephen and Churms, 2006).

Development of films from water-soluble gel-forming polysaccharides has brought a surge of new types of coatings for extending the shelf-life of a wide variety of food products. Polysaccharide coatings are generally poor moisture barriers, but in contrast they have moderately low oxygen permeability and, at the same time, selective permeability to O_2 and CO_2 (Lacroix and LeTien, 2005). Therefore, polysaccharide-based coatings, like most hydrocolloids, have been applied very often to fruits and vegetables, either fresh or minimally processed, to reduce their respiration by creating modified atmosphere conditions inside the product, provide a partial barrier to moisture, improve mechanical handling properties, carry additives, as well as contribute to the retention and even the production of volatile compounds (Nisperos-Carriedo and Baldwin, 1990; Olivas and Barbosa-Cánovas, 2005).

Water-soluble polysaccharides are long-chain polymers that dissolve or disperse in water to give a thickening or viscosity-building effect (Nussinovitch, 1997). These compounds serve numerous diverse roles such as adhesiveness, gel-forming ability, and mouthfeel (Whistler and Daniel, 1990). Attention has been drawn on a wide variety of polysaccharides also because of their wide availability, low cost, and nontoxicity. As other coating materials, because they are an integral part of the edible portion of food products, they should follow all regulations required for food additives or ingredients. The foremost governmental regulations concerning food additives are the Food and Drug Act (FDA), the European Union standards, and the Codex Alimentarius, which constitutes the Food and Agriculture Organization (FAO)/World Health Organization (WHO) joint regulatory body (Raju and Bawa, 2006). The U.S. Food and Drug Administration (FDA) classifies these compounds as either food additives or generally recognized as safe (GRAS) substances. The Federal Drug and Cosmetic Act states that any substance that is intentionally added to food is a food additive that is subject to premarket review and approval by FDA, unless the substance is GRAS, among qualified experts, as having been adequately shown to be safe under the conditions of its intended use, or unless the use of the substance is otherwise excluded from the definition of a food additive (FDA, 2006). In the specific case of polysaccharide gums, most are approved in the Code of Federal Regulations (Title 21, Volume 3) as general use additives, including carrageenan, furcelleran, and their corresponding salts, gellan, xanthan, and locust bean gums. In Europe, these substances are majorly regarded as food additives and are listed within the list of additives for general purposes, although pectins, Acacia, and karaya gums are mentioned apart for coating applications (European Directive, 1995). In any case, the use of polysaccharide gums is permitted, provided that the *quantum satis* principle is observed.

4.2 Properties of polysaccharide edible films and coatings

4.2.1 Cellulose and derivatives

Cellulose is the world's most abundant polysaccharide, rivaled only by chitin. It is the major building block of plant cell walls, but it is also contained in the cell wall structure of green algae and in membranes of fungi. Cellulose is composed of linear chains of $(1\rightarrow 4)$-β-D-glucopyranolsyl units (Krassig et al., 1986). The abundance of hydroxyl groups and concomitant tendency to form intra- and intermolecular hydrogen bonds result in the formation of linear aggregates, which contributes to the strength shown by cellulose-containing structures in plants and also to the virtual insolubility of cellulose in common solvents, particularly water (Goffey et al., 2006), though cellulose is a hygroscopic material, insoluble but able to swell in water, dilute acid, and most solvents. Solubility can be achieved in concentrated acid solutions but at the expense of considerable degradation through glycosidic hydrolysis. Alkali solutions lead to considerable swelling and dissolution of present hemicelluloses. The degree of crystallinity greatly affects cellulose solubility. Thus, highly crystalline celluloses have a solvation energy threshold that does not allow the swelling of the polymer in water. Alkaline solutions penetrate cellulose polymers and interfere between polymer chains by placing substituents that block the formation of crystalline units (Dinand et al., 2002). This process is used for cellulose ether production. Despite the wide variety of cellulose derivatives that can be obtained, only a few of the cellulose ethers find application in foodstuffs (Majewicz and Polas, 1993; Murray, 2000), especially methylcellulose (MC), ethylmethylcellulose (EMC), hydroxypropyl methylcellulose (HPMC), hydroxypropylcellulose (HPC), and sodium carboxymethylcellulose (CMC). Changing the level of methoxyl, hydroxypropyl, and carboxymethyl substitution affects a number of physical and chemical properties such as water retention, sensitivity to electrolytes and other solutes, dissolution temperatures, gelation properties, and solubility in nonaqueous systems.

4.2.1.1 Microcrystalline cellulose

Microcrystalline cellulose (MCC) is prepared by controlled acid hydrolysis of native cellulose to dissolve the amorphous regions of the polysaccharide, leaving behind the less reactive crystalline regions as fine crystals. MCC dispersions have been shown to exhibit both thixotropic and pseudoplastic behavior (Brownsey and Redout, 1985). It has been used in a number of low-caloric foods where it acts as a bulking agent and provides stabilization as a result of its ability to form a gel structure. Commercial MCC is presented with the brand name of Avicel and incorporates a 10% CMC.

Addition of CMC to coating formulations has been reported to substantially increase film strength.

4.2.1.2 Microfibrillated cellulose

Microfibrillated cellulose (MFC) is obtained by passing a cellulose slurry through a small orifice under conditions of high shear force and great pressure differential, which results into the disruption of cellulose into microfibrillar fragments. MFC has considerably more water-retention capability than normal-grade materials. Suspensions of MFC are shear thinning (pseudoplastic), exhibit slight thixotropic behavior, and apparently do not suffer a viscosity drop on heating, while electrolyte tolerance is similar to that of other commonly used cellulosics (Goffey et al., 2006).

4.2.1.3 Nonionic cellulose ethers

MC, HPMC, and HPC are nonionic water-soluble ethers with good film-forming properties for edible coatings. These materials are also used in edible coatings for thickening, modification of surface activity, and the ability to form thermal gels that melt upon cooling. Ethylmethylcellulose (EMC) is a cellulose ether with rheological and physical properties similar to those of MC.

MCs compose a group of cellulose ethers in which methyl substitution occurs with or without additional functional substituents. This category also includes HPMC. The amount of substituent groups on the anhydroglucose units of cellulose can be designated by weight percent or by the average number of substituent groups attached to the ring, which is known as degree of substitution (DS). Each glucose unit has three available positions for substitution. Hence, DS can vary from 0 to 3.

Nonionic cellulose ether solutions are stable at pH 2 to 11. They are compatible with surfactants, other water-soluble polysaccharides, and with certain salts depending on the concentration used. If relatively high concentrations of dissolved salts are used, these cellulose derivatives tend to form finely divided and highly swollen precipitates. Namely, MC and HPMC gel upon the addition of sufficient coagulative cosolutes such as phosphate, sulfate, and carbonate salts. Firm gels can be formed at room temperature by adding 3% trisodium polyphosphate to a 2% solution of MC. Nonionic cellulose ethers are good film formers, capable of yielding tough and flexible transparent films owing to the linear structure of the polymer backbone (Krumel and Lindsay, 1976). Nonionic cellulose ethers precipitate from aqueous solution when heated, whereas re-solution occurs on cooling. Dissolution of nonionic cellulose ethers may be considered a two-step process: dispersion and hydration. Incomplete dissolution of particles may lead to the formation of agglomerates with a swollen outer skin. The solution behavior of the nonionic MC family is markedly different from that of the ionic CMC. When dissolved in water, MCs give clear, smooth-flowing pseudoplastic solutions. DS does not affect the rheology of MC solutions. These solutions at >1.5 wt% form gels when heated, and then on cooling return to the solution state at their original viscosity (Goffey et al., 2006).

Three basic procedures are recommended for the preparation of lump-free solutions of these compounds. The first procedure consists of adding the powder to the vortex of well-agitated water at room temperature. The rate of addition must be slow enough to allow particles to separate in water so that surfaces become individually wetted and agitation must continue until complete dissolution. Inclusion of other soluble ingredients into the solution may slow the solution rate of cellulose ethers because they compete for the solvent. The second procedure is dry blending the solvent with an inert or nonpolymeric soluble material, which reduces the tendency to lump. The third procedure is by dispersing the cellulose ethers in a water miscible nonsolvent such as glycerin, ethanol, or propylene glycol and then adding the slurry to water.

Nonionic cellulose ethers yield tough and flexible transparent films that are soluble in water and resistant to fats and oils. The determinant property affecting thermal gelation is the concentration of methyl groups and the methyl hydroxypropyl ratio. As the methyl concentration increases, the gels formed on heating become firmer. Contrarily, the addition of hydroxypropyl groups to MC tends to diminish the rigidity of the gel and increase the temperature at which gelation occurs. Besides, gel strength does not appear to be favored by an increase in molecular weight, but by polymer chemistry and thermal kinetics. Hence, MC films are the least hydrophilic of the cellulose ethers and produce films with relatively high water vapor permeability. MC films are tough and flexible, while the internal plasticizing effects of the high level of hydroxypropyl substitution in HPC result in a film of lower tensile strength and greater elongation. HPC also shows the unique property of being a water-soluble thermoplastic that is capable of injection molding and extrusion. Insolubility of MC and HPMC films in water can be obtained by cross-linking with melamine formaldehyde resins or with other multifunctional resins. With HPC, insolubility can be attained by incorporating zein, shellac, or ethyl cellulose using common solvents. Plasticity of the gels can be improved by adding polyglycols, glycerin, or propylene glycol.

4.2.1.4 Anionic cellulose ethers

CMC, also known as cellulose gum, is an anionic, linear, water-soluble cellulose derivative that is produced by reacting alkali cellulose with sodium monochloroacetate under controlled conditions. Although the sodium salt is the most common for food applications, a free acid, water-insoluble, form is also available. CMC is commercially available in the DS range of 0.4 to 1.5, because 0.4 is the minimum substitution needed to place the gum in solution. At low DS, hydroxyl sites are poorly substituted, thus resulting in a high number of interactions between chains and, in turn, water solubility. A DS above 1 ensures that little or no interchain association occurs, because hydroxyl sites are almost completely substituted. Hence, the DS of CMC has a profound impact on the physical properties of film-forming solutions. Käistner et al. (1997) report the solution properties of a range of commercial CMCs.

The procedure for the preparation of lump-free, clear CMC solutions is similar to that of the nonionic cellulose ethers. Commercial food-grade CMCs offer 1% aqueous solution viscosities within the range of 20 to 4000 mPa. Viscosity is a critical factor conditioning the choice for a food application. However, it is worth mentioning that CMC at typical concentrations does not gel even in the presence of cross-linking ions. As with most polyelectrolytes, the pH of the environment affects the viscosity of a CMC solution, even though, in general, little effect can be observed between pH 5 and 9. Below pH 4, the less-soluble free acid CMC predominates, viscosity is largely increased, and even precipitation of the polymer may occur. Above pH 10, a slight decrease in viscosity occurs, and cellulose degradation becomes important. The effect of added salts on viscosity can be marked and depends on the type and concentration of the salt, DS, and order of addition. The problem of viscosity reduction can be solved by first dispersing the polymer in water and allowing hydration. Once it is hydrated, the salt can be added to the desired concentration. In general, monovalent salts form soluble salts of CMC, whereas the use of divalent and trivalent cations dramatically decreases the quality of the solution (Mulchandani and Mahmoud, 1997). CMCs can also interact with proteins, as long as the pH of the food system is greater than the isoelectric point of the protein. In such a case, ionic interaction between the anionic CMC chain and the cationic protein chains generates a higher viscosity than is otherwise expected (Klemm et al., 2005).

4.2.2 Starches and derivatives

4.2.2.1 Raw starches

Starches are among the most abundant of plant products and constitute the reserve polysaccharide of most plants, as well as a naturally occurring bulk nutrient and low-cost energy source for human nutrition. Starch is the most commonly used food hydrocolloid because of its low cost relative to alternatives, and also the wide range of functional properties it can provide in its native and modified forms, including improvement in film formation of coatings.

Since the earliest times, cultivation of plants containing high proportions of starch has been of the utmost importance to mankind. Cereal grains, legume seeds, tubers, and certain fruits contain 30% to 85% starch in dry basis (Zobel and Stephen, 2006). In the native state, starch exists as insoluble granules with characteristic shape and some crystallinity.

Starch is defined as a mix of two different polysaccharide fractions: amylose and amylopectin. Both are composed of glucose but differ in size and shape. Amylose is the smaller of the two fractions and is defined as a linear molecule of $(1\rightarrow 4)$-α-D-glucopyranosyl units, with degree of polymerization (DP) up to 5000, but it is today well established that some molecules are slightly branched by $(1\rightarrow 6)$-α-linkages (Buléon et al., 1998). Amylopectin is a multiply branched macromolecule produced by transglycosylation of various lengths of $(1\rightarrow 4)$-linked chains of α-D-glucose units to O-6 of existing chains. Starches contain 18% to 30% amylose, except for waxy corn types that are virtually all amylopectin. Each raw starch has its own characteristic viscosity/temperature profile as a result of its particular granular composition and structure (Oates, 1997). Palviainen et al. (2001) evaluated the film formation ability and mechanical properties of aqueous native maize starches. Maize starches with a higher amylose content exhibit potential film-forming capacity and produce stronger films than those starches that contain less amylose. Thus, a high proportion of amylose is required for film forming and for the preparation of strong gels.

Films developed from amylose are described as isotropic, odorless, tasteless, colorless, nontoxic, and biologically absorbable. They exhibit physical characteristics, chemical resistance, and mechanical properties similar to those of plastic films (Wolff et al., 1951). Nevertheless, unmodified starches have very limited use in the food industry because raw starches do not have the functional properties demanded by processors. On the one hand, it is difficult to disperse raw starch in water, and at the same time, gels are formed and retrograde very rapidly once cooling is initiated. Therefore, there are few works in literature reporting the application of amylose and native starches as film-forming agents. Combinations of amylose and ethylcellulose, both as aqueous-based (Milojevic et al., 1996a, 1996b) and organic-based coatings (Siew et al., 2000) have been studied for microencapsulation in order to facilitate colon drug delivery.

4.2.2.2 Modified starches

Modification of native starch by disruption of hydrogen bonding through reduction of molecular weight or chemical substitution leads to lower gelatinization temperatures and reduced tendency for retrogradation. Starch hydrolysis by acid treatment shortens the chain length of the amylose fraction and converts some of the branched amylopectin into linear amylose units. The properties of acid-converted starches vary according to the base starch type and processing. They are all characterized by weakened granules that no longer swell, imbibing large volume of water in the way native starches do, but develop radial fissures and disintegrate on heating in water (Wurzburg, 2006). The hot-paste viscosity of acid-converted starches is considerably lower than that of unmodified starches. As a result, acid-converted starches may be dispersed at higher concentrations than unmodified starch. The concentration at which the starch may be dispersed increases as the fluidity rises. On cooling, the sols of acid-converted starches cloud and still form opaque gels because the amylose units, even though they are degraded, can still effectively associate and retrograde. Molecular weight measurements and gelling properties of various acid-modified starches are reported by Pessa et al. (1992).

Another way of converting starches is bleaching by exposure to treatments with certain specified oxidizing agents such as sodium hypochlorite. The term *chlorinated starch* refers to those starches that are not chlorinated but simply oxidized with hypochlorite. Oxidized starches cover roughly the same fluidity range as acid-converted starches, but on cooling they are more resistant to thickening and forming gels or pastes than acid-converted starches (Han, 2002).

Furthermore, dextrins obtained by pyroconversion of starches, namely, white dextrins, yellow dextrins, and British gums, have physical properties covering a very wide range. The film properties of these dextrins are important factors in governing their usage as binders, adhesives, and coatings. Films prepared from sols of unmodified starch have higher tensile strength than those made from dextrins. The advantage of dextrins is that they can be heated at much higher concentrations than unmodified starch. As a result, films formed from their sols contain higher proportions of solids so that they dry faster, can form thicker films, and have faster tack and greater ability to stick to surfaces shortly after application.

Chemical cross-linking of starch polymers stabilizes the granules and ensures that they do not overswell or rupture, especially in cases where food processing demands stability to high shear and low pH. Cross-linking reinforces the hydrogen bonds already present in the granules. When the hydrogen bonds are broken during heating or shearing, the chemical cross-links remain intact. The net result is that the granules of cross-linked starches are much more robust and retain their swollen integrity under conditions of excessive heat, shear, and reduced pH. Cross-linking may be varied to provide modified starches with optimal viscosity for specific applications under a wide range of acidities and treatment temperatures. It also increases the resistance of the swollen granules to shear so that viscosity of cross-linked starch solutions does not break down under mechanical agitation. Cross-linked starches are extensively used throughout the food industry to thicken, stabilize, and texturize food systems (Wurzburg, 2006), but no food coating applications have been developed because of the difficulty in developing coatings with good film-forming properties. Consistently, Rioux et al. (2002) reported hydrophilic and brittle stress–strain behavior for epichlorohydrin cross-linked starches, thus concluding that flexibility in normal environmental conditions needs to be improved in order to produce high-quality films from this source.

It is also possible to react the hydroxyl groups of starch to produce ester such as acetates of phosphates, and ethers such as the hydroxypropyl ether derivative. The introduction of substituents on the hydroxyl groups weakens the hydrogen bonding in the starch, lowering the gelatinization temperature and improving the freeze–thaw stability as well as the clarity by interfering with the ability of the linear amylose polymer to associate. The ether linkage tends to be more stable than the ester linkage. Hence, water-soluble, transparent films have been produced from hydroxypropylated (1.1%) amylomaize starch having an apparent amylose content of 71% (Roth and Mehltretter, 1967). This film exhibited very low oxygen permeability. Hydroxypropylation reduced the dry tensile strength of amylomaize starch film but considerably increased bursting strength and elongation. This hydroxypropyl derivative may be compounded with other ingredients in order to improve coating pliability, to enable speed setting upon cooling or drying, and to control rate or resolubility (Jokay et al., 1967). Comprehensive information on structural and mechanical properties of hydroxypropylated pea starch films as influenced by casting conditions are presented by Lafargue et al. (2007).

4.2.3 Pectins and derivatives

Pectins are a family of complex polysaccharides present in all plant cell walls. The backbone of pectin contains (1→4)-α-D-galacturonic acid units interrupted by single (1→2)-α-L-rhamnose residues (Ridley et al., 2001). Actually, three pectic polysaccharides (homogalacturonan, rhamnogalacturonan, and substituted galacturonans) have been isolated from primary cell walls and structurally characterized. The carboxyl groups of the galacturonic acid units are partly esterified by methyl groups. The degree of esterification (DE), corresponding to the number of methyl-esterified galacturonic acid residues versus the total number of galacturonic acid units, as well as the distribution of the methylated groups along the polysaccharide chains exert a determinant influence on the solubility and gelation properties of pectins (Axelos et al., 1991).

Currently, apple pomace and citrus peel are the main sources of commercial pectin used as film-forming agents, while other potentially valuable sources remain unused because of certain undesirable structural properties. As extracted from these sources, pectins have a typical DE of 55% to 75%. Pectins with lower DE are produced by controlled de-esterification during the production process. High-methoxyl (HM) pectins require a minimum amount of soluble solids (55% to 80%) and a low pH (2.5 to 3.5) in order to form gels (Lopes da Silva and Rao, 2006). Unlike most polysaccharide gels, the structure of HM pectin-sugar gels is considered to be irreversible on heating (gels do not melt). As the DE of HM pectin decreases, a lower pH is required for gelation. In low-methoxyl (LM) pectins, the mechanism for gel formation is based on ionic mediation through divalent cations, especially calcium, in a similar way to the *egg-box* model proposed for alginates (Figure 4.1).

For both HM and LM pectins, the gel modulus is determined by the number of effective junction zones formed between pectin chains. Therefore, the higher the molecular weight and the concentration, the stronger is the gel. Pectin gels have high water vapor permeabilities; thus, they can prevent dehydration only by acting as sacrificial agents.

4.2.4 Seaweed extracts

4.2.4.1 Alginates

Alginate is a generic term for the salts and derivates of alginic acid. Alginic acid is a high molecular weight polysaccharide consisting of varying proportions of D-mannuronic acid and L-guluronic acid; the variation being dependent principally upon the seaweed species from which the alginic acid was isolated (King, 1983). This polysaccharide is considered a family of unbranched binary copolymers

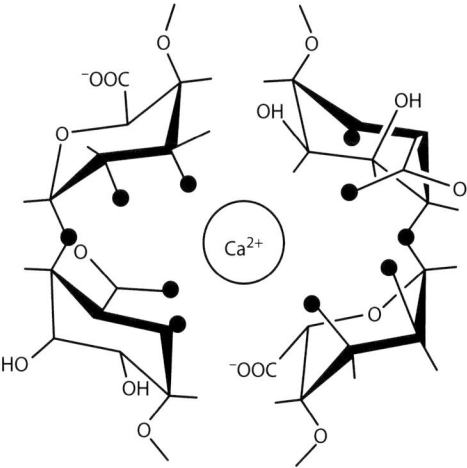

Figure 4.1 The egg-box model for the gelation of pectin-calcium systems.

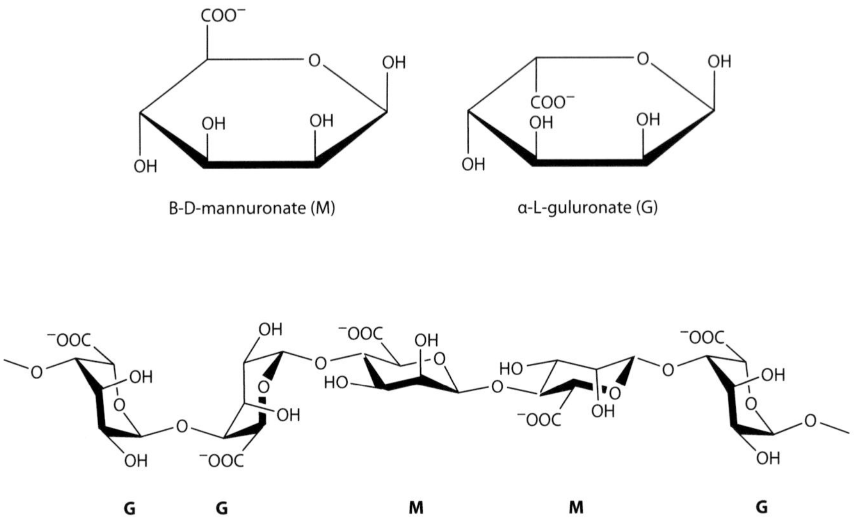

β-D-mannuronate (M)

α-L-guluronate (G)

G G M M G

Figure 4.2 Structure and conformation of alginates: alginate monomers and chain conformation.

of (1→4)-linked β-D-mannuronic acid (M) and α-L-guluronic acid (G) residues of widely varying composition and sequential structure (Figure 4.2).

Commercial alginates are produced mainly from brown algae *Macrocystis pyrifera, Laminaria hyberborea, Laminaria digitata, Ascophyllum nodosum, Laminaria japonica, Edonia maxima, Lessonia nigrescens, Durvillea Antarctica,* and *Sargassum* spp. (Draget, 2005). *A. vinelandii* (soil bacteria) has been evaluated as a source for industrial application, but at present all commercial alginates are extracted from algal sources (Draget et al., 2006).

Alginates' gel-forming properties are mainly due to their capacity to bind a number of divalent ions like calcium and are strongly correlated with the proportion and length of the guluronic acid blocks (G-blocks) in their polymeric chains. A very rapid and irreversible binding reaction of multivalent cations is typical for alginates; a direct mixing of these two components therefore rarely produces homogeneous gels.

Following the addition of calcium ions, alginate undergoes conformational changes, giving rise to the well-known egg-box model of alginate gelation. This is based on chain dimerization and eventually further aggregation of the dimers (Moe, 1995).

The industrial applications of alginates are linked to its ability to retain water, and its film-forming, gelling, viscosifying, and stabilizing properties. Also, alginates possess good film-forming properties, which make them particularly useful in food applications. In addition to food uses, the other major outlets are pharmaceutical application, technical uses as print paste in the textile industry, and biotechnological applications. The last are based either on specific biological effects of the alginate molecule or on its unique, gentle, and almost temperature-independent sol/gel transition in the presence of multivalent cations, which makes alginate highly suitable as an immobilization matrix for living cells (Draget et al., 2005). Alginates are used as food additives to improve, modify, and stabilize the texture of foods. This is valid for such properties as viscosity enhancement, gel-forming ability, and stabilization of aqueous mixtures, dispersions, and emulsions. In addition, alginate gels can be heated without melting and for this reason are used in baking creams.

Because of its simplicity, alginate also can be used in restructured food based on Ca-alginate gels (pimento olive fillings, onion rings, crabsticks). Restructuring of foods is based on binding together a flaked, sectioned, chunked, or milled foodstuff to make it resemble the original (Draget et al., 2006). For applications in jams, jellies, fruit fillings, and so forth, the synergistic gelation between alginates high in guluronate and highly esterified pectins may be utilized (Toft et al., 1986). In effect, the alginate/pectin system can give thermoreversible gels in contrast to the purely ionically cross-linked alginate gels. This gel structure is almost independent of sugar content, in contrast to pectin gels, and therefore may be used in low caloric products. Several researchers have reported that mixed gel systems containing alginates and pectins give firm and cuttable gels at low pH (Mancini and McHugh, 2000).

4.2.4.2 Carrageenans

Carrageenan is a generic name for a family of natural water-soluble, sulfated galactans that are isolated from red seaweeds, namely from various species of the *Rhodophyta* (De Ruiter and Rudolph, 1997). Carrageenan is a complex mixture of several polysaccharides. The three main commercial carrageenans are ι-, κ-, and λ-carrageenan, whose names specify the major substitution pattern present in the galactan backbone. The structure of carrageenans is mostly composed of repeating dimers of an α(1→4)–linked D-galactopyranose or 3,6-anydro-D-galactopyranose residue and a β(1→3)–linked D-galactopyranose residue. Sulfate hemiester substituents and, to a lesser extent, methoxy and pyruvate groups, are covalently coupled to the carbon atoms C-2, C-4, or C-6 of individual galactose residues. κ-Carrageenan contains the lowest number of sulfate groups and the highest concentrations of the 3,5-anhydro-α-D-galactopyranosyl units, whereas λ-carrageenan has the highest amount of sulfate groups, thus resulting in highly negatively charged macromolecules. ι-Carrageenan differs from κ-carrageenan with an additional sulfate group at the 2 position (van de Velde, 2008). λ-Carrageenan does not form gels because typical helical conformation is not formed by this structure but can be used in combination with other gelling carrageenans to reduce brittleness and decrease the tendency for syneresis. The strongest gels form with K^+ and κ-carrageenan, and with Ca^{2+} and ι-carrageenan at polysaccharide levels ranging from 0.5% to 1%, syneresis occurring with the former combination (Piculell, 2006). Commercial carrageenans are mixtures of these three fractions, and gelling properties may substantially vary depending on the selected type. A well-studied gelling carrageenan is furcellaran, obtained from *Furcellaria lumbricalis*. Furcellaran is currently considered to be a type of κ-carrageenan, although it contains other repeating polysaccharide structures (Friedenthal et al., 2001).

Carrageenans, which are film-formers, are used mainly in the food industry as texturizing agents (Bixler, 1996; De Ruiter and Rudolph, 1997). In dairy products, gels form with κ-carrageenan at far lower polysaccharide concentration than in the absence of caseins due to the strongly anionic character of the polysaccharide

(Piculell, 2006). However, carrageenans can also interact with substances other than caseins in processed foods (e.g., with sodium or calcium phosphates to provide improved texture) (de Vries, 2002). Hence, potential use of carrageenans covers applications as coating agents that control transfer of moisture, gases, flavors, and lipids in diverse food systems, but also manufacturing soft capsules and especially nongelating capsules (Karbowiak et al., 2006). Another promising emerging application of carrageenan-based coatings is their use as antimicrobial agent carriers (Choi et al., 2005). For most of these applications, plasticizers are needed in order to improve their mechanical and structural properties. The rigid gels formed from the κ-carrageenan may be turned more elastic by incorporating other hydrocolloids. Locust bean gum forms an outstanding cooperative association with the double helix structure of carrageenans, although this may give gels with poor clarity.

4.2.4.3 Agar

Agar is the word referring to members of a class of galactan polysaccharides that occur as intercellular matrix material in numerous species of red seaweeds (Selby and Whistler, 1993). Agars are linear polysaccharides made up of disaccharides composed of ß(1,3)- and α(1,4)-linked galactose residues. The sol-gel transitions of agars and carrageenans typically exhibit hysteresis as the melting temperatures are higher than the gelling temperatures. Agar gels and films are subject to syneresis, or separation of water from the gel on aging. This is attributed to aggregation of double helices, resulting in contraction of the polymer network, and thus decreasing the interstitial space available for holding water (Whytock and Finch, 1991). Therefore, it is best known as a culture medium and is not used to a great extent in food applications. Syneresis is less evident for agars high in ester sulfate, which confers hydrophilicity. For food applications, multicomponent gel systems incorporating agar are frequently formulated. Sucrose incorporation into the systems can increase gel strength, as well as the addition of polyols such as ethylene glycol and glycerol (Nakahama, 1966; Watase et al., 1992). The incorporation of other gums into an agar gel system generally does not alter its gelling properties in any useful way. Locust bean gum is capable of enhancing the strength of agar gels and films, which are typically brittle (Deuel and Neukom, 1954).

4.2.5 Exudate gums

4.2.5.1 Acacia gums

The terms *gum arabic* and *Acacia gum* have been traditionally used as synonyms, although only the gums from *Acacia senegal* and *A. seyal* are approved as additives for food use (Williams and Phillips, 2000). *Acacia gums* are the secretions or exudates from the stems of *Acacia* species in response to injury to the plant tissue. The gums are hand sorted and often converted to a spray-dried powder after being dissolved in hot water and clarified, preferably by centrifugation (Whistler, 1993). In general, structural variations among gums from different *Acacia* species are not large but could be relevant in the development of other gums for specific purposes.

Though, the extremely high cost of developing new food additives and having them approved for public use is a serious deterrent to the wider exploitation of these gums (Williams et al., 2006). Among the different gums from *Acacia* species, only gum arabic is approved for food use in most countries. Gum arabic is a neutral or slightly acidic salt of a complex polysaccharide containing calcium, magnesium, and potassium ions. The gum is a complex and variable structure of arabinogalactan oligosaccharides, also containing glucuronic acid and rhamnose units, and a small amount of glycoproteins.

Gum arabic is the least viscous and the most soluble of the hydrocolloids. Therefore, its use is mostly based upon its action as a protective colloid or stabilizer and the adhesiveness of its water solutions. It produces stable emulsions with most oils over a wide pH range. Electrolytes generally reduce the viscosity of the gum arabic solutions. Gum arabic solutions are subject to bacterial attack and can be preserved using 0.1% concentrations of sodium benzoate or benzoic acid.

One of the major food applications of gum arabic is encapsulation of volatile compounds for aroma and flavor retention. This is not only because of its film-forming properties but is also due to the existence of hydrophobic regions in its structure. This hydrophobicity gives rise to emulsifying power, which is exploited in the fixation of flavors and aromas by microencapsulation. Oil-in-water emulsions containing flavor compounds and gum arabic are spray-dried, so that water is removed rapidly with exposure to a minimum of heat, resulting in little damage even to heat-sensitive compounds. Dry mixes for puddings, cakes, soups, or beverages often contain flavor and aroma compounds encapsulated within a protective film of gum arabic. Other applications reported for encapsulation by gum arabic include the production of clouding agents for beverages by spray-drying emulsions containing gum arabic and hydrogenated vegetable oils. Protective coatings for oil-soluble vitamins may also be prepared using this method (Glicksman, 1983).

4.2.5.2 Gum tragacanth

Gum tragacanth, the exudate from *Astragalus* spp., is fundamentally a mixture of two polysaccharide components. About 30% to 40% of the gum is a typical branched water-soluble arabinogalactan (tragacanthin) including galactose units and terminal arabinose units. The bulk of the gum is a water-insoluble pectic polymer (tragacanthic acid) consisting of a (1,4)-linked α-D-galacturonan substituted by xylose groups or short side chains in which fucose and galactose units are attached to the xylose (Williams et al., 2006).

As noted above, the tragacanthin fraction is water soluble, which is probably due to its highly branched structure. On the other hand, the gelling properties of the tragacanthic acid water-insoluble fraction could be linked to the calcium that can be found associated in the structure, among other ions. Gum tragacanth swells rapidly in either cold or hot water. Aqueous dispersions of more than 2% of the gum form thick pastes having the texture of soft gels (Glicksman, 1983). At room temperature, viscosity of gum tragacanth dispersions remains almost unchanged up to pH 10 and decreases by only 30% as the pH falls to 2 (Stauffer and Andon, 1975). It also exhibits an unusual stability to microbial attack. When preservatives are needed, benzoic acid or a combination of the methyl and propyl esters of p-hydroxybenzoic acid have been reported as effective (Meer et al., 1973). Because of this unusual stability toward both microbial attack and acidic conditions, it has been extensively used for maintaining viscosity in low acidic food products such as salad dressings and sauces. Nevertheless, xanthan gum is commonly used instead. Furthermore, the gum is an excellent emulsion stabilizer, not only due to its water thickening properties, but also because it substantially lowers the interfacial tension between oil and water.

4.2.5.3 Gum ghatti

Gum ghatti, or Indian gum, is an amorphous translucent exudate obtained from the species *Anogeissus latifolia*, a deciduous tree of the family Combretaceae, which is native to India (Glicksman, 1983). The GRAS status of gum ghatti is tempered by very low maximum usage allowances. Nevertheless, the cost is favorable in comparison with gum arabic.

The gum is a water-soluble, complex polysaccharide in which complex arrays of neutral sugars (galactopyranosyl, arabinofuranosyl, arabinopyranosyl) and glucuronic acid are attached to a molecular core of alternating ß-D-glucuronic acid and D-mannopyranosyl residues (Williams et al., 2006). The gum is not completely soluble in water at concentrations of 5% or more but swells to form colloidal dispersions. As a whole, the gum is a nongelling polysaccharide but allows dispersion in hot or cold water to give a solution for which maximum viscosity develops at pH between 5 and 7. However, the viscosity of dispersions is very variable from one gum to another because of the presence of two fractions, one soluble in cold water and the other forming a dispersible gel, which is more viscous than the soluble fraction. The proportion of the latter fraction, which has been found in amounts varying between 8% and 23% (w/w), strongly conditions the viscosity and gelling properties of the solutions. Gum ghatti closely resembles the viscosity and emulsifying properties of gum arabic. It has excellent emulsifying properties, and its high viscosity in solution makes it an outstanding stabilizer for dense emulsions and suspensions. Films formed from ghatti as the only polymer are relatively water soluble and brittle and are not considered very useful.

4.2.5.4 Gum karaya

Gum karaya is the exudate obtained from trees of the genus *Sterculia* and *Cochlospermum*, which are found mostly in central and northern India. Structurally, gum karaya is a partially acetylated polysaccharide of the substituted rhamnogalacturonoglycan (pectic) type. The composition of the gum may differ depending on the species, but the presence of an uninterrupted acetylated chain with galactose units occurring only in side chains has been demonstrated (Williams et al., 2006). Gum karaya is the least soluble of the commercial exudates gums but swells to form colloidal dispersions of high viscosity at concentrations up to 5%. The rate of swelling depends on the particle size of the gum. Thus, finely ground powders hydrate more rapidly than more coarsely ground samples (Glicksman, 1983). At concentrations above 2% to 3%, very thick, nonflowing pastes resembling spreadable gels are formed.

In the past, the gum has been used as an emulsifier, stabilizer, or binder in frozen desserts, dairy products, salad dressing, or meat products. It is currently used in foods such as mayonnaises and dressings, but its main use is in medical and pharmaceutical applications such as adhesives for colostomy bags and dentures or as a bulk laxative (Williams et al., 2006). Gum karaya forms smooth films that can be plasticized with compounds such as glycols to reduce brittleness.

4.2.6 Seed gums

The cell walls of the storage tissues of many seeds contain large deposits of polysaccharides, which are mobilized after germination and are similar in structure to individual hemicellulosic or pectic components of the normal plant cell wall matrix (Gidley and Reid, 2006). Most of these compounds are usually used in the food industry as thickeners or stabilizers or to impart special rheologies to food products. The most important polysaccharides are the galactomannans, guar gum and locust bean gum, both of which are film formers. Another of importance is xyloglucans, namely, tamarind seed polysaccharide.

The endosperm of palm seeds, such as date, and of leguminous seeds have thickened cell walls with massive deposits of mannans and galactomannans, respectively (Reid, 1985). Galactomannans from locust bean or carob, guar, and tara are widely used commercially. Carob or locust bean gum is processed from the seeds of *Ceratonia siliqua*, which is indigenous to Mediterranean countries, whereas guar gum is obtained from the ground endosperm of *Cyamopsis tetragonolobus*, a plant that is cultivated commercially mostly in India and Pakistan for human and animal consumption. Tara (*Caesalpinia spinosa*) is a small leguminous tree or thorny shrub native to Peru. Galactomannans are similar in overall structure to mannans but are largely substituted with galactose. These galactomannans are made up of a (1→4)-β-D-mannopyranosyl backbone having single-unit (1→6)-α-D-galactopyranosyl side-chain residues (Dey, 1978). Generally, the galactose to mannose (G–M) ratio varies roughly from 1:2 to 1:4, respectively, for the three gums. The distribution of galactose units on the mannan backbone is not yet fully understood, but it is believed to be important for the functional properties of these polysaccharides (Savitha et al., 2006).

Locust bean gum is insoluble in cold water but totally soluble when warmed at 80°C, with a peak viscosity when the gum is dispersed at 95°C and then cooled. Guar gum disperses and swells almost completely in cold or hot water to form a sol of high viscosity. Both carob and guar gums are insoluble in organic solvents but compatible with most other plant hydrocolloids, chemically modified starches or cellulose, and water-soluble proteins. Locust bean gum presents synergy with β-carrageenan and xanthan gums, forming elastic gels that are very cohesive and relatively free from syneresis (Gidley and Reid, 2006), which does not occur with guar gum. By interacting in Maillard-type reactions, galactomannans have been shown to improve the quality of proteins in various ways, as antioxidants, emulsifiers, or even as bactericidal agents (when conjugated with lysozyme). They also can be used as drug deliverers, thickeners or viscosity modifiers, binders of free water, suspending agents or stabilizers for a wide array of food applications, including dairy, bakery, and meat products, sauces, and beverages.

4.2.7 Microbial fermentation gums

4.2.7.1 Xanthan gum

Xanthan gum is the major bacterial polysaccharide used by the food industry, obtained from the secretion of a wide range of bacteria of the genus *Xanthomonas*. It is commercially produced from the microorganism *Xanthomonas campestris*. Xanthan gum is a pentasaccharide, and its structure is identical to that of cellulose. Trisaccharide side chains contain a D-glucuronic acid unit between two D-mannose units linked at the O-3 position of every other glucose residue in the main chain (García-Ochoa et al., 2000).

Xanthan gum is readily dispersed in water and has film-forming properties. Aqueous solutions of xanthan gum are thixotropic, thus exhibiting high viscosity at low shear rates. Because of its unique molecular conformation and structure, it is regarded as the most acid-stable gum. It is as well compatible with salts at high concentrations and tolerates well the incorporation of water-miscible solvents. Although xanthan gum alone forms weak gels, mixtures of xanthan gum with certain galactomannans such as locust bean gum or tara gum lead to interesting thermoreversible gels. Xanthan can also be gelled by the addition of trivalent cations or borate anions (Morris, 2006). The interaction with other polysaccharides offers a potential to improve the release of compounds in protein systems. As a polyanion, xanthan gum forms macromolecular complexes with polycationic compounds such as chitosan (Dumitriu and Chornet, 1998), thus constituting a porous network with fibrillar structures that allows the stabilization of proteins as well as the lodging of enzymes and their substrates, or any substance with certain activity that is wanted immobilized. Xanthan gum gels are more efficient than alginate and carrageenan gels in recovering their porous structure after compression. Therefore, gels have increased elasticity in comparison to other polymers and are very suitable to be used for the controlled release of drugs and other bioactive compounds (Coviello et al., 2007).

The unexpected rheological behavior of aqueous xanthan solutions has led to a wide array of applications for a number of important reasons, including emulsion stabilization, temperature stability, compatibility with food ingredients, and its pseudoplastic rheological properties. Because of its thickening properties, it has been used as a dispersing agent and stabilizer of emulsions and suspensions in pharmaceutical formulations, cosmetics, or agricultural products (García-Ochoa et al., 2000).

4.2.7.2 Gellan gum and related exopolysaccharides

Gellan gum, a film former, is a comparatively new gum produced by aerobic submerged fermentation of a broth culture by the Gram-negative bacterium *Sphingomonas paucimobilis* (formerly *Pseudomoas elodea*) (Morris, 2006). The bacterium was first isolated from a water plant (elodea) by Kelco, while trying to identify new polysaccharide bacterial sources. The broth is pasteurized to kill bacteria, and gellan gum is isolated by alcohol precipitation. The gum has a linear, anionic heteropolysaccharide based on a tetrasaccharide repeating unit (O'Neill et al., 1983), which consists of (1→3)-ß-D-glucopyranosyl, (1→4)-ß-D-glucopyranosyluronic acid, (1→4)-ß-D-glucopyranosyl, and (1→4)-α-L-rhamnopyranosyl units with C-2 linked L-glycerate and about 50% C-6 linked acetated substituents. Extraction conditions can greatly affect the degree of esterification of the final product. Other related biopolymers, such as welan or rhamsan gums, possess a similar linear structure in which there is at least an identical trisaccharide sequence with the same anomeric configuration as in gellan. The main structural differences are in the nature and location of the monosaccharide and disaccharide side-chain grouping and in some of the polymers, the presence of L-mannose as an alternative to L-rhamnose in the backbone chain of the polymer (Banik et al., 2000). A reevaluation of the biochemical and physiological properties of the bacteria producing these gellan-type polysaccharides has shown that all of them belong to genus *Sphingomonas*.

Gellan gels are formed by dispersing the gum in water, heating, adding cross-linking cations, and then cooling to set. The degree of acylation of the polysaccharide and the type and concentration of cations affect the performance of the gels formed (Crescenzi et al., 1990). The mechanism of gelation involves the formation of a three-dimensional network, which in turn is formed by double helical junction segments (Takahashi et al., 2004). Partial or complete de-acylation of gellan strongly affects the mechanical properties of the gels. Acylated gellan gels are soft and elastic, whereas de-acylated gels are hard and brittle (Moorhouse et al., 1981). On the other hand, several studies considering various monovalent and divalent cations on the mechanical properties of gellan gum gels suggest that gelation may be sensitive to the cation type as well as cation valency (Morris, 1991; Singh and Kim, 2005). Generally, at equivalent ionic strength, divalent cations give rise to stronger gels with higher elastic moduli.

4.2.8 Chitosan

Chitosan is obtained by deacetylation of chitin, which is the structural element in the exoskeleton of crustaceans as well as a cell wall constituent of fungi and green algae. Chitin and chitosan are two of the most abundant polymers in nature after cellulose and its derivatives (Sandford and Hutchings, 1987). Chitin is a ß-1,4 linked linear polymer of 2-acetamido-2-deoxy-D-glucopyranosyl residues (BeMiller, 1965). Chitin does not exist alone but is found in close association with calcium carbonate and protein and other organic substances, which make the compound stable to most reagents. It is known that heterogeneous conditions during chitin deacetylation provide a block-wise distribution of acetyl groups, whereas under homogeneous conditions a random distribution of acetyl groups appears in chitosan (Harish Prashanth and Tharanathan, 2007). Net cations as well as presence of reactive functional groups allows manipulation of the molecule for preparing a broad spectrum of derivatives for specific end-use applications in diversified areas.

Chitosan has been extensively studied for coating applications because of its film-forming properties. Chitosan films do not dissolve in water and exhibit good wet tensile strength. The methods of preparation of chitosan gels can be broadly divided into four groups: solvent evaporation, neutralization, cross-linking, and ionotropic gelation. The first of the methods is mainly used for the preparation of films. A solution of chitosan in organic acid is cast onto a plate and allowed to dry, if possible, at elevated temperature. Upon drying, the film is usually neutralized with a dilute NaOH solution and cross-linked to avoid disintegration in acidic solutions (Krajewska, 2004). Film formation also may be followed by precipitation after increasing the pH of a chitosan solution with an alkali. Another possibility is the use of a cross-linking agent such as glutaraldehyde in order to facilitate gel formation. So far, no simple ionic and non-toxic cross-linking agents have been reported providing reproducible chitosan gels at low concentrations, such as calcium ions in the case of alginates. Eventually, by virtue of the attraction of oppositely charged molecules, chitosan, owing to its cationic polyelectrolyte nature, spontaneously forms water-insoluble complexes with anionic polyelectrolytes such as alginate, carrageenan, xanthan, various polyphosphates, and organic sulfates. This method is frequently used for enzyme immobilization into gel beads, which is achieved by adding an anionic polyelectrolyte solution containing the enzyme into an acidic chitosan solution.

Chitosan can form semipermeable coatings, thereby delaying ripening and decreasing water migration. It also presents some positive effects against fungi and phytopathogens (El Ghaouth et al., 1991; Hirano and Nagao, 1989).

4.3 Applications of polysaccharide coatings

Because of the large number of different applications, polysaccharide-based films and coatings are regarded as the most versatile. These applications span from traditional technical applications in foods to biomedicine (Table 4.1).

Polysaccharide films and coatings may be used to preserve the quality of several food commodities. High-fat meat and fish products, such as sausages, jerky, and fillets, can be protected from oxidation by coating them with appropriate polysaccharide coatings. The oxygen and moisture barrier properties of these coatings can protect fresh fruit and vegetables from dehydration and, in some cases, even retard their respiration rate. Polysaccharide coatings can also be used to prevent oxidation of lipid ingredients, or to reduce loss of food colors and flavors. Some of the most common applications of edible coatings for improving the quality and extending the shelf life of foods, as well as its use with other purposes are discussed next.

4.3.1 Fruits and vegetables

When used to coat fresh-cut fruits and vegetables, edible films may reduce the deleterious effects concomitant with minimal processing. The first documented examples of the use of polysaccharide edible coatings on fresh-cut fruits was reported by Bryan (1972), who observed that a carrageenan-based coating applied on cut grapefruit halves resulted in less shrinkage, leakage, or deterioration of taste after 2 weeks of storage at 4°C. In the last years, polysaccharide edible coatings have been evaluated in order to improve the quality and shelf life of some fresh-cut produce. Recently, Eissa (2007) suggested that the use of a coating containing chitosan (2%) could be beneficial for extending shelf life, maintaining quality, and, to some extent, controlling decay of fresh-cut mushroom (Eissa 2007). Likewise, Pen and Jiang (2003) reported that chitosan retarded the development of browning, maintained sensory quality, and retained levels of total soluble solids, acidity, and ascorbic acid

Table 4.1 Main food applications of polysaccharide coatings and their functions

Foods	Coatings	Function of coating	Reference
Fresh-cut fruits: *Apple, pear, strawberry, papaya, mango, red pitayas*	Carrageenan, alginate, gellan gum, apple puree + pectin, maltodextrin, cellulose derivatives, chitosan	Antioxidant, antimicrobial, and nutraceutical carrier; gases and moisture barrier; antifungal protection; color, flavor, and texture improvement	Assis and Pessoa, 2004; Baldwin et al., 1996; Brancoli and Barbosa-Cánovas, 2000; Chien et al., 2007a, 2007b; Del-Valle et al., 2005; Lee et al., 2003; McHugh and Senesi, 2000; Olivas et al., 2003; Rojas-Graü et al., 2007a, 2007b; Tapia et al., 2007
Whole fruits and vegetables: *Apple, grape, cherry, plum, strawberry, baby carrot, mango, banana, kiwifruit, zucchini*	Cellulose derivatives Aloe vera gel, chitosan, cactus-mucilage, xanthan gum, Semperfresh™, pullulan	Gases, lipid and moisture barrier; antifungal protection; color, flavor, and texture improvement; retention of functional compounds; antimicrobial and functional compounds carrier	Baldwin et al., 1999; Chien et al., 2006, 2007c; Conforti and Totty, 2007; Diab et al., 2001; Han et al., 2004; Kaynas and Ozelkok, 1999; Martínez-Romero et al., 2006; Mei et al., 2002; Serrano et al., 2006; Valverde et al., 2005; Yaman and Bayoindirli, 2002
Meats, poultry, fish and seafood	Carrageenan, alginate, cellulose derivatives, chitosan, pullulan	Oxygen, lipid and moisture barrier; texture improvement; antioxidant and antimicrobials carrier; bacterial growth prevention	Hargens-Madsen, 1995; Holownia et al., 2000; Jeon et al., 2002; Ouattara et al., 2000; Oussalah et al., 2006; Sathivel, 2005; Wu et al., 2001
Nuts and cereals	Starch, dextrin, cellulose derivatives	Adhesive of seasonings, oxygen and moisture barrier	Laohakunjit and Kerdchoechuen, 2007; Noznick and Bundus, 1976; Roudaut et al., 2002
Other applications: *Pharmaceutical industry*	Pectin, alginate, chitosan	Drug release, carrier, cell encapsulation	Efentakis and Buckton, 2002; Liew et al., 2006; Sriamornsak and Kennedy, 2007; Sriamornsak and Nunthanid, 1999; Sriamornsak and Puttipipatkhachorn, 2004

in sliced Chinese water chestnuts. Olivas et al. (2007) reported that alginate coatings extended the shelf life of fresh-cut 'Gala' apples without causing anaerobic respiration (Olivas et al. 2007). McHugh and Senesi (1999) developed a novel method (apple wraps) to extend the shelf life and improve the quality of fresh-cut apples. These wraps were made from apple puree containing various concentrations of fatty acids, fatty alcohols, beeswax, vegetal oil, and high methoxyl pectin. Fruit and vegetable wraps can be used to extend the shelf life of fresh-cut fruits and vegetables, as well as to enhance their nutritional value and increase their consumer appeal. McHugh and Senesi (2000) have also suggested that fruit and vegetable wraps can be used as barriers for other food systems, such as nuts, baked goods, and confectionary products (McHugh and Senesi, 2000).

On the other hand, the use of several polysaccharide edible coatings to deliver functional compounds in order to improve the safety, nutritional, and quality properties of fresh-cut fruits has been proposed in recent studies. Tapia et al. (2007) developed the first edible films for probiotic coatings on fresh-cut apple and papaya, observing that both fruits were successfully coated with alginate or gellan film-forming solutions containing viable bifidobacteria. Values higher than 10^6 cfu/g *Bifidobacterium lactis Bb*-12 were maintained for 10 days during refrigerated storage of both papaya and apple pieces, thus demonstrating the feasibility of these polysaccharide coatings to carry and support viable probiotics on fresh-cut produce. Rojas-Graü et al. (2007b) applied alginate- and gellan-based coatings to fresh-cut 'Fuji' apples, proving that these coatings were good carriers of antioxidant agents such as cysteine and glutathione, which helped to maintain color of cut fruits during storage. Furthermore, the incorporation of antimicrobial agents into edible coatings is gaining importance as a way of reducing the deleterious effects imposed by minimal processing on fresh-cut produce. In this way, Rojas-Graü et al. (2007a) combined the efficacy of polysaccharide edible coatings (alginate and gellan gum) with the antimicrobial effect of plant essential oils (lemongrass, oregano oil, and vanillin) to prolong shelf life of fresh-cut apples.

In the case of whole fruits and vegetables, edible coatings are used as a protective barrier to reduce respiration and transpiration rates through surfaces, retard microbial growth and color changes, prevent weight loss, and improve the appearance, texture, and quality of fruits (Kester and Fennema, 1986). For instance, cellulose-based edible coatings have been used to delay ripening in some climacteric fruits such as mangoes, papayas, and bananas. Furthermore, the application of the same formulation on sliced mushrooms has been shown to significantly reduce enzymatic browning (Nisperos-Carriedo et al., 1991). Conforti and Totty (2007) developed three coatings using a vegetable oil with various combinations of hydrocolloids (maltodextrin, locust bean gum, gum arabic, algin, and sodium CMC) to be applied on 'Golden Delicious' apples. They observed that coated apples maintained low internal oxygen levels, had very little starch breakdown, minimum flesh color development, and better maintained firmness, crispness, and juiciness throughout the storage period compared with noncoated apples. Garcia et al. (2000) also found that the incorporation of plasticizers and sunflower oil into polysaccharide coatings significantly improved the loss of moisture in apples.

Chitosan-based coatings have been studied for their application on some fresh commodities, including peaches (Li and Yu, 2001), fresh berries (El Ghaouth et al.,

1991; Garcia et al., 1998; Han et al., 2004; Hernández-Muñoz et al., 2006; Park et al., 2005; Vargas et al., 2006; Zhang and Quantick, 1998), mangoes (Chien et al., 2007a; Kittur et al., 2001; Wang et al., 2007), table grapefruits (Meng et al., 2008), banana (Kittur et al., 2001), citrus (Chien and Chou, 2006; Chien et al., 2007c), carrots (Durango et al., 2006), and lettuce (Devlieghere et al., 2004). In every case, reductions in the respiration rate and ethylene production of the produce were observed. In addition, coatings were effective in controlling decay and firmness loss.

Others have proposed the use of a semipermeable coating containing CMC and sucrose fatty acid esters for extending shelf life of fresh fruits and vegetables (Lowings and Cutts, 1982). These coatings reduce oxygen uptake without causing an equivalent increase in carbon dioxide level in internal atmospheres of fruit or vegetal tissues, preventing the occurrence of anaerobic respiration. As a result, a coating manufactured by Agricoat Industries Ltd. as Semperfresh™ is formulated with sucrose esters of fatty acids, mono- and diglycerides, and the sodium salt of CMC. It forms an invisible coating that is odorless and tasteless, and creates a barrier that is differentially permeable to oxygen and carbon dioxide. Semperfresh coating has been applied successfully to extend the shelf life and preserve important flavor compounds of fresh fruits and vegetables, such as apples (Bauchot et al., 1995; Drake et al., 1987; Soria et al., 1999; Sumnu and Bayindirli, 1995), bananas (Banks, 1984), mangoes (Carrillo-Lopez et al., 2000; Dhalla and Hanson, 1988), and tomatoes (Tasdelen and Bayindirli, 1998). Yaman and Bayindirli (2001, 2002) studied the effect of Semperfresh coating on the shelf life and quality of fresh sweet cherries. They demonstrated that coating application reduced weight loss and increased firmness, ascorbic acid content, titratable acidity, and skin color of cherries during storage, thus extending shelf life by 21% at 30°C and by 26% at 0°C without perceptible losses in quality. Yurdugul (2005) observed that a triple combination of Semperfresh, ascorbic acid, and cold storage provided high microbial, chemical, and sensorial qualities as well as an extension in the shelf life of quinces, a native fruit of Persia. Also, application of Semperfresh in fully matured tomato fruits delayed ripening and decreased physiological weight loss (Tasdelen and Bayindirli, 1998). Besides, Nova Chem (Halifax, Canada) developed Nutrisave™, a commercial formulation containing chitosan with antimicrobial, barrier, and film-forming properties. The coating was successfully applied to reduce respiration rates and water loss of whole apples and pears by creating special conditions inside the coated products (Elson et al., 1985).

Alternatively, some studies have reported the use of polysaccharide-based edible coatings as carriers of functional ingredients for whole fruits and vegetables. Several researchers have endeavored to incorporate calcium (Han et al., 2004; Hernández-Muñoz et al., 2006; Mei et al., 2002), vitamin E (Han et al., 2005; Mei et al., 2002), and potassium (Park et al., 2005), or oleic acid (Vargas et al., 2006) into polysaccharide coating formulations to prolong shelf life as well as to enhance fruit quality and nutritional value. For instance, Mei et al. (2002) developed xanthan gum coatings containing high concentrations of calcium and vitamin E with the purpose of enhancing nutritional and sensory qualities of fresh baby carrots. The results of this study showed that calcium and vitamin E contents in the coated carrots increased from 2.6% to 6.6% and from 0 to about 67% of the Dietary Reference Intake (DRI) values per serving (85 g), respectively, without affecting the fresh aroma, fresh flavor, sweetness, crispness, and β-carotene level. Additionally, baby carrots coated with xanthan gum exhibited less white surface discoloration and greater orange color intensity ratings than uncoated samples. In another study, a chitosan-based coating was shown to perform well as carrier of vitamin E and calcium in order to improve storability and nutritional properties of fresh and frozen strawberries and red raspberries (Han et al., 2004). From the point of view of storability, the drip loss due to freezing–thawing in chitosan-coated raspberries was reduced by at least 24%, whereas firmness consistently rose by about 25% in comparison with uncoated fruits.

Moreover, coatings containing antimicrobials can inhibit microbial growth and extend the shelf life of produce. Franssen and Krochta (2003) significantly reduced the populations of *Salmonella Montevideo* on tomatoes when incorporating citric, sorbic, or acetic acids in HPMC coatings. The coating alone resulted in a 2 log CFU/g reduction in the counts of the pathogenic strain, but addition of 0.4% sorbic acid led to a significantly higher inactivation.

4.3.2 Meats, poultry, fish, and seafood

Meat and meat products are highly susceptible to lipid oxidation, which leads to rapid development of rancid or warmed-over flavor. Certain polysaccharide films and coatings may provide effective protection against oxidation of lipids and other food components on foods of animal origin. Gennadios et al. (1997) described some potential benefits of using edible coatings on meat, poultry, and fish products, which provide moisture loss reduction during storage of fresh or frozen meats; prevention of juice dripping when these products are packaged in retail plastic trays; decrease in myoglobin oxidation in meats; reduction of the load of spoilage and pathogenic microorganisms on meat, poultry, and seafood products; delay of meat rancidity, discoloration, and microbial growth when antimicrobials or antioxidant compounds are incorporated into the polymer matrix; and even improvement of the nutritional quality of the coated foods.

From an industrial point of view, edible films and coatings can increase the effectiveness of some food processing unit operations. The ability of some water-soluble polysaccharides to form thermally induced gelatinous coatings can be used to reduce oil absorption during frying. Dziezak (1991) reported that cellulose-based films and coatings reduce moisture loss as well as the amount of oil absorbed by fried foods. For instance, Balasubramaniam et al. (1997) demonstrated that fat absorption was substantially reduced (33.7%) and moisture retention significantly improved (16.4%) when meatballs prepared from ground chicken breast were fried after coating with HPMC. Holownia et al. (2000) observed that incorporating HPMC into a breading mix for whole chicken strips resulted in lower fat absorption and reduced oil degradation in comparison to uncoated samples or samples coated before or after breading.

Polysaccharide-based formulations may also contribute to retard moisture loss from meat products during short-term storage. They act as sacrificial moisture barriers, so that the water content of the food can be maintained (Kester and Fennema, 1986). Baker et al. (1994) observed that MC and HPMC can reduce moisture loss during cooking of poultry and seafood. Williams et al. (1978) reported that alginate coatings did not affect cooking losses, flavor, odor, or overall acceptability of beef and poultry products. Some authors have reported that alginate coatings can be used to retard the development of off-flavors due to lipid oxidative phenomena in precooked beef patties and cooked pork chops, thus improving flavor and juiciness in both products (Hargens-Madsen, 1995; Wu et al., 2001). Some polysaccharide films (e.g., pullulan-based films), are used in Japan to coat meat products, ham and poultry, before smoking or steaming processes. The film is dissolved during the process, and the coated meat exhibits improved yield, structure, and texture, as well as reduced moisture loss (Labell, 1991; Stollman et al., 1994). These pullulan films can be also used to entrap flavors, colors, and other active ingredients and are available in several colors, with spices and seasonings included (Guilbert and Gontard, 2005).

Furthermore, edible films and coatings can carry compounds with different functionality, thus providing a novel way to improve the safety and quality shelf life of food systems. Several studies have demonstrated that antimicrobial edible films can reduce bacterial levels on meat products. For instance, antimicrobial chitosan films containing acetic or propionic acid reportedly inhibited growth of *Enterobacteriaceae* and *Serratia liquefaciens* on bologna, cooked ham, and pastrami (Ouattara et al., 2000). Oussalah et al. (2006) applied alginate-based edible films containing Spanish oregano or Chinese cinnamon essential oils on beef muscle slices to control the growth of *Escherichia coli* O157:H7 and *Salmonella typhimurium*. The incorporation of essential oils into films helped to reduce the growth of both bacterial species on beef muscle during 5 days of storage.

On the other hand, fish and seafood are highly susceptible to quality deterioration due to lipid oxidation of unsaturated fatty acids, catalyzed by the presence of high concentrations of hematin compounds and metal ions in the fish muscle (Decker and Hultin, 1992). Jeon et al. (2002) studied the effect of chitosan on shelf-life extension of fresh fillets of Atlantic cod (*Gadus morhua*) and herring (*Clupea harengus*) over 12 days of refrigerated storage. They observed that chitosan coatings helped to significantly reduce lipid oxidation, chemical spoilage, and growth of microorganisms in both fishes compared to the uncoated controls. Sathivel (2005) showed the potential of chitosan for reducing moisture loss and delaying lipid oxidation of salmon fillets through 3 months of storage.

4.3.3 Nuts and cereal-based products

Edible coatings are applied to the surface of snack foods and crackers to serve as a base or adhesive for seasonings. Fats and oils have traditionally been used for adhering seasonings and flavorings to the surface of cereal-based snacks. However, as a result of the recent market demand for low-fat snack foods, many companies have introduced the use of other edible coatings, which are especially indicated because of their low-caloric value. An alternative to oil to achieve adherence is the use of an aqueous starch solution that acts like a glue for the dry seasoning (Clark, 2007). For instance, oil-roasted and dry-roasted peanuts require an adhesive to act as a coating or bonding agent for salting and seasonings. In this application, modified starches in combination with corn syrup, water, and glycerin are used to make an adhesive solution. This solution is applied to the peanuts during tumbling. After the peanuts have been coated with adhesive, the seasoning is added. Likewise, a gluten-dextrin coating was used to coat dry-roasted peanuts prior to the addition of salt (Noznick and Bundus, 1967).

On the other hand, polysaccharide coatings have been used to reduce the loss of crispness and subsequent softening in low-moisture bakery and extruded products such as biscuits, snacks, and breakfast cereals that have a crispy texture (Roudaut et al., 2002; Sauvageot and Blond, 1991). Bravin et al. (2006) evaluated the effectiveness of an edible coating composed of corn starch, MC, and soybean oil in controlling moisture transfer of coating crackers, a low a_w-type cereal food. They observed that coated crackers had longer shelf life and higher water vapor transmission rates than uncoated crackers and confirmed the potential of the edible layer to become an integral part of the food, thus reducing the kinetics of hydration in high a_w environments.

In the last few years, there has been a renewed interest in using the coating material as a flavor carrier in cereals. Laohakunjit and Kerdchoechuen (2007) coated nonaromatic milled rice with 30% sorbitol-plasticized rice starch containing 25% natural pandan leaf extract (*Pandanus amaryllifolius* Roxb.), which is mainly responsible for the jasmine aroma of aromatic rice. These authors established that rice-starch coating containing natural pandan extract produced nonaromatic rice with aroma compounds similar to those of aromatic rice. Additionally, the coating treatment reduced the *n*-hexanal content of the stored grains. This coating technique is a promising approach for improving cereal aroma and, at the same time, for reducing detrimental processes occurring during grain storage, such as lipid oxidation.

4.3.4 Other applications

Because of their nontoxicity, polysaccharides are increasingly finding applications in several pharmaceutical and biotechnology sectors (Coviello et al., 2007). The potential of pectin to be cross-linked by calcium ions has been used to manufacture various food products as well as controlled release dosage forms. Because pectin is a readily available, highly innocuous product, its use as a carrier matrix for entrapment of living cells is strongly recommended. In fact, several oral delivery systems based on calcium pectinate gels (mainly matrix gel beads) have already been investigated (Sriamornsak and Nunthanid, 1999). Sriamornsak and Kennedy (2007) examined the effect of pellet size, pectin type, pectin concentration, and dissolution medium on the swelling and drug release behavior of spherical pellets containing theophylline and coated with two different calcium pectinates. They concluded that the application of calcium pectinate gel coats onto spherical pellets containing theophylline markedly slowed drug release (compared with uncoated pellets). In another work, Sriamornsak and Puttipipatkhachorn (2004) studied chitosan-pectin composite gel spheres in which pectin solution containing indomethacin, a model drug, was extruded into a mixture of chitosan and calcium chloride. In this study, authors discussed the influence of different factors affecting release behavior, such as pectin type, chitosan molecular weight, cross-linking time, and release medium. Their results suggest that composite gel spheres of pectin and chitosan can be used as a controlled release drug delivery carrier.

Alginate is another polysaccharide coating that is being widely studied within the pharmaceutical field. In fact, alginates have been used for decades in various human-health-related applications, including wound dressings, dental impression materials, and formulations to prevent gastric reflux. Because of the ability of alginic sodium salts to rapidly form viscous solutions and gels in aqueous media, they have been used as excipient in tablets to modulate drug delivery dosage (Tonnesen and Karlsen, 2002). Moreover, matrices incorporating alginate salts or a combination of alginate with other polymers have been employed to successfully prolong release of many drugs (Efentakis and Buckton, 2002; Giunchedi et al., 2000b; Liew et al., 2006; Sriamornsak and Sungthongjeen, 2007; Veski and Marvola, 1993). Some have studied the influence of alginate grades on the drug release properties of drug delivery tablets (Efentakis and Buckton, 2002; Liew et al., 2006). Due to the intrinsic properties of calcium alginate (biocompatibility, mucoadhesion, porosity, and ease of manipulation), much attention has recently been devoted to the delivery of proteins, cell encapsulation, and tissue regeneration (Coviello et al., 2007).

4.4 Future trends

Polysaccharide-based films and coatings have the potential to be used in a variety of applications. The attractive, nongreasy, and low-caloric polysaccharide films and coatings can be used to extend the shelf life of fruits, vegetables, meats, seafood, and bakery products by preventing dehydration, oxidative rancidity, surface browning, and oil diffusion, or even by modifying the internal atmosphere of metabolically active food products. Furthermore, edible films and coatings are promising systems for the improvement of food quality, shelf life, safety, and functionality. Several studies have demonstrated that the incorporation of active compounds such as antimicrobials, antioxidants, or nutraceuticals in edible films and coatings can improve safety and shelf life and can be used to better keep quality throughout storage. However, when active compounds are added to polysaccharide edible films and coatings, mechanical properties can be dramatically affected. Studies in this subject are still very limited, and more information is needed in order to understand these changes.

Several factors such as cost, lack of new materials with desirable functionality, investment for the installation of new film production or coating equipment, technical requirements for implementing production processes, and regulatory issues are currently limiting the development of new applications. In spite of these limitations, polysaccharide coatings meet the demands of food processors, who aim to find materials that can be used on a broad spectrum of applications and add value to their products, increase shelf life, and reduce packaging. Nevertheless, more research on an industrial scale and under commercial conditions should be conducted with the objective of obtaining more realistic and practical information for actual commercialization of coated products.

References

Assis, O.B., and J.D. Pessoa. 2004. Preparation of thin films of chitosan for use as edible coatings to inhibit fungal growth on sliced fruits. *Brazilian Journal of Food Technology* 7:7–22.

Axelos, M.A.V., Lefebvre, J., Qiu, C.G., and M.A. Rao. 1991. Rheology of pectin dispersions and gels. In *The Chemistry and Technology of Pectin*, ed. R.H. Walter, 228. San Diego: Academic Press.

Baker, R.A., Baldwin, E.A., and M.O. Nisperos-Carriedo. 1994. Edible coatings and films for processed foods. In *Edible Coatings and Films to Improve Food Quality*, eds. J.M. Krochta, E.A. Baldwin, and M.O. Nisperos-Carriedo, 89–105. Lancaster, PA: Technomic.

Balasubramaniam, V.M., Chinnan, M.S., Mallikarjunan, P., and R.D. Phillips. 1997. The effect of edible film on oil uptake and moisture retention of a deep-fat fried poultry product. *Journal of Food Process Engineering* 20:17–29.

Baldwin, E.A., Burns, J.K., Kazokas, W., Brecht, J.K., Hagenmaier, R.D., Bender, R.J., and E. Pesis. 1999. Effect of two edible coatings with different permeability characteristics on mango *(Mangifera indica* L.) ripening during storage. *Postharvest Biology and Technology* 17:215–226.

Baldwin, E.A., Nisperos, M.O., Chen, X., and R.D. Hagenmaier. 1996. Improving storage life of cut apple and potato with edible coating. *Postharvest Biology and Technology* 9:151–163.

Banik, R.M., Kanari, B., and S.N. Upadhyay. 2000. Exopolysaccharides of the gellan family: prospects and potential. *World Journal of Microbiology and Biotechnology* 16:407–414.

Banks, N.H. 1984. Studies of the banana fruit surface in relation to the effects of TAL prolong coating on gaseous exchange. *Scientia Horticulturae* 24:279–286.

Bauchot, A.D., John, P., Soria, Y., and I. Recasens. 1995. Sucrose ester-based coatings formulated with food-compatible antioxidants in the preservation of superficial scald in stored apples. *Journal of the American Society for Horticultural Science* 120:491–496.

BeMiller, J.N. 1965. Chitosan. In *Methods in Carbohydrate Chemistry. General Polysaccharides,* Vol. V, ed. R.L. Whistler, 103–106, New York: Academic Press.

Bixler, H.J. 1996. Recent developments in manufacturing and marketing carrageenan. *Hydrobiologia* 326/327:35–37.

Brancoli, N., and G.V. Barbosa-Cánovas. 2000. Quality changes during refrigerated storage of packaged apple slices treated with polysaccharide films. In *Innovations in Food Processing*, eds. G.V. Barbosa-Cánovas, and G.W. Gould, 243–254. Lancaster, PA: Technomic.

Bravin, B., Peressini, D., and A. Sensidoni. 2006. Development and application of polysaccharide-lipid edible coatings to extend shelf-life of dry bakery products. *Journal of Food Engineering* 76:280–290.

Brownsey, G.J., and M.J. Redout. 1985. Rheological characterization of microcrystalline cellulose dispersions: Avicel RC 591. *Journal of Food Technology* 20:231.

Bryan, D.S. 1972. Prepared citrus fruit halves and method of making the same. U.S. patent 3,707,383.

Buléon, A., Colonna, P., Planchot, V., and S. Ball. 1998. Starch granules: structure and biosynthesis. *International Journal of Biological Macromolecules* 23:85–112.

Carrillo-Lopez, A., Ramirez-Bustamante, F., Valdez-Torres, J.B., Rojas-Villegas, R., and Yahia, E.M. 2000. Ripening and quality changes in mango fruit as affected by coating with an edible film. *Journal of Food Quality* 23:479–486.

Chien, P.J., and C.C. Chou. 2006. Antifungal activity of chitosan and its application to control post-harvest quality and fungal rotting of Tankan citrus fruit (*Citrus tankan* Hayata). *Journal of the Science of Food and Agriculture* 86:1964–1969.

Chien, P.J., Sheu, F., and H.R. Lin. 2007b. Quality assessment of low molecular weight chitosan coating on sliced red pitayas. *Journal of Food Engineering* 79:736–740.

Chien, P.J., Sheu, F., and H.R. Lin. 2007c. Coating citrus (*Murcott tangor*) fruit with low molecular weight chitosan increases postharvest quality and shelf life. *Food Chemistry* 100:1160–1164.

Chien, P., Sep, F., and F. Yang. 2007a. Effects of edible chitosan coating on quality and shelf life of sliced mango fruit. *Journal of Food Engineering* 78:225–229.

Choi, J.H., Choi, W.Y., Cha, D.S., Chinnan, M.J., Park, H.J., Lee, D.S., and J.M. Park. 2005. Diffusivity of potassium sorbate in κ-carrageenan based antimicrobial film. *Lebensmittel-Wissenschaft und -Technologie* 38:417–423.

Clark, P. 2007. Applying seasonings and coatings. *Food Technology* 11:72–74.

Conforti, F.D., and J.A. Totty. 2007. Effect of three lipid/hydrocolloid coatings on shelf life stability of Golden Delicious apples. *International Journal of Food Science and Technology* 42:1101–1106.

Coviello, T., Matricardi, P., Marianecci, C., and F. Alhaique. 2007. Polysaccharide hydrogels for modified release formulations. *Journal of Controlled Release* 119:5–24.

Crescenzi, V., Dentini, M., and T. Coviello. 1990. Solution and gelling properties of microbial polysaccharides of industrial interest: the case of gellan. In *Novel Biodegradable Microbial Polymers*, ed. E.A. Dawes, 227–284. Boston: Kluwer Academic.

De Ruiter, G.A., and B. Rudolph. 1997. Carrageenan biotechnology. *Trends in Food Science and Technology* 8:389–429.

De Vries, J.A. 2002. Interaction of carrageenan with other ingredients in dairy dessert gels. In *Gums and Stabilizers for the Food Industry*, Vol. 11, ed. P.A. Williams, and G.O. Phillips, 201–210. Cambridge: The Royal Society of Chemistry.

Decker, E.A., and H.O. Hultin. 1992. Lipid oxidation in muscle foods via redox iron. In *Lipid Oxidation in Food*, ed. A.J. Angelo, 33–54. Washington, DC: American Chemical Society.

Del-Valle, V., Hernández-Muñoz, P., Guarda, A., and M.J. Galotto. 2005. Development of a cactus-mucilage edible coating (*Opuntia ficus indica*) and its application to extend strawberry (*Fragaria ananassa*) shelf-life. *Food Chemistry* 91:751–756.

Deuel, H., and H. Neukom. 1954. Some properties of locust bean gum. In *Natural Plant Hydrocolloids*, 51–61. Washington, DC: American Chemical Society.

Devlieghere, F., Vermeulen, A., and J. Debevere. 2004. Chitosan: antimicrobial activity, interactions with food components and applicability as a coating on fruit and vegetables. *Food Microbiology* 21:703–714.

Dey, P.M. 1978. Biochemistry of plant galactomannans. *Advances in Carbohydrate Chemistry and Biochemistry* 35:341–376.

Dhalla, R., and S.W. Hanson. 1988. Effect of permeable coatings on the storage life of fruits. II. Pro-long treatment of mangoes (*Mangifera indica* L. cv. Julie). *International Journal of Food Science and Technology* 23:107–112.

Diab, T., Biliaderis, C.G., Gerasopoulos, D., and E. Sfakiotakis. 2001. Physicochemical properties and application of pullulan edible films and coatings in fruit preservation. *Journal of the Science of Food and Agriculture* 81:988–1000.

Dinand, E., Vignon, M., Chanzi, H., and L. Heux. 2002. Mercerization of primary cell wall cellulose and its implication for the conversion of cellulose I → cellulose II. *Cellulose* 9:7–18.

Draget, K.I., Moe, S.T., Skjåk-Broek, G., and O. Smidsrød. 2006. Alginates. In *Food Polysaccharides and Their Applications*, eds. A.M. Stephen, G.O. Phillips, and P.A. Williams, 290–328. Boca Raton, FL: Taylor & Francis.

Draget, K.I., Smidsrød, O., and G. Skjåk-Broek. 2005. Alginates from algae. In *Polysaccharides and Polyamides in the Food Industry. Properties, Production, and Patents*, eds. A. Steinbüchel, and S.K. Rhee, 1–30. Weinheim: Wiley-VCH.

Drake, S.R., Fellman, J.K., and J.W. Nelson. 1987. Postharvest use of sucrose polyesters for extending the shelf-life of stored 'Golden delicious' apples. *Journal of Food Science* 52:685–690.

Dumitriu, S., and E. Chornet. 1998. Inclusion and release of proteins from polysaccharide-based polyion complexes. *Advanced Drug Delivery Reviews* 31:223–246.

Durango, A.M., Soares, N.F.F., and N.J. Andrade. 2006. Microbiological evaluation of an edible antimicrobial coating on minimally processed carrots. *Food Control* 17:336–341.

Dziezak, J.D. 1991. Special report: A focus on gums. *Food Technology* 45:116–132.

Efentakis, M., and G. Buckton. 2002. The effect of erosion and swelling on the dissolution of theophylline from low and high viscosity sodium alginate matrices. *Pharmaceutical Development and Technology* 7:69–77.

Eissa, H.A.A. 2007. Effect of chitosan coating on shelf life and quality of fresh-cut mushroom. *Journal of Food Quality* 30:623–645.

El Ghaouth, A., Arul, J., Ponnampalan, R., and M. Boutlet. 1991. Chitosan coating effect on storability and quality of fresh strawberries. *Journal of Food Science* 56:1618–1620.

Elson, C.W., Hayes, E.R., and P.D. Lidster. 1985. Development of the differentially permeable fruit coatings 'Nutri-Save' for the modified atmosphere storage of fruit. In Proceedings of the Fourth National Controlled Atmosphere Research Conf., S.M. Blankenship, ed., Raleigh, NC: Dept. of Hort. Sci., North Caroline State Univ., Rpt. 126:248.

European Parliament and Council Directive No. 95/2/EC. (1995). On food additives other than colors and sweeteners. http://ec.europa.eu/food/fs/sfp/addit_flavor/flav11_en.pdf.

FDA (U.S. Food and Drug Administration). 2006. Food additives permitted for direct addition to food for human consumption 21CFR172, subpart C. Coatings, Films and Related Substances.

Franssen, L.R., and J.M. Krochta. 2003. Edible coatings containing natural antimicrobials for processed foods. In *Natural Antimicrobials for Minimal Processing of Foods*, ed. S. Roller, 250–262. Boca Raton, FL: CRC Press.

Friedenthal, M., Eha, K., Viitak, A., Lukas, A., and E. Siimer. 2001 Effects of drying on the gel strength and cation mobility of furcellaran. *Innovative Food Science and Emerging Technologies*, 1:275–279.

García, M.A., Martino, M.N., and N.E. Zaritzky. 2000. Lipid addition to improve barrier properties of edible starch-based films and coatings. *Journal of Food Science* 65:941–947.

García, M.A., Martino, M.N., and N.E. Zaritzky. 1998. Starch-based coatings: effect on refrigerated strawberry (*Fragaria ananassa*) quality. *Journal of the Science of Food and Agriculture* 76:411–420.

García-Ochoa, F., Santos, V.E., Casas, J.A., and E. Gómez. 2000. Xanthan gum: production, recovery, and properties. *Biotechnology Advances* 18:549–579.

Gennadios, A., Hanna, M.A., and L.B. Kurth. 1997. Application of edible coatings on meats, poultry and seafoods: a review. *Lebensmittel-Wissenschaft und -Technologie* 30:337–350.

Gidley, M.J., and J.S.G. Reid. 2006. Galactomannans and other cell wall storage polysaccharides in seeds. In *Food Polysaccharides and Their Applications*, eds. A.M. Stephen, G.O. Phillips, and P.A. Williams, 181–215. Boca Raton, FL: Taylor & Francis.

Giunchedi, P., Gavini, E., Bonacucina, G., and G.F. Palmieri. 2000a. Tabletted polylactide microspheres prepared by a w/o emulsion-spray drying method. *Journal of Microencapsulation* 17:711–720.

Giunchedi, P., Gavini, E., Domenico, M., Moretti, L., and G. Pirisino. 2000b. Evaluation of alginate compressed matrices as prolonged drug delivery systems. *AAPS PharmSciTech* 1(3):19.

Glicksman, M. 1983. Gum ghatti (Indian gum). In *Food Hydrocolloids*, Vol. II, ed. M. Glicksman, 111–113. Boca Raton: CRC Press.

Goffey, D.G., Bell, D.A., and A. Henderson. 2006. Cellulose and cellulose derivatives. In *Food Polysaccharides and Their Applications*, eds. A.M. Stephen, G.O. Phillips, and P.A. Williams, 147–179. Boca Raton, FL: Taylor & Francis.

Guilbert, S., and N. Gontard. 2005. Agro-polymers for edible and biodegradable films: review of agricultural polymeric materials, physical and mechanical characteristics. In *Innovation in Food Packaging*, ed. J.H. Han, 263–276. Oxford: Elsevier.

Han, C., Lederer, C., McDaniel, M., and Y. Zhao. 2005. Sensory evaluation of fresh strawberries (*Fragaria ananassa*) coated with chitosan-based edible coatings. *Journal of Food Science* 70:S172–S178.

Han, C., Zhao, Y., Leonard, S.W., and M.G. Traber. 2004. Edible coatings to improve storability and enhance nutritional value of fresh and frozen strawberries (*Fragaria x ananassa*) and raspberries (*Rubus ideaus*). *Postharvest Biology and Technology* 33:67–78.

Han, J.S. 2002. Changes of dynamic viscoelastic properties of oxidized corn starch suspensions during heating and cooling. *Food Science and Biotechnology* 11:231.

Hargens-Madsen, M.R. 1995. Use of edible coatings and tocopherols in the control of warmed-over flavour. M.S. thesis, University of Nebraska.

Harish Prashanth, K.V., and R.N. Tharanathan. 2007. Chitin/chitosan: modifications and their unlimited application potential: an overview. *Trends in Food Science and Technology* 18:117–131.

Hernández-Muñoz, P., Almenar, E., Ocio, M.J., and R. Gavara. 2006. Effect of calcium dips and chitosan coatings on postharvest life of strawberries (*Fragaria×ananassa*). *Postharvest Biology and Technology* 39:247–253.

Hirano, S., and N. Nagao. 1989. Effect of chitosan, pectic acid, lysozyme and chitinase on the growth of several phytopathogens. *Agricultural and Biological Chemistry* 53:3065.

Holownia, K.I., Chinnan, M.S., Erickson, M.C., and P. Mallikarjunan. 2000. Quality evaluation of edible film-coated chicken strips and frying oils. *Journal of Food Science* 65:1087–1090.

Jeon, Y.J., Kamil, J.Y.V.A., and F. Shahidi. 2002. Chitosan as an edible invisible film for quality preservation of herring and Atlantic cod. *Journal of Agricultural and Food Chemistry* 50:5167–5178.

Jokay, L., Nelson, G.E., and E.L. Powell. 1967. Development of edible amylaceous coatings for foods. *Food Technology* 21:12–14.

Käistner, U., Hoffmann, H., Dönges, R., and J. Hilbig. 1997. Structure and solution properties of sodium carboxymethyl cellulose. *Colloids and Surfaces A* 123–124:307–328.

Karbowiak, T., Hervet, H., Leger, L., Champion, D., Debeaufort, F., and A. Voilley. 2006. Effect of plasticizers (water and glycerol) on the diffusion of a small molecule in iota-carrageenan biopolymer films for edible coating application. *Biomacromolecules* 7:2011–2019.

Kaynas, K., and I.S. Ozelkok. 1999. Effect of Semperfresh on postharvest behavior of cucumber (*Cucumis sativus* L.) and summer squash (*Cucurbita pepo* L.) fruits. *Acta Horticulturae* 492:213–220.

Kester, J.J., and O. Fennema. 1986. Edible films and coatings: a review. *Food Technology* 40:47–59.

King, A. 1983. Brown seaweed extracts (alginates). In *Food Hydrocolloids*, ed. M. Glicksman, 116–188. Boca Raton, FL: CRC Press.

Kittur, F.S., Saroja, N., Habibunnisa, and R.N. Tharanathan. 2001. Polysaccharide-based composite coating formulations for shelf-life extension of fresh banana and mango. *European Food Research and Technology* 213:306–311.

Klemm, D., Heublein, B., Fink, H.-P., and A. Bohn 2005. Cellulose: fascinating biopolymer and sustainable raw material, *Angewandte Chemie International Edition* 44:3358–3393.

Krajewska, B. 2004. Application of chitin- and chitosan-based materials for enzyme immobilizations: a review. *Enzyme Microbiology and Technology* 35: 126–139.

Krassig, H., Steadman, R.G., Schliefer, K., and W. Albrecht. 1986. *Ullmann's Encyclopedia of Organic Compounds*, 5th ed., A5, p. 325, New York: VCH.

Krumel, K.L., and T.A. Lindsay. 1976. Nonionic cellulose ethers. *Food Technology* 30:36–43.

Labell, F. 1991. Edible packaging. *Food Process Engineering* 52:24.

Lacroix, M., and C. LeTien. 2005. Edible films and coatings from non-starch polysaccharides. In *Innovation in Food Packaging*, ed. J.H. Han, 338–361. New York: Elsevier Academic Press.

Lafargue, D., Pontoire, B., Buléon, A., Doublier, J.L., and D. Lourdin. 2007. Structure and mechanical properties of hydroxypropylated starch films. *Biomacromolecules* 8:3950–3958.

Laohakunjit, N., and O. Kerdchoechuen. 2007. Aroma enrichment and the change during storage of non-aromatic milled rice coated with extracted natural flavour. *Food Chemistry* 101:339–344.

Lee, J.Y., Park, H.J., Lee, C.Y., and W.Y. Choi. 2003. Extending shelf-life of minimally processed apples with edible coatings and antibrowning agents. *Lebensmittel-Wissenschaft und -Technologie* 36:323–329.

Li, H.Y., and T. Yu. 2001. Effect of chitosan on incidence of brown rot, quality and physiological attributes of postharvest peach fruit. *Journal of the Science of Food and Agriculture* 81:269–274.

Liew, C.V., Chan, L.W., Ching, A.L., and P.W.S. Heng. 2006. Evaluation of sodium alginate as drug release modifier in matrix tablets. *International Journal of Pharmaceutics* 309:25–37.

Lopes da Silva, J.A., and M.A. Rao. 2006. Pectins: structure, functionality, and uses. In *Food Polysaccharides and Their Applications*, eds. A.M. Stephen, G.O. Phillips, and P.A. Williams, 353–411. Boca Raton, FL: Taylor & Francis.

Lowings, P.H., and D.F. Cutts. 1982. The preservation of fresh fruit and vegetables. Paper presented at the Annual Symposium of the Institute of Food Science and Technology, Nottingham.

Majewicz, T.G., and T.J. Polas. 1993. Cellulose ethers. In *Kirk-Othmer Encyclopaedia of Chemical Technology*, 4th ed., Vol. 5, 529–540. New York: Wiley.

Mancini, F., and T.H. McHugh. 2000. Fruit–alginate interactions in novel restructured products. *Nahrung* 44:152–157.

Martínez-Romero, D., Alburquerque, N., Valverde, J.M., Guillén, F., Castillo, S., Valero, D., and M. Serrano. 2006. Postharvest sweet cherry quality and safety maintenance by Aloe Vera treatment: a new edible coating. *Postharvest Biology and Technology* 39:93–100.

McHugh, T.H., and E. Senesi. 2000. Apple wraps: a novel method to improve the quality and extend the shelf life of fresh-cut apples. *Journal of Food Science* 65:480–485.

McHugh, T.H., and E. Senesi, inventors; USDA-ARS-WRRC assignee. Filed 1999 June 11. Fruit and vegetable edible film wraps and methods to improve and extend the shelf life of foods. U.S. Patent Application Serial No. 09/330,358.

Meer, G., Meer, W.A., and T. Gerard. 1973. Gum tragacanth. In *Industrial Gums*, 2nd ed., eds. R.L. Whistler and J.N. BeMiller, 289–299. New York: Academic Press.

Mei, Y., Zhao, Y., Yang, J., and H.C. Furr. 2002. Using edible coating to enhance nutritional and sensory qualities of baby carrots. *Journal of Food Science* 67:1964–1968.

Meng, X., Li, B.Q., Liu, J., and S.P. Tian. 2008. Physiological responses and quality attributes of table grape fruit to chitosan preharvest spray and postharvest coating during storage. *Food Chemistry* 106:501–508.

Milojevic, S., Newton, J.M., Cummings, J.H., Gibson, G.R., Botham, R.L., Ring, S.G., Stockham, M., and M.C. Allwood. 1996a. Amylose as a coating for drug delivery to the colon: preparation and in vitro evaluation using 5-aminosalicylic acid pellets. *Journal of Controlled Release* 38:75–84.

Milojevic, S., Newton, J.M., Cummings, J.H., Gibson, G.R., Botham, R.L., Ring, S.G., Stockham, M., and M.C. Allwood. Amylose as a coating for drug delivery to the colon: preparation and in vitro evaluation using glucose pellets. *Journal of Controlled Release* 38:85–94.

Moe, S.T., Draget, K.I., Skjåk-Broek, G., and O. Smidsrød. 1995. In *Food Polysaccharides and Their Applications*, ed. A.M. Stephen, 245–286. New York: Marcel Dekker.

Moorhouse, R., Cosgrove, G.T., Sandford, P.A., Baird, J., and K.S. Kang. 1981. PS-60: a new gel-forming polysaccharide. In *Solution Properties of Polysaccharides*, ed. D.A. Brant, 111. Washington: American Chemical Society.

Morris, V.J. 1991. Weak and strong polysaccharide gels. In *Food Polymers, Gels and Colloids*, ed. E. Dickinson, 310. London: Royal Society of Chemistry.

Morris, V.J. 2006. Bacterial polysaccharides. In *Food Polysaccharides and Their Applications*, eds. A.M. Stephen, G.O. Phillips, and P.A. Williams, 413–454. Boca Raton, FL: Taylor & Francis.

Mulchandani, R.P., and M.I. Mahmoud. 1997. Liquid nutritional product containing improved stabilizer composition of carrageenan/microcrystalline cellulose/carboxymethylcellulose. *PCT International Application WO/1997/025878*

Murray, J.C.F. 2000. Cellulosics. In *Handbook of Hydrocolloids*, eds. G.O. Phillips, and P.A. Williams, 219–230. Cambridge: Woodhead.

Nakahama, N. 1966. Rheological studies on the agar-agar gel. *Journal of Home Economics* 17:197.

Nisperos-Carriedo, M.O., and E.A. Baldwin. 1990. Edible coatings for fresh fruits and vegetables. *Subtropical Technology Conference Proceedings*, October 18, Lake Alfred, FL.

Nisperos-Carriedo, M.O., Baldwin, E.A., and P.E. Shaw. 1991. Development of an edible coating for extending postharvest life of selected fruits and vegetables. *Proceedings of the Florida State Horticultural Society* 104:122–125.

Noznick, P., and R. Bundus. 1967. Dry roasted peanuts. Patent GB1096490.

Nussinovitch, A. 1997. *Hydrocolloid Applications: Gum Technology in the Food and Other Industries.* Berlin: Springer.

O'Neill, M.A., Selevendran, R.R., and V.J. Morris. 1983. Structure of the extracellular polysaccharide produced by *Pseudomonas elodea*, *Carbohydrate Research* 124:123–133.

Oates, C.G. 1997. Towards an understanding of starch granule structure and hydrolysis. *Trends in Food Science and Technology* 8(11): 375–382.

Olivas, G.I., Rodríguez, J.J., and G.V. Barbosa-Cánovas. 2003. Edible coatings composed of methylcellulose, stearic acid, and additives to preserve quality of pear wedges. *Journal of Food Processing and Preservation* 27:299–320.

Olivas, G.I., and G.V. Barbosa-Cánovas. 2005. Edible coatings for fresh-cut fruits. *Critical Reviews in Food Science and Nutrition.* 45:657–670.

Olivas, G.I., Mattinson, D.S., and G.V. Barbosa-Canovas. 2007. Alginate coatings for preservation of minimally processed 'Gala' apples. *Postharvest Biology and Technology* 45:89–96.

Ouattara, B., Simard, R.E., Piette, G., Begin, A., and R.A. Holley. 2000. Inhibition of surface spoilage bacteria in processed meats by application of antimicrobial films prepared with chitosan. *International Journal of Food Microbiology* 62:139–148.

Oussalah, M., Caillet, S., Salmieri, S., Saucier, L., and M. Lacroix. 2006. Antimicrobial effects of alginate-based film containing essential oils for the preservation of whole beef muscle. *Journal of Food Protection* 69:2364–2369.

Palviainen, P., Heinämäki, J., Myllärinen, P., Lahtinen, R., Yliruusi, J., and P. Forssell. 2001. Corn starches as film formers in aqueous-based film coating. *Pharmaceutical Development and Technology* 6:351–359.

Park, S.I., Stan, S.D., Daeschel, M.A., and Y.Y. Zhao. 2005. Antifungal coatings on fresh strawberries (*Fragaria* × *ananassa*) to control mold growth during cold storage. *Journal of Food Science* 70:M202–M207.

Pen, L.T., and Y.M. Jiang. 2003. Effects of chitosan coating on shelf life and quality of fresh-cut Chinese water chestnut. *Lebensmittel-Wissenschaft und -Technologie-Food Science and Technology* 36:359–364.

Pessa, E., Sueortti, T., Autio, K., and K. Poutanen. 1992. Molecular weight characterization and gelling properties of acid-modified starches. *Starch/Staerke* 44:64–69.

Piculell, L. 2006. Bacterial polysaccharides. In *Food Polysaccharides and Their Applications*, eds. A.M. Stephen, G.O. Phillips, and P.A. Williams, 239–287. Boca Raton, FL: Taylor & Francis.

Raju, P.S., and A.S. Bawa 2006. Food additives in fruit processing. In *Handbook of Fruits and Fruit Processing*, ed. Y.H. Hui, 145–170. Ames, IA: Blackwell.

Reid, J.S.G. 1985. Cell wall storage carbohydrates in seeds: biochemistry of the seed "gums" and "hemicelluloses." *Advances in Botanical Research* 11:125–155.

Ridley, B.L., O'Neill, M.A., and D. Mohnen. 2001. Pectins: structure, biosynthesis, and oligogalacturonide-related signaling. *Phytochemistry* 57:929–967.

Rioux, B., Ispas-Szabo, P., Aït-Kadi, A., Mateescu, M.-A., and J. Juhász. 2002. Structure-properties relationship in cross-linked high amylose starch cast films. *Carbohydrate Polymers* 50:371–378.

Rojas-Graü, M.A., Raybaudi-Massilia, R.M., Soliva-Fortuny, R.C., Avena-Bustillos, R.J., McHugh, T.H., and O. Martín-Belloso. 2007a. Apple puree-alginate edible coating as carrier of antimicrobial agents to prolong shelf-life of fresh-cut apples. *Postharvest Biology and Technology* 45:254–264.

Rojas-Graü, M.A., Tapia, M.S., Rodríguez, F.J., Carmona, A.J., and O. Martín-Belloso. 2007b. Alginate and gellan based edible coatings as support of antibrowning agents applied on fresh-cut Fuji apple. *Food Hydrocolloids* 21:118–127.

Roth, W.B., and C.L. Mehltretter. 1967. Some properties of hydroxypropylated amylomaize starch films. *Food Technology* 21:72–74.

Roudaut, G., Dacremont, C., Pamies, B.M., Colas, B., and M. Le Meste. 2002. Crispness: a critical review on sensory and material science approaches. *Trends in Food Science and Technology* 13:217–227.

Sandford, P.A., and G.P. Hutchings. 1987. Chitosan—a natural, cationic biopolymer. In *Industrial Polysaccharides: Genetic Engineering, Structure/Property Relations and Applications*, ed. M. Yalpani, 363–376. Amsterdam: Elsevier.

Sathivel, S. 2005. Chitosan and protein coatings affect yield, moisture loss, and lipid oxidation of pink salmon (Oncorhynchus gorbuscha) fillets during frozen storage. *Journal of Food Science* 70:E455–459.

Sauvageot, F., and G. Blond. 1991. Effect of water activity on crispness of breakfast cereals. *Journal of Texture Studies* 22:423–442.

Savitha Prashantha, M.R., Parvathya, K.S., Susheelammab, N.S., Harish Prashantha, K.V., Tharanathana, R.N., Chac, A., and G. Anilkumarc. 2006. Galactomannan esters: a simple, cost-effective method of preparation and characterization. *Food Hydrocolloids* 20:1198–2120.

Selby, H.H., and R.L. Whistler. 1993. Agar. In *Industrial Gums*, 2nd ed., eds. R.L. Whistler and J.N. BeMiller, 145–180. New York: Academic Press.

Serrano, M., Valverde, J.M., Guillén, F., Castillo, S., Martínez-Romero, D., and D. Valero. 2006. Use of aloe vera gel coating preserves the functional properties of table grapes. *Journal Agricultural and Food Chemistry* 54:3882–3886.
Siew, L.F., Basit, A.W., and J.M. Newton. 2000. The potential of organic-based amylose–ethylcellulose film coatings as oral colon-specific drug delivery systems. *AAPS PharmSciTech* 1(3):22.
Singh, B.N., and K.H. Kim. 2005. Effects of divalent cations on drug encapsulation efficiency of deacetylated gellan gum. *Journal of Microencapsulation* 22(7):761–771.
Sriamornsak, P., and R.A. Kennedy. 2007. Development of polysaccharide gel-coated pellets for oral administration: swelling and release behavior of calcium pectinate gel. *Aaps Pharmscitech* 8(3):79.
Sriamornsak, P., and J. Nunthanid. 1999. Calcium pectinate gel beads for controlled release drug delivery. II. Effect of formulation and processing variables on drug release. *Journal of Microencapsulation* 16:303–313.
Sriamornsak, P., and S. Puttipipatkhachorn. 2004. Chitosan-pectin composite gel spheres: effect of some formulation variables on drug release. *Macromolecular Symposia* 216:17–21.
Sriamornsak, P., and S. Sungthongjeen. 2007. Modification of theophylline release with alginate gel formed in hard capsules. *AAPS Pharmscitech* 8(3):51.
Stauffer, K.R., and S.A. Andon. 1975. Comparison of the functional properties of two grades of gum tragacanth. *Food Technology* 29:46.
Stephen, A.M., and S.C. Churms. 2006. Introduction. In *Food Polysaccharides and Their Applications*, eds. A.M. Stephen, G.O. Phillips, and P.A. Williams, 1–24. Boca Raton, FL: Taylor & Francis.
Stollman, U., Hohansson, F., and A. Leufven. 1994. Packaging and food quality. In *Shelf Life Evaluation of Foods*, eds. C.M.D. Man, and A.A. Jones, 52–71. New York: Blackie Academic and Professional.
Sumnu, G., and L. Bayindirli. 1995. Effects of coatings on fruit quality of Amasya apples. *Lebensmittel-Wissenschaft und -Technologie* 28:501–505.
Takahashi, R., Tokunou, H., Kubota, K., Ogawa, E., Oida, T., Kawase, T. and K. Nishinari (2004). Solution properties of gellan gum: Change in chain stiffness between single- and double-stranded chains. *Biomacromolecules* 5:516–523.
Tapia, M.S., Rojas-Graü, M.A., Rodríguez, F.J., Ramírez, J., Carmona, A., and O. Martín-Belloso. 2007. Alginate- and gellan-based edible films for probiotic coatings on fresh-cut fruits. *Journal of Food Science* 72:E190–E196.
Tasdelen, O., and L. Bayindirli. 1998. Controlled atmosphere storage and edible coating effects on storage life and quality of tomatoes. *Journal of Food Processing and Preservation* 22:303–320.
Toft, K., Grasdalen, H., and O. Smidsrød. 1986. Synergistic gelation of alginate and pectins. *Journal of the American Chemical Society* 10:117–132.
Tonnesen, H.H., and J. Karlsen. 2002. Alginate in drug delivery systems. *Drug Development and Industrial Pharmacy* 28:621–630.
Valverde, J.M., Valero, D., Martínez-Romero, D., Guillen, F., Castillo, S., and M. Serrano. 2005. Novel edible coating based on Aloe vera gel to maintain table grape quality and safety. *Journal of Agricultural and Food Chemistry* 53:7807–7813.
Van de Velde, F. 2008. Structure and function of hybrid carrageenans. *Food Hydrocolloids* 22:727–734.
Vargas, M., Albors, A., Chiralt, A., and C. Gonzalez-Martinez. 2006. Quality of cold-stored strawberries as affected by chitosan-oleic acid edible coatings. *Postharvest Biology and Technology* 41:164–171.

Veski, P., and M. Marvola. 1993. Sodium alginates as diluents in hard gelatin capsules containing ibuprofen as a model-drug. *Pharmazie* 48:757–760.

Wang, J., Wang, B., Jiang, W., and Y. Zhao. 2007. Quality and shelf life of mango (*Mangifera indica* L. cv. 'Tainong') coated by using chitosan and polyphenols. *Food Science and Technology International* 13:317–322.

Watase, M., Kohyama, K., and K. Nishinari. 1992. Effects of sugars and polyols on the gel-sol transition of agarose by differential scanning calorimetry. *Thennochimica Acta* 2006:163.

Whistler, R.L. 1993. Exudate gums. In *Industrial Gums*, eds. R.L. Whistler, and J.N. BeMiller, 295–303. San Diego: Academic Press.

Whistler, R.L., and J.R. Daniel. 1990. Functions of polysaccharides in foods. In *Food Additives*, eds. A.L. Branen, P.M. Davidson, and S. Salminen, 395–424. New York: Marcel Dekker.

Whytock, S., and J. Finch. 1991. The substructure of agarose gels as prepared for electrophoresis. *Biopolymers* 31:1025–1028.

Williams, P.A., and G.O. Phillips. 2000. Gum arabic. In *Handbook of Hydrocolloids*, eds. G.P. Phillips, and P.A. Williams, 155–158. Cambridge: Woodhead.

Williams, P.A., Phillips, G.O., Stephen, A.M., and S.C. Churms. 2006. Gums and mucilages. In *Food Polysaccharides and Their Applications*, eds. A.M. Stephen, G.O. Phillips, and P.A. Williams, 455–495. Boca Raton, FL: Taylor & Francis.

Williams, R.B.H., Oblinger, J.L., and R.L. West. 1978. Evaluation of a calcium alginate film for use on beef cuts. *Journal of Food Science* 43:292–296.

Wolff, I.A., Davis, H.A., Cluskey, J.E., Gundrum, L.J., and C.E. Rist. 1951. Preparation of films from amylose. *Industrial and Engineering Chemistry* 43:915–919.

Wu, Y., Weller, C.L., Hamouz, F., Cuppett, S., and M. Schnepf. 2001. Moisture loss and lipid oxidation for precooked ground-beef patties packaged in edible starch-alginate-based composite films. *Journal of Food Science* 66:486–493.

Wurzburg, O.B. 2006. Modified starches. In *Food Polysaccharides and Their Applications*, eds. A.M. Stephen, G.O. Phillips, and P.A. Williams, 87–118. Boca Raton, FL: Taylor & Francis.

Yaman, O., and L. Bayindirli. 2001. Effects of an edible coating, fungicide and cold storage on microbial spoilage of cherries. *European Food Research and Technology* 213:53–55.

Yaman, O., and L. Bayindirli. 2002. Effects of an edible coating and cold storage on shelf-life and quality of cherries. *Lebensmittel-Wissenschaft und -Technologie* 35:146–150.

Yurdugul, S. 2005. Preservation of quinces by the combination of an edible coating material, Semperfresh, ascorbic acid and cold storage. *European Food Research and Technology* 220:579–586.

Zhang, D.L., and P.C. Quantick. 1998. Antifungal effects of chitosan coating on fresh strawberries and raspberries during storage. *Journal of Horticultural Science and Biotechnology* 73:763–767.

Zobel, H.F., and A.M. Stephen. 2006. Starch: structure, analysis, and application. In *Food Polysaccharides and Their Applications*, eds. A.M. Stephen, G.O. Phillips, and P.A. Williams, 25–85. Boca Raton, FL: Taylor & Francis.

5 Gas-exchange properties of edible films and coatings

Robert D. Hagenmaier

Contents

5.1	Introduction	137
	5.1.1 Permeance	137
	5.1.2 Effusion	138
	5.1.3 Mass transfer	139
5.2	Permeance equations	139
	5.2.1 Homogeneous barriers	139
	5.2.2 Layered coatings and films	140
	5.2.3 Moisture gradients	141
	5.2.4 Units for permeability and permeance	141
	5.2.5 Measurement of barrier properties	142
	5.2.5.1 Methods	142
	5.2.5.2 Equipment	144
	5.2.5.3 Analytical errors	144
	5.2.5.4 Coating permeability	145
	5.2.6 Permeability values	146
	5.2.7 Effusion compared with permeance	147
	5.2.8 Fruit coatings as test of permeance equations	148
	5.2.9 Nitrogen and mass transfer in fresh fruit	150
References		153

5.1 Introduction

The gas exchange between atmosphere and food products is often described as permeation of gases through a coating barrier (Amarante and Banks, 2001). This chapter will discuss and demonstrate the shortcomings of that approach. In order to do that, it seems appropriate to begin with the definition of the word *permeance*.

5.1.1 Permeance

Permeance is the mechanism by which a gas migrates through a continuous barrier, for example, the skin of a balloon, a bubble, or a sealed plastic bag. This process consists of the gas dissolving into one side of the barrier, diffusing through that barrier by random motion, and escaping out the other side of the barrier. The driving force is the difference in partial pressure of the gas on the two surfaces of the barrier.

The equations for this model are well known and are regularly used to describe and predict the barrier properties of flexible, uniform, factory-made packaging films (Crank, 1964).

Unfortunately, however, the process just described often does not accurately describe how gases migrate through the many edible films and coatings that are not continuous, but rather contain holes for gases to pass through without needing to dissolve in the film. Therefore, discussion of the standard equations for permeance, let us first consider effusion and mass transfer.

5.1.2 Effusion

Effusion is the diffusive passage of gas molecules through small holes in a barrier. Equations 5.1 and 5.2 both show formulas for the effusion rate (E) based on the kinetic theory of gases for the case where the orifices are sufficiently small in diameter and length that gas molecules pass through the barrier relatively unhampered by gas collisions, where gas flow is not turbulent (Glasstone, 1946).

$$E\left(kg \times m^{-2} s^{-1}\right) = \sqrt{\frac{p(Pa) \times density\left(kg \times m^{-3}\right)}{2\pi}} \qquad (5.1)$$

$$E\left(m^3 \times m^{-2} s^{-1}\right) = \sqrt{\frac{p(Pa)}{2\pi \times density\left(kg \times m^{-3}\right)}} \qquad (5.2)$$

Here pressure (p) is the partial pressure of the diffusing gas, and density is the contribution of that gas to total density. Two forms of the equation are shown to make the point that it is ambiguous to state that efflux increases with decreasing molecular weight. In fact, the weight of efflux decreases, but its volume increases as molecular weight decreases.

The assumptions used to derive these equations from the kinetic theory of gases are very limiting. For example, effusion of O_2 at one atmosphere difference in partial pressure would produce a calculated velocity of 106 m/s through holes in the barrier. More importantly, the units of effusion are rate of flow per unit area, and holes in a barrier are generally imperfections with unknown area.

Nevertheless, Equation 5.2 is well established for comparing effusion rates of different gases through the same barrier, having been used since 1833, when Thomas Graham showed that the effusion volumes of several gases were inversely proportional to gas density with less than 1% error for a 5-mm-thick barrier of plaster-of-Paris. In Graham's experiments, considerable care was given to maintaining the same total pressure on both sides of the plaster barrier, in order to avoid mass transfer of gases through the barrier (Graham, 1833).

For the present study, the exchange rate of CO_2 through O_2 coatings and films is of much importance. To anticipate that discussion, it is important to know that CO_2 permeates through most barriers three to four times as fast as O_2, whether the permeation rate is expressed in units of volume or mass. By contrast, as shown by Graham, the effusion rates for these two gases are similar. CO_2 effuses 1.17 times as fast as O_2 at the same difference in partial pressure when E is expressed as weight of gas, and 0.85 times as fast if expressed in terms of volume of gas.

5.1.3 Mass transfer

Mass transfer is passage of gas through a barrier because of a difference in total pressure across the barrier, a mechanism. Mass transfer through food coatings occurs when something happens to change internal gas pressure (e.g., temperature change of a rigid food product, causing interior gas to expand or contract). Less obvious are pressure changes that result from respiration, to be discussed in detail in the last section of this chapter.

Finally, there is the important and difficult question of whether mass transfer through a barrier changes the composition of interior gases as do permeance and effusion, because the permeance and effusion rates vary for different gases. In case of very large holes in the coating, it would seem obvious that gas concentrations on both sides of the barrier are the same. However, for transfer through very small holes in an edible coating or through complicated barriers consisting of edible coating and product skin, some change might be possible. After all, Graham's plaster barrier changed the gas composition, even though his experimental setup was very different from the simple holes assumed for development of the diffusion equations. No attempt here will be made to address that difficult question, which is beyond the scope of this chapter and the knowledge of this author.

5.2 Permeance equations

To repeat, *permeance* here refers only to the process by which gases pass through a barrier the same way these diffuse through a plastic bag with an airtight seal. The word as used here does not include effusion and mass transfer.

5.2.1 Homogeneous barriers

The equations for permeance through a homogeneous barrier are based on Fick's laws of diffusion (Comyn, 1985; Crank, 1964; Park, 1999). Equation 5.3 predicts the steady-state one-dimensional flow rate (F) passing through a barrier of uniform thickness (t), when the diffusion constant (D) is uniform throughout a barrier separating two different concentrations of the diffusing gas. The concentrations of the diffusion gas just within the two walls of the barrier film are C_1 and C_2.

$$F = D \times (C_1 - C_2)/t \qquad (5.3)$$

Compatible SI units are kg m^{-2}s^{-1} for F, m^{2}s^{-1} for D, kg m^{-3} for C_1 and C_2, and m for the thickness. Equation 5.3 is not easily used because the gas concentrations inside the two surfaces of the barrier are not readily measurable. However, gas pressures in the air on the two sides of the barrier are much more easily measured, and according to Henry's law are proportional to C; that is,

$$C_i = S \times p_i \tag{5.4}$$

where S is the solubility of the gas in the barrier, and p_i is the partial pressure of the gas in the atmosphere in contact with the barrier. Combining Equations 5.3 and 5.4 gives

$$F = D \times S \times (p_1 - p_2)/t \tag{5.5}$$

Methods for measuring D and S are discussed in the first edition of this book (McHugh and Krochta, 1994b). Here, however, we will follow a simpler approach. Equation 5.6 is the permeance equation rewritten to eliminate the diffusion constant (D) and solubility (S) and instead include the permeability constant (P.C.), defined as $D \times S$, and *Permeance* (P), defined as P.C./thickness:

$$F = P.C. \times (p_1 - p_2)/t = P \times (p_1 - p_2) \tag{5.6}$$

Unfortunately, however, the words *permeance* and *permeability* are often used interchangeably in the literature. Sometimes it is obvious which is meant from the units used (see Section 5.2.4).

The permeance (P), rather than permeability (P.C.), is the more useful parameter for barriers of unknown thickness, and it is also useful for dealing with composite barriers, as shown in the next section. After simplification, the SI units of P are s/m (again, see Section 5.2.4).

5.2.2 Layered coatings and films

As mentioned, homogeneity of the coatings and films was assumed for development of Equations 5.3 through 5.6. However, edible coatings and films often lack homogeneity. Nevertheless, one form of nonhomogeneity that readily lends itself to modeling is the case of a film made up of two or more homogeneous films layered one upon the other. From Equation 5.6, it follows that the drop in partial pressure of gas across each film layer is

$$(p_1 - p_2) = F/P \tag{5.7}$$

Because the films are layered, the flow through each layer is the same. From Equation 5.6, the pressure drop across each layer of a layered film is F/P for that layer. The total pressure drop across the composite film is the sum of the pressure drops across each layer equal to the pressure drop across the layered film; therefore,

$$F/P_{layered} = F(1/P_a + 1/P_b + \cdots + 1/P_n) \tag{5.8}$$

where $P_{layered}$ is permeance of the layered coating, P_a is the permeance of layer a, P_b is the permeance of layer b, and so forth. Now divide both sides of the equation by F. For a barrier with two layers,

$$1/P_{layered} = 1/P_a + 1/P_b \quad (5.9)$$

The reciprocal of permeance is sometimes called resistance (R), which helps those familiar with Ohm's law to make parallels between electrical current flowing through series resistors and gas permeating through layered barriers. In applying Equation 5.9 to fresh fruit coatings, the two layers would be noncoated peel and coating (Hagenmaier and Shaw, 1992). Thus,

$$R_{CoatedFruit} = R_{Coating} + R_{Fruit} \quad (5.10)$$

5.2.3 Moisture gradients

Probably the most common departure from homogeneity for edible coatings is the case when relative humidity is different on the two sides of the barrier (Batdorf, 1989; Lebovits, 1966; Stannet, 1985). For most coatings and films, P.C. becomes higher with increasing relative humidity, especially for hydroscopic films.

In case of a relative humidity (R.H.) gradient across a barrier, the effective permeability can be calculated by integration:

$$P.C._{mean} = (R.H._{.2} - R.H._{.1})^{-1} \times \int_{1}^{2} P \times d(R.H.) \quad (5.11)$$

Crank (1964) gives a similar equation for the mean diffusion constant. Not having a mathematical expression for P.C. as a function of R.H., it seems best to do the integration graphically from measurements of P.C. at various levels of R.H. on the two sides of the coating or film. The difficulty here is that at high levels of R.H., some coatings swell or undergo a glass-state transition, in which case steady-state flow is reached very slowly (Crank, 1964; Frisch and Stern, 1981; Marom, 1985). Such cases are beyond the scope of this chapter.

5.2.4 Units for permeability and permeance

Unfortunately, the units for permeability and permeance are so awkward that it is necessary to discuss them. Had diffusion units been used as a guide, the situation could be much simpler, From Equation 5.3, it is evident that the units for diffusion (D) could have been chosen to be thickness × flow/concentration, for example, $m \times (kg/m^2 s) \div (kg/m^3)$. Instead, by custom, the individual terms are *canceled out*, leaving only m^2/s as the units for diffusion.

Had that same simple approach been used for permeability, then from Equation 5.6, the SI units for permeability would be just s (seconds) as shown by the following:

$$P.C. = thickness \times flowrate/pressure$$
$$= m \times (kg/m^2 s) / [kg \times m/(s^2 m^2)] = s \tag{5.12}$$

Unfortunately, the tradition for permeability is to employ units for thickness, flow, and pressure, and many different choices have been made for each of these components. For example, flow rates for fixed gases are commonly given in volume or moles per day or per second; thickness is expressed in mm, cm, m, or even mils (1 mil = 0.001 inch). Barrier area is given in units of cm^2, m^2, or 100 $inches^2$. Gas pressure is indicated in terms of mmHg, atmospheres, bars, or (for water vapor) as % relative humidity. Suppliers of packaging film often use $mL/(m^2$ day atm) for oxygen permeability and $g/(m^2$ day) at 100% RH (www.Dupont.com, www.sealedair.com). In the scientific literature, there is some preference for SI units, although even those differ because quantity of gas can be expressed in m^3 or moles (Banks et al., 1993). The wide variety of units in use makes it frequently necessary to convert from one set of units to another (Table 5.1). Thus, it has become the convention in discussions of permeability to present a table of conversion factors. However, such tables are understandably too small. For example, with just 12 permeability units, 66 different conversion factors would be needed for each gas. A simpler method for converting permeability to other units is presented in Table 5.1. The first step is to convert from the units given to the simplest set of units, namely, seconds. A second calculation is then performed to convert to any of the other units listed. With this approach, only 12 factors are needed to convert between that many permeability units for all gases.

Finally, the units for permeance are those used for permeability divided by thickness.

5.2.5 Measurement of barrier properties

The methods and equipment currently used for measurement of barrier properties are almost exclusively designed for packaging materials, which are manufactured in great quantities and demand routine quality-control measurements of permeability. Methods used for packaging films can sometimes be used unchanged to determine the properties of edible films, although somewhat different methods and equipment are sometimes more appropriate.

5.2.5.1 Methods

The American Society for Testing and Materials (www.astm.org) publishes many methods for measuring permeability, including E96/E96M (gravimetric water vapor), F1249 (water vapor with infrared sensor), F3985 and F1307 (oxygen with coulometric sensors), and F1927 (oxygen with coulometric detector). The German standards are DIN 53380-1 (volumetric method), DIN 53380-2 (manometric method), DIN53380-3 (carrier gas method), and DIN 53380-4 (infrared absorption method).

Table 5.1 Conversion factors with a confectionary shellac[a] coating as an example

Units	Conversion factor	Permeability of confectionary shellac		
		O_2	CO_2	H_2O vapor
kg m/(m² s Pa), or kg/m s Pa, or s[b]	1	9.52×10^{-19}	4.57×10^{-18}	1.12×10^{-14}
pg/m s Pa	1.00×10^{-15}	9.52×10^{-4}	4.57×10^{-3}	11.2
g/m d atm	1.14×10^{-3}	8.34×10^{-6}	4.00×10^{-5}	0.0984
µg/m d Pa, or g mm/m² d kPa	1.16×10^{-4}	8.23×10^{-5}	3.95×10^{-4}	97.2
g mil/m² d atm	2.90×10^{-18}	0.328	1.58	3.88×10^{3}
g mil/m² d mmHg	2.21×10^{-15}	4.32×10^{-4}	2.07×10^{3}	5.1
fL/m s Pa	$4.46M \times 10^{-20}$	0.67	2.33	1.40×10^{4}
mol/m s Pa	$M/1000$	2.97×10^{-17}	1.04×10^{-16}	6.24×10^{-13}
pL/m d Pa, or mL µm/m² d kPa	$5.16M \times 10^{-22}$	57.6	201	1.21×10^{6}
mL/m d atm	$5.10M \times 10^{-18}$	5.84×10^{-3}	2.04×10^{-2}	123
mL mil/m² d atm, or mL/m² d atm at 25.4 µm thickness	$1.29M \times 10^{-22}$	230	803	4.83×10^{6}
mL mil/(100 in²) d atm	$2.01M \times 10^{-21}$	14.8	51.8	$3.11M \times 10^{5}$

Notes: To convert into units of s (seconds), multiply by the conversion factor. To convert from s, divide by the factor. M = g/mol; pg = 10^{-12} gram; fL = 10^{-15} liter; mil = 1/1000 inch.

[a] See Hagenmaier, R., and P. Shaw, *J. Agric. Food Chem.* 39, 825–829, 1991a. The confectionary shellac coating was made by drying an ethanol solution of shellac.

[b] These three options are identical in value. The first shows all terms. For the second, a common factor (m) in numerator and denominator was removed. For the third, all possible common factors were removed, leaving only s (seconds).

[c] Polymer films are often made with thickness of 1 mil, and their barrier properties are here expressed as permeance for a film of that thickness, namely, 25.4 µm. Because permeability equals permeance × thickness, reporting the CO_2 permeance of a 1-mil film as 803 mL/m² d is equivalent to saying it's 803 mL mil/m² d atm. What is reported is actually permeance, unfortunately often labeled as permeability, which makes it seem as if the units are incorrect.

The Japanese Industrial Standards are JIS K7129 for water vapor and K7126 for oxygen. The ISO standard for oxygen is CD15105-2. The Technical Association of the Pulp and Paper Industry (TAPPI) standard for water vapor is T557. Detailed discussion of these methods is beyond the scope of this chapter. Some of the standard methods were, however, ably addressed in some detail by McHugh and Krochta (1994b) in the first edition of this book.

5.2.5.2 Equipment

The standards are incorporated into some of the equipment designed to measure permeability. In alphabetical order, here are some companies that sell equipment to measure gas permeabilities of barriers:

- Hanatek, East Sussex, United Kingdom (www.hanatekinstruments.com)
- Illinois Instruments, Johnsburg, Illinois (www.illinoisinstruments.com)
- Mocon, Minneapolis, Minnesota (www.mocon.com)
- PBI-Dansensor, Ringsted, Denmark (www.pbi-dansensor.com)
- Versaperm, Sittingbourne, Kent, United Kingdom (www.versaperm.com)

In general, instruments made to measure permeability hold the film in a cell divided into two chambers by the test film. Pure or diluted test gas (usually water vapor or oxygen) is circulated through the first chamber. An inert carrier gas (usually nitrogen) is circulated through the second chamber. The concentration of the test gas in the second chamber is measured at different times. Testing is complete when the test gas concentration no longer increases with time. The instruments measure permeance. Permeability is determined by multiplying permeance by barrier thickness.

5.2.5.3 Analytical errors

The equipment and methods are designed to deal with the major issues involved in measuring permeability, namely, gas leaks; concentration gradients; control of temperature, pressure, and relative humidity; and analysis of gas concentrations.

Gas leaks are the most common problem and are also the most serious. Leaks may occur not only across the film, but also between cell and atmosphere. If there are holes in the film, then what is being measured is not just permeability but rather a combination of permeability, effusion, and—if the barometric pressures of the two cells are not perfectly matched—mass transfer. Leaks between equipment and atmosphere are particularly a problem when measuring the permeability of oxygen and nitrogen because of their high concentration in the atmosphere. Therefore, it is important to routinely test for leaks. The most common leaks are holes in the film, inadequate seals between film and cell, and poor tubing connections.

One method for testing for leaks is to increase total pressure of the test gas on one side of the barrier and monitor change in its concentration on the low-pressure side. Leakage from the atmosphere can be tested by monitoring N_2, O_2, or water vapor while circulating the carrier gas through both sides of the cell. Another method is to monitor concentration of the target gas when the test film has been replaced by an impermeable barrier, for example, a Saran™-aluminum foil laminate. Holes in the barrier can sometimes be detected with microscopic examination of the film.

Circulation of the carrier gas should be sufficient to avoid concentration gradients within the cells, which is important because the barrier is usually some distance removed from the analytical device.

Control of temperature is important for two reasons. First, knowledge of the temperature is needed in order to calculate gas concentration. Second and most importantly, permeability is dependent on temperature. For example, the oxygen permeability of a polyethylene wax coating has activation energy of about 5 kcal/mol (Hagenmaier and Shaw, 1991b), which means that oxygen permeability increases about 35% with a 10°C rise in temperature.

It is important to control relative humidity on both sides of the barrier for three reasons. First, for measurement of water vapor permeability, the difference in relative humidity is the driving force and therefore must be known. Second, some gas analytical methods are adversely affected by high relative humidity of the gas stream. Third, and most importantly, the permeability of most barriers, particularly food coatings and films, is highly dependent on relative humidity (more below).

The choice of analytical method is important because of the very big range in permeability values of different barriers. This wide range challenges almost all analytical methods on the high or low side of their capability; therefore, different pieces of equipment often have quite different ranges of permeability.

5.2.5.4 Coating permeability

Measurements of permeability of coatings have all the problems encountered with films and others in addition. Thus, if a coating can be cast as a film sturdy enough to maintain its integrity and structure during handling and analysis, then it is best to measure its permeability in that form. However, many coatings disintegrate when removed from the surface on which they are prepared.

For such coatings, the method of determining coating permeability in our laboratory was based on measurements of layered barriers consisting of the coating spread onto packaging film (Hagenmaier and Shaw, 1991a, 1991b, 1992).

In order to determine coating permeability, measurements of permeability are required for noncoated film, and also for coated film. The coating permeability is then calculated from Equation 5.9.

There are at least four major sources of error for this method. First, there are the difficulties in preparation of a uniformly coated film, which is particularly troublesome for liquid coatings that tend to bead up and form bare spots on the film early in the drying process. This difficulty can be lessened by coating the *treated* side of a packaging film, which is the side on which the label is to be printed. Second, there is the problem of increased variance caused by subtracting two numbers that are sometimes rather close together in value. To minimize this problem, the bare film should have the highest possible permeability. Third, when exposed to high relative humidity, some coatings become sticky enough to adhere to the cell walls or to any screens used to hold the film in place. Therefore, it is best to use a cell designed so that nothing contacts the coated side film. Cell surgery was sometimes required. Finally, the calculations of permeability can be difficult for the case when there is a relative humidity gradient across the barrier, because the gradients across the coating are difficult to estimate (see Equation 5.11).

Finally, good control of total pressure on the two sides of the film is necessary in order to avoid distorting the film. Packaging films tend to be more elastic than coatings, with the result that uneven pressure can cause separation of coating from the film.

Table 5.2 Examples of edible films and packaging films with relatively high and low values of permeability

Gas	Type of film	High or low	Film description	Permeability constant(s)
Oxygen	Packaging	High	Polyethylene vinyl acetate[a]	530×10^{-19}
		Low	NC coated cellophane[a]	0.4×10^{-19}
	Edible	High	Methylcellulose[b]	170×10^{-19}
		Low	Chitosan[c]	0.2×10^{-19}
Water vapor	Packaging	High	Cellulose acetate[a]	120×10^{-15}
		Low	Polyvinylidene[a]	0.3×10^{-15}
	Edible	High	Lactic acid casein and glycerol[d]	630×10^{-15}
		Low	Candelilla wax[e]	0.2×10^{-15}

Notes: Water vapor permeabilities were measured with 90% to 100% R.H. on one side of the film, 0% on the other. Oxygen permeabilities were measured at 0% R.H.

[a] The packaging film permeabilities are limited to those included in data from Sacharow, S., *Principles of Food Packaging*, AVI, Westport, CT, 1980.
[b] Park, J.J., and M.S. Chinnan. *J. Food Engr.*, 25, 497–507, 1995.
[c] Gontard et al., 1996
[d] Chick, J., and Z. Ustonol, *J. Food Sci.*, 63, 1024–1027, 1998. 0/90% R.H.
[e] Greener, I.K., *Physical properties of edible films and their components*. Ph.D. thesis, University of Wisconsin, Madison, 1992; 25°C, 0/100% R.H. (see first edition, p. 179).

5.2.6 Permeability values

The range of values for permeability to water vapor and oxygen for edible films and coatings is comparable to the range for polymeric packaging films (Table 5.2). Thus, there is good opportunity to find an edible film or coating suitable for any given application.

The study of the permeability of polymers used as packaging films has shown that many factors contribute to permeability of polymers (Rogers, 1985). Low density contributes to high permeability because loosely packed polymers provide sites for diffusing molecules to occupy when moving through the polymer. For example, it is well known that low density polyethylene film has higher gas permeability than polyethylene with higher density.

The presence of certain functional chemical groups in the barrier tends to result in low oxygen permeability. In decreasing order of permeability, these are $-OH$, $-CN$, $-Cl$, $-F$, and $-COOCH_3$ (Ashley, 1985).

It has been frequently noted that the gas permeabilities of various polymers tend to have fairly consistent G values, where G for oxygen is the ratio of its oxygen permeability to its nitrogen permeability:

$$G_{oxygen} = P_{oxygen}/P_{nitrogen} \qquad (5.13)$$

For example, Ashley (1985) reported the mean values of G_{oxygen} for 11 polymers to be 3.6 (range 2.8 to 5.6) despite the fact that measured CO_2 permeability differed 50,000-fold for the different polymers. Stannet (1985) reported similar findings.

However, no findings consistently apply to the relationship between permeability of water vapor and the permanent gases.

Published values of permeability, even if very accurately measured, might not be useful to predict barrier properties. One reason is that relatively small differences in coating composition can result in big differences in permeability. For example, McHugh and Krochta (1994a) found that the oxygen permeability of whey protein isolate films increased fourfold when the amount of glycerol plasticizer was increased from 15% to 30% of the mixture. They found a 140-fold increase for a similar film in which sorbitol replaced the glycerin.

Another problem is that permeability of coatings and films can change with time. First, volatile components of the film will slowly dissipate. Second, for a film or coating in contact with a food product, some components of the coating, especially plasticizers, can diffuse into the food product or into the air.

Finally, and most importantly, permeability is very dependent on relative humidity (R.H.). For example, in the same study, McHugh and Krochta found about 18-fold increase in water vapor permeability of a film consisting of 62% whey protein isolate and 38% glycerin when permeability was measured at 65% R.H. instead of 11% to 65% R.H. on the high-humidity side of the film. In both cases, the R.H. was 0% on the low-humidity side of the barrier.

However, large as the prediction error may be because of effects of composition and humidity, the unpredictability resulting from effusion may be even larger.

5.2.7 Effusion compared with permeance

Consider a film or coating with hole area amounting to one billionth of the total area, equivalent to one pinhole (8 micron diameter) for a 500 cm² package area, about the size a typical consumer package. Table 5.3 shows the amount of gas passing through the hole as a fraction of that permeating through the intact coating or film. For O_2 the ratio is very high, less so for CO_2 because its permeation rate is higher than that of O_2. However, even for CO_2, a large error can result from ignoring the effusion rate.

An admitted shortcoming of Table 5.3 is that it shows values for permeance flow rates that are measured and effusion flow rates that are calculated. It would be better, it might seem, to compare experimental flow rates through the same barriers with

Table 5.3 Effusion flow rate divided by permeance flow rate through a barrier consisting of 25-μm-thick coating with open holes with area one billionth of the total barrier area, at 0°C, 75% R.H.

Coating or film	Effusion flow/permeance flow[a]		
	O_2	CO_2	Water vapor
Shellac[b]	390	9.6	0.0024
Candelilla Wax[c]	13	3.6	0.016

[a] Effusion was calculated from Equation 5.1 and permeance from Equation 5.7.
[b] The shellac permeabilities are from Table 5.1.
[c] For candelilla wax, the O_2 and W.V. permeabilities are 2.9×10^{-18} s and 1760×10^{-18} s, respectively (Greener, 1992). The CO_2 permeability is 12.2×10^{-18} s (Hagenmaier unpublished).

and without holes, attributing difference to effusion. Unfortunately, however, that does not seem possible with the equipment used to measure permeance, because, as discussed, these require flows of gas on both sides of the barrier to minimize pressure gradients. Because of gas flowing through the system, pressure gradients are inevitable, leading almost certainly to mass transfer though any holes in the barrier.

The passage of gas through holes in the barrier is very important for fruit coatings, which will now be considered as a real-life example of errors involved in using the permeance equations to predict barrier properties.

5.2.8 Fruit coatings as test of permeance equations

Fruit coatings are probably the most common type of edible coating. Almost all fresh apples and citrus fruit sold to consumers in the United States, about 2.4 and 2.8 million M.T. per year, are coated. These are coated at a typical application rate of roughly 1 kg coating per M.T. with coatings made up of about 20% total solids and 80% volatiles, mostly water. Therefore, annual U.S. usage of edible citrus and apple coatings is roughly 5000 and 1000 M.T., wet weight and dry weight, respectively. Total world consumption of fresh citrus and apples is about 90 to 100 million M.T. per year, but it is difficult to determine what percentage of that fruit is coated.

The primary reason coatings are applied to citrus and apples is to increase gloss and in that way improve marketability. However, it is a well-known fact that the flavor quality of coated apples and oranges is sometimes adversely affected by coatings that overly restrict gas exchange (Ahmad and Khan, 1987; Bai et al. 2002c; Cohen et al., 1990). Therefore, it is very important for the fresh apple and citrus industry to be aware of just how various coatings affect their gas exchange.

Therefore, testing of the permeance equations for gas exchange of fruits is doubly important. First, it is an opportunity to understand how the application of coatings affects gas exchange. Second, because gas exchange can be directly measured with and without the coatings, this exercise serves as a test of the value of the equations for fruit coatings.

Equation 5.10 is our basis for verifying the accuracy of the permeance-only theory of gas exchange, because all terms can be independently measured. Values for $R_{CoatedFruit}$ and R_{Fruit}, the resistance of coated and noncoated fruit, respectively, are determined by measurement of the gas exchange rate, surface area, and gas concentrations. The CO_2 and H_2O exchange rates are determined from measurement of respiration rate and weight loss, respectively. The values for $R_{Coating}$ are measured as already described in the discussion of Equation 5.9.

Tables 5.4 through 5.7 contain data that show how poorly the gas flow of coated fruit was predicted from Equation 5.10. For example, for oranges, the CO_2 resistance of oranges with carnauba wax coating was more than twice the value predicted from Equation 5.10 (18×10^{10} m/s instead of 7.7×10^{10} m/s) (Table 5.4). For shellac-coated apples, it was much lower—namely 27.2×10^{10} m/s instead of 43.1×10^{10} m/s (Table 5.6). Further, the water vapor resistances were also not well predicted by the equations (Tables 5.5 and 5.7). In fact, the errors were sufficiently large that the equations are not useful.

These shortcomings are not surprising considering that the skin of fruit has many holes that allow effusion to occur. For example, the skin of Shamouti orange fruit has

Table 5.4 Increase in CO_2 resistance of Valencia oranges when coated[a]

	How determined	Polyethylene wax	Carnauba wax	High-gloss resin
R_{Fruit}	Measured	5.5	5.5	5.5
$R_{Coating}$	Measured	0.1	2.2	25
$R_{CoatedFruit}$	Measured	9.5	17.9	23.7
$R_{CoatedFruit}$	Equation 5.10	5.6	7.7	30.5

[a] Resistance values are in units of 10^{10} m/s. Based on data from Hagenmaier, R.D., and R.A. Baker, *J. Agric. Food Chem.*, 41, 283–287, 1993b; Hagenmaier, R., and R.A. Baker, *Proc. Fla. State Hort. Soc.*, 107, 261–265, 1994; Hagenmaier, R., and R.A. Baker, *HortScience*, 30, 296–298, 1995; Hagenmaier, R., *Postharvest Biol. Tech.*, 37, 56–64, 2005; Hagenmaier, R., *Proc. Fla. State Hort. Soc.*, 6, 418–423, 2003.

Table 5.5 Increase in H_2O resistance of Valencia oranges when coated[a]

	How determined	Polyethylene wax	Carnauba wax	High-gloss resin
R_{Fruit}	Measured	0.07	0.07	0.07
$R_{Coating}$	Measured	0.09	0.06	0.02
$R_{CoatedFruit}$	Measured	0.13	0.18	0.11
$R_{CoatedFruit}$	Equation 5.10	0.16	0.13	0.09

[a] Resistance values are in units of 10^{10} m/s. Based on data from Hagenmaier, R.D., and R.A. Baker, *J. Agric. Food Chem.*, 41, 283–287, 1993b; Hagenmaier, R., and R.A. Baker, *Proc. Fla. State Hort. Soc.*, 107, 261–265, 1994; Hagenmaier, R., and R.A. Baker, *HortScience*, 30, 296–298, 1995; Hagenmaier, R., *Postharvest Biol. Tech.*, 37, 56–64, 2005; Hagenmaier, R., *Proc. Fla. State Hort. Soc.*, 6, 418–423, 2003.

Table 5.6 Increase in CO_2 resistance of 'Delicious' apples when coated[a]

	How determined	Shellac	Carnauba wax
R_{Fruit}	Measured	8.1	8.1
$R_{Coating}$	Measured	35	2.2
$R_{CoatedFruit}$	Measured	27.2	12.8
$R_{CoatedFruit}$	Equation 5.10	43.1	10.3

[a] Resistance values are in units of 10^{10} m/s. Based on data from Bai et al., 2002b; Hagenmaier, R., *Postharvest Biol Tech.*, 19, 147–154, 2000; Hagenmaier, R., *Postharvest Biol. Tech.*, 37, 56–64, 2005; Hagenmaier, R., *Proc. Fla. State Hort. Soc.*, 6, 418–423, 2003.

Table 5.7 Increase in H_2O resistance of 'Delicious' apples when coated[a]

	How determined	Shellac	Carnauba wax
R_{Fruit}	Measured	0.40	0.40
$R_{Coating}$	Measured	0.02	0.06
$R_{CoatedFruit}$	Measured	0.50	0.72
$R_{CoatedFruit}$	Equation 5.10	0.42	0.46

[a] Resistance values are in units of 10^{10} m/s. Based on data from Bai et al., 2002b; Hagenmaier, R., *Postharvest Biol. Tech.*, 37, 56–64, 2005; Hagenmaier, R., *Proc. Fla. State Hort. Soc.*, 6, 418–423, 2003.

about 7000 stomata/cm^2 (Rokach, 1953). Orange fruit stomata, which have diameter of roughly 10 μm, are partially blocked with natural or applied wax (Ben-Yehosha et al., 1983, 1985). It has been estimated that applied coatings completely block the openings if coating thickness is at least 20 μm thick (Brusewitz and Singh, 1985), but that is about 10 times the thickness of fruit coatings (Trout, 1953). The holes in the skin of apple fruit consist of lenticels, stem end, and calyx openings (Bai et al., 2002b). Like citrus and apples, papaya has many stomata, about 2000/cm$_2$ (Paull and Chen, 1989). Air passes through the skin of the tomato primarily through the stem scar (Cameron, 1982). Notably, stomata are the major pathway for gas exchange through leaves (Ball, 1987).

Knowing that the fruit peel has holes makes it easy to understand why the layer formula, Equation 5.10, can give a result that is either too high or too low. When a fruit is coated, some of the holes in the peel can be blocked, thus reducing the gas flow through that layer and giving lower gas flow than expected (Bai et al., 2002b; Hagenmaier and Baker, 1993a, 1993b). On the other hand, the added coating might have holes aligned with those in the peel, thus making the coating layer a more porous barrier than the continuous film used to measure coating permeability. With luck, these effects may cancel one another. If not, the calculated gas exchange rate result may be higher or lower than that measured. However, a suitable equation is not useful if it depends on luck.

Therefore, it seems that to be useful, mathematical models for gas exchange in fruit must include gas exchange through holes by effusion. This is admittedly difficult, especially considering the fact that application of coatings blocks some of the holes. Must a useful model of gas exchange for fruit also include mass transfer through the holes? That is the question next addressed.

5.2.9 Nitrogen and mass transfer in fresh fruit

Consider now some data that suggest mass transfer is important in the steady-state gas exchange of some fresh fruit, based on a new look at previously published work in our laboratory. The N_2 concentrations were omitted from the publications because we failed to realize their importance. We thought of N_2 as of no importance in showing how fruit quality is affected by coatings.

See Figure 5.1 for the ratio (interior/exterior) of N_2 concentrations of Valencia oranges. The data also include the sum of measured concentrations of N_2, O_2, CO_2, and Ar. It is important to note that this sum was calculated by adding the individual measured values of these gases, not by measuring total pressure, T. The gas samples were injected into the gas chromatograph at atmospheric pressure, not at the pressure inside the fruit. The mean sum of the concentrations of N_2, O_2, CO_2, and Ar for the Valencia gas samples in Figure 5.1 was 98.7 kPa, with 0.07 kPa S.D. of mean. The expected partial pressure of water vapor inside the fruit at 20°C is 2.3 kPa. Therefore, the partial pressures of N_2, O_2, CO_2, Ar, and water vapor added up to 101 kPa, namely, one atmosphere. The point of this discussion is that the difference in measured internal and external N_2 cannot be attributed to higher pressure inside the fruit. Measurements with other fruit show the same pattern observed with Valencia

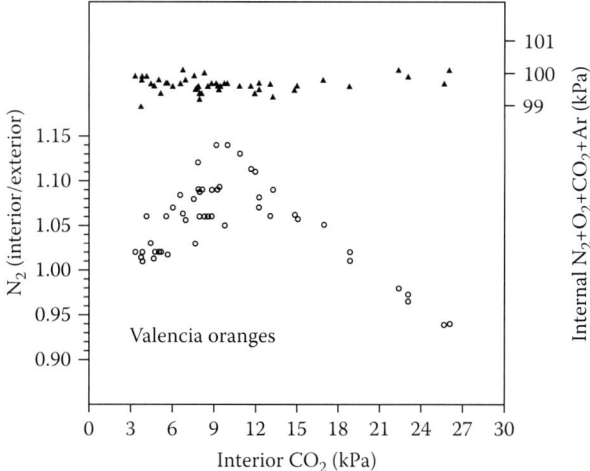

Figure 5.1 The ratio of N_2 concentrations (interior/exterior) for coated and noncoated 'Valencia' oranges stored at 20°C. (See Hagenmaier, R., *Postharvest Biol. Tech.* 37, 56–64, 2005.)

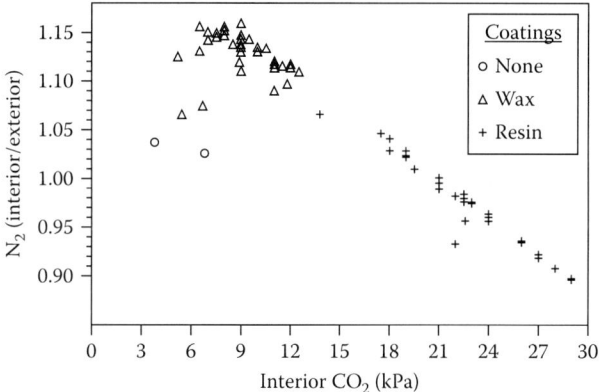

Figure 5.2 The ratio of N_2 concentrations (interior/exterior) for coated and noncoated 'Honey' tangerines stored at 20°C. (See Hagenmaier, R., *Proc. Fla. State Hort. Soc.*, 114, 170–173, 2001; Hagenmaier, R., *Postharvest Biol. Tech.*, 24, 79–87, 2002.

oranges (Figures 5.2, 5.3, and 5.4). Fruit with low internal CO_2 had about equal partial pressures of N_2 inside and outside the fruit. Apples or citrus fruit with about 3 to 18 kPa internal CO_2 had elevated internal N_2, with maximum at roughly 8 kPa internal CO_2, and many had internal N_2 partial pressures above 85 kPa, 10% higher than the N_2 concentration in the air outside of the fruit (about 77 kPa, depending on temperature and relative humidity). At elevated values of CO_2, usually indicative of anaerobic respiration, the N_2 partial pressure was lower inside the fruit than outside.

Consider now those fruit with internal N_2 concentrations above ambient values. First, remember that effusion and permeance of nitrogen are both driven by difference in its partial pressure. Second, it is well known that internal gas pressures in

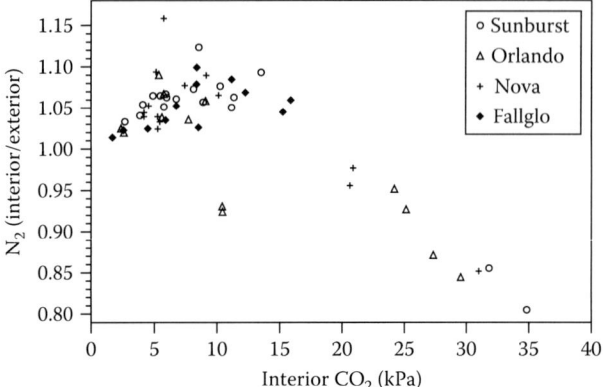

Figure 5.3 The ratio of N_2 concentrations (interior/exterior) for coated and noncoated tangerines stored at 20°C. (See Hagenmaier, R., *Postharvest Biol. Tech.*, 24, 79–87, 2002.)

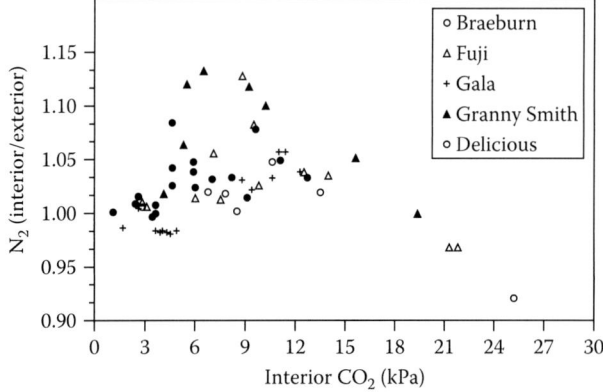

Figure 5.4 The ratio of N_2 concentrations (interior/exterior) for coated and noncoated apples stored at 20°C. (See Bai, J., E.A. Baldwin, and R.D. Hagenmaier, *HortScience*, 37, 559–563, 2002b; Bai, J., R.D. Hagenmaier, and E.A. Baldwin, *Postharvest Biol. Tech.* 28, 381–390, 2002a; Bai, J., V. Alleyne, R.D. Hagenmaier et al., *Postharvest Biol. Tech.*, 28, 259–268, 2003; Hagenmaier, R., *Postharvest Biol. Tech.*, 37, 56–64, 2005.)

citrus fruit, which does not ripen much during storage, are approximately steady state soon after the fruit is coated (Hasegawa and Iba, 1980). In other words, under steady-state conditions, nitrogen permeates and effuses out of the fruit. But if the fruit constantly lost that gas, how did it manage to maintain steady-state internal N_2 partial pressure? I propose as a hypothesis that the total internal pressure was slightly lower than atmospheric pressure, at least for some portions of the peel.

As already noted, mass transfer normally does not separate gases; therefore, if nitrogen flowed into the fruit because of difference in total pressure, so did O_2. Therefore, if the pressure was lower inside the fruit, this inward flow supplied some of the oxygen needed for respiration and maintenance of quality, and is therefore of some importance.

Finally, to briefly summarize this section, a big hole in the prevailing theory of how edible coatings affect gas exchange is its failure to include the gases passing through holes by effusion and mass transfer.

In summary, understanding coating permeability to gases and water vapor is essential to the successful design of commercial coating and film products.

References

Ahmad, M., and I. Khan. 1987. Effect of waxing and cellophane lining on chemical quality indices of citrus fruit. *Plant Foods Human Nutr.* 37:47–57.

Amarante, C., and N.H. Banks. 2001. Postharvest physiology and quality of coated fruits and vegetables. *Hort. Rev.* 26:161–238.

Ashley, R.J.G. 1985. Permeability and Plastics Packaging. In *Polymer Permeability*, J. Comyn, ed. New York: Elsevier, pp. 269–307.

Bai, J., V. Alleyne, R.D. Hagenmaier, J.P. Mattheis and E.A. Baldwin. 2003. Formulation of zein coatings for apples (*Malus domestica* Borkh). *Postharvest Biol. Tech.* 28:259–268.

Bai, J., E.A. Baldwin, and R.D. Hagenmaier. 2002a. Alternatives to shellac coatings provide comparable gloss, internal gas modification, and quality for 'Delicious' apple fruit. *HortScience* 37:559–563.

Bai, J., R.D. Hagenmaier, and E.A. Baldwin. 2002b. Coating selection for 'Delicious' and other apples. *Postharvest Biol. Tech.* 28:381–390.

Bai., J., R.D. Hagenmaier, and E.A. Baldwin. 2002c. Volatile response of four apple varieties and different coatings during marketing at room temperature. *J. Agric. Food Chem.* 50:7660–7668.

Ball, J.T. 1987. Calculations related to gas exchange. In *Stomatal Function*. E. Zeiger, G.D. Farquhar, and I.R. Cowan, eds. Stanford, CA: University Press, pp. 445–473.

Banks, N.H., D.J. Cleland, A.C. Cameron, R.M. Beaudry, and A.A. Kader. 1993. Proposal for a rationalised system of units for postharvest research in gas exchange. Unpublished recommendations.

Batdorf, V. 1989. Effect of relative humidity on permeance of coatings. *J. Testing Eval.* 17:299–306.

Ben-Yehoshua, S., S.P. Burg, and R. Young. 1985. Resistance of citrus fruit to mass transport of water vapor and other gases. *Plant Physiol. (Bethesda)* 79:1048–1053.

Ben-Yehoshua, S., S.P. Burg, and R. Young. 1983. Resistance of citrus fruit to C_2H_4, O_2, CO_2 and H_2O mass transport. *Proc. Plant Growth Reg. Soc. Am.* 10:145–150.

Brusewitz, G.H., and R.P. Singh. 1985. Natural and applied wax coatings on oranges. *J. Food Proc. Preserv.* 9:1–9.

Cameron, A.C. 1982. Gas diffusion in bulky plant organs. PhD Diss., University of California, Davis. *Diss. Abstr. Intl.* 43:929B–930B.

Chick, J., and Z. Ustonol. 1998. Mechanical and barrier properties of lactic acid and rennet precipitated casein-based edible films. *J. Food Sci.* 63:1024–1027.

Cohen, E., Y. Shalom, and E. Rosenberger. 1990. Postharvest ethanol buildup and off-flavor in 'Murcott' tangerine fruit. *J. Amer. Soc. Hort. Sci.* 115:775–778.

Comyn, J. 1985. Introduction to polymer permeability and the mathematics of diffusion. In *Polymer Permeability*. J. Comyn, ed. New York: Elsevier, pp. 1–10.

Crank, J. 1964. *The Mathematics of Diffusion*. Glasgow: Oxford at the Clarendon Press.

Frisch, H.L., and S.A. Stern. 1981. Diffusion of small molecules in polymers. *CRC Crit. Rev. Solid State Mater Sci.* II:123–187.

Glasstone, S. 1946. *Textbook of Physical Chemistry*. New York: D. Van Nostrand.

Graham, T. 1833. On the law of the diffusion of gases. Republished in 1995, *J. Membrane Sci.* 100:12–21.

Greener, I.K. 1992. *Physical properties of edible films and their components.* Ph.D. thesis, University of Wisconsin, Madison.

Hagenmaier, R. 2005. A comparison of ethane, ethylene and CO_2 peel permeance for fruit with different coatings. *Postharvest Biol. Tech.* 37:56–64.

Hagenmaier, R. 2003. Methods for measuring internal gases of citrus fruit and determining peel permeance. *Proc. Fla. State Hort. Soc.* 6:418–423.

Hagenmaier, R. 2002. The flavor of mandarin hybrids with different coatings. *Postharvest Biol. Tech.* 24:79–87.

Hagenmaier, R. 2001. Ethanol content of 'Mucrott' tangerines harvested at different times and treated with coatings of different O_2 permeability. *Proc. Fla. State Hort. Soc.* 114:170–173.

Hagenmaier, R. 2000. Evaluation of a polyethylene-candelilla coating for 'Valencia' oranges. *Postharvest Biol Tech.* 19:147–154.

Hagenmaier, R., and R.A. Baker. 1995. Layered coatings to control weight loss and preserve gloss of citrus fruit. *HortScience* 30:296–298.

Hagenmaier, R., and R.A. Baker. 1994. Internal gases, ethanol content and gloss of citrus fruit coated with polyethylene wax, carnauba wax, shellac or resin at different application levels. *Proc. Fla. State Hort. Soc.* 107:261–265.

Hagenmaier, R.D., and R.A. Baker. 1993a. Cleaning method affects shrinkage rate of citrus fruit. *HortScience* 28:824–825.

Hagenmaier, R.D., and R.A. Baker. 1993b. Reduction in gas exchange of citrus fruit by wax coatings. *J. Agric. Food Chem.* 41:283–287.

Hagenmaier, R., and P. Shaw. 1992. Gas permeability of fruit coating waxes. *J. Amer. Soc. Hort. Sci.* 117:105–109.

Hagenmaier, R., and P. Shaw. 1991a. Permeability of shellac coatings to gases and water vapor. *J. Agric. Food Chem.* 39:825–829.

Hagenmaier, R., and P. Shaw. 1991b. Permeability of coatings made with emulsified polyethylene wax. *J. Agric. Food Chem.* 39:1705–1708.

Hasegawa, Y., and Y. Iba. 1980. The effects of coating with wax on citrus fruit; characteristics of various wax coatings. *Bull. Fruit Tree Res. Sta., Okitsu, Japan.* 7:85–97.

Lebovits, A. 1966. Permeability of polymers to gases, vapors, and liquids. *Modern Plastics* March, 139–213.

Marom, G. 1985. The role of water in composite materials. In *Polymer Permeability*, J. Comyn, ed. New York: Elsevier, pp. 341–374.

McHugh, T.H., and J.M. Krochta. 1994a. Sorbitol- vs. glycerol-plasticized whey protein edible films: integrated oxygen permeability and tensile property evaluation. *J. Agric. Food Chem.*, 42(4):841–845.

McHugh, T.H., J.F. Aujard, and J.M. Krochta. 1994b. Permeability properties of edible films. In *Edible Coatings and Films to Improve Food Quality*, J.M. Krochta, E.A. Baldwin, and M.O. Nisperos-Carriedo, eds. Lancaster, PA: Technomic, pp. 139–187.

Park, H.J. 1999. Development of advanced edible coatings for fruits. *Trends Food Sci. Technol.* 10:254–260.

Park, J.J., and M.S. Chinnan. 1995. Gas and water vapor barrier properties of edible films from protein and cellulosic materials. *J. Food Engr.* 25:497–507.

Paull, R.E., and N.J. Chen. 1989. Waxing and plastic wraps influence water loss from papaya fruit during storage and ripening. *J. Amer. Soc. Hort. Sci.* 114:937–942.

Rogers, C.E. 1985. Permeation of gases and vapours in polymers. In *Polymer Permeability*, J. Comyn, ed. New York: Elsevier, pp. 11–73.

Rokach, A. 1953. Water transfer from fruits to leaves in the Shamouti orange tree and related topics. *Palestine J. Botany.* 8:146–151.

Sacharow, S. 1980. *Principles of Food Packaging.* Westport, CT: AVI.

Stannett, V.T. 1985. The permeability of plastic films and coated papers to gases and vapors. *Converting & Packaging*, September 22–26.

Tock, W. 1983. Permeabilities and water vapor transmission rates for commercial polymer films. *Adv. Polym. Technol.* 3(3):223–231.

Trout, S.A., E.G. Hall, and S.M. Sykes. 1953. Effects of skin coatings on the behavior of apples in storage. *Australian J. Agric. Res.* 4:57–81.

6 Role of edible film and coating additives

Roberto de Jesús Avena-Bustillos and Tara H. McHugh

Contents

6.1 Introduction	157
6.2 Plasticizers	158
6.2.1 Plasticizers in protein films and coatings	159
6.2.2 Plasticizers in polysaccharide films and coatings	166
6.3 Emulsifiers	167
6.4 Antimicrobial agents	170
6.5 Antioxidants and antibrowning agents	173
6.6 Nutrients, flavors, and colorants	175
6.7 Nanoparticles	176
References	177

6.1 Introduction

Edible films and coatings offer potential to extend the shelf life and improve the quality of virtually any food system. Edible films and coatings can control moisture, oxygen, carbon dioxide, flavor, and aroma transfer between food components and the atmosphere surrounding the food. Several reviews are available on edible film and coating systems (Gennadios et al., 1997; Krochta and De Mulder-Johnston, 1997; Arvanitoyannis and Gorris, 1999; Baldwin, 1999). The functional, sensory, nutritional, barrier, and mechanical properties of an edible coating can be altered by the addition of various chemicals in minor amounts (Krochta and De Mulder-Johnston, 1997; Debeaufort et al., 1998; Min and Krochta, 2005; Lin and Zhao, 2007). Films are usually regarded as stand alone, formed separate of any eventual intended use. Coatings involve formation of film directly on the surface of the object they intend to protect or enhance in some manner (Krochta, 2002). One of the unique functions of both edible films and coatings is the capability to incorporate functional ingredients into the film or coating matrix to enhance its functionality. This may include

1. Improving basic film or coating functionality, such as plasticizers for improved mechanical properties or emulsifiers for increased stability and better adhesion
2. Enhancing food quality, stability, and safety by incorporating antioxidants, antimicrobials, nutraceuticals, flavors, and color agents

Edible films and coatings have received increasing interest because films and coatings can carry a diversity of functional ingredients. Plasticizers, such as glycerol, acetylated monoglycerides, polyethylene glycol, and sucrose, are often used to modify the mechanical properties of the film or coating. Incorporation of these additives may cause significant changes in the barrier properties of the film. For example, the addition of hydrophilic plasticizers usually increases the water vapor permeability of the film. Other types of film additives often found in edible film formulations are antimicrobial agents, vitamins, antioxidants, flavors, and pigments (Krochta et al., 1994). A greater emphasis on safety features associated with the addition of antibacterial agents is an active area of research in edible film and coating technology (Cha and Chinnan, 2004). Application of nanoparticles to edible films and coatings is another new research area. In this chapter, we will discuss intended and unintended effects of food additives on film and coating properties.

A food additive was defined by Congress as "any substance, the intended use of which results directly or indirectly, in its becoming a component or otherwise affecting the characteristics of food." Excluded from that are substances that are generally recognized, among experts qualified by scientific training and experience to evaluate their safety, as having been adequately shown through scientific procedures (or, in the case of a substance used in food before January 1, 1958, through either scientific procedures or through experience based on common use in food) to be safe under the conditions of their intended use. This is the generally recognized as safe (GRAS) exemption. Importantly, it is the use of a substance, rather than the actual substance, that is eligible for the GRAS exemption. In addition, it is well settled that a mere showing that use of a substance is "safe" is not sufficient to exempt the substance from the act's definition of "food additive." Instead, the substance must be shown to be "generally recognized" as safe under the conditions of its intended use (FDA, 1997). The coating additives discussed in this chapter include a number of GRAS substances.

6.2 Plasticizers

A plasticizer, in most cases, is required for making edible films and coatings, especially for polysaccharide- and protein-based films because the structure of these is often brittle and stiff due to extensive interactions between polymer molecules (Krochta, 2002; Gennadios, 2004). Plasticizers are added to polymers to reduce brittleness. They decrease intermolecular forces between polymer chains, thus increasing flexibility and extendibility. Plasticizers are added to protein film-forming solutions to increase the flexibility and "soften" the structure of formed films. Generally, plasticizers act by entering between polymeric molecular chains, physicochemically associating with the polymer, reducing cohesion within the film network, and effectively extending and softening the film structure (Banker, 1966; Arnold, 1968; Guilbert, 1986). Examples of food-grade plasticizers incorporated into edible films are polyols (e.g., glycerol, sorbitol, and polyethylene glycol), sugars (e.g., glucose and honey), and lipids (e.g., monoglycerides, phospholipids, and surfactants). Commonly used plasticizers are liquid organic compounds such as polyols, mono-, di-, oligosaccharides, lipids, and lipid derivatives (Reiners et al., 1973; Padua and Wang, 2002). Glycerol, acetylated monoglyceride, polyethylene glycol, and sucrose are some of the most common plasticizers incorporated into the polymeric matrix to decrease glass transition temperature and increase film or coating flexibility (Guilbert and Gontard, 1995). Plasticizers are usually hygroscopic and attract water molecules. Water can also function as plasticizer, but it is easily lost due to dehydration in low relative humidity environments (Guilbert and Gontard, 1995). In addition to improving mechanical properties, plasticizers also affect the resistance of coatings to the permeation of vapors and gases (Sothornvit and Krochta, 2000, 2001), where hydrophilic plasticizers usually increase the water vapor permeability of the coating or film. Unfortunately, plasticizers generally also decrease the ability of coatings and films to act as barriers to the transport of moisture, gases, and aroma compounds (Krochta, 2002).

6.2.1 Plasticizers in protein films and coatings

Protein films and coatings need to have good strength and flexibility to prevent cracking during handling and storage (Torres, 1994). Therefore, plasticizers of relatively low molecular weight are typically added to protein film-forming preparations to increase the flexibility and "soften" the structure of formed films.

The amount of plasticizer added into protein film-forming preparations varies widely within the range of 10% to 60% by weight of the protein. Plasticizers that are commonly used with protein films and coatings are summarized in Table 6.1. The polyols, glycerol, and sorbitol, are the most commonly used plasticizers for protein-based films (Gennadios et al., 1994). Other hydrophilic molecules that have been used as protein film plasticizers include triethylene glycol, polyethylene glycol, propylene glycol, glucose, and sucrose (Gennadios, 2004). Also, fatty acids, such as oleic, palmitic, stearic, and linoleic acids, have been used to plasticize zein films (Lai et al., 1997; Santosa and Padua, 1999). In fact, water depresses the glass transition temperature of protein films acting as an effective plasticizer (Gontard and Ring, 1996). Thus, film moisture content, as affected by the relative humidity (RH) of the surrounding environment, largely affects protein film properties (Krochta, 2002).

Protein films and coatings are often quite stiff and brittle due to extensive interactions between protein chains through hydrogen bonding, electrostatic forces, hydrophobic bonding, and disulfide cross-linking (Krochta, 2002). Relatively small molecular weight hydrophilic plasticizers are often added, which mainly compete for hydrogen bonding and electrostatic interactions with the protein chains. The result of plasticizer addition is a reduction in protein chain-to-chain interactions, a lowering of the protein glass transition temperature(s), and an improvement in film flexibility (lowering of film elastic modulus). Also, film elongation (stretchiness or ductility) increases, and film strength decreases when plasticizers are added (Krochta, 2002). Unfortunately, plasticizers generally also decrease the film's ability to act as a barrier to moisture, oxygen, aroma, and oils.

Glycerol has been used to plasticize zein films. However, it tends to migrate from the bulk of the film matrix to the surface (Gennadios, 2004). According to Park et al. (1994a), the interaction between protein molecules and glycerol is weak, and excess glycerol easily migrates through the film matrix. Cast zein films are initially transparent, but after only a few hours they appear greasy and "cloudy" because of glycerol

Table 6.1 Edible films and coating plasticizers allowed for use in foods

Food additive	Classification, effect, or use	21 CFR #	Allowed use
Sugars			
Sucrose	Nutritive sweetener	184.1854	GRAS, GMP
Polyols			
Glycerol	Plasticizer, food flavoring	182.1320	GRAS/FS, GMP
Polyethylene glycol (MW 2200-9500)	Coating component on fresh citrus fruits, binder, plasticizer, lubricant, resinous/polymeric coatings	172.210	GMP, REG
Propylene glycol	Plasticizer, solvent, thickener, component of resinous and polymeric coating	184.1666	GRAS/FS
Sorbitol	Plasticizer, component of resinous and polymeric coatings	184.1835	GRAS, GMP, REG
Lipids			
Acetylated monoglycerides	Emulsifiers, coating component	172.828	GMP, REG
Lauric acid	Coating for fresh fruits	172.210	REG
Linoleic acid	Nutrient, dietary supplement	182.5065, 184.1065	GRAS, GMP
Oleic acid	Lubricant, binder, defoaming agent, and as excipient of other food-grade additives	172.840	GMP, REG
Palmitic acid	Lubricant, binder, defoaming agent, and as excipient of other food-grade additives; coating for fresh citrus fruits; antifoaming in food processing	172.860, 172.210, 173.340	GMP, REG
Stearic acid	Lubricant, binder, chewing gum component	172.860, 184.1090	GRAS, GMP, REG

Notes: GRAS: Generally recognized as safe. Substances in this category are by definition, under Sec. 201(s) of the FD&C Act, not food additives. Most GRAS substances have no quantitative restrictions as to use, although their use must conform to good manufacturing practices. GRAS/FS: Substances generally recognized as safe in foods but limited in standardized foods where the standard provides for its use. GMP: Good manufacturing practices. REG: Food additive for which a petition has been filed and a regulation issued.

"sweating out." Plasticizer migration results in loss of film flexibility (Park et al., 1992). Park et al. (1994a) used a mixture of glycerol and polyethylene glycol 400 (PEG with average molecular weight 400). They observed that migration rates of glycerol-PEG mixtures in zein-based films were slower than that of glycerol alone, and such mixtures could slow the deterioration of mechanical properties during film storage.

Hydrophilic plasticizers (e.g., glycerol) act as humectants, retaining the moisture that plasticizes zein. However, they are not effective at temperatures below freezing or in low RH environments. As water freezes or evaporates, zein reverts to a stiff and brittle material. Fatty acids have been used to plasticize zein films (Reiners et al., 1973; Park et al. 1994b; Lai et al., 1997; Lai and Padua, 1998). Unsaturated fatty acids remain effective plasticizers at temperatures below freezing and in low RH environments. However, oleic acid-plasticized zein films became brittle and almost colorless over a storage period of 6 weeks in a light stability chamber. Property changes were attributed to oleic acid oxidation (Padua and Wang, 2002).

Water vapor transfer rate increased with increasing moisture content and plasticizer concentration of caseinate films (Arvanitoyannis et al., 1996; Siew et al., 1999; Lima-Lima et al., 2006). Chick and Ustunol (1998) reported an increase in water vapor permeability (by 8% to 32%) when the protein-to-plasticizer ratio decreased from 1.4:1 to 1:1 for films made from lactic acid casein or rennet casein with either glycerol or sorbitol as plasticizer.

Different plasticizers may impart different effects on film water barrier properties (Chen, 2002). Glycerol-plasticized casein films had lower WVP values than films plasticized with polyethylene glycol (PEG) (Siew et al., 1999). A correlation between protein conformation and WVP was found. The caseinate films plasticized with glycerol had a higher percentage of helical structure and a lower percentage of random coil structure than films plasticized with PEG. It was perceived that the random coil structure formed a more "open" molecular network that facilitated water transmission. The higher WVP of the PEG-plasticized films compared to the glycerol-plasticized films was also attributed to the relatively higher hydration rate of PEG. Sorbitol-plasticized casein films were more effective moisture barriers than glycerol-plasticized films (Chick and Ustunol, 1998). Sorbitol is less hygroscopic than glycerol.

The mechanical strength of casein films is directly proportional to the ratio of structural materials and fillers. The higher is the ratio of protein to plasticizer, the greater is the tensile strength of the films (Chick and Ustunol, 1998; Tomasula et al., 1998). Siew et al. (1999) reported that plasticizers were effective within a range (i.e., too little plasticizer resulted in brittle films, and too much plasticizer rendered the films too tacky to yield reliable mechanical data).

The molecular characteristics of plasticizers affect film strength significantly (Chen, 2002). Casein films plasticized with sorbitol had higher tensile strength than those plasticized with glycerol (Chick and Ustunol, 1998). This was attributed to the less hydrophilic nature of sorbitol compared to glycerol, therefore giving casein films with lower affinity to bind water at equivalent RH. However, no equilibrium moisture content data were reported to support this statement. The higher tensile strength of sorbitol-plasticized films might have also been partially due to the fact that the number of glycerol molecules was about double the number of sorbitol molecules at the same plasticizer weight as that used in the film-forming formulations. Glycerol is a much smaller molecule than sorbitol; therefore, it interrupts more hydrogen bonds within the protein matrix of casein films where an equal weight of plasticizer is applied. With less hydrogen bonding between protein strands, films have lower tensile strength.

Siew et al. (1999) compared the effects of glycerol and PEG 400 plasticizers on the tensile strength of sodium caseinate films and noticed a "cross-over" phenomenon. Specifically, films plasticized with glycerol had higher tensile strength at low plasticizer concentration, but lower tensile strength at higher plasticizer concentration than films plasticized with PEG. Glycerol caused a larger change in film tensile strength (about 20 MPa), while PEG resulted in a smaller change (about 10 MPa). The transition from glassy to rubbery state was observed at lower plasticizer concentration for glycerol-plasticized films compared to PEG-plasticized films. Furthermore, the transition in the glycerol films occurred over a broader plasticizer range, suggesting that the PEG system had a more homogeneous bonding distribution. Smaller plasticizer molecules, such as glycerol, can access the hydrophilic sites on the caseinate chain more easily than larger plasticizer molecules. Glycerol can be easily inserted between the protein strands through hydrogen bonding, thus reducing intermolecular protein interactions, increasing intermolecular spacing, and lowering the mechanical strength of casein films.

The elasticity (E) of caseinate films generally showed an increase, plateau, and then decrease with increasing moisture and plasticizer content (Chick and Ustunol, 1998; Tomasula et al., 1998; Siew et al., 1999). Water molecules have a plasticizing effect on protein films. Insertion of plasticizer molecules as lubricants facilitates the movement of protein chains, thus increasing film flexibility and stretchability. However, too much plasticizer dilutes the cohesion of the protein network as the number of plasticizer–plasticizer bonds increases.

Depending on the type of proteins and plasticizers, as well as on the ratio of protein to plasticizer, one, two, or all three relationships between plasticizer concentration and E may be observed. The data reported by Chick and Ustunol (1998) serve as the best examples—that is film E only increases with increasing plasticizer content (lactic acid casein with sorbitol); film E increases to a peak value and then declines with increasing plasticizer content (lactic acid casein with glycerol); and film E decreases with increasing plasticizer content (rennet casein with either glycerol or sorbitol).

Addition of different plasticizers to the casein film matrix resulted in different elasticity values (Chen, 2002). For example, glycerol-plasticized films were generally more stretchable than sorbitol-plasticized films (Chick and Ustunol, 1998). Glycerol, being more hygroscopic, absorbs more moisture, which serves as an additional plasticizer.

Arvanitoyannis (2002) discussed properties of collagen and gelatin films. Glycerol-plasticized collagen films had lower moduli than gelatin films. The effect of gelatin plasticization was more pronounced (3.5 times higher moduli values than nonplasticized gelatin films) when plasticizer content varied within 30% to 50% (w/w). In the diluent range of 60% to 75% (w/w) plasticizer content, gelatin films were tough rubbers extending up to 700%. Gelatin coatings and films have limited resistance to water vapor transmission and require significant amounts of plasticizers (e.g., glycerol and sorbitol) to impart flexibility (Gennadios et al., 1997).

Plasticizer addition is necessary to reduce brittleness and increase flexibility of whey protein isolate films. Plasticizers achieve these changes by interacting with polymer chains, thus reducing intermolecular forces among them. Plasticizers act mainly by disrupting hydrogen bonding between neighboring polymer chains, thus increasing chain mobility (Guilbert, 1986; Kester and Fennema, 1986). Glycerol, sorbitol, and polyethylene glycol have been commonly used to plasticize whey protein isolate films. The increase in film flexibility is also accompanied by an increase in film permeability, which depends on the type of plasticizer (McHugh et al., 1994; McHugh and Krochta, 1994a).

Plasticized whey protein films are transparent, bland, and flexible and have excellent oxygen, aroma, and oil barrier characteristics. However, these films are poor moisture barriers due to their hydrophilic character. Incorporation of lipid materials improves film moisture barrier properties by increasing hydrophobicity. The lipid materials are homogenized at temperatures above their melting points into the aqueous film-forming solutions of heat-denatured whey protein and plasticizer. The barrier efficiency of resulting composite films strongly depends on polarity of film components and distribution of lipid material in the film matrix (Kamper and Fennema, 1985; Debeaufort et al., 1993). Lipids that have been added to whey protein films include acetylated monoglycerides, waxes, fatty alcohols, and fatty acids. All protein-plasticizer solutions, either alone or emulsified with a lipid, are degassed until no air bubbles are observed, to ensure accurate permeability values.

Plasticizer and relative humidity (RH) have the most significant effects on permeability and mechanical properties of whey protein films. As mentioned, plasticizers reduce internal hydrogen bonding in films, thereby increasing film flexibility while also increasing film permeability. For this reason, simultaneous comparison of film permeability and mechanical properties is necessary for complete assessment of the effect of plasticizer type and amount (Krochta, 1998). McHugh et al. (1994) studied the effect of various plasticizers, including glycerol, sorbitol, and polyethylene glycol of different chain lengths, on water vapor permeability of whey protein films. Sorbitol provided the highest flexibility increase per unit increase in water vapor permeability among the studied plasticizers at comparable concentrations and RH conditions.

In a later study, McHugh and Krochta (1994a) compared the oxygen permeability and mechanical properties of whey protein films. Sorbitol was more effective than glycerol as a plasticizer. In films of equal tensile strength, Young modulus, or elasticity values, oxygen permeability values were reduced to a lesser degree when plasticized with sorbitol rather than glycerol. These results suggested that at the same mass content levels in films, lower molecular weight plasticizers might result in films with higher permeability, flexibility, and elongation, with lower strength. This is understandable considering that the molar content of the lower molecular weight plasticizer in the film is greater at constant plasticizer mass content (Krochta, 1998).

Maté and Krochta (1996) studied the effect of glycerol amount on permeabilities of water vapor and oxygen whey protein and β-lactoglobulin films. Again, increased glycerol levels gave higher permeability values. However, permeabilities of whey protein and β-lactoglobulin films were not significantly different at each glycerol content studied. Despite the different protein molecular structures in whey protein films compared to β-lactoglobulin films and the resulting differences in hydrogen bonds, hydrophobic interactions, and S–S bonds, no differences between whey protein and β-lactoglobulin permeabilities were detected. Similarly, Anker et al. (1998) reported no significant differences in water vapor permeability and mechanical properties between whey protein and β-lactoglobulin films using multivariate analysis.

Due to the hydrophilic nature of whey proteins, moisture plasticizes whey protein films. Therefore, increased permeability as film moisture content increases is expected. McHugh et al. (1994) and McHugh and Krochta (1994a) studied the effect of RH on water vapor and oxygen permeabilities of whey protein films. The results suggested an exponential-type relationship between permeabilities of and RH for whey protein films. Interestingly, the effect of RH on water vapor permeability was more pronounced for glycerol-plasticized films compared to sorbitol-plasticized films, in spite of higher whey protein:glycerol ratio compared to whey protein:sorbitol ratio.

Plasticizers are added to soy protein film-forming solutions to increase the flexibility and "soften" the structure of formed films. Glycerol and sorbitol are the most commonly used plasticizers for soy protein-based films (Gennadios et al., 1994). Extruding soy protein formulations into films is also possible. Such an extrusion process, using soy protein, plasticizer (e.g., polyhydric alcohols), and water as the main formulation ingredients, was described by Park et al. (2002). Pressures of 5 to 50 kg/cm^2 and temperatures of 110 to 180°C were used in the extruder to knead and melt the raw material.

Compression molding and extrusion are technologies used for the thermoplastic processing of proteins for film formation (Hernandez-Izquierdo and Krochta, 2008). Plasticizers are generally added to the protein matrix to improve processability during thermoplastic processing and to modify the properties of the final structure. As opposed to "internal" plasticizers, which are copolymerized or reacted with the polymer, "external" plasticizers consist of low molecular weight, low volatility substances that interact with the polymer chains producing swelling (Sothornvit and Krochta, 2005).

Water is the most effective plasticizer in biopolymer materials, enabling them to undergo the glass transition, facilitating deformation, and processability of the biopolymer matrix. Without water addition, the temperature region of thermal degradation would be easily reached before films could be formed (Tolstoguzov, 1993). However, an excessive amount of water during protein extrusion would decrease melt viscosity, which, in turn, would lead to low motor torque and specific mechanical energy input, resulting in low product temperature that could reduce the degree of protein transformation and interactions. Besides water, common plasticizers for edible films include monosaccharides, oligosaccharides, polyols, lipids, and derivatives (Sothornvit and Krochta, 2005). Plasticizer composition, size, shape, and ability to attract water have been shown to affect solution-cast whey protein film barrier properties (Sothornvit and Krochta, 2000).

Glycerol is a low molecular weight, hydrophilic plasticizer that has been widely used in the thermoplastic processing of proteins (Hernandez-Izquierdo and Krochta, 2008). Its high plasticizing effect has been attributed to the ease with which glycerol can insert and position itself within the three-dimensional biopolymer network (di Gioia and Guilbert, 1999). Sucrose and sorbitol have also been studied for their plasticizing effects, including plasticizing fish myofibrillar proteins to produce biopackaging materials by thermal compression molding (Cuq et al., 1997). Pommet et al. (2005) tested several compounds with different chemical functions, number of functional groups, and degree of hydrophobicity as wheat gluten plasticizers. From this study, the critical factors for a good plasticizer were found to be low melting point, low volatility, and protein compatibility. In addition to these characteristics, permanence in the film and amount of plasticizer needed should be taken into account when choosing a good plasticizer (di Gioia and Guilbert, 1999; Sothornvit and Krochta, 2001). The efficiency with which a plasticizer affects specific mechanical and barrier properties can be quantified (Sothornvit and Krochta, 2001). The relative effect on mechanical and barrier properties can vary a great amount among different plasticizers. Table 6.2 summarizes the plasticizers that have been used in the thermoplastic processing of proteins.

Table 6.2 Plasticizers used in the thermoplastic processing of proteins

Plasticizer	Protein studied	Processing method	Reference
1,4-Butanediol	Wheat gluten	Compression molding	Pommet et al. (2005)
DATEM[a]	Corn gluten meal (rich in zein)	Compression molding	di Gioia and Guilbert (1999)
Dibutyl tartrate	Corn gluten meal (rich in zein)	Compression molding	di Gioia and Guilbert (1999)
Dibutyl phthalate	Corn gluten meal (rich in zein)	Compression molding	di Gioia and Guilbert (1999)
Glycerol	Corn gluten meal (rich in zein)	Compression molding	di Gioia and Guilbert (1999)
	Wheat gluten	Extrusion	Redl et al. (1999)
	Soy protein	Compression molding	Cunningham et al. (2000)
	Soy protein	Extrusion	Zhang et al. (2001)
	Wheat gluten	Extrusion	Pommet et al. (2003)
	Wheat gluten	Compression molding	Pommet et al. (2003)
	Wheat gluten	Compression molding	Pommet et al. (2005)
	Whey protein	Compression molding	Sothornvit et al. (2003)
	Whey protein	Extrusion	Pommet et al. (2005)
Lactic acid	Wheat gluten	Compression molding	Hernandez-Izquierdo (2007)
Octanoic acid	Corn gluten meal (rich in zein)	Compression molding	Pommet et al. (2005)
	Wheat gluten	Compression molding	di Gioia and Guilbert (1999)
Palmitic acid	Corn gluten meal (rich in zein)	Compression molding	Pommet et al. (2005)
Sorbitol	Myofibrillar proteins	Compression molding	Cuq et al. (1997)
Sucrose	Myofibrillar proteins	Compression molding	Cuq et al. (1997)
Water	Corn gluten meal (rich in zein)	Compression molding	di Gioia and Guilbert (1999)
	Wheat gluten	Compression molding	Pommet et al. (2005)

Source: Hernandez-Izquierdo, V.M., and Krochta, J.M. *Journal of Food Science* 73(2), R30–R39, 2008. With permission.

[a] DATEM, diacetyl tartaric acid ester of monodiglycerides.

6.2.2 Plasticizers in polysaccharide films and coatings

Water-soluble polysaccharides in edible films serve numerous roles such as providing hardness, crispness, compactness, viscosity, adhesiveness, mouthfeel, gelling, and thickening (Nisperos-Carriedo, 1994). Polysaccharide films often require plasticizers for the same reasons they are used in protein-based films. Nonionic cellulose ethers, such as methylcellulose, hydroxypropyl methylcellulose, and hydroxypropylcellulose are good film-formers, yielding tough and flexible transparent films due to the linear structure of the polymer backbone. Plasticity can be further improved by adding polyglycols, glycerin, or propylene glycol (Nisperos-Carriedo, 1994).

Plasticization of the biopolymer structure is often required to obtain satisfactory mechanical film properties, but it strongly affects aroma transfer. For example, addition of polyethylene glycol 400 (PEG 400) in the methylcellulose network substantially increased the transfer rate of 1-octen-3-ol (Debeaufort et al., 2002). This was due to film swelling in the presence of PEG 400, decrease of film density and crystallinity (Greener-Donhowe and Fennema, 1993), and high solubility of the aroma compound in PEG. Most plasticizers used in biopolymer film formulations, such as polyols, are potent solvents for aroma compounds and are widely used as liquid support in the flavoring industry (Reineccius, 1994).

Film plasticization can occur during aroma transfer depending on the aroma concentration, as observed by Quezada-Gallo et al. (1999, 2000) for permeation of 2-heptanone and 2-pentanone through methylcellulose-based films. The ketone group of the flavor compound interacts with hydroxyl groups on methylcellulose chains as revealed by infrared spectroscopy determination. The characteristic wave numbers of OH and CO in the film containing 2-heptanone were notably decreased compared to those of pure methylcellulose and 2-heptanone. This flavor compound forms weak hydrogen bonds with methylcellulose. This likely widens the spaces among polymer chains resulting in swelling and plasticization of the film network. This mechanism strongly favors the transfer of volatiles.

Anionic cellulose ethers such as sodium carboxy-methylcellulose, or cellulose gum, form strong, oil-resistant films of great importance in many food applications. However, they are seldom used to prepare free or unsupported films. Plasticizers such as ethanolamine, glycerol, polyglycols, or propylene glycol have been reported to be effective with cellulose gum (Nispero-Carriedo, 1994). Amylose, the linear raw starch polymer, produces transparent film with physical, chemical, and mechanical properties similar to those of polymeric films. The addition of plasticizers and absorption of water molecules by the hydrophilic plasticizers can increase polymer chain mobility and gas permeability (Banker, 1966).

Alginates possess good film-forming properties, which makes them particularly useful in food applications to reduce water loss or oxidation (Onsøyen, 2001). Alginate films tend to be quite brittle when dry but may be plasticized by the inclusion of glycerol (Glickman, 1983). Gum Karaya forms smooth films that can be plasticized with compounds such as glycerol to reduce brittleness (Nispero-Carriedo, 1994).

Use of sorbitol rather than propylene glycol and glycerol as a plasticizer was generally shown to be better in reducing weight loss and preserving the albumen and yolk quality of chitosan-coated eggs during storage periods (Kim et al., 2008).

Caner et al. (1998) used PEG 400 in chitosan at 0.25% and 0.5% concentrations and reported higher water vapor and oxygen permeabilities and elongation and lower tensile strength at the higher concentration of plasticizer used.

6.3 Emulsifiers

An emulsion consists of two immiscible liquids (usually oil and water), with one of the liquids dispersed as small spherical droplets in the other. In most foods, the diameters of the droplets usually lie somewhere between 0.1 and 100 µm (McClements, 2005). For many wax coatings, including almost all of those used for apples or citrus fruit, the droplet diameters are too small to scatter light, and therefore do not appear milky. These are called microemulsions or nanoemulsions, depending on whether their mean droplet diameters are more or less than about 0.1 µm, respectively. Emulsion technology is particularly suited for the design and fabrication of delivery systems for encapsulating bioactive lipids, such as carotenoids, tocopherols, flavonoids, and polyphenols (McClements et al., 2007). The major food-grade structural components that can be used to construct delivery systems for bioactive components are lipids, including flavor oils such as lemon and orange; surfactants; and biopolymers, including globular and flexible proteins and nonionic, anionic, and cationic polysaccharides, some of them originated from fruits and vegetables, such as soy, pea, starch, cellulose, and pectin (McClements et al., 2007). An emulsion-based coating delivery system should be compatible with the food matrix in which it is going to be incorporated; that is, it should not adversely affect the appearance, texture, stability, or taste. Some of the most important bulk physicochemical characteristics of emulsions used for coatings are viscosity, optical properties, stability, molecular distribution, and release characteristics (McClements et al., 2007). Milky emulsions are prepared by high-energy procedures using high-pressure homogenizers or microfluidizers. Clear emulsions (micro- or nanoemulsions) are prepared by low-energy methods, including catastrophic phase inversion (Hernandez, 1994; Bilbao-Sáinz et al., 2010b).

Composite (microemulsion or nanoemulsion) coatings are promising in improving moisture and oxygen barrier properties of hydrophilic coating materials and improving coating adhesion, durability, and delivery of active compounds. Research on identifying the most compatible material combinations, the lipid particle size, and stability of the emulsion system needs to be continuously investigated for meeting the specific needs of coating processed foods and fresh and minimally processed fruits and vegetables (Lin and Zhao, 2007).

Emulsifiers are surface-active compounds with both polar and nonpolar molecular structures, which absorb at the water-lipid or water-air interface reducing surface tension (Krochta, 2002; Lin and Zhao, 2007). Emulsifiers are essential for the formation of protein or polysaccharide coatings containing lipid emulsion particles. Polysaccharides and proteins are polymeric and hydrophilic in nature, and thus are good film-formers with excellent oxygen, aroma, and lipid barriers at low relative humidity. However, they are poor moisture barriers compared to synthetic moisture barrier films such as low-density polyethylene. On the other hand, lipids are hydrophobic with better moisture-barrier properties than those of polysaccharides and proteins. However, their nonpolymeric nature limits their cohesive film-forming capacity (Krochta, 1997). In composite films and coatings, the polysaccharide or protein provides the film integrity and entraps the lipid component, and the lipid component imparts the moisture-barrier property (Krochta, 1997). Without the presence of an effective emulsifier, the lipid often separates from the polysaccharide or protein film.

Composite film/coating can be categorized as a bilayer or a stable emulsion. For bilayer composite films/coatings, a lipid generally forms an additional layer over the polysaccharide or protein layer, while the lipid in the emulsion composite films/coatings is dispersed and entrapped in the matrix of protein or polysaccharide. Emulsified films improve moisture barrier properties and can be manufactured in a single step in contrast to a layer-by-layer process for multilayer films (Hambleton et al., 2009).

The amphiphilic character of proteins enables proteins to stabilize the protein-lipid emulsions through the balance between forces, primarily electrostatic and hydrophobic. Polysaccharides stabilize emulsions by strongly attaching to the surface of the lipid and significantly protruding into the continuous phase to form a polymeric layer or a network of appreciable thickness (Callegarin et al., 1997).

To produce protein-lipid or polysaccharide-lipid composite films from aqueous solution, it is often necessary to add an emulsifier to allow dispersion of the hydrophobic lipid material in the solution to improve emulsion stability and increase the particle distribution in composite emulsion-based edible films (Debeaufort and Voilley, 1995). Also, for some food-coating applications, addition of a surface-active agent to a coating formulation may be necessary to achieve satisfactory surface wetting and spreading with the coating formulation and then adhesion of the dry coating (Krochta, 2002).

There are numerous emulsifiers and surfactants, and new ones are developed constantly. Table 6.3 lists some of the common emulsifiers that can be used for edible films and coating formulations. The U.S. Code of Federal Regulations provides the status of protein, polysaccharide, lipid, resin, plasticizer, emulsifier, preservative, and antioxidant materials related to acceptable use (Baldwin, 1999). Selection of emulsifiers for formulation of composite films can be facilitated using the hydrophilic-lipophilic balance and phase inversion temperature criteria (Hernandez, 1994; McClements, 2005).

The barrier and mechanical properties of the composite films/coatings are affected by the composition and distribution of the hydrophobic substances in the film/coating matrix (Kamper and Fennema, 1985; Debeaufort et al., 1993). In general, bilayer films/coatings are more effective water vapor barriers than emulsion films/coatings due to the existence of a continuous hydrophobic phase in the matrix, and their moisture-barrier property can be improved by increasing the degree of lipid saturation and chain length of fatty acids (Kamper and Fennema 1984a, 1984b; Hagenmaier and Shaw, 1990). For emulsion composite films/coatings, the type of lipid, location, volume fraction, polymorphic phase, and drying conditions significantly impact the moisture-barrier property (Gontard et al., 1994).

Moisture barriers of whey protein-lipid emulsion films/coatings are improved when the hydrocarbon chain length of fatty acid alcohols and monoglycerides increased from 14 to 18 carbon atoms (McHugh and Krochta, 1994b). Beeswax and fatty acids are more effective in reducing water vapor permeability of whey protein isolate (WPI)–based emulsion films/coatings than fatty acid alcohols due to the lipid polarity.

Table 6.3 Edible films and coating emulsifiers allowed for use in foods

Component	Classification	21 CFR #	Allowed use
Acetylated monoglycerides	Emulsifier, coating agent	172.828	REG, GMP
Ammonium lauryl sulfate	Emulsifier, scald agent	172.822	GMP
Ethylene glycol monostearate	Emulsifier, confections	73.1	GMP
Glycerol monostearate	Coating agent, emulsifier	184.1324	GRAS/FS, GMP
Oleic acid	Emulsifier, binder, lubricant, coating for citrus fruits	172.840, 172.860, 172.863, 172.210	REG, GMP
Linoleic acid	Emulsifier, nutrient, dietary supplement	182.5065, 184.1065	GRAS/GMP
Potassium oleate	Emulsifier, stabilizer	172.863	REG, GMP
Propylene glycol monostearate	Emulsifier, stabilizer	172.856	REG/FS, GMP
Sodium alkyl sulfate	Emulsifier, scald agent	172.210	REG
Sodium oleate	Packaging coating, emulsifier, anticaking agent	172.863	REG
Sorbitan monostearate	Emulsifier, defoamer, stabilizer, coating fresh fruit	172.842	REG, GMP
Lecithin	Emulsifier, antioxidant	172.814	REG, GMP
Stearic acid	Emulsifier	184.1090	GRAS, GMP
Sucrose stearate	Emulsifier, texturizer, and component of fruit coatings	172.859	REG, GMP

Notes: GRAS: Generally recognized as safe. Substances in this category are by definition, under Sec. 201(s) of the FD&C Act, not food additives. Most GRAS substances have no quantitative restrictions as to use, although their use must conform to good manufacturing practices. GRAS/FS: Substances generally recognized as safe in foods but limited in standardized foods where the standard provides for its use. GMP: Good manufacturing practices. REG: Food additive for which a petition has been filed and a regulation issued. REG/FS: Food additive regulated under the FAA and included in a specific food standard.

To produce protein-lipid or polysaccharide-lipid composite films from aqueous solution, it is often necessary to add an emulsifier to allow dispersion of the lipid material in the solution. Also, for some food-coating applications, addition of a surface-active agent to a coating formulation may be necessary to achieve satisfactory surface wetting and spreading with the coating formulation and then adhesion of the dry coating. Emulsifiers also modify surface energy to control adhesion and wettability of the coating surfaces (Krochta, 2002).

Some proteins are sufficiently surface active that no emulsifier is necessary to form well-dispersed composite films or provide good surface wetting and adhesion (Krochta, 2002). Milk proteins are important functional ingredients. Their solubility in aqueous solutions and unique surface characteristics (the balance of hydrophilic and hydrophobic forces) make them excellent emulsifiers. Hence, whey proteins are excellent candidates for developing composite or emulsion coatings with improved moisture barrier property (Lin and Zhao, 2007). Addition of an emulsifier into whey protein coatings increases the hydrophilicity and coatability of peanut surfaces, thus improving the oxygen barrier of the coatings (Lin and Krochta, 2005).

The β- and κ-caseins are amphipathic proteins having hydrophobic and hydrophilic ends and, thus, are especially suitable for use as emulsifiers. This feature helps the formation of stable composite protein-lipid emulsions for coating wet surfaces (Chen, 2002).

The amphiphilic antioxidant ascorbyl palmitate worked as a surfactant and emulsifier incorporated into 10% whey protein isolate coating solution containing 6.67% glycerol (Han and Krochta, 2007).

Valencia-Chamorro et al. (2008) prepared films from edible composite emulsions combining the hydrophilic phase (hydroxyproyl methylcellulose) and the lipid phase (beewax and shellac) suspended in water. Glycerol and stearic acid were used as plasticizer and emulsifier, respectively.

Some cellulose ethers are used in conjunction with other stabilizers as emulsifiers in fluid systems that contain fat. The ability of cellulose ethers such as hydroxypropylcellulose, hydroxypropyl-methylcellulose, and methylcellulose to accumulate at oil droplet interfaces and prevent oil droplet coalescence is important to prevent oiling-off during storage. Celluloses and cellulose ethers fall under regulations stemming from EEC directive 74/329/EEC and its amendments, covering emulsifiers, stabilizers, thickeners, and gelling agents (Coffey et al., 2006).

Sucrose stearate, an emulsifier promoting oil-in-water emulsions, was added (5% w/w of fatty acid) to soybean protein isolate film-forming solutions. Fatty acid–containing soybean protein isolate films had notably lower water vapor permeability values than control soy protein films (Park et al., 2002).

Glycerol monostearate emulsifier was mixed to acetic acid ester of mono- and diglycerides blended with 20% w/w beeswax prior to homogenization to encapsulate the aroma compound n-hexanal (Hambleton et al., 2009). In the films containing fat, this amphiphilic emulsifier is usually present at the fat globule surface to reduce surface hydrophobicity and surface energy.

Alginates act as thickeners, stabilizers, and emulsifiers, giving improved body and prolonged shelf life to a variety of products. The effect obtained depends on the formulation, whether it is neutral or acid, emulsion or a suspension (Onsøyen, 2001).

6.4 Antimicrobial agents

Edible coating and film research has undergone rapid expansion in the past 20 years, in part due to increased consumer interest in food safety and environmental issues. In addition to improving food safety, antibacterial films offer opportunities to reduce the need for synthetic packaging materials and improve their recyclability by simplifying their structure (Krochta and De Mulder-Johnston, 1997; Arvanitoyannis and Gorris, 1999). Incorporating antimicrobial compounds into edible films or coatings provides a novel way to improve the safety and shelf life of ready-to-eat foods (Cagri et al., 2004). Some of the more commonly used antimicrobials include benzoic acid, sorbic acid, lysozyme, lactoferrin, bacteriocins (nisin and pediocin), and plant-derived secondary metabolites, such as essential oils and phytoalexins.

Common antimicrobial agents used in food systems, such as benzoic acid, sodium benzoate, sorbic acid, potassium sorbate, and propionic acid, may be incorporated into edible films and coatings. Starch-based coatings containing potassium sorbate were applied on the surface of fresh strawberries for reducing microbial growth and extending storage life of the fruit (Garcia et al., 1998). A hydroxypropyl methylcellulose coating containing ethanol was effective in inactivating *Salmonella Montevideo* on the surface of fresh tomatoes (Zhuang et al., 1996). Lysozyme was incorporated into chitosan coatings for enhancing the antimicrobial activity of chitosan against *Escherichia coli* and *Streptococcus faecalis* (Park et al., 2004). In addition, chitosan coatings containing potassium sorbate were shown to increase antifungal activity against the growth of *Cladosporium* and *Rhizopus* on fresh strawberries (Park et al., 2005). A new patented edible film including organic acids, protein, and glycerol (e.g., 0.9% glycerol, 10% soy protein, and 2.6% malic acid) can inhibit pathogen growth, including *L. monocytogens*, *S. gaminara*, and *E. coli* O157:H7. Such a film also provides a method for coating edible products with edible films without masking the color but increasing the shelf life (Hettiarachchy and Satchithanandam, 2007).

Recent studies have shown that essential oils of oregano, thyme, cinnamon, lemongrass, and clove are among the most active against strains of *E. coli* (Smith-Palmer et al., 1998; Hammer et al., 1999; Dorman and Deans, 2000; Friedman et al., 2002). Although the effectiveness of all these compounds has been widely reported, carvacrol (a major component of the essential oils of oregano and thyme) appears to have received the most attention from investigators. Carvacrol is GRAS and used as a flavoring agent in baked goods, sweets, ice cream, beverages, and chewing gum (Fenaroli, 1995). Some plant essential oils and their components are compatible with the sensory characteristics of fruits and have been shown to prevent bacterial growth. Among the complex constituents of citrus essential oils, the terpene citral is known to have strong antifungal properties (Rodov et al., 1995). In addition, cinnamon oil and its active compound (cinnamaldehyde) also have been tested for their inhibitory activity against *E. coli* (Helander et al., 1998; Friedman et al., 2002; Friedman et al., 2004b). In addition to those mentioned earlier, related published studies by Roller and Seedhar (2002) showed that carvacrol and cinnamaldehyde were effective at reducing the viable count of the natural flora on kiwifruit when 0.15 µL/mL was used in the dipping solution, but it was less effective on honeydew melon. Treatment of fresh kiwifruit and honeydew melon with 1 mM carvacrol or cinnamic acid has been found to delay spoilage without causing adverse sensory changes (Roller and Seedhar, 2002). Lanciotti et al. (1999) added citrus essential oils to fresh sliced fruit mixtures (apple, pear, grape, peach, and kiwifruit) to inhibit the proliferation of naturally occurring microbial population. Because many of the oils are derived from citrus products and cinnamon oil is already widely used in food (Friedman et al., 2000), it may be safe to include them in small amounts in apple puree edible films. Cinnamaldehyde and some of the other antibacterials are GRAS-listed (Adams et al., 2004; Burt, 2004). Phenolic compounds present in teas (Friedman et al., 2005; Friedman, 2006), pigmented rice brans (Nam et al., 2006), and in most fruits and vegetables (Friedman et al., 2003; Shahidi and Naczk, 2004; Friedman et al., 2005) are also reported to exhibit antibacterial effects. Some of these have been incorporated in edible films and coatings (Cagri et al., 2004).

Du et al. (2011) reported that an apple skin extract added to an apple-based film inhibited the growth of *Listeria monocytogenes*. Quinones et al. (2009) reported that grape seed and grape pomace extracts inactivated the Shiga toxin produced by *E. coli* O157:H7.

Friedman et al. (2006) evaluated the antibacterial activities of seven green tea catechins and four black tea theaflavins as well as aqueous infusions of 36 black, green, oolong, white, and herbal teas against the foodborne pathogen *Bacillus cereus*. The results show that the activities of some tea catechins and theaflavins were greater than those of medical antibiotics such as vancomycin and tetracycline and that added tea and apple skin extracts facilitated thermal destruction of *E. coli* O157:H7 in cooked ground beef (Juneja et al., 2009).

Friedman et al. (2002) screened 120 naturally occurring plant-derived oils and oil compounds for their antibacterial activities against four species of foodborne pathogens. The most active oils in terms of BA_{50} values (percent of oil in phosphate buffer that killed 50% of the bacteria) are

Campylobacter jejuni (BA_{50}, 0.003% to 0.009%): Marigold, ginger root, jasmine, patchouli, Gardenia, cedarwood, carrot seed, celery seed, mugwort, spikenard, and orange bitter
Escherichia coli O157:H7 (BA_{50}, 0.046% to 0.14%): Oregano, thyme, cinnamon, palmarosa, bay leaf, clove bud, lemon grass, and allspice
Listeria monocytogenes (BA_{50}, 0.057% to 0.092%): Gardenia, cedarwood, bay leaf, clove bud, oregano, cinnamon, allspice, thyme, and patchouli
Salmonella enterica (BA_{50}, 0.045% to 0.14%): Thyme, oregano, cinnamon, clove bud, allspice, bay leaf, palmarosa, and marjoram

Recently, Friedman et al. (2009) discovered that carvacrol in apple films inhibited the growth of *E. coli* O157:H7 on the surfaces of raw chicken breast; inhibited the growth of *Clostridium perfringens* during chilling of cooked ground beef (Juneja et al., 2006); and inactivated *E. coli* O157:H7 and inhibited the formation of carcinogenic heterocyclic amines during grilling of hamburger beef patties (Friedman et al., 2009).

Friedman et al. (2000, 2004a) showed that carvacrol, oregano, and cinnamaldehyde were effective antibacterials against antibiotic-resistant *Bacillus cereus, Campylobacter jejuni, E. coli, Salmonella enterica*, and *Staphylococcus aureus*. These compounds are candidates for incorporation into film and coating formulations to reduce both nonresistant as well as antibiotic-resistant pathogens in human foods.

Du et al. (2009) studied the effect of allspice, cinnamon, and clove bud oil in apple film, as well as allspice, garlic, and oregano oil in tomato film on antimicrobial activities against *E. coli* O157:H7, *Salmonella enterica*, and *Listeria monocytogenes* using overlay and vapor phase methods. The results of the study show that these plant-derived essential oils can be used to prepare fruit and vegetable-based antimicrobial edible films with good physical properties for food applications by both direct contact and indirectly by vapors emanating from the films.

Rojas-Graü et al. (2007a) evaluated the antimicrobial activity of essential oils and active compounds in apple films against target pathogen microorganisms. They found that the order of antibacterial activities was as follows in apple-based edible films: carvacrol > oregano > citral > cinnamaldehyde > lemon grass > cinnamon oil. Evaluation of physicochemical properties of films made from apple slurries revealed no adverse effect of the additives on water vapor permeability properties. The incorporation of these bactericidal compounds caused a significant increase in tensile strength, percentage elongation, and elastic modulus of the films. Antimicrobials commonly used with edible films and coatings are summarized in Table 6.4.

6.5 Antioxidants and antibrowning agents

Antioxidants can be added into the coating matrix to protect against oxidative rancidity, degradation, and discoloration of certain foods. Nuts were coated with pectinate, pectate, and zein coatings containing BHA, BHT, and citric acid to prevent rancidity and maintain their texture (Swenson et al., 1953).

Ascorbic acid was incorporated into edible coatings to reduce enzymatic browning in sliced apples and potatoes (Baldwin et al., 1996). Xanthan gum coatings mixed with α-tocopherol enhanced nutritional quality and improved the surface color of peeled baby carrots (Mei et al. 2002). Carrageenan and whey protein coatings containing antibrowning agents such as ascorbic acid and citric acid effectively prolonged the shelf life of apple slices (Lee et al., 2003). Methylcellulose-based coatings containing ascorbic acid and sorbic acid were able to retard browning and to

Table 6.4 Classification and regulation of some antimicrobial agents allowed for use in food coatings

Food additive	Classification	21 CFR #	Allowed use
Benzoic acid	Preservative	None	GRAS
Clove bud oil	Essential oil	184.1257	GRAS
Potassium sorbate	Preservative	182.3640	GRAS/FS
Propionic acid	Preservative	184.1081	GRAS/FS, GMP
Sodium benzoate	Preservative for fruit jellies and citrus juices	150.141, 150,161, 184.1733	GRAS/FS, GMP
Sorbic acid	Preservative	182.3089	GRAS, GMP

Notes: GRAS: Generally recognized as safe. Substances in this category are by definition, under Sec. 201(s) of the FD&C Act, not food additives. Most GRAS substances have no quantitative restrictions as to use, although their use must conform to good manufacturing practices. GRAS/FS: Substances generally recognized as safe in foods but limited in standardized foods where the standard provides for its use. GMP: Good manufacturing practices. REG: Food additive for which a petition has been filed and a regulation issued.

enhance texture of cut pear wedges (Guadalupe et al., 2003). Chitosan-based coating containing α-tocopheryl acetate significantly delayed the color change of fresh and frozen strawberries (Han et al., 2004b). Antioxidants that are commonly used with edible films and coatings are summarized in Table 6.5.

Cutting of fruits and vegetables can induce undesirable changes in color and appearance of these products during storage and marketing. The phenomenon is usually caused by the enzyme polyphenol oxidase, which in the presence of oxygen, converts phenolic compounds into dark-colored pigments.

Antioxidant constituents of fruit such as phenolic compounds and ascorbic acid are related to enzymatic browning. Phenolic compounds are oxidized to highly unstable quinones that are later polymerized to brown, red, and black pigments (Nicolas et al. 1994; Martínez and Whitaker, 1995). Ascorbic acid is extensively used to avoid enzymatic browning of fruit due to the reduction of the o-quinones, generated by the action of the polyphenol oxidase (PPO) enzymes, back to the phenolic substrates (McEvily et al., 1992).

Table 6.5 Classification and regulation of some antioxidants allowed for use in food coatings

Food additive	Classification	21 CFR #	Allowed use
Ascorbic acid	Antioxidant, preservative, color stabilizer, nutrient	182.3013	GRAS, GMP
Ascorbyl palmitate	Antioxidant, preservative, color stabilizer	182.3149	GRAS
Butylated hydroxyanisole (BHA)	Antioxidant	172.110	GRAS, FS
Butylated hydroxytoluene (BHT)	Antioxidant	137.350, 172.115	GRAS, REG, FS
Citric acid	Sequestrant, buffer and neutralizing agent	182.1033, 182.6033	GRAS/FS, GMP
Propyl gallate	Antioxidant	184.1660	GRAS
Tertiary butylhydroquinone (TBHQ)	Antioxidant	172.185	REG
Tocopherols	Preservative, dietary supplement, nutrient	182.3890, 184.1890, 182.8890	GRAS, GMP

Notes: GRAS: Generally recognized as safe. Substances in this category are by definition, under Sec. 201(s) of the FD&C Act, not food additives. Most GRAS substances have no quantitative restrictions as to use, although their use must conform to good manufacturing practices. GRAS/FS: Substances generally recognized as safe in foods but limited in standardized foods where the standard provides for its use. GMP: Good manufacturing practices. REG: Food additive for which a petition has been filed and a regulation issued. FS: Substance permitted as optional ingredient in a standardized food.

However, ascorbic acid is oxidized to dehydroascorbic acid after a certain time (Luo and Barbosa-Canovas, 1997; Rojas-Graü et al., 2008), thus allowing the accumulation of o-quinones (Sapers, 1993). The incorporation of antioxidant agents containing sulfur-containing amino acids such as *N*-acetylcysteine and glutathione into alginate- and gellan-based coatings offered an alternative approach to prevent fresh-cut apples and papayas from browning (Oms-Oliu et al., 2008; Rojas-Graü et al., 2007b, 2008).

6.6 Nutrients, flavors, and colorants

Edible films and coatings are excellent vehicles to enhance the nutritional value of fruits and vegetables by delivering basic nutrients and nutraceuticals that are lacking or are present in only low quantity in fruits and vegetables. Xanthan gum coating was utilized to carry a high concentration of calcium and vitamin E, for not only preventing moisture loss and surface whitening, but also to significantly increase the calcium and vitamin E contents of the carrots (Mei et al., 2002). The development of chitosan coatings containing high concentrations of calcium, zinc, or vitamin E also provides alternative ways to fortify fresh fruits and vegetables that otherwise could not be accomplished with common processing approaches (Park and Zhao, 2004). This application was successfully demonstrated on fresh and frozen strawberries (Han et al., 2004b). Flavor and coloring agents may also be added to edible coatings to improve the sensory quality of coated products. However, very little has been reported regarding this application. Nutrients, flavors, and colorants commonly used with edible films and coatings are summarized in Table 6.6.

Table 6.6 Classification and regulation of some miscellaneous additives allowed for use in foods

Food additive	Classification	21 CFR #	Allowed use
Calcium chloride	Antimicrobial agent, firming agent	184.1193	GRAS/FS
Carotene (β-carotene)	Nutrient, dietary supplement, coloring agent	182.5245, 182.8245	GRAS, GMP
Essential oils	Natural flavorings	182.20	GRAS
Silicon dioxide	Anticaking agent, component of microcapsules for flavoring oils	172.480 172.230	REG, GMP

Notes: GRAS: Generally recognized as safe. Substances in this category are by definition, under Sec. 201(s) of the FD&C Act, not food additives. Most GRAS substances have no quantitative restrictions as to use, although their use must conform to good manufacturing practices. GRAS/FS: Substances generally recognized as safe in foods but limited in standardized foods where the standard provides for its use. GMP: Good manufacturing practices. REG: Food additive for which a petition has been filed and a regulation issued.

6.7 Nanoparticles

An ongoing challenge for composite edible films is the relatively high water vapor permeability and poor mechanical behavior of the hydrocolloid fractions. By using nanoscience, new forms of nanocomposites dispersed with nanoparticles, nanofibers, or nanoemulsions can be developed to minimize migration of water through hydrocolloid films as well as to improve mechanical properties.

Nanocomposite edible films have been developed by incorporating microcrystalline cellulose nanofibers (Dogan and McHugh., 2007; Azeredo et al., 2009; De Moura et al., 2009; Bilbao-Sáinz et al., 2010a; De Moura et al., 2010), chitosan nanoparticles (De Moura et al., 2008), and nanoemulsions (Bilbao-Sáinz et al., 2010b), into edible films. Microcrystalline cellulose nanofibers and chitosan nanoparticles significantly improved the tensile strength of composite films formulated with hydroxypropyl methylcellulose, chitosan, and fruit-based edible films. This was attributed to the stronger interfacial adhesion between the polar groups of the fillers and the hydrophilic groups of the hydrocolloids and fruit matrixes. Incorporation of microcrystalline cellulose nanofibers in chitosan (Azeredo et al., 2010), hydroxypropyl methylcellulose (Dogan and McHugh., 2007; De Moura et al., 2009; Bilbao-Sáinz et al., 2010a; De Moura et al., 2010), or fruit-based films (De Moura et al., 2008) exhibited water barrier properties that improved with increasing concentration of nanofibers. Lipid coating of the microcrystalline cellulose nanofibers resulted in further improvement in the water barrier properties of the hydroxypropyl methylcellulose films. The water vapor permeability values of the hydroxypropyl methylcellulose films also decreased significantly when chitosan nanoparticles were included in the film matrix. In addition, the water barrier properties of hydrophilic films can be improved by the addition of hydrophobic nanoemulsions into the film matrix (Bilbao-Sáinz et al., 2010b). Because of their extremely small oil droplet size, droplet coalescence was avoided, achieving sufficient short-term stability to minimize creaming during film casting.

Nanocomposites are promising to expand the use of edible films and coatings, because the addition of nanoparticles has been related to improvements in overall performance of films and coatings. However, there are important safety concerns about nanotechnology applications in food systems. On the one hand, the properties and safety of most edible films and coatings starting materials in their bulk form are well understood. On the other hand, nano-sized counterparts frequently exhibit different properties from those found at the macroscale, because the very small sizes of the former allow them to move through the body more freely than larger particles, while their high surface area increases their reactivity. Few studies have been conducted to assess the risks associated with the presence of such extremely small particles, some of them biologically active, in the human body or dispersed in the environment. Hence, significant research is still required to evaluate the potential toxicity of nanotechnology products, as well as the environmental safety of their use.

In conclusion, edible films and coatings contain many useful ingredients besides the basic film-forming biopolymeric matrix. These ingredients generally serve to improve coating structural and functional properties. For example, increasing elasticity of coatings with plasticizers improves coating structure, while antimicrobial agents improve coating function. These often small molecules are essential to successful applications of coatings.

References

Adams, T.B., Cohen, S.M., Doull, J., Feron, V.J., Goodman, J.I., Marnett, L.J., Munro, I.C., Portoghese, P.S., Smith, R.L., Waddell, W.J., and Wagner, B.M. 2004. The FEMA GRAS assessment of cinnamyl derivatives used as flavor ingredients. *Food and Chemical Toxicology* 42:157–185.

Anker, M., Stading, M., and Hermansson, A.-M. 1998. Mechanical properties, water vapor permeability, and moisture contents of β-lactoglobulin and whey protein films using multivariate analysis. *Journal of Agricultural and Food Chemistry* 46:1820–1829.

Arnold, L.K. 1968. *Introduction to Plastics*. Ames, IA: Iowa State University Press.

Arvanitoyannis, I., Psomiadou, E., and Nakayama, A. 1996. Edible films made from sodium caseinate, starches, sugars or glycerol. Part 1. *Carbohydrate Polymers*, 31:179–192.

Arvanitoyannis, I., and Gorris, L.G.M. 1999. Edible and biodegradable polymeric materials for food packaging or coating. In *Processing foods: quality optimization and process assessment*. F.A.R. Oliveira and J.C. Oliveira (Eds.), CRC Press, Boca Raton, FL, pp. 357–371.

Arvanitoyannis, I. 2002. Formation and properties of collagen and gelatin films and coatings. In *Protein-based films and coatings*. A. Gennadios (Ed.), CRC Press, Boca Raton, FL, pp. 275–304.

Azeredo, H.M.C., Mattoso, L.H.C., Wood, D., Williams, T.G., Avena-Bustillos, R.J., and McHugh, T.H. 2009. Nanocomposite edible films from mango puree reinforced with cellulose nanofibers. *Journal of Food Science* 74(5):N31–N35.

Azeredo, H.M.C., Mattoso, L.H.C., Avena-Bustillos, R.J., Filho, G.C., Munford, M.L., Wood, D., and McHugh, T.H. 2010. Nanocellulose reinforced chitosan composite films as affected by nanofiller loading and plasticizer content. *Journal of Food Science* 75(1):N1–N7.

Baldwin, E.A., Níspero-Carriedo, M.O., Chen, X., and Hagenmaier, R.D. 1996. Improving storage life of cut apple and potato with edible coating. *Postharvest Biology and Technology* 9:151–163.

Baldwin, E.A. 1999. Surface treatments and edible coatings in food preservation. In *Handbook of food preservation*. M.S. Rahman (Ed.), Marcel Dekker, New York, pp. 577–609.

Banker, G.S. 1966. Film coating theory and practice. *Journal of Pharmaceutical Sciences* 55:81–89.

Bilbao-Sáinz, C., Avena-Bustillos, R.J., Wood, D., Williams, T.G., and McHugh, T.H. 2010a. Composite edible films based on hydroxypropyl methylcellulose reinforced with microcrystalline cellulose nanoparticles. *Journal of Agricultural and Food Chemistry* 58:3753–3760.

Bilbao-Sáinz, C., Avena-Bustillos, R.J., Wood, D.F., Williams, T.G., and McHugh, T.H. 2010b. Nano-emulsions prepared by a low-energy emulsification method applied to edible films. *Journal of Agricultural and Food Chemistry* 58:11932–11938.

Burt, S. 2004. Essential oils: their antibacterial properties and potential applications in foods. A review. *International Journal of Food Microbiology* 94:223–253.

Cagri, A., Ustunol, Z., and Ryser, E.T. 2004. Antimicrobial edible films and coatings. *Journal of Food Protection* 67:833–848.

Callegarin, F., Quezada Gallo, J.A., Debeaufort, F., and Voilley, A. 1997. Lipids and biopackaging. *Journal of the American Oil Chemists' Society* 74:1183–1192.

Caner, C., Vergano, P.J., and Wiles, J.L. 1998. Chitosan film mechanical and permeation properties as affected by acid, plasticizer, and storage. *Journal of Food Science* 63(8):1049–1053.

Cha, D.S., and Chinnan, M.S. 2004. Biopolymer-based antimicrobial packaging: a review. *Critical Reviews in Food Science and Nutrition* 44(4):223–237.

Chen, H. 2002. Formation and properties of casein films and coatings. In *Protein-based films and coatings*. A. Gennadios (Ed.), CRC Press, Boca Raton, FL, pp. 181–211.

Chick, J., and Ustunol, Z. 1998. Mechanical and barrier properties of lactic acid and rennet precipitated casein-based edible films. *Journal of Food Science* 63:1024–1027.

Coffey, D.G., Bell, D.A., and Henderson, A. 2006. Cellulose and cellulose derivatives. In *Food polysaccharides and their applications* (2nd ed.). A.M. Stephen, G.O. Phillips, and P.A. Williams (Eds.), CRC Press, Boca Raton, FL, pp. 147–179.

Cunningham, P., Ogale, A.A., Dawson, P.L., and Acton, J.C. 2000. Tensile properties of soy protein isolate films produced by a thermal compaction technique. *Journal of Food Science* 65(4):668–671.

Cuq, B., Gontard, N., and Guilbert, S. 1997. Thermoplastic properties of fish myofibrillar proteins: application to biopackaging fabrication. *Polymer* 38(16):4071–4078.

Debeaufort, F., Martin-Polo, M., and Voilley, A. 1993. Polarity homogeneity and structure affect water vapor permeability of model edible films. *Journal of Food Science* 58:426–429, 434.

Debeaufort, F., and Voilley, A. 1995. Effect of surfactants and drying rate on barrier properties of emulsified edible films. *International Journal of Food Science and Technology* 30:183–190.

Debeaufort, F., Quezada-Gallo, J.A., and Voilley, A. 1998. Edible films and coatings: tomorrow's packagings: a review. *Critical Reviews in Food Science and Nutrition* 38(4):299–313.

Debeaufort, F., Voilley, A., and Guilbert, S., 2002. The procedures of product stabilization due to films barrier. In *Water in Foods*. Tec&Doc Lavoisier, Paris, pp. 549–622.

de Moura, M.R., Avena-Bustillos, R.J., McHugh, T.H., Krochta, J.M., and Mattoso, L.H.C. 2008. Properties of novel hydroxypropyl methylcellulose films containing chitosan nanoparticles. *Journal of Food Science* 73(7):N31–N37.

de Moura, M.R., Aouada, F.A., Avena-Bustillos, R.J., McHugh, T.H., Krochta, J.M., and Mattoso, L.H.C. 2009. Improved barrier and mechanical properties of novel hydroxypropyl methylcellulose edible films with chitosan/tripolyphosphate nanoparticles. *Journal of Food Engineering* 92:448–453.

de Moura, M.R., Avena-Bustillos, R.J., McHugh, T.H., Wood, D.F., Otoni, C.G., and Mattoso, L.H.C. 2010. Miniaturization of cellulose fibers and effect of addition on the mechanical and barrier properties of hydroxypropyl methylcellulose films *Journal of Food Engineering*, doi:10.1016/j.jfoodeng.2010.12.008.

di Gioia, L., and Guilbert, S. 1999. Corn protein-based thermoplastic resins: effect of some polar and amphiphilic plasticizers. *Journal of Agricultural and Food Chemistry* 47:1254–1261.

Dogan, N., and McHugh, T.H. 2007. Effects of microcrystalline cellulose on functional properties of hydroxyl propyl methyl cellulose microcomposite films. *Journal of Food Science* 72(1):E16–E22.

Dorman, H.J., and Deans, S.G. 2000. Antimicrobial agents from plants: antibacterial activity of plant volatile oils. *Journal of Applied Microbiology* 88:308–316.

Du, W., Olsen, C.W., Avena-Bustillos, R.J., McHugh, T.H., Levin, C.E., and Friedman, M. 2009. Effects of allspice, cinnamon, and clove bud essential oils in edible apple films on physical properties and antimicrobial activities. *Journal of Food Science* 74:M372–M378.

Du, W., Olsen, C.W., Avena Bustillos, R.J., Friedman, M., and McHugh, T.H. 2011. Physical and antibacterial properties of edible films formulated with apple skin polyphenols. *Journal of Food Science* 76(2):M149–M155.

FDA (U.S. Food and Drug Administration). 1997. Substances Generally Recognized as Safe. *Federal Register*. 62(74):18938–18964. Available at www.gpo.gov/fdsys/pkg/FR-1997-04-17/pdf/97-9706.pdf. Accessed December 13, 2010.

Fenaroli, G. 1995. *Fenaroli's handbook of flavor ingredients*, 3rd ed. Boca Raton, FL: CRC Press.

Friedman, M., Kozukue, N., and Harden, L.A. 2000. Cinnamaldehyde content in foods determined by gas chromatography-mass spectrometry. *Journal of Agricultural and Food Chemistry* 48:5702–5709.

Friedman, M., Henika, P.R., and Mandrell, R.E. 2002. Bactericidal activities of plant essential oils and some of their isolated constituents against *Campylobacter jejuni*, *Escherichia coli* O157:H7, *Listeria monocytogenes*, and *Salmonella enterica*. *Journal of Food Protection* 65:1545–1560.

Friedman, M., Henika, P.R., and Mandrell, R.E. 2003. Antibacterial activities of phenolic benzaldehydes and benzoic acids against *Campylobacter jejuni*, *Escherichia coli* O157:H7, *Listeria monocytogenes*, and *Salmonella enterica*. *Journal of Food Protection* 66:1811–1821.

Friedman, M., Buick, R., and Elliott, C.T. 2004a. Antibacterial activities of naturally occurring compounds against antibiotic-resistant *Bacillus cereus* vegetative cells and spores, *Escherichia coli*, and *Staphylococcus aureus*. *Journal of Food Protection* 67:1774–1778.

Friedman, M., Henika, P.R., Levin, C.E., and Mandrell, R.E. 2004b. Antibacterial activities of plant essential oils and their components against *Escherichia coli* O157:H7 and *Salmonella enterica* in apple juice. *Journal of Agricultural and Food Chemistry* 52:6042–6048.

Friedman, M., Kim, S.Y., Lee, S.J., Han, P.G., Han, J.S., Lee, K.R., and Kozukue, N. 2005. Distribution of catechins, theaflavins, caffeine, and theobromine in 77 teas consumed in the United States. *Journal of Food Science* 70:C550–C559.

Friedman, M. 2006. Structure-antibiotic activity relationships of plant compounds against nonresistant and antibiotic-resistant foodborne pathogens. In *Advances in microbial food safety*, V.K. Juneja, J.P. Cherry, and M.H. Tunick (Eds.), ACS Symposium Series. American Chemical Society, Washington, DC.

Friedman, M. Henika, P.R., Levin, C.E., Mandrell, R.E., Kozukue, N. 2006. Antimicrobial activities of tea catechins and theaflavins and tea extracts against *Bacillus cereus*. *Journal of Food Protection* 69:354–361.

Friedman, M., Zhu, L., Fienstein, Y., and Ravishankar, S. 2009. Carvacrol facilitates heat induced inactivation of *Escherichia coli* 0157:H7 and inhibits formation of heterocylic amines in grilled ground beef patties. *Journal of Agricultural and Food Chemistry* 57:1848–1853.

Garcia, M., Martino M., and Zaritzky, N. 1998. Plasticized starch based coatings to improved strawberry (*Fragaria ananassa*) quality and stability, *Journal of Agricultural and Food Chemistry* 46:3758–3767.

Gennadios, A., McHugh, T.H., Weller, C.L., and Krochta, J.M. 1994. Edible coatings and films based on proteins. In *Edible coatings and films to improve food quality*. J.M. Krochta, E.A. Baldwin, M.G. Nispero-Carriedo (Eds.), Technomic, Lancaster, PA, pp. 201–277.

Gennadios, A., Hanna, M.A., and Kurth, L.B. 1997. Application of edible coatings on meats, poultry and seafoods: a review. *Lebensmittel-Wissenschaft und -Technologie* 30:337–350.

Gennadios, A. 2004. Edible films and coatings from proteins. In *Proteins in food processing*. R. Yada (Ed.), Woodhead, Cambridge, UK, pp. 442–467.

Glickman, M. 1983. *Food hydrocolloids,* Vol. II. Boca Raton, FL: CRC Press.

Gontard, N., Duchez, C., Cuq, B., and Guilbert, S. 1994. Edible composite films of wheat gluten and lipids: water vapor permeability and other physical properties. *International Journal of Food Science and Technology* 44:1064–1069.

Gontard, N., and Ring, S. 1996. Edible wheat gluten film: influence of water content on glass transition temperature. *Journal of Agricultural and Food Chemistry* 44:3474–3478.

Greener-Donhowe, I., and Fennema, O. 1993. The effects of plasticizers on crystallinity, permeability, and mechanical properties of methylcellulose films. *Journal of Food Processing and Preservation* 17:247–257.

Guadalupe, I.O., Rodriguez, J.J., and Barbosa-Canovas, G.V. 2003. Edible coatings composed of methylcellulose, stearic acid, and additives to preserve quality of pear wedges. *Journal of Food Processing and Preservation* 27:299–320.

Guilbert, S. 1986. Technology and application of edible protective films. In *Food packaging and preservation: theory and practice.* M. Mathlouthi (Ed.), Elsevier Applied Science, New York, pp. 371–394.

Guilbert, S., and Gontard, N. 1995. Edible and biodegradable food packaging. In *Foods and packaging materials—chemical interactions.* P. Ackermann, M. Jägerstad, and T. Ohlsson (Eds.). The Royal Society of Chemistry, Cambridge, UK, pp. 159–168.

Hagenmaier, R.D., and Shaw, P.E. 1990. Moisture permeability of edible films made with fatty acid and (hydroxypropyl) methyl cellulose. *Journal of Agricultural and Food Chemistry* 38:1799–1803.

Hambleton, A., Fabra, M.J., Debeaufort, F., Dury-Brun, C., and Voilley, A. 2009. Interface and aroma barrier properties of iota-carrageenan emulsion–based films used for encapsulation of active food compounds. *Journal of Food Engineering* 93:80–88.

Hammer, K.A., Carson, C.F., and Riley, T.V. 1999. Antimicrobial activity of essential oils and other plant extracts. *Journal of Applied Microbiology* 86:985–990.

Han, C., Zhao, Y., Leonard, S.W., and Traber, M.G. 2004b. Edible coatings to improve storability and enhance nutritional value of fresh and frozen strawberries (*Fragaria×ananassa*) and raspberries (*Rubus ideaus*). *Postharvest Biology and Technology* 33:67–78.

Han, J.H., and Krochta, J.M. 2007. Physical properties of whey protein coating solutions and films containing antioxidants. *Journal of Food Science* 72: E308–E314.

Helander, I.M., Alakomi, H.L., Latva-Kala, K., Mattila-Sandholm, T., Pol, I., Smid, E.J., Gorris, L.G., and Von Wright, A. 1998. Characterization of the action of selected essential oil components on Gram-negative bacteria. *Journal of Agricultural and Food Chemistry* 46:3590–3595.

Hernandez, E. 1994. Edible coatings from lipids and resins. In *Edible coatings and films to improve food quality.* J.M. Krochta, E.A. Baldwin, and M.G. Nispero-Carriedo (Eds.), Technomic, Lancaster, PA, pp. 279–303.

Hernandez-Izquierdo, V.M. 2007. Thermal transitions, extrusion, and heat-sealing of whey protein edible films. Ph.D. dissertation. University of California–Davis. 110 p.

Hernandez-Izquierdo, V.M., and Krochta, J.M. 2008. Thermoplastic processing of proteins for film formation—a review. *Journal of Food Science* 73(2):R30–R39.

Hettiarachchy, N.S., and Satchithanandam, E. 2007. Organic acids incorporated edible antimicrobial films. U.S. patent 7,160,580.

Juneja, V.K., Thippareddi, H., and Friedman, M. 2006. Control of *Clostridium perfringens* in cooked ground beef by carvacrol, cinnamaldehyde, thymol, or oregano oil during chilling. *Journal of Food Protection* 69:1546–1551.

Juneja, V.K., Bari, M.L., Inatsu, Y., Kawamoto, S., and Friedman, M. 2009. Thermal destruction of *Escherichia coli* O157:H7 in sous-vide cooked ground beef as affected by tea leaf and apple skin powders. *Journal of Food Protection* 72:860–865.

Kamper, S.L., and Fennema, O. 1984a. Water vapor permeability of edible bilayer films. *Journal of Food Science* 49:1478–1481, 1485.

Kamper, S.L., and Fennema, O. 1984b. Water vapor permeability of edible, fatty acid, bilayer film. *Journal of Food Science* 49:1482–1485.

Kamper, S.L., and Fennema, O. 1985. Use of an edible film to maintain water vapor gradients in food. *Journal of Food Science* 50:382–384.

Kester, J.J., and Fennema, O.R. 1986. Edible films and coatings: a review. *Food Technology* 40(12):47–59.

Kim, S.H., No, H.K., and Prinyawiwatkul, W. 2008. Plasticizer types and coating methods affect quality and shelf life of eggs coated with chitosan. *Journal of Food Science* 73(3):S111–S117.

Krochta, J.M., Baldwin, E.A., and Nisperos-Carriedo, M.O. 1994. *Edible coatings and films to improve food quality*. Technomic, Lancaster, PA. 379 p.

Krochta, J.M. 1997. Edible protein films and coatings. In *Food proteins and their applications in foods*. S. Damodaran and A. Paaraf (Eds.), Marcel Dekker, New York, pp. 529–549.

Krochta, J.M., and De Mulder-Johnston, C. 1997. Edible and biodegradable polymer films: challenges and opportunities. *Food Technology* 51(2):61–74.

Krochta, J.M. 1998. Whey protein interactions: effects on edible film properties. In *Functional properties of proteins and lipids*. J.R. Whitaker, F. Shahidi, A.L. Munguia, R.Y. Yada, and G. Fuller (Eds.), American Chemical Society, Washington, DC, pp. 158–167.

Krochta, J.M. 2002. Proteins as raw materials for films and coatings: definitions, current status, and opportunities. In *Protein-based films and coatings*. A. Gennadios (Ed.), CRC Press, Boca Raton, FL, pp. 1–41.

Lai, H.M., and Padua, G.W. 1998. Water vapor barrier properties of zein films plasticized with oleic acid. *Cereal Chemistry* 75:194–199.

Lai, H.M., Padua, G.W., and Wei, L.S. 1997. Properties and microstructure of zein sheets plasticized with palmitic and stearic acids. *Cereal Chemistry* 74:83–90.

Lanciotti, R., Anese, M., Sinigaglia, M., Severini, C., and Massini, R. 1999. Effects of heated glucose-fructose-glutamic acid solutions on the growth of *Bacillus stearothermophilus*. *Food Science and Technology* 32:223–230.

Lee, J.Y., Park, H.J., Lee, C.Y., and Choi, W.Y. 2003. Extending shelf-life of minimally processed apples with edible coatings and antibrowning agents. *Lebensmittel-Wissenschaft und -Technologie* 36:323–329.

Lima-Lima, E., Altamirano-Romo, S., Rivas-Araiza, R., Luna-Bárcenas, G., and Pérez-Pérez, C. 2006. Effect of nutraceuticals on physico-chemical properties of sodium caseinate films plasticized with glycerol. In *Water properties of food, pharmaceutical and biological materials*. M.P. Buera, J. Welti-Chanes, P.J. Lillford, and H.R. Corti (Eds.), CRC Press, Boca Raton, FL, pp. 445–454.

Lin, S.Y., and Krochta, J.M. 2005. Whey protein coating efficiency on surfactant-modified hydrophobic surfaces. *Journal of Agricultural and Food Chemistry* 53:5018–5023.

Lin D., and Zhao Y. 2007. Innovations in the development and application of edible coatings for fresh and minimally processed fruits and vegetables. *Comprehensive Reviews in Food Science and Food Safety* 6(3):60–75.

Luo, Y., and Barbosa-Cánovas, G.V. 1997. Enzymatic browning and its inhibition in new apple cultivars slices using 4-hexylresorcinol in combination with ascorbic acid. *International Journal of Food Science and Technology* 3:195–201.

Martínez, M.V., and Whitaker, J.R. 1995. The biochemistry and control of enzymatic browning, *Trends in Food Science and Technology* 6:195–200.

Maté, J.I., and Krochta, J.M. 1996. Comparison of oxygen and water vapor permeabilities of whey protein isolate and β-lactoglobulin edible films. *Journal of Agricultural and Food Chemistry* 44:3001–3004.

McClements, D.J. 2005. *Food emulsions: principles, practice, and techniques.* Boca Raton, FL: CRC Press.

McClements, D.J., Decker, E.A., and Weiss, J. 2007. Emulsion-based delivery systems for lipophilic bioactive components. *Journal of Food Science* 72(8):R109–R124.

McEvily, A.J., Iyengar, R., and Otwell, W.S. 1992. Inhibition of enzymatic browning in foods and beverages. *Critical Reviews in Food Science and Nutrition* 32:253–273.

McHugh, T.H., Aujard, J.-F., and Krochta, J.M. 1994. Plasticized whey protein edible films: water vapor permeability properties. *Journal of Food Science* 59:416–419, 423.

McHugh, T.H., and Krochta, J.M. 1994a. Sorbitol- vs glycerol-plasticized whey protein edible films: integrated oxygen permeability and tensile property evaluation. *Journal of Agricultural and Food Chemistry* 52:841–845.

McHugh, T.H., and Krochta, J.M. 1994b. Permeability properties of edible films. In *Edible coatings and films to improve food quality*. J.M. Krochta, E.A. Baldwin, and M.G. Nispero-Carriedo (Eds.), Technomic, Lancaster, PA, pp. 139–187.

Mei, Y., Zhao, Y., Yang, J., and Furr, H.C. 2002. Using edible coating to enhance nutritional and sensory qualities of baby carrots. *Journal of Food Science* 67:1964–1968.

Min, S., and Krochta, J.M. 2005. Antimicrobial films and coatings for fresh fruit and vegetables. In *Improving the safety of fresh fruit and vegetables*. W. Jongen (Ed.), Woodhead, Cambridge, UK, pp. 454–492.

Nam, S.H., Choi, S.P., Kang, M.Y., Koh, H.J., Kozukue, N., and Friedman, M. 2006. Antioxidative activities of bran extracts from twenty one pigmented rice cultivars. *Food Chemistry* 94:613–620.

Nicolas, J.J., Richard-Forget, F.C., Goupy, P.M., Amiot, M.J., and Aubert, S.Y. 1994. Enzymatic browning reactions in apple and products. *Critical Reviews in Food Science and Nutrition* 34:109–157.

Nispero-Carriedo, M.G. 1994. Edible coatings and films based on polysaccharides. In *Edible coatings and films to improve food quality*. J.M. Krochta, E.A. Baldwin, and M.G. Nispero-Carriedo (Eds.), Technomic, Lancaster, PA, pp. 305–335.

Oms-Oliu, G., Soliva-Fortuny, R., and Martín-Belloso, O. 2008. Edible coatings with antibrowning agents to maintain sensory quality and antioxidant properties of fresh-cut pears. *Postharvest Biology and Technology* 50:87–94.

Onsøyen, E. 2001. Alginate. Production, composition, physicochemical properties, physiological effects, safety, and food applications. In *Handbook of dietary fiber*. M.L. Dreher and S.S. Cho (Eds.), CRC Press, Boca Raton, FL, pp. 659–674.

Padua, G.W., and Wang, Q. 2002. Formation and properties of corn zein films and coatings. In *Protein-based films and coatings*. A. Gennadios (Ed.), CRC Press, Boca Raton, FL, pp. 43–67.

Park, H.J., Weller, C.L., Vergano, P.J., and Testin, R.F. 1992. Factors affecting barrier and mechanical properties of protein-based edible, degradable films. Paper No. 428, presented at the Annual Meeting of the Institute of Food Technologists, June 20–24, New Orleans, LA.

Park, H.J., Bunn, J.M., Weller, C.L., Vergano, P.J., and Testin, R.F. 1994a. Water vapor permeability and mechanical properties of grain protein-based films as affected by mixtures of polyethylene glycol and glycerin plasticizers. *Transactions of the American Society of Agricultural Engineering* 37:1281–1285.

Park, J.W., Testin, R.F., Park, H.J., Vergano, P.J., and Weller, C.L. 1994b. Fatty acid concentration effect on tensile strength, elongation, and water vapor permeability of laminated edible films. *Journal of Food Science* 59:916–919.

Park, S.K., Hettiarachchy, N.S., Ju, Z.Y., and Gennadios, A. 2002. Formation and properties of soy protein films and coatings. In *Protein-based films and coatings*. A. Gennadios (Ed.), CRC Press, Boca Raton, FL, pp. 123–137.

Park, S.I., Daeschel, M.A., and Zhao, Y. 2004. Functional properties of antimicrobial lysozyme–chitosan composite films. *Journal of Food Science* 69:M215–M221.

Park, S.I., and Zhao, Y. 2004. Incorporation of a high concentration of mineral or vitamin into chitosan-based films. *Journal of Agricultural and Food Chemistry* 52:1933–1939.

Park, S.I., Stan, S.D., Daeschel, M.A., and Zhao, Y. 2005. Antifungal coatings on fresh strawberries (*Fragaria×ananassa*) to control mold growth during storage. *Journal of Food Science* 70:M202–M207.

Pommet, M., Redl, A., Morel, M.H., Domenek, S., and Guilbert, S. 2003. Thermoplastic processing of protein-based bioplastics: chemical engineering aspects of mixing, extrusion and hot molding. *Macromolecular Symposia* 197:207–217.

Pommet, M., Redl, A., Guilbert, S., and Morel, M.H. 2005. Intrinsic influence of various plasticizers on functional properties and reactivity of wheat gluten thermoplastic materials. *Journal of Cereal Science* 42:81–91.

Quezada-Gallo, J.A., Deveaufort, F., and Voilley, A. 1999. Interactions between aroma and edible films. 1. Permeability of methylcellulose and polyethylene films to methyl ketones. *Journal of Agricultural and Food Chemistry* 47:108–113.

Quezada-Gallo, J.A., Deveaufort, F., and Voilley, A. 2000. Mechanism of aroma transport through edible and plastic packagings. In *Food packaging: testing methods and applications*, ACS Symposium Series No. 753. S. Rish (Ed.), American Chemical Society, Washington, DC, pp. 125–140.

Quinones, B., Massey, S., Friedman, M., Swimley, M.S., and Teter, K. 2009. Novel cell-based method to detect Shiga Toxin 2 from *Escherichia coli* O157:H7 and inhibitors of toxin activity. *Applied Environmental Microbiology* 75:1410–1416.

Redl, A., Morel, M.H., Bonicel, J., Vergnes, B., and Guilbert, S. 1999. Extrusion of wheat gluten plasticized with glycerol: influence of process conditions on flow behavior, rheological properties, and molecular size distribution. *Cereal Chemistry* 76(3):361–370.

Reineccius, G.A. 1994. Flavor encapsulation. In *Edible coatings and films to improve food quality*. J.M. Krochta, E.A. Baldwin, and M.G. Nispero-Carriedo (Eds.), Technomic, Lancaster, PA, pp. 105–120.

Reiners, R.A., Wall, J.S., and Inglett, G.E. 1973. Corn proteins: potential for their industrial use. In *Industrial uses of cereals*. Y. Pomeranz (Ed.), St. Paul, MN: American Association of Cereal Chemists, pp. 285–298.

Rodov, V., Ben-Yehoshua, S., Fang, D., Kim, J., and Ashkenazi, R. 1995. Performed antifungal compounds of lemon fruit: citral and its relation to disease resistance. *Journal of Agricultural and Food Chemistry* 43:1057–1061.

Rojas-Graü, M.A., Avena-Bustillos, R.J., Olsen, C.W., Friedman, M., Henika, P.R., Martín-Belloso, O., Pan, Z., and McHugh, T.H. 2007a. Effects of plant essential oils and oil compounds on mechanical, barrier and antimicrobial properties of alginate-apple puree edible films. *Journal of Food Engineering* 81:634–641.

Rojas-Graü, M.A., Grasa-Guillem, R., and Martín-Belloso, O. 2007b. Quality changes in fresh-cut Fuji apple as affected by ripeness stage, antibrowning agents and storage atmosphere. *Journal of Food Science* 72:S36–S43.

Rojas-Graü, M.A., Soliva-Fortuny, R., and Martín-Belloso, O. 2008. Effect of natural antibrowning agents on color and related enzymes in fresh-cut Fuji apples as an alternative to the use of ascorbic acid. *Journal of Food Science* 6:267–272.

Roller, S., and Seedhar, P. 2002. Carvacrol and cinnamic acid inhibit microbial growth in fresh-cut melon and kiwifruit at 4°C and 8°C. *Letters in Applied Microbiology* 35:390–394.

Santosa, F.X.B., and Padua, G.W. 1999. Tensile properties and water absorption of zein sheets plasticized with oleic and linoleic acids. *Journal of Agricultural and Food Chemistry* 47:2070–2074.

Sapers, G.M. 1993. Browning of foods—control by sulfites, antioxidants, and other means. *Food Technology* 47:75–84.

Shahidi, F., and M. Naczk. 2004. *Phenolics in food and nutraceuticals.* Boca Raton, FL: CRC Press.

Siew, D.C.W., Heilmann, C., Easteal, A.J., and Cooney, R.P. 1999. Solution and film properties of sodium caseinate/glycerol and sodium caseinate/polyethylene glycol edible coating systems. *Journal of Agricultural and Food Chemistry* 47:3432–3440.

Smith-Palmer, A., Stewart, J., and Fyfe, L. 1998. Antimicrobial properties of plant essential oils and essences against five important food-borne pathogens. *Letters in Food Microbiology* 26:118–122.

Sothornvit, R., and Krochta, J.M. 2000. Plasticizer effect on oxygen permeability of β-lactoglobulin films. *Journal of Agricultural and Food Chemistry* 48:6298–6302.

Sothornvit, R., and Krochta, J.M. 2001. Plasticizer effect on mechanical properties of β-lactoglobulin films. *Journal of Food Engineering* 50:149–155.

Sothornvit, R., Olsen, C.W., McHugh, T.H., and Krochta, J.M. 2003. Formation conditions, water vapor permeability, and solubility of compression-molded whey protein films. *Journal of Food Science* 68:1985–1989.

Sothornvit, R., and Krochta, J.M. 2005. Plasticizers in edible films and coatings. In *Innovations in food packaging.* J.H. Han (Ed.), Elsevier Academic Press, San Diego, CA, pp. 403–433.

Swenson, H.A., Miers, J.C., Schultz, T.H., and Owens, H.S. 1953. Pectinate and pectate coatings II. Applications to nuts and fruit products. *Food Technology* 7:232–235.

Tolstoguzov, V.B. 1993. Thermoplastic extrusion—the mechanism of the formation of extrudate structure and properties. *Journal of the American Oil Chemists' Society* 70(4):417–424.

Tomasula, P.M., Parris, N., Yee, W., and Coffin, D. 1998. Properties of films made from CO_2-precipitated casein. *Journal of Agricultural and Food Chemistry* 46:4470–4474.

Torres, J.A. 1994. Edible films and coatings from proteins. In *Protein functionality in food systems.* N.S. Hettiarachchy and G.R. Ziegler (Eds.), Marcel Dekker, New York, pp. 467–507.

Valencia-Chamorro, S.A., Palou, L., Del Río, M.A., and Pérez-Gago, M.B. 2008. Inhibition of *Penicillium digitatum* and *Penicillium italicum* by hydroxypropyl methylcellulose-lipid edible composite films containing food additives with antifungal properties. *Journal of Agricultural and Food Chemistry* 56:11270–11278.

Zhang, J., Mungara, P., and Jane, J. 2001. Mechanical and thermal properties of extruded soy protein sheets. *Polymer* 42:2569–2578.

Zhuang, R., Beuchat, L.R., Chinnan, M.S., Shewfelt, R.L., and Huang, Y.W. 1996. Inactivation of *Salmonella Montevideo* on tomatoes by applying cellulose-based films. *Journal of Food Protection* 59:808–812.

7 Coatings for fresh fruits and vegetables

Jinhe Bai and Anne Plotto

Contents

7.1	Introduction		186
7.2	Postharvest physiology		188
	7.2.1	Respiration	188
		7.2.1.1 Effect of oxygen and carbon dioxide on respiration	188
		7.2.1.2 Climacteric and nonclimacteric fruit	191
		7.2.1.3 Other factors affecting respiration	191
	7.2.2	Transpiration	192
		7.2.2.1 Effect of relative humidity and temperature on transpiration	192
		7.2.2.2 Influence of the type of produce on water loss	193
7.3	Postharvest deterioration and effect of coatings		193
	7.3.1	Dehydration and shrinkage	193
	7.3.2	Softening, mealy texture, or hardening	194
	7.3.3	Discoloration and reduction in gloss	195
	7.3.4	Decreased nutritional quality	197
	7.3.5	Flavor loss	197
	7.3.6	Physiological disorders	198
	7.3.7	Disease and physical injury	199
	7.3.8	Quarantine treatments	201
7.4	Properties of edible coatings on fresh commodities		202
	7.4.1	Lipids and resins	202
	7.4.2	Polysaccharides	206
	7.4.3	Proteins	212
	7.4.4	Composites and bilayer coatings	212
7.5	Responses of commodities to edible coatings		213
	7.5.1	Apples and pears	213
	7.5.2	Citrus fruit	215
	7.5.3	Stone fruits	217
	7.5.4	Grapes	218
	7.5.5	Strawberries and other berries	219
	7.5.6	Kiwifruit	219
	7.5.7	Melons	220
	7.5.8	Bananas	220
	7.5.9	Mangoes and other tropical fruits	221

7.5.10	Tomatoes and other fruit vegetables	223
7.5.11	Root and tuber crops	224
7.5.12	Leafy vegetables	225
7.6	Conclusion	225
References		226

7.1 Introduction

Coatings applied to the surfaces of fruit and vegetables are commonly called waxes, whether or not any component thereof is actually a wax. The application of coatings to apples, citrus, stone fruits, avocadoes, tomatoes, and cucumbers prior to marketing is standard practice in the United States and many other countries. The purpose of coatings on fruit and vegetables is to reduce water loss, slow senescence and aging, impart shine, and allow for better quality and marketing price. Furthermore, coatings create a modified atmosphere inside the produce and protect the product from pathogens and contaminants. Coatings become more important for fruit that are washed prior to storage and shipping, because during that process, brushes partially remove the natural wax on the produce surface, which aggravates water loss and increases deterioration (Baldwin, 1994; Hagenmaier and Baker, 1993c). When coatings are formulated, special attention needs to be paid to the produce on which it will be used to avoid an excess modification of internal O_2 and CO_2, which can cause anaerobic respiration with consequent ethanol production, off-flavor, and other physiological disorders (Bai et al., 2002b; Baldwin et al., 1999a; Hagenmaier, 2002). Finally, when making a coating, companies need to take into account applicable food laws. Some ingredients, such as polyethylene, are approved for citrus in the United States by the U.S. Food and Drug Administration (FDA) but are not approved for pome fruit (Baldwin, 1994). Likewise, morpholine, the basic portion that allows waxes to get into emulsions, is approved in the United States but not in Europe.

Although coating citrus fruit by wax had been used in China since the twelfth and thirteenth centuries, extended commercial use of coatings for fruit and vegetables did not occur until the twentieth century (Baldwin, 1994). The development of modern-day waxes began in the United States in the 1930s, where melted paraffin waxes were applied onto oranges and apples by brushes (Hardenburg, 1967; Hitz and Haut, 1938; Hitz and Haut, 1942). These waxes were utilized to control weight loss (from water losses) occurring in storage. "Solvent waxes" replaced paraffin waxes in the late 1940s and remained the predominant type for more than 40 years (Hall, 1981a). The solvent waxes were composed primarily of either synthetic resin (coumarone-indene) or wood rosin dissolved in petroleum solvents. Plasticizers and other compounds were added to improve shine and handling characteristics (Petracek et al., 1999). These coatings not only decreased weight loss, but also imparted an attractive shine to the fruit surface. Solvent waxes are no longer in use in the United States for safety and environmental reasons (Hagenmaier, 1998). Nevertheless, their development has initiated a marketing trend that emphasizes fruit shininess (Hall, 1981a; Kaplan, 1986), which is being continued on today's citrus and apple commercial coatings.

Carnauba wax microemulsions, the first commercially available water-based waxes, were introduced for coating fresh fruits and vegetables in the late 1950s

(Kaplan, 1986). Carnauba-based waxes are valued for decreasing water loss, but they produce less gloss compared with the solvent waxes. Candelilla-based waxes give excellent protection against water loss but offer even less gloss than carnauba waxes. Polyethylene waxes offer moderate protection against water loss, and because of their high oxygen and CO_2 permeability, they are preferred for commodities like tangerines that develop off-flavors with shellac or resin coatings. However, polyethylene-based coatings are not used in the United States for fruits whose peel is normally eaten (CFR, 1990). High-shine, water-soluble waxes based on shellac and alkali-soluble resin were introduced in the early 1960s and have become commonly used in packing houses. High-solids waxes, or concentrated waxes, were introduced in the late 1970s as high-shine waxes that required less drying prior to waxing.

In the mid 1980s, several water-soluble polysaccharide coatings containing carboxymethyl cellulose and sucrose fatty acid esters became available. These generally produce poor gloss and have limited effect on water loss but have advantages for creating suitable modified atmosphere (MA), and these are easily removed by washing before consumption (Drake et al., 1987; Petracek et al., 1999; Smith and Stow, 1984). They are preferentially used on fruit other than citrus and apples. Chitosan, another polysaccharide, has attracted research attention as a new coating since the early 1980s because it not only performs well as a semipermeable film to slow senescence of products, but it also has potential as a natural food preservative due to its antimicrobial activity against a wide range of foodborne filamentous fungi, yeast, and bacteria (Sagoo et al., 2002). Chitosan has been developed as a commercial fruit coating (Elson et al., 1985) but is not yet approved for use in the United States.

In the last decade, new natural materials such as alginate (Falcão-Rodrigues et al., 2007; Olivas et al., 2007; Rojas-Graü et al., 2007; Tapia et al., 2007), gellan (Tapia et al., 2007), pullulan (Chlebowska-Śmigiel et al., 2007; Diab et al., 2001; Xu et al., 2001), *Aloe vera* (Martínez-Romero et al., 2006; Valverde et al., 2005), and turmeric starch (Jagannath et al., 2006) have shown potential to be new coatings for fresh and fresh-cut fruit. For satisfying consumers' preference for "all natural" ingredients, one new approach for coating development has been to use fruit purees as a base to develop coatings and films (McHugh et al., 1996). Those have properties of polysaccharides, and depending on the type and amount of lipid component, desired properties may be obtained. For instance, McHugh and Senesi (2000) made edible films from apple puree with various concentrations of fatty acids, fatty alcohols, beeswax, and vegetable oil that effectively maintained apple slice texture and color. Sothornvit and Pitak (2007) and Sothornvit and Rodsamran (2008) developed banana and mango films from banana flour and mango puree, respectively, and used them as wraps on whole or fresh-cut fruit. These films have mostly been experimented with fresh-cut commodities, and with whole fruit to a lesser extent. A new approach to use nanoparticles as food coating components has been discovered. Jin et al. (2008) found that ZnO nanoparticles possess antimicrobial powers against some food-borne microbes including *Escherichia coli* O157:H7 in culture media. The antimicrobial activity was decreased when ZnO was suspended in a polyvinylpyrrolidone (PVP) coating in comparison with the direct use of the material (Jin et al., 2008). An et al. (2008) used a silver nanoparticles–PVP coating on green asparagus and found that the coating decreased the growth of microorganisms, retarded tissue hardening, and significantly extended storage life.

At present, "organic" fruits are usually not treated with any coating. Regular fruit coatings are not allowed because of some commonly used ingredients, for example, morpholine. Only a few coatings with beeswax- or carnauba-based wax are available for organic products (OMRI, 2008). There is need for organic coatings research and development.

Table 7.1 summarizes edible coatings studied for fruits and vegetables. More information on edible coatings for fruits and vegetables can be found in the following reviews: Baldwin (1994), Petracek et al. (1999), Park (1999), Amarante and Banks (2001), Farber et al. (2003), Olivas et al. (2008), and Vargas et al. (2008).

7.2 Postharvest physiology

7.2.1 Respiration

Respiration is the process by which plants take in oxygen and give out carbon dioxide. During that process, carbohydrates in the plant are metabolized into carbon dioxide and water. This reaction produces energy to maintain life of the produce (Saltveit, 2004).

Respiration is a continuing process in the growing plant as the leaves uptake atmospheric CO_2 and make carbohydrates through photosynthesis. However, after harvest, the plant parts—leaves, fruit, roots—cannot replace carbohydrates or water. Thereafter, respiration uses stored starch, sugars, and acids, and only stops doing so when these reserves are exhausted, although marketability and edibility end long before complete exhaustion of those reserves (Saltveit, 2004).

7.2.1.1 Effect of oxygen and carbon dioxide on respiration

Respiration rates, like rates of any chemical reaction, decrease with falling concentrations of reactants and increasing concentrations of reaction products. Thus, respiration slows when the environmental oxygen level falls below the ambient value of about 20.7% or the environmental carbon dioxide exceeds the ambient level of about 0.04%. Controlled atmosphere (CA) storage and modified atmosphere packaging (MAP) of fruit and vegetables are two widely used techniques that make use of low O_2 and high CO_2 in the environment to reduce respiration rate and thereby extend storage life (Mir and Beaudry, 2004). Edible coatings can extend storage life of produce by a similar mechanism as CA and MAP. Coatings restrict gas exchange through peel of produce and thus lead to a modified internal atmosphere and an extended storage life of produce (Baldwin, 1994).

However, it is important to recognize that while atmosphere modification can improve the storability of some fruit and vegetables, it also has the potential to induce undesirable effects. Fermentation and off-flavors may develop if decreased O_2 levels cannot sustain aerobic respiration (Petracek et al., 1999). Similarly, physiological injury will occur if CO_2 exceeds tolerable levels (Bai et al., 2002b; Baldwin et al., 1995; Cisneros-Zevallos and Krochta, 2003; Hagenmaier, 2001a). Ranges of

Table 7.1 Edible coatings studied for fruits and vegetables

Main ingredient	Fruit	Reference
Beeswax	Cut apples, plum, cherry, mandarins	Pérez-Gago et al., 2002, 2005, 2006; Rojas-Argudo et al., 2005
Candelilla	Apple, citrus	Bai et al., 2003b; Hagenmaier and Baker, 1996
Carnauba	Apple, cut apple, pear, citrus, tomato	Amarante et al., 2001b; Bai et al., 2003b; Chiumarelli and Ferreira, 2006; Hagenmaier and Baker, 1994b; Pérez-Gago et al., 2005
Shellac	Apple, cherry, plum, citrus	Bai et al., 2002a; Hagenmaier, 2002; Hagenmaier and Shaw, 1991b; Pérez-Gago et al., 2003a; Rojas-Argudo et al., 2005
Soybean oil/vegetable oil	Apple, pear	Conforti and Totty, 2007; Ju and Curry, 2000a 2000b; Ju et al., 2000
Fatty acids	Cut pear, cut apples, apricots, green peppers, kiwifruit	Ayranci and Tunc, 2004; Olivas et al., 2003, 2007; Xu et al., 2001
Polyethylene	Citrus	Hagenmaier, 2000; Hagenmaier and Shaw, 1991a
Polyvinyl acetate	Citrus, apple	Bai et al., 2002a; Hagenmaier and Grohmann, 1999
Alginate	Cut apple, cut papaya, cut fruit, apple	Falcão-Rodrigues et al., 2007; Olivas et al., 2007; Rojas-Graü et al., 2007; Tapia et al., 2007
Aloe vera	Cherry, grapes	Martínez-Romero et al., 2006; Valverde et al., 2005
Cactus-mucilage	Strawberry	Del-Valle et al., 2005
Carboxymethyl cellulose	Shelled pecan, cut mango, cut apple, apple	Baldwin and Wood, 2006; Nisperos-Carriedo and Baldwin, 1994
Carrageenan	Strawberry	Ribeiro et al., 2007
Chitosan	Cucumber, grapes, citrus, carrot, pepper, strawberry, apple, cut fruit, grape	Chien et al., 2007b; Choi et al., 2002; El Ghaouth et al., 1991b; Fornes et al., 2005; Hernández-Muñoz et al., 2006; Ratanachinakorn et al., 2005; Ribeiro et al., 2007; Romanazzi et al., 2002; Vargas et al., 2006, 2009
Dextrans	Guava, apricot	Quezada Gallo et al., 2005
Gellan	Cut apple, cut papaya	Tapia et al., 2007
Hydroxypropyl methyl cellulose	Cut apple, plum, tomato, citrus	Pérez-Gago et al., 2002, 2003a, 2005; Zhuang and Huang, 2003
Locust bean gum	Cherry, apple	Conforti and Totty, 2007; Rojas-Argudo et al., 2005

(*Continued*)

Table 7.1 Edible coatings studied for fruits and vegetables (Continued)

Main ingredient	Fruit	Reference
Maltodextrin	Apple	Conforti and Totty, 2007
Methyl Cellulose	Cut pear, apricots, green peppers, cut apple	Ayranci and Tunc, 2004; Brancoli and Barbosa-Canovas, 2000; Olivas et al., 2003
Pullulan	Apple, kiwifruit, strawberry	Chlebowska-Śmigiel et al., 2007; Diab et al., 2001; Xu et al., 2001
Semperfresh	Cherry, tomato, apple	Drake et al., 1987; Sümnü and Bayindirli, 1995a, 1995b; Tasdelen and Bayindirli, 1998; Yaman and Bayoindirli, 2002
Starch	Apple, strawberry	Bai et al., 2002a; Ribeiro et al., 2007
Xanthan gum	Baby carrots	Mei et al., 2002
Casein/milk protein	Apple, cherry	Certel et al., 2004; Le Tien et al., 2001; Shon and Haque, 2007
Soy protein	Apple	Eswarandam et al., 2006; Shon and Haque, 2007
Whey protein concentrate	Cut apple, cut persimmon	Le Tien et al., 2001; Pérez-Gago et al., 2005, 2006
Whey protein isolate	Cut apple	Pérez-Gago et al., 2003b, 2005
Zein	Apple, tomato	Bai et al., 2003a; Park et al., 1994

Sources: Adapted from Olivas, G.I., Davila-Avina, J.E., Salas-Salazar, N.A., and Molina, F.J., Stewart Postharvest Review, 4, 1, 2008; Vargas, M., Pastor, C., Chiralt, A., McClements, D.J., and Gonzalez-Martinez, C., Critical Reviews in Food Science and Nutrition, 48, 496, 2008.

nondamaging O_2 and CO_2 levels have been published for a number of fruits and vegetables (Beaudry, 1999, 2000; Mir and Beaudry, 2004). Horticultural crops differ in their tolerance for O_2 and CO_2. Most fruit and vegetables do not tolerate environmental O_2 <2%, or CO_2 >1% to 15%. However, some commodities, such as mushrooms, strawberries, and blueberries, have a tolerance level >20% CO_2.

It seems important here to point out that the "environmental" O_2 and CO_2 values just referred to for CA and MAP apply to gas samples taken from the space outside of the fruit but inside the man-made barriers that separate fruit from the environment (building walls for CA and packaging for MAP). Edible coatings also act as barriers between fruit and atmosphere, but there is no space between fruit and coating from which to take gas samples. Instead, the measureable O_2 and CO_2 concentrations are those of gas samples taken from cavities inside the fruit, namely, the "internal" gas, whose composition is often much different from that of the environment. Therefore, the environmental gas concentrations just cited, however valid they are for CA and MAP, are not the internal O_2 and CO_2 values critical for safe storage of coated fruit.

7.2.1.2 Climacteric and nonclimacteric fruit

There are two distinct patterns of ripening that can be identified, and these are termed *climacteric* and *nonclimacteric* types. In nonclimacteric fruit, the process of maturation and ripening is a continuous but gradual process. In contrast, climacteric fruit undergo a rapid ripening phase when triggered by changes in hormonal composition. The onset of climacteric ripening is thus a well-defined event marked by rapid increase in the rate of respiration and the natural evolution of ethylene gas by the fruit at a point in its development known as the respiratory climacteric (Baldwin, 2004). An example of the importance of the respiratory climacteric can be seen in a fruit such as banana: bananas may be held at a moderate temperature when in the green state, but as they begin to ripen, they will rapidly increase their respiration rate and generate much more heat. The consequence may be that this heat cannot be controlled and even more respiration will occur in an inflationary spiral rapidly leading to spoilage of the fruit in a very short time. Once climacteric fruit start to ripen, there is very little that can be done except to market them for immediate consumption.

For climacteric-type fruits, additional attention should be paid when applying the coating at a preclimacteric stage. One should be aware that once the climacteric initiates onset, respiration rate increases dramatically; a tight gas barrier coating may cause anaerobic fermentation and physiological disorder (Bai et al., 2003b).

7.2.1.3 Other factors affecting respiration

Temperature control is the most important single factor for postharvest storage and handling of produce. All plant metabolisms, including respiration, are accelerated at "physiological temperature" between 20 and 25°C. The maximal rate of respiration for most fruit and vegetable products undergoes a four- to sixfold increase when temperature changes from 0 to 15°C (Beaudry et al., 1992; Cameron et al., 1994; Mir and

Beaudry, 2004). However, the increase of coating permeability to oxygen and CO_2 was only two times when raising temperature from 0°C to 25°C for apples coated with a carnauba-shellac mixture (Amarante and Banks, 2001). This means that a fruit coating that allows adequate respiration at low temperature may cause anaerobic problems should the storage temperature rise. When respiratory demand for O_2 increases faster than O_2 permeation as temperature increases, O_2 levels decline and may pose a risk to product quality.

Ethylene gas is produced in most plant tissues and is known to be an important factor in increasing respiration rate and inducing ripening and senescence of produce. The ethylene production is increased when produce is injured or attacked by microorganisms. The major undesirable effects caused by ethylene include shortened shelf life, yellowing of some crops like cucumber and green apples, sprouting of potatoes, softening and core-browning of apples, internal breakdown of watermelon and kiwifruit, and increased sensitivity to chilling injury in avocado and grapefruit (Baldwin, 2004). Many horticultural products are sensitive to ethylene. Their exposure to ethylene must therefore be minimized. Ethylene-producing commodities should not be mixed with ethylene-sensitive commodities during storage and transport. Some coatings might inhibit ethylene production because of decreased oxygen and increased carbon dioxide inside of the produce, although ethylene is usually accumulated at a higher concentration in the coated produce (Bai et al., 2002b).

7.2.2 Transpiration

Most fresh produce contain from 65% to 85% water when harvested. Within growing plants there is a constant flow of water. After harvest, the water supply is broken but loss of water continues. It was found that 97% of weight loss in apples stored at 20°C and 60% relative humidity (RH) was from transpiration, in comparison with about 3% of weight loss from respiration (Maguire et al., 2000). Normally, if weight loss exceeds 5% of weight, produce will appear too shrunken to be salable. Water loss from the plant occurs through the entire produce surface, especially through openings including stomata, lenticels, wound openings, and blossom and stem scars.

Water loss can be limited by the natural waxy or corky layer on the plant surface. All primary aerial surfaces of plants are covered with a hydrophobic cuticle consisting of epicuticular wax embedded in a cutin matrix composed largely of cross-linked hydroxyl and hydroxylepoxy fatty acids. One of the major functions of the cuticle is the formation of an efficient barrier against unregulated water loss (Kerstiens, 1996). Pieniazek (1944) determined that lenticular transpiration accounted for 21% of total transpiration in 'Golden Delicious' and 'Baldwin,' 19% in 'Canada Red,' and 8% in 'Turley' apples. This route represented a significant proportion of water loss in some cultivars, but still 5 to 10 times more water was being lost directly through the cuticle. The barrier properties of the cuticle compose the main constraint to transpiration (Maguire et al., 2001).

7.2.2.1 Effect of relative humidity and temperature on transpiration

The important determinants of transpiration rate are the difference between the water vapor pressure inside and outside the plant, air movement, packaging or coating, and surface area. The relative humidity in the produce is almost 100% versus 90% to

95% in a well-maintained storage room and 30% to 80% on the market shelves. To prevent water loss from fresh produce, it must be kept in a moist atmosphere as much as possible. Also, the faster the surrounding air moves, the quicker water is lost. High temperature raises activity of water molecules and thus increases water loss from produce. The ideal storage condition for preventing water loss from produce is a low temperature (as low as possible above chilling injury temperature) and relative humidity of 90% or above inside the storage room or packaging.

7.2.2.2 Influence of the type of produce on water loss

The water loss rate varies with the type of produce. Leafy green vegetables, especially spinach and leaf lettuce, lose water quickly because they have a high surface-area-to-volume ratio and a thin waxy cuticle with many pores. Others, such as tomatoes, which have a thick cuticle layer without pores (Cameron and Yang, 1982), have a much lower rate of water loss. Some produce, such as the multilayered structure of onions and iceberg lettuce, are carrying an inherent protection because water transport between leaves occurs via the dwarf stem. Some produce are more sensitive to water loss than others. For example, at room temperature without coating or packaging, tangerines lose water very quickly and become unmarketable within 3 to 4 days; however, apples store much longer in the same environment. Varieties of the same fruit also can have highly different transpiration rates. For example, Maguire et al. (2000) reported that mean water vapor permeance for 'Braeburn,' 'Pacific Rose,' 'Cripps Pink,' and 'Granny Smith' apples were 44, 35, 20, and 17 nmol s^{-1} m^{-2} Pa^{-1}, respectively. Research has also shown that the water vapor permeance in apples largely increased with harvest date, from 21 nmol s^{-1} m^{-2} Pa^{-1} at first harvest to 46 nmol s^{-1} m^{-2} Pa^{-1} at final harvest, a 219% increase during 10 weeks, on average, across all four cultivars. There are other factors contributing to total variation in water vapor permeance, such as fruit-to-fruit differences (22% to 25%), tree (4%), and orchard (Maguire et al., 2000). Edible hydrophobic coatings with good water vapor barrier properties can offer resistance to water loss of fresh produce (Baldwin, 1994).

7.3 Postharvest deterioration and effect of coatings

7.3.1 Dehydration and shrinkage

Water loss control was the original target of coating application for preserving fresh produce. Singh (1971) studied optimum storage temperature, storage life, and fruit quality of mandarins and sweet lemon. During storage, the loss of moisture from the peel is continuously replenished by the movement of the moisture from the pulp. If this loss, due to combined effects of respiration and transpiration, goes on unchecked, the fruit shrivels up and becomes unmarketable. Washing, which is a standard practice for today's packing houses, removes part of the protective materials on the skin and thereby increases water loss in citrus (Hagenmaier and Baker, 1993b). Washing grapefruits, oranges, and other citrus with brushes increased the shrinkage rate by 40% to 50%. Thus, edible coating application becomes a necessity in order to protect the freshness of produce during subsequent storage, transportation, and marketing.

Figure 7.1 Bubbles emitted from apples submerged in water when air is pushed through the seed cavity for noncoated 'Granny Smith' (left), noncoated 'Delicious,' and shellac-coated 'Delicious.' (From Jinhe Bai, USDA-ARS. With permission.)

Waxing decreases produce permeance to water vapor by adding a layer through which the water molecules must move, and by blocking pores on the skin (Figure 7.1) (Bai et al., 2002a; Baldwin, 1994; Hagenmaier and Shaw, 1991b). Lipid coatings have the least water vapor permeability and best protect against water loss followed by resin, protein, and polysaccharides (Bai et al., 2002a; Hagenmaier, 2002).

7.3.2 Softening, mealy texture, or hardening

Plant cell walls play a central role in determining texture of fruits and vegetables. There are three major categories of wall polysaccharides: pectic polysaccharides, hemicelluloses, and cellulose. Cellulose provides the microfibrillar component and composes the single most abundant polysaccharide component of vegetables (Waldron et al., 2003). Pectic polysaccharides are present throughout all primary cell walls and form a gel matrix interspersing the cellulose-hemicellulose network (Steele et al., 1997). During fruit softening, pectins (Fischer and Bennett, 1991) and hemicelluloses (Wakabayashi, 2000) typically undergo solubilization and depolymerization that are thought to contribute to cell wall loosening and disintegration. In most fruit, firmness diminishes as the degree of ripening increases due to the action of pectolytic enzymes (Muramatsu et al., 1996). In overripe and mealy fruits, the high level of cell separation results in poor juice release and a dry mouthfeel (Waldron et al., 2003). The loss of firmness in apples and tomatoes represents a major deterioration for those fruits. Some fruit, such as pears and mangoes, need proper ripening prior to consumption; however, once the ripening process is triggered, fruit have a very short shelf-life (Amarante et al., 2001a). In apples, Link et al. (2004) reported that the relationship between firmness and mass loss, or shrinkage, was dependent on cultivar. Firmness was significantly and linearly related to mass loss and to shrinkage in 'Delicious' apples. Hence, it is possible to predict firmness of 'Delicious' apples under regular air storage conditions by tracking mass loss or shrinkage.

Edible coatings slowed softening and thus extended shelf life of many fruits, such as apples (Bai et al., 2002a, 2003a), pears (Amarante et al., 2001a), tomatoes (Park et al., 1994), mangoes (Baldwin et al., 1999b), pepino dulce (Schreiner et al., 2003), radish (Schreiner et al., 2003), and sweet cherries (Yaman and Bayoindirli, 2002). Edible coatings maintain fruit firmness by similar mechanisms as do CA and MAP—by decreasing respiration and transpiration, slowing ripening and senescence, and retarding degradation of cell wall (Bai et al., 2009a; Baldwin, 1994). For example, a sucrose fatty acids ester coating maintained firmness by inhibiting pectin degradation of pepino dulce, and starch coatings prevented the loss of soluble pectins in radish (Schreiner et al., 2003).

Postharvest hardening occurs in some vegetables, such as cauliflower and asparagus. This is usually attributed to the continued development of vascular and support tissues involving natural wall thickening with associated lignification (Waldron and Selvendran, 1992). Modified atmosphere packaging combined with cold temperature slow the toughening process (Heyes et al., 1998). A silver nanoparticles-polyvinylpyrroloidone coating decreased water loss and toughening of asparagus (An et al., 2008).

7.3.3 Discoloration and reduction in gloss

Color is considered to be one of the most important external factors of fruit and vegetable quality, as the appearance greatly influences consumers. Yellowing of fruit and vegetables is often a result of the disappearance of chlorophylls which allows the yellow pigments and carotenes to become more visible (Shewfelt, 1993). A sucrose fatty acid ester coating was effective in retaining higher contents of chlorophyll to maintain green color in green peppers (Özden and Bayindirli, 2002). The relationship between color and level of maturation has been widely studied in tomatoes (Choi et al., 1995), peaches, and nectarines (Luchsinger and Walsh, 1993; Mitchell, 1987). Along the same line, Mercado-Silva et al. (1998) identified L^*, a^*, and hue angle values as being the best parameters for differentiating the different stages of the maturation of guava. In citrus, Jimenez-Cuesta et al. (1981) proposed the use of the formula $1000a^*/(L^*b^*)$ as "Color Index" for recording the process of orange degreening.

Some coatings suppressed yellowing in apples (Bai et al., 2003b), sour citrus (Yamauchi et al., 2008), asparagus (An et al., 2008), and mango (Baldwin et al., 1999a) due to the formation of modified atmosphere in the fruit. In addition, some specific ingredients in the coatings may influence discoloration. For example, laurates de-esterified from sucrose laurate ester inhibited activities of chlorophyllase and a chlorophyll-degrading peroxidase, thus suppressing the degreening of citrus (Yamauchi et al., 2008). However, other fatty acid salts, such as caprylate, caprate, myristate, palmitate, or stearate, had much less effect on those enzymes (Yamauchi et al., 2008).

Another example of use of coating to maintain color (anthocyanins) is with litchi fruit. Discoloration of litchi generally is caused by browning and disappearing of red color in the pericarp. Chitosan with high concentration of citric acid maintained litchi color (Ducamp-Collin et al., 2008; Jiang et al., 2005) by blocking the activity of polyphenol oxidase (PPO) (Jiang et al., 2005) and peroxidase (Zhang et al., 2005), which degrade the anthocyanins. This method is a proposed alternative to SO_2, which was a principal postharvest treatment but is now banned by many importing countries for most foods, including litchi (Holcroft and Mitcham, 1996).

Without coatings, apples and oranges have lower gloss, particularly after storage. Coatings impart fruit gloss and protect against its loss (Table 7.2) (Bai et al., 2002a; Hagenmaier and Baker, 1995). Many coatings lose gloss during storage, especially

Table 7.2 Gloss (Gloss Unit) of apples and citrus coated with different coatings. fruit were stored at 20°C after coating for up to 14 days

Coating	Apple					Citrus		
	Delicious[1]		Granny Smith[2]		Gala[3]	Valencia Orange[4]		Marsh Grapefruit[4]
	Day 1	Day 14	Day 1	Day 14	Day 14	Day 7		Day 7
Control	3.7	3.1	5.9		5.2	1.1		1.1
Candelilla (Can)	10.3	9	8.5					
Carnauba (Car)					7.2			
Shellac (Sh)	11.2	10.1	10.9		9.5	5.4		6.4–7.3
Polyethylene (Pe)			10.9			2.3–4.7		4.2
Polyvinyl acetate	8.9	7.3						
Starch	12.2	7.4						
Zein	7.6	7.5			8.9			
Car-Sh[a]			10.1					
Car-Can						2.2		
Car-polysaccharide	7.5	7.4						
Pe/Sh[b]						7.4		

Sources: Data from [1] Bai, J., Baldwin, E.A., and Hagenmaier, R.D., *HortScience*, 37, 559, 2002; [2] Bai, J., Hagenmaier, R.D., and Baldwin, E.A., *Postharvest Biol. Technol.*, 28, 381, 2003; [3] Bai, J., Alleyne, V., Hagenmaier, R.D., Mattheis, J.P., and Baldwin, E.A., *Postharvest Biol. Technol.*, 28, 259, 2003; [4] Hagenmaier, R.D., and Baker, R.A., *HortScience*, 30, 296, 1995.

[a] Composite coating.
[b] Layered coatings with polyethylene as first layer and shellac second.

when the fruit encounters water, such as condensation when removed from cold storage (Bai et al., 2003a; Baldwin, 1994; Hagenmaier and Baker, 1995; Hall, 1981b). Shellac coatings give high gloss, while carnauba coatings give less gloss than shellac (Hall, 1981b). Layered coatings preserve gloss of citrus (Hagenmaier and Baker, 1993a, 1995). Edible coatings also can act as carriers of colorants to cover surface blemish and improve fruit appearance (Baldwin, 1999; Cuppett, 1994). Colored waxes were used on the red-skinned varieties of potatoes and sweet potatoes (Hartman and Isenberg, 1956). Oranges are also commonly dyed with Citrus Red No. 2 (1-(2,5-dimethoxyphenylazo)-2-naphthol, synthetic, 21CFR74.302) before shellac coating is applied (Plotto and Baker, 2004).

However, improper coating application may cause cosmetic discoloration. "Whitening" is a typical problem caused by shellac coating. When commodities go through several "sweats" due to humidity and temperature changes, fruit surface shows white stains (Baldwin, 1994). A compromised method is to decrease shellac and increase carnauba content, thus the coating offers less gloss but has less risk to "whiten." Zein coating also has whitening problems; however, by regulating the formulation, whitening was alleviated (Bai et al., 2003a). Coatings can also detract from appearance by coming off, which can happen, for example, if the coating is sticky, causing the fruit to stick together or to the container.

7.3.4 Decreased nutritional quality

Fruits and vegetables lose some nutritional value during storage, even under optimal conditions. A major benefit from a higher intake of fruits and vegetables may be the increased consumption of vitamins (vitamin C, vitamin A, vitamin B_6, thiamin, and niacin), minerals, dietary fiber, and antioxidant compounds such as carotenoids, flavonoids, and other phenolics (Kevers et al., 2007). In most fruits and vegetables, levels of minerals, phenolics, and antioxidant capacity are generally stable during storage (Kevers et al., 2007). However, contents of ascorbic acid and dietary fiber consistently decrease during storage (Dong et al., 2008).

'Hami' melons coated with bilayer coatings of chitosan and polyethylene wax had significantly higher retention of ascorbic acid content (Cong et al., 2007). A sucrose fatty acid ester coating was effective in retaining higher contents of ascorbic acids in green peppers (Özden and Bayindirli, 2002), sweet cherries (Yaman and Bayoindirli, 2002), and apples (Sümnü and Bayindirli, 1995a). A methyl cellulose coating containing ascorbic acid or citric acid as antioxidant and acidulant lowered the ascorbic acid loss in apricots and green peppers (Ayranci and Tunc, 2004).

7.3.5 Flavor loss

Coatings influence flavor in two ways. One is by directly accumulating volatile compounds in the fruit as a result of reduced skin permeability; the other way is by altering metabolism of fruit tissues, such as by retarding ripening and deterioration to extend the flavor life of commodities, or by excessively reducing internal O_2 and elevating CO_2 to cause fermentation, abnormal accumulation of ethanol and ethyl acetate, and off-flavor (Baldwin, 1994).

Nisperos-Carriedo et al. (1990) found that the use in coatings of beeswax and a sucrose fatty acid ester, carboxymethylcellulose (CMC) sodium salt and mono-/diglycerides) on oranges increased the levels of the volatile components acetaldehyde, ethyl acetate, ethyl butyrate, methyl butyrate, ethanol, and methanol in oranges stored at 21°C for up to 12 days. An informal tasting suggested that these coatings improved the fruit quality without causing off-flavors. The coating altered the aroma profile and increased the ethanol content (Baldwin et al., 1995).

Concentrations of ethanol and ethyl esters increased remarkably for tangerines with shellac and resin-based coatings; however, the concentration increases were relatively less for fruit coated with candelilla or polyethylene wax-based coatings (Hagenmaier and Shaw, 2002). Hagenmaier (2002) reported that with tangerine coatings having oxygen permeability equal to or less than 1.1×10^{-16} mol m s^{-1} m^{-2} Pa^{-1}, the fruit had less fresh flavor. Such coatings often resulted in internal $O_2 < 4$ kPa, and internal $CO_2 > 14$ kPa at 21°C.

Bai et al. (2002b, 2003b) applied several coatings with a very wide range of gas permeabilities to apples. 'Granny Smith' coated with shellac had very low internal O_2 (<2 kPa), and 'Braeburn' had very high internal CO_2 (25 kPa). These excessive modifications of internal gas induced an unusually high accumulation of ethanol and ethyl esters in the fruit and caused off-flavor. Candelilla and carnauba-shellac coatings had higher permeabilities that maintained better internal O_2 and CO_2 and better quality for 'Fuji' and 'Braeburn'; however, these acceptable coatings may still present too much of a gas barrier for 'Granny Smith' apples. In general, the gas permeabilities of the coatings are useful as indicators of differences in coating barrier properties, but do not account for differences in pore blockage, which is expected to play an important role in the gas exchange for the whole fruit (Bai et al., 2002b, 2003b).

7.3.6 Physiological disorders

Postharvest physiological disorders are caused by nonpathological factors such as improper temperature, humidity, and gas combinations, which affect the functioning of the plant system. The symptoms of physiological disorders may appear as scald, pitting, core flush, browning, and other disease-like forms, depending on the fruit.

> Superficial scald: After extended storage, some apple and pear cultivars develop scald and other disorders (Mir et al., 1998). Many reports have shown that coatings may alleviate disorders such as superficial scald and senescent blotch on 'Granny Smith' apples, a susceptible cultivar (Bai et al., 2009a, 2009b; Ju and Curry, 2000a, 2000b). Spots and soft scald on 'Jonathan' apples were controlled by carnauba coatings, with increased coating concentrations providing a better control (Farooqi and Hall, 1973); waxed apples had 20% less scald than the noncoated control (Smock, 1935); a sucrose fatty acid ester coating and a chitosan-based coating reduced the incidence of scald in apples (Lau and Meheriuk, 1994). However, other reports showed little or opposite results. For example, Semperfresh® increased scald incidence in apples (Kerbel et al., 1989) but had little effect in other studies (Bauchot and

John, 1996; Bauchot et al., 1995). Generally, the above coatings influence disorders by the same mechanisms as CA storage, which inhibits ripening and senescence in fruit, and restrict O_2 entering into fruit, thus reducing oxidative metabolism. In addition to atmosphere modification, coatings may influence the disorders by other mechanisms. A coating application with 5% to 10% soybean or corn oil emulsions inhibited senescent scald and core breakdown in Chinese pears (Ju et al., 2000). Corn oil emulsion coatings also retarded ethylene and α-farnesene production rates, and controlled superficial scald in 'Granny Smith' apples and 'Anjou' pears (Ju and Curry, 2000b). There is evidence that these oil coatings inhibited ethylene production independently from atmosphere modification (Ju and Curry, 2000b; Ju et al., 2000). Absorption of α-farnesene by the coating may also play a role in the decrease of superficial scald (Ju and Curry, 2000b). Bauchot et al. (1995) applied sucrose fatty acid ester coatings containing antioxidants, ascorbyl palmitate, and propyl gallate on 'Red Chief' and 'Golden Delicious' apples, and controlled the superficial scald for 4 months. Because the gas combinations in coated fruit depend on cultivar, fruit quality, ripening stage, storage temperature, humidity, and coating characteristics, the influence of coatings on these disorders was not always the same.

Chilling injury: Many tropical fruits and vegetables, such as tomatoes, cucumbers, citrus, mangoes, and papayas, exhibit chilling injury when exposed to low temperatures, usually less than 13°C. The typical symptom is surface pitting, although some commodities show different symptoms. For example, chilling injury in citrus is characterized by an initial collapse of peel tissue in discrete spots that become darkened and depressed, sometimes appearing in a circular form (Dou, 2004). Many reports show that coating applications suppress chilling injury in grapefruit (Bajwa and Anjum, 2007; Davis and Harding, 1960; Dou, 2004), honeydew melons (Edwards and Blennerhassett, 1994), and breadfruit (Maharaj and Sankat, 1990). Dou (2004) reported that polyethylene wax, a low gas barrier coating, was less effective at preventing chilling injury in grapefruit; however, shellac, a high gas barrier coating, prevented chilling injury more effectively.

Pitting of citrus: Petracek et al. (1998, 1999) showed that postharvest pitting of waxed grapefruit stored occurred at 10°C or higher temperature. Shellac coating caused higher CO_2 and lower O_2 inside the fruit and caused more severe postharvest pitting in comparison to fruit coated with carnauba wax. Bajwa and Anjum (2007) also showed that polyethylene coating, which has high gas permeability, suppressed pitting incidence in mandarin.

7.3.7 Disease and physical injury

Fresh produce can become infected before or after harvest by diseases spread in the air, soil, and water. Some diseases are able to penetrate the unbroken skin of produce; others require an injury in order to cause infection. Damage from postharvest infection is probably the major cause of loss of fresh produce. Careless handling of fresh produce causes internal bruising, which results in abnormal physiological damage or splitting and breaks in the peel, thus rapidly increasing water loss and the rate of normal physiological breakdown. Breaks in the peel also provide sites for infection by disease organisms causing decay. Coatings can form a physical barrier against pathogenic infection, reducing incidence of postharvest diseases. Sanitation followed by coating application in the packing process may decrease microbial load on fruit surface and protect produce from microbial infection (Pao et al., 1999). Coatings also reduce decay by delay of ripening and water loss. Both of these processes lead to senescence, making the commodity more prone to pathogenic infection as a result of loss of cellular integrity and natural tissue defense mechanisms (Akagi and Stotz, 2007; Amarante and Banks, 2001). Adding antimicrobials, such as organic acids and their salts (Franssen et al., 2004; Vojdani and Torres, 1990), parabens (Chung et al., 2001), essential oils, or natural antimicrobials (Min and Krochta, 2007; Rojas-Grau et al., 2006), into edible films and coatings has been effective in delaying the growth of contaminating microorganisms during storage or distribution of fresh horticultural products.

Claypool (1939) observed a reduction of postharvest decay in waxed deciduous fruits. Farooqi and Hall (1973) reported that coated apples and pears had less decay. Bananas coated with polyethylene-based wax had delayed and reduced decay, particularly of the cut surfaces of fruit (Ben-Yehoshua, 1966). Baldwin et al. (1997) reported that cucumber coated with 1% to 2% hydroxylpropyl cellulose (HPC), with or without the addition of carnauba wax, had less decay than controls. However, guavas coated with 2% to 4% HPC and 5% carnauba did not show change in the incidence of decay in comparison with the control (McGuire and Hallman, 1995). Edward and Blennerhassett (1994) reported that Citruseal wax did not reduce breakdown caused by bacteria or fungi in honeydew melon. Coating oranges with shellac decreased the incidence of decay; however, methylcellulose-based coating did not have any beneficial or negative effect (Potjewijd et al., 1995).

Chitosan coatings have been shown to be effective against decay by inhibiting fungal, yeast, and bacterial activities (El Ghaouth et al., 1991b, 1992; No et al., 2007). The mechanism of the antimicrobial activity of chitosan has not yet been fully clarified; however, the most feasible hypothesis is a change in cell permeability due to interaction between the positively charged chitosan molecules and the negatively charged microbial cell membranes. This interaction leads to the leakage of proteinaceous and other intracellular constituents (Papineau et al., 1991; Young et al., 1982). Chitosan is also believed to elicit plant defense mechanism, such as chitinase and β-1,3-glucanase or a combination (El Ghaouth et al., 1991a, 1991b). Badawy and Rabea (2009) found that the antifungal effects of chitosan were concentration and molecular weight dependent. Smaller molecular weight and higher concentration in chitosan resulted in higher antifungal activity. However, a larger molecule, 5.7×10^4 g mol^{-1} had the highest biological activity, as measured by enhanced total protein and phenolic compounds, and decreased polyphenol oxidase activity (Badawy and Rabea, 2009).

Coatings can serve as vehicle for postharvest fungicides. Fungicides are often incorporated into citrus coatings for simplicity and economy. However, effectiveness of decay control is often reduced when fungicides are dissolved into the coating. A two- to threefold increase in fungicide concentration must be used to obtain the same decay control as that of precoating aqueous solution (Eckert and Brown, 1986). Fresh market tomatoes treated with a fungicidal wax containing 2.5% o-phenylphenol (Jung et al., 2000) in a commercial packinghouse had lower incidence of decay than fruit treated with plain wax, and almost no chemical residue of OPP could be detected (Hall, 1989). Incorporation of biocontrol organisms into coatings to restore surface populations of beneficial microorganisms can provide an opportunity for biological control of postharvest decay pathogens (McGuire and Baldwin, 1994; Potjewijd et al., 1995).

A new type of antimicrobials, nanoparticles, has recently been assessed for food and agricultural use (Weiss et al., 2006). Jin et al. (2008) evaluated the antimicrobial activity of zinc oxide (ZnO) nanoparticles against *Listeria monocytogenes*, *Salmonella Enteritidis*, and *Escherichia coli* O157:H7, and found that both ZnO powder and ZnO suspended in polyvinylpyrroloidone gel showed significant antimicrobial activities against the food-borne pathogens, with ZnO-PVP coating having less inhibitory effect than the direct addition of ZnO. No antimicrobial activities of the film made of ZnO bound in a polystyrene were observed (Jin et al., 2008). An et al. (2008) coated asparagus using silver nanoparticles-polyvinylpyrroloidone. The results showed that the coating reduced the growth of total aerobic psychrotrophic count, yeast and molds on asparagus stored at 2 and 10°C. ZnO is currently listed as generally recognized as safe by the U.S. Food and Drug Administration (21CFR182.8991). The U.S. Environmental Protection Agency standard for the secondary maximum contaminant level of silver ion in drinking water is less than 0.10 mg L^{-1} (U.S. Environmental Protection Agency, 2006).

7.3.8 Quarantine treatments

Wax coatings on fruits and vegetables can be an effective method of killing all stages of insects by suffocation (Amarante and Banks, 2001; Baldwin, 1994; Fields and White, 2002). As an alternative to methyl bromide fumigation, the effects of wax application on *Brevipalpus chilensis* quarantine on various fruits and vegetables were studied in Chile (Gonzalez, 1997). The study was based on the hypothesis that if a wax cover was spread over the body of the mite, it would affect the respiratory system of the organism, causing asphyxiation and eventually death. Many tests were carried out in all of the developing stages of the mite, and it was demonstrated that the theoretical supposition was true, at least on cherimoyas and lime (Gonzalez, 1997). The method has been approved by the U.S. Department of Agriculture (USDA) for cherimoyas and lime, and the code T102 (b) was assigned for products admissible from Chile with import permit (Gonzalez, 1997). Some coatings gave a gas barrier large enough to control fruit fly (*Anastrepha suspensa*) in fruits (Hallman et al., 1994, 1995). Combination of coating with hot air (Hallman et al., 1994) or insecticide (Hallman and Foos, 1996) achieved even better insect control in comparison with coating, hot air, or insecticide alone. No larvae emerged from coated grapefruit treated with 48°C hot air for 60 minutes, whereas 24% survived in noncoated fruit (Hallman et al., 1994). Use of coatings in combination with other quarantine treatments such as hot water dips, fumigation, and extended cold storage was investigated. Quarantine treatments may result in some damage to fruit peel and possibly flavor, thereby reducing fruit quality and value. There is a possibility that the use of coatings could replace or reduce the severity of these established procedures, and improve fruit quality (Amarante and Banks, 2001).

7.4 Properties of edible coatings on fresh commodities

7.4.1 Lipids and resins

Most or many commercial coatings are wax or lipid based. Their use is rather extensive on citrus, apples, mature green tomatoes, rutabagas, and cucumbers (Baldwin, 1994). Those coatings are made with natural or synthetic waxes (carnauba, beeswax, and polyethylene), oils (vegetable and mineral oils), rosin and resin (wood rosin, shellac, and coumarone indene resin), fatty acids, emulsifiers, plasticizers, antifoam agents, surfactants, and preservatives. Many coatings are emulsions, microemulsions, or solutions in water.

Oxidized polyethylene is defined as the basic resin produced by the mild air oxidation of polyethylene, a petroleum by-product. Several grades of polyethylene wax are available to obtain desired emulsion properties (Eastman Chemical Inc., 1984, 1990). Polyethylene-based water waxes have been approved for use on citrus because the coated peels are generally not consumed. Coatings made with polyethylene have low-to-medium water-vapor and high CO_2 and O_2 permeabilities. These characteristics are desired to maintain fruit quality (limit water loss and do not create anaerobic internal atmosphere), but polyethylene coatings provide limited gloss (shine) (Bai et al., 2003b; Baldwin, 1994). On the contrary, shellac and rosin coatings impart the most gloss (Table 7.2) and have low or very low permeability to CO_2 and O_2 and moderate permeability to water vapor (Table 7.3) (Bai et al., 2002a; 2003b; Baldwin, 1994; Hagenmaier, 2002; Hagenmaier and Goodner, 2002). Coatings made with carnauba and other waxes have low to moderate permeability to O_2 and CO_2, low permeability to water vapor (Figure 7.2), and provide moderate gloss (Bai et al., 2003b; Hagenmaier, 2002; Hagenmaier and Baker, 1997).

Permeability to O_2 is also affected by the relative humidity and polar components used to raise pH and solubilize the polymer (Hagenmaier and Shaw, 1992). Polymers that contain hydroxy, ester, and other polar groups tend to have lower O_2 permeability than those with hydrocarbon and other nonpolar groups (Ashley, 1985). Water vapor permeability of coatings is also sensitive to relative humidity, especially for waxes that contain polar ingredients (Hagenmaier and Shaw, 1992). The characteristics of some commercial coatings are shown in Table 7.4, along with recommendations for their use on fresh produce.

Coatings made from mixtures of shellac and carnauba wax or polyethylene wax are increasingly used. Depending on the ratio of ingredients, these share the benefits and limitation of wax and shellac coatings, namely gloss, gas and water vapor permeability, and water damage from condensation. Our experience showed that the shellac/carnauba ratio should be no greater than 1:4 to avoid "whitening" (unpublished data).

Table 7.3 Water vapor, O_2, and CO_2 permeability of edible coatings for fruits and vegetables

	Water vapor permeability			O_2 and CO_2 permeability				
	(10^{-12} g m^{-1} s^{-1} Pa^{-1})^z			(10^{-10} L m^{-1} d^{-1} Pa^{-1})^z				
Coating	WV	HR gradient (%/%)	Temperature (°C)	O_2	CO_2	RH (%)	Temperature (°C)	Reference
Lipid								
Shellac	1.46–7.71	0/100^z	30	0.96	0.34–1.31	60	20	Bai et al., 2003b; Hagenmaier, 2000; Hagenmaier and Baker, 1995
Beeswax	0.58	0/100^z	25	1.06		0	25	Hagenmaier and Baker, 1997
Candelilla	0.17	0/100^z	25	6.22	23.61	60	30	Bai et al., 2003b; Guibert, 2000; Hagenmaier and Baker, 1997
Carnauba	0.33–0.61	0/100^z	25	0.19	2.62–16.89	0	25	Guibert, 2000; Hagenmaier and Shaw, 1992; Martin-Polo et al., 1992
Polyethylene	0.65			8.30	26.1			Park, 1999
Polypropylene				0.05				Park, 1999
PVA	2.98	0/85	30	10.89		75	30	Hagenmaier and Grohmann, 1999
Polysaccharide								
Methylcellulose	76–92	0/75	25	2.17–12.96	69–743	0	30	Garcia et al., 2004; Kester and Fennema, 1986; Park, 1999; Pinotti, 2007

(*Continued*)

Table 7.3 Water vapor, O_2, and CO_2 permeability of edible coatings for fruits and vegetables (Continued)

Coating	Water vapor permeability (10^{-12} g m^{-1} s^{-1} Pa^{-1})			O_2 and CO_2 permeability (10^{-10} L m^{-1} d^{-1} Pa^{-1})				Reference
	WV	HR gradient (%/%)	Temperature (°C)	O_2	CO_2	RH (%)	Temperature (°C)	
Hydroxypropyl cellulose	110			3.57–11.69	144–718	0	30	Garcia et al., 2004; Hagenmaier and Shaw, 1991; Kester and Fennema, 1986; Park, 1999; Pinotti, 2007
Hydroxypropyl methylcellulose	105	0/85	27	0.12–1.16		50	25	Hagenmaier and Shaw, 1990; Krochta and De-Mulder-Johnston, 1997
Starch	2170	74/50	23	1591	29209	63.8	20	Garcia et al., 2000
Chitosan	490	100/50	25	0.0014		93	25	Garcia et al., 2000; Park et al., 2001
Pectin				6.6–29.5	472	96	25	Gontard et al., 1996; Liu et al., 2006
Xanthan gum, low-density	0.7–0.97	0/90	38	22.22	120	90	25	Farber et al., 2003
Xanthan gum, high-density	0.24	0/90	38	7.43	21.70	90	25	Farber et al., 2003
Sucrose polyester	0.42			2.1				Park, 1999
Protein								

Material							References
Gluten	43–616	0/50	23	0.2–28.6	2.1–811	91	25	Gontard et al., 1992, 1996; Guilbert et al., 1996; Hernandez-Munoz et al., 2004; Park, 1999
Soy	3540	100/50	25	0.775		50	25	Cho and Rhee, 2004; Cho et al., 2007; Gennadios and Weller, 1991
Whey	616–4170	100/55	25	0.012	2.13	50	25	Krochta and De-Mulder-Johnston, 1997; Mate et al., 1996; McHugh and Krochta, 1994; Mei and Zhao, 2003
Sodium caseinate	425	0/81	25	8.80	52.78	77	25	Dangaran et al., 2006; Guilbert et al., 1996; Khwaldia et al., 2004
Zein	89–132	0/85 [a]	21	1.79–3.81	2.7–13.1	60	20	Bai et al., 2003a; Gennadios and Weller, 1990; McHugh and Krochta, 1994; Park, 1999

Sources: Adapted from Farber, J.N., Harris, L.J., Parish, M.E., Beuchat, L.R., Suslow, T.V., Gorney, J.R., and Garrett, E.H. *Comprehensive Reviews in Food Science and Food Safety,* 2, 142, 2003; Vargas, M., Pastor, C., Chiralt, A., McClements, D.J., and Gonzalez-Martinez, C., *Critical Reviews in Food Science and Nutrition,* 48, 496, 2008; Park, H.J., *Trends in Food Science and Technology,* 10, 254, 1999.

[a] The values are based on polyethylene carrier file plus coating, because the coatings alone are too fragile to form a film.

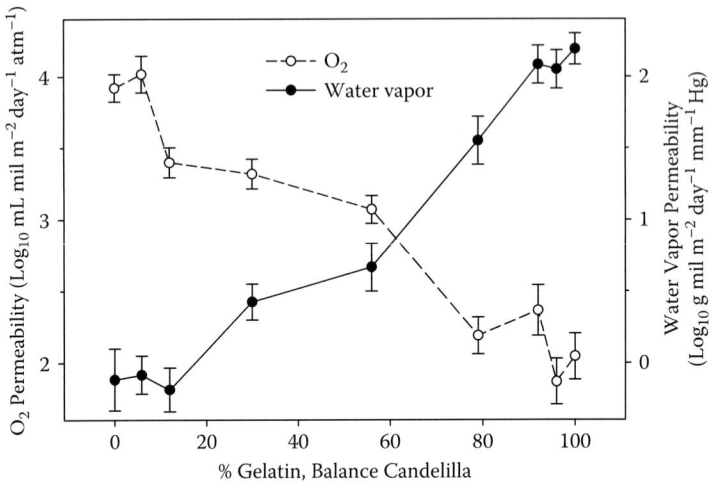

Figure 7.2 Permeability to oxygen and water vapor at 30°C, 60% RH of coatings made from mixtures of gelatin and candelilla. (Reprinted from Hagenmaier, R.D., and Baker, R.A., *J. Food Sci.*, 61, 562, 1996. With permission.)

7.4.2 Polysaccharides

Edible coatings based on polysaccharides have been extensively applied to delay loss of quality in fresh products such as apples, pears, tomatoes, cherries, fresh beans, strawberries, mangoes, and bananas (Ayranci and Tunc, 2004; Baldwin, 1999; Kester and Fennema, 1986; Kittur et al., 2001; Krochta and De Mulder-Johnston, 1997; Yaman and Bayoindirli, 2002; Zhuang and Huang, 2003). Polysaccharides show effective gas barrier properties, although they are highly hydrophilic, and they also show high water vapor permeability in comparison with lipid and rosin coatings (Figure 7.2).

The polysaccharides used in edible coatings are starch and derivatives, cellulose derivatives, chitosan, gums, pectin, and other compounds. Table 7.3 shows the barrier properties of the major polysaccharide coatings. Cellulose derivatives, including carboxymethylcellulose (CMC), methylcellulose (MC), hydroxypropylcellulose (HPC), and hydroxypropyl methylcellulose (HPMC) are widely used materials for edible coating development. CMC and sucrose fatty acid esters have been used in commercial coating formulations, such as TAL Pro-long (a coating similar to Semperfresh and no longer made) and Semperfresh (AgriCoat NatureSeal Ltd., Berkshire, UK). These coatings are hydrophilic; thus, they are poor barriers to water vapor (Bai et al., 2002a; Banks, 1984, 1985a; Drake et al., 1987; Kester and Fennema, 1986). O_2 is limited in entering the fruit more than CO_2 is from escaping. Therefore, they are potentially good coatings for some varieties of apples (such as 'Breaburn') and pears (such as 'd'Anjou') that are susceptible to high CO_2 (Bai et al., 2003b).

One problem with polysaccharides is that they are quite permeable to water vapor. A typical starch coating exhibited a water vapor permeability of 2170 g m^{-1} s^{-1} Pa^{-1} at 23°C with a relative humidity (RH) differential of 24% (Vargas et al., 2008). A chitosan coating was 490 to 3600 g m^{-1} s^{-1} Pa^{-1}. These values of water vapor permeability

Table 7.4 Some commercial edible coatings for fruits and vegetables

Coating	Formulation base	Intended function	Suggested application
JBT FoodTech, Lakeland, Florida			
Sta-Fresh MP	Carnauba	High shine, high blush control	Citrus, tropical fruits, pome fruits, pomegranate
Sta-Fresh 123	Shellac	—	Citrus
Sta-Fresh 151	Oil	Moderate shine	Vegetables
Sta-Fresh 216	Shellac/resin	Good shine, but lacks shell color control	Citrus, pome fruits, pomegranate, pineapple
Sta-Fresh 223HS	Shellac/resin	High shine/blush resistant	Citrus
Sta-Fresh 227	Shellac/resin	High shine/blush resistant	Citrus
Fresh-Cote 241HL	Shellac	High shine	Citrus, pome fruits, pomegranate
Sta-Fresh 410	Shellac/resin	Shine	Citrus
Sta-Fresh 590HS	Shellac/carnauba	High gloss	Citrus
Sta-Fresh 711	Oil	Drying speed, preservative	Lime, stone fruits, potato, melon, vegetables, pineapple
Sta-Fresh 819F	Carnauba	High shine, high blush control	Pome fruits, pomegranate
Sta-Fresh 890H	Shellac	High shine	Citrus
Sta-Fresh 2109	Carnauba/resin	Higher blushing potential	Citrus
Sta-Fresh 2210	Shellac/carnauba	High shine on mandarins	Citrus
Sta-Fresh 2505	Carnauba	Fast drying, high blushing potential	Citrus, pome fruits, pomegranate
Sta-Fresh 2862OR	Carnauba/resin	High shine, organic	Citrus
Sta-Fresh 2952	—	Protective coating	Pineapple
Sta-Fresh 2981	Vegetable oil	Protective coating	Pineapple
Sta-Fresh 4201	Polyethylene	High shine, low odor, shine duration	Citrus, tropical fruits

(*Continued*)

Table 7.4 Some commercial edible coatings for fruits and vegetables (Continued)

Coating	Formulation base	Intended function	Suggested application
Sta-Fresh 7051	Paraffin	Internal quality control	Pineapple
Sta-Fresh 7055	Polyethylene	Internal quality control	Pineapple
Sta-Fresh 7100	Polyethylene	Internal quality control	Citrus, pineapple
Pace International, Seattle, Washington			
CitraShine™ 585 Ft	—	Premium high solids citrus coating	Citrus
Lustre Dry	Wood rosin and shellac	High gloss, quick-drying citrus coating	Citrus
Lustre Dry w/ TBZ	—	Fast drying, high gloss citrus coating containing thiabendazole fungicide	Citrus
Natural Shine™ 320OR	—	Premium organic citrus coating	Citrus
Natural Shine 505	Carnauba	Organic fruit coating	Citrus
Natural Shine 960	Carnauba	High shine for domestic and export use	Citrus
Natural Shine 990	Carnauba	High shine for domestic and export use	Citrus
Natural Shine 2015	Vegetable wax and shellac	Coating for prolonged storage	Citrus
Natural Shine 4000	Vegetable wax	Control ripening	Pear
Natural Shine 6000	—	Prolong storage	Lemon
Natural Shine 9000	Carnauba	Carnauba citrus coating	Citrus
Natural Shine TFC 210	Carnauba	Carnauba tropical fruit coating	Tropical fruits
PacRite® 425	Vegetable, resin and shellac waxes	Premium high-gloss shellac coating	Citrus
PacRite Durafresh	Petroleum wax	Protective coating	Stone fruits, vegetables
PrimaFresh® 45	Carnauba-based	Carnauba-based protective coating	Stone fruits
PrimaFresh 50-V	Vegetable oil–based	Protective coating	Stone fruits, vegetables
PrimaFresh 200	Petroleum wax	Protective mineral oil emulsion coating	Stone fruits, vegetables
PrimaFresh FD	Carnauba	Premium high shine and fast drying	Citrus
PrimaFresh Gold	Carnauba	Resist "whitening"	Apple

PrimaFresh Ultra	Vegetable and shellac waxes	Ultra high shine	Pome fruits
Shield-Brite® AP-40	Shellac	Premier high-shine shellac coating	Apple
Shield-Brite AP-431FD	Shellac	Fast-drying high-shine shellac coating	Apple
Shield-Brite AP-747	Shellac	Maximum shine shellac coating	Apple
Shield-Brite Melon Coating	Carnauba	Carnauba coating	Melon
Decco Cerexagri Inc., Monrovia, California			
Apl-Lustr® 221	Shellac	Fast drying, serves as a carrier for fungicides	Apple
Apl-Lustr 225	Shellac	Morpholine free, fast drying, serves as a carrier for fungicides	Apple
Apl-Lustr 229	Neutral lac resin	Fast drying	Apple
Apl-Lustr 231	Carnauba	Reducing weight loss and serving as a compatible carrier for most fungicides	Apple, pear, nuts, root crops, and other
Apl-Lustr 275	Alkali-soluble resins	High shine, high solids, reduce water loss	Apple
Apple Lustr® Premium	Shellac	Fast drying, reduces weight loss, and serves as a carrier for fungicides	Apple
Citrus Lustr® 204	Polyethylene	Reduces weight loss and serves as a carrier for most fungicides	Citrus
Citrus Lustr 209	Resin	High shine, high solids, packout coating	Citrus
Citrus Lustr 267	Alkali-soluble resins, fatty acids, propylene glycol, silicone antifoam and propyl paraben as a wax preservative	The coating reduces weight loss and serves as a compatible carrier for most fungicides	Citrus
Citrus Lustr 287	Resin	High shine, high solids, packout coating	Citrus

(Continued)

Table 7.4 Some commercial edible coatings for fruits and vegetables (Continued)

Coating	Formulation base	Intended function	Suggested application
Citrus Lustr 402	Alkali-soluble natural lac resins, fatty acids, propylene glycol, silicone antifoam, and emulsifiers	High shine, reduces weight loss, and serves as a carrier for most fungicides	Citrus
Citrus Lustr 525	Resin	High shine	Citrus
Decco Lustr® 202	Natural and synthetic waxes, fatty acids, and silicon antifoam	Concentrated, water miscible, controlling shrinkage	Citrus, cantaloupe, cherry, mango, avocado, stone fruits
Decco Lustr 231	Carnauba	Prevents dehydration, imparts natural gloss, and serves as a carrier for most fungicides	Citrus, pome fruits, stone fruits, vegetables, pomegranate, apple, roots, mango, papaya
Decco Lustr 295	Food-grade mineral hydrocarbons	High gloss, enhances appearance, and reduces weight loss	Stone fruits
Decco Lustr 505 Org	Carnauba	Organic fruit coating	Citrus, pome fruit, stone fruit
Decco Lustr Carnauba Plus	Carnauba	Reduces weight, enhances appearance, and serves as a carrier for fungicides	Citrus, vegetables, apple, pear, mango, papaya, sweet potato
Decco Pearl Lustr®	Soluble natural lac resins, fatty acids soaps, propylene glycol, food-grade emulsifiers and silicone antifoam	High shine, reduces weight loss, and serves as a carrier for fungicides	Citrus
Peach, Nectarine & Plum Lustr® 251	Food-grade mineral hydrocarbons, paraffin and carnauba wax, fatty acid soaps and surfactants	Improves fruit appearance, reduces weight loss, and serves as a carrier for fungicides	Stone fruits, vegetables, potato, mango, cherry
Peach, Nectarine & Plum Lustr 255	Vegetable oil	Coating for domestic and export markets	Stone fruits, pineapple, cherry, avocado, cantaloupe
Peach, Nectarine and Plum Lustr 282	Mineral oil	To improve fruit appearance, reduce weight loss, and act as a carrier for fungicides	Stone fruits, vegetables, cherry, mango, potato

Pineapple Lustr® 444	—	Concentrated; reduce exterior and interior deterioration of fruit	Pineapple
Sweet Potato Lustr®	Carnauba	Reduce weight loss	Red, white, and sweet potato
Vegetable Lustr® 227F	Mineral oil, petrolatum, and paraffin	Enhances appearance and reduces weight loss due to evaporation	Vegetables
AgriCoat NatureSeal Limited, Berkshire, UK/Mantrose-Haeuser Co., Inc., Westport, Connecticut			
SemperFresh	Sucrose fatty acid ester, sodium carboxymethyl cellulose, and vegetable oils	Prolong storage and marketing life, cover surface blemishes, and minimize scuffing and rub marking	Most fruits and vegetables
NatureSeal	Vitamin–mineral premix	Mainly for quality maintenance of fresh-cut product	Sliced apple, pear, avocado, and banana, avocado, cantaloupe, mango, and papaya
Fresh Seal	Cellulose derivatives and emulsifiers	Coatings for melons to maintain firmness and skin color and limit sunken areas	Melon (honeydew and cantaloupe)
Fruit Coating Resins	Natural resins	Enhances gloss and appearance, increases shelf life, delays ripening	Apple, citrus

Source: Modified from Olivas, G.I., Davila-Avina, J.E., Salas-Salazar, N.A., and Molina, F.I., *Stewart Postharvest Review*, 4, 1, 2008.

are much higher than those for wax and oil coatings, such as shellac (1.5 to 7.7 g m^{-1} s^{-1} Pa^{-1}), carnauba (0.3 to 0.6 g m^{-1} s^{-1} Pa^{-1}), and candelilla (0.17 g m^{-1} s^{-1} Pa^{-1}) (Table 7.3) (Vargas et al., 2008).

Chitosan is another polysaccharide widely used in the postharvest coating of intact and fresh-cut fruits and vegetables (El Ghaouth et al., 1991b, 1992; Kittur et al., 2001; No et al., 2007). Nutri-Save®, a chitosan-based commercial coating, has been used in different fruits (Lau and Meheriuk, 1994; Meheriuk and Lau, 1988; Worrell et al., 2002). As a result of chitosan coating, a reduction in the respiration rate and ethylene production, control of decay, and retention of firmness have been reported for apples (Hwang et al., 1998), bananas (Kittur et al., 2001), citrus (Chien et al., 2007a), litchi (Jiang et al., 2005; Zhang et al., 2005); mangoes (Chien et al., 2007b; Kittur et al., 2001), peaches (Li and Yu, 2001), strawberries (El Ghaouth et al., 1991a; Han et al., 2004), carrots (Durango et al., 2006), lettuce (Devlieghere et al., 2004), and tomatoes (El Ghaouth et al., 1992). Chitosan coating is likely to modify the internal atmosphere without causing anaerobic respiration, because chitosan films are more selectively permeable to O_2 than to CO_2 (No et al., 2007). There is ample evidence that chitosan coatings have the potential to prolong storage life and control decay of fruits (see Section 7.3.7).

7.4.3 Proteins

Proteins for edible coatings may be derived from maize, wheat, soybeans, and collagen (gelatin). Protein coatings, similar to carbohydrates, due to their hydrophilic nature, are not effective in reducing water loss (Table 7.3) (Gennadios et al., 1994; Gennadios and Weller, 1990; Kester and Fennema, 1986; Park and Chinnan, 1995). Certain protein materials such as soy protein and casein, which contain higher levels of hydrophobic amino acids, present more effective moisture barriers, especially in combination with lipids. Protein coatings are more effective for ripening control via creation of a gentle MA. Only the corn protein, zein, can result in a high-gloss appearance (Gennadios and Weller, 1990) that equals that of resin-based coatings (Bai et al., 2003a).

7.4.4 Composites and bilayer coatings

The advantage of the good water barrier properties of lipid coatings and the good gas permeability properties and nongreasy texture of polysaccharide or protein coatings can also be combined to form edible composite or bilayer coatings. One such film, consisting of a layer of stearic and palmitic acids and a layer of hydroxypropyl cellulose, substantially reduced the transfer of water (Kamper and Fennema, 1985). A film composed of 53% hydroxypropyl methylcellulose and 45% stearic acid (not a bilayer) had a water vapor permeability of 0.17 g mil m^{-2} day^{-1} mm Hg^{-1} at 85% RH compared to 48 g water mil m^{-2} day^{-1} mm Hg^{-1} for hydroxypropyl methylcellulose alone (Hagenmaier and Shaw, 1990). Rojas-Argudo et al. (2005) studied the effect of composite coatings with different hydrophobic and hydrophilic ratios and solids contents on storability of cherries. They found that increasing coating hydrophobicity (higher shellac or beeswax to locust bean gum content ratio) decreased weight and firmness loss of cherries. Conforti and Totty (2007) reported that 'Golden Delicious' apples coated with lipid-polysaccharide coatings maintained a better quality in firmness, crispness, acidity, and juiciness than uncoated fruit. In gelatin-starch coatings, increases in the starch content and pH resulted in higher CO_2 permeability but lower film puncture strength (Aguilar-Méndez et al., 2008). Jagannath et al. (2006) developed an edible coating by blending casein and turmeric starch as the main component (67%). Carrots coated with this coating had less microbial contamination, probably due to the turmeric, and maintained 7 days longer shelf life at ambient temperature.

Hagenmaier and Baker (1996a) reported that the addition of protein or polysaccharide components into candelilla wax formulations improved gloss but decreased O_2 permeability and increased water vapor permeability of the films. Such characteristics are not favorable for fruit coatings, because commodities coated with such combinations may be rendered anaerobic and still have high water loss. Navarro-Tarazaga and coworkers optimized the types of fatty acids (stearic, palmitic, or oleic) (Navarro-Tarazaga et al., 2008a), plasticizers (Navarro-Tarazaga et al., 2008b), and solid content ratios in HPMC:beeswax composite coatings. These studies illustrate how changing one type of ingredient can affect properties of the entire coating.

7.5 Responses of commodities to edible coatings

7.5.1 Apples and pears

Freshly harvested apples have their own waxy cuticle, but when apples are washed to remove dust and chemical residues, the process removes about half of the original apple wax, which creates a need for coating. Imparting shine and covering blemishes is the foremost purpose for commercial coatings of apples, although there are other benefits from coatings: protection from water loss and shrinkage, and modification of an internal atmosphere that results in slowing fruit metabolism and senescence (Bai et al., 2002a, 2002b; Baldwin, 1994). Major commercial apple coatings are carnauba, shellac, or the mixture of the two materials. Shellac-based coatings, which provide the highest gloss, are used in the United States for the domestic market. Carnauba-based coatings, on the other hand, give adequate gloss with a low risk of "whitening" and are used for the international market. Coatings can be applied before or after storage. The gloss decreases during storage, and anaerobic risk may arise if fruit are stored in controlled atmosphere when the coating is applied before storage.

Bai et al. (2002a, 2003b) applied several experimental coatings, with a wide range of gas permeabilities, to four apple varieties. The shellac coating resulted in maximum fruit gloss, lowest internal O_2, highest CO_2, and least loss of flesh firmness for all of the varieties. 'Granny Smith' with shellac coating had very low internal O_2 (<2 kPa), and the freshly harvested 'Braeburn' had very high internal CO_2 (25 kPa). These excessive modifications of internal gas induced an abrupt rise of the respiratory quotient and a prodigious accumulation of ethanol in both 'Braeburn' and 'Granny Smith,' and flesh browning at the blossom end of 'Braeburn' (Bai et al., 2003b). In addition, the shellac coating gave an unusual accumulation of ethanol in freshly harvested and 5-month stored 'Fuji.' Candelilla and carnauba-shellac coatings maintained more optimal internal O_2 and CO_2 and better quality for 'Fuji,' 'Braeburn,' and 'Granny Smith'; however, even these coatings may present too much of a gas barrier for 'Granny Smith.' In general, the gas permeabilities of the coatings were useful as an indicator of differences in coating barrier properties but did not account for gas exchange through pores, which can play an important role (Bai et al., 2002a, 2003b). An example is that candelilla wax coating, which had a lower permeability than the carnauba-shellac, resulted in higher values in internal O_2 and lower CO_2 (Bai et al., 2003b). We can speculate that the carnauba-shellac coating had more tendency to block pores in the fruit skin, but we have insufficient evidence to conclude that was the case. Figure 7.1 shows density of pores for 'Delicious' and 'Granny Smith' apples, and the effects of coating.

Sucrose ester-carboxymethyl cellulose formulations retarded color development and retained acids and firmness compared to controls when tested on apples (Banks, 1985a; Drake et al., 1987; Smith and Stow, 1984) and pears (Meheriuk and Lau, 1988). This was due to the effect of decreased ethylene production, increased internal fruit CO_2, and decreased O_2. Different storage temperatures seem to affect coating performance in terms of increasing internal CO_2 and especially decreasing internal O_2 concentrations. The lower storage temperatures (3 to 10°C) resulted in minor increases in internal CO_2 and negligible decreases in internal O_2 concentrations in the same variety of apples, but not at the higher storage temperature (20°C) (Bai et al., 2002a; Banks, 1985a). These differences in internal gas changes due to storage temperature are probably related to the effect of temperature on fruit respiration.

Apples coated by a specially formulated carnauba-polysaccharide (CPS) coating caused low internal O_2 and CO_2 concentration that were quite different from apples with other coatings (Figure 7.3) (Bai et al., 2002a). Unlike other coatings, for which the internal $O_2 + CO_2 \approx 20\%$, this sum for CPS was only 12% to 14%. These observations suggest that the CPS coating tended to block pores in the fruit more than other coating treatments.

According to the Kupferman (1998) survey in Washington, Oregon, and California, which produce more than 90% of U.S. pears, edible coatings were applied on 71% 'd'Anjou,' 25% of 'Bartlett,' and 26% of 'Bosc' fruits. Carnauba-based waxes are the major coatings for pears, but sucrose fatty acid ester–based Semperfresh and shellac were also used by some packers in some cultivars. Carnauba waxes used in pears are often especially formulated for that commodity, but diluted apple waxes are also used. Pear wax has less total solids than apple wax because pears are more susceptible to a high gas barrier that may cause anaerobic metabolism and disturb fruit ripening. Reducing shrinkage and improving appearance are the major purposes for pear coatings. Pears, especially after long-term storage, are susceptible to friction discoloration (Amarante et al., 2001b; Bai et al., 2006). High concentration carnauba coatings reduced this discoloration (Amarante et al., 2001b).

Amarante and Banks (2001) showed that the difference in skin structure of pears significantly affected coating performance. For varieties without lignified cells in the skin ('Bartlett,' 'Comice,' and 'Packham's Triumph'), the diffusion of water vapor, CO_2, and O_2 decreased remarkably by a small increase in total solids in carnauba wax because the coating covered the entire surface of the peel, including lenticels. However, for 'Bosc'

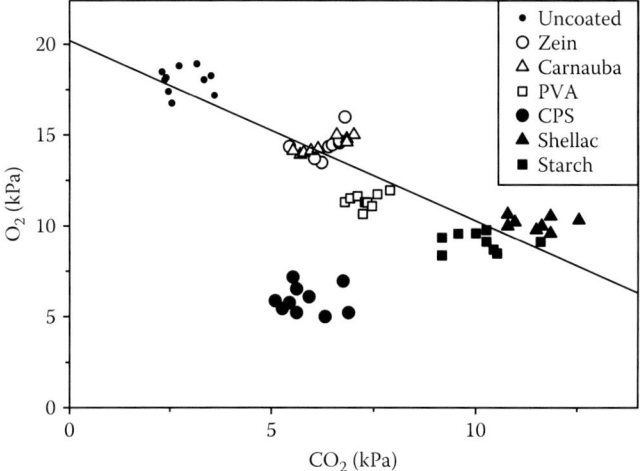

Figure 7.3 Relationship between internal O_2 and CO_2 of 'Delicious' apples coated with zein-, carnauba-, polyvinyl acetate (see Ngarmsak, M., Delaquis, P., Toivonen, P., Ngarmsak, T., Ooraikul, B., and Mazza, G., J. Food Protect., 69, 1724–1727, 2006)-, carnauba-polysaccharide (CPS)-, shellac-, and starch-based coatings or uncoated control. Fruit were stored for 14 days at 2°C followed by 14 days at 21°C. The linear regression line is for formulations other than the CPS, with a slope of –0.99 and an intercept of 20.2 ($r^2 = 0.8264$). (Reprinted from Bai, J., Baldwin, E.A., and Hagenmaier, R.H., *HortScience,* 37, 559, 2002. With permission.)

pears, which are covered by lignified cells in the peel, the diffusion of CO_2 and O_2 decreased, but water vapor diffusion only slightly decreased due to increasing the total solids content. This is because O_2 and CO_2 exchange mostly occurs through the lenticels and for 'Bosc,' and the coating effectively covered them. On the other hand, increasing the coating concentration was not effective in covering the lignified cells in the epidermis, which is the major route for water vapor exchange.

Elson et al. (1985) reported that Nutri-Save, a chitosan-based coating, reduced the incidence of scald and core flush in apples. Plant oil emulsions inhibited ethylene production independent of internal CO_2 and O_2 concentrations and controlled superficial scald of 'Anjou' pears and 'Granny Smith' apples (Ju and Curry, 2000a, 2000b).

7.5.2 Citrus fruit

Water loss and off-flavor development are important factors limiting citrus postharvest quality. Edible coatings can reduce water loss and slow the senescence by providing a barrier to moisture and gas transfer. Edible coatings also reduce decay incidence (Pao et al., 1999; Waks et al., 1985) and suppress degreening (Yamauchi et al., 2008), chilling injury (Dou, 2004), and pitting (Petracek et al., 1998, 1999). However, a high gas barrier caused by coatings can induce off-flavor in the fruit (Baldwin, 1994; Baldwin et al., 1995; Hagenmaier, 1998, 2001b, 2002) and pitting (Petracek et al., 1998, 1999). The flavor quality of tangerines and tangerine hybrids are generally more sensitive to high gas resistance coatings that cause off-flavor (Hagenmaier and Shaw, 2002).

Citrus peel consists of albedo and flavedo. Flavedo is the outer, pigmented portion of the peel which restricts gas exchange between inside and outside of the fruit. Albedo is the inner, spongy, white portion of the peel which is composed primarily of parenchyma cells. Gas can move through this layer relatively freely. The outermost layer of the peel is covered by a cuticle that is the barrier to water vapor and gas exchange (Lendzian, 1982). Stomata are holes on the continuous cuticle and epidermal cells. Turrell and Klotz (1940) estimated that there are 200,000 stomata found in the peel of 'Navel' oranges. Although the cuticle of the guard cells often plug the stomata in mature citrus fruit (Albrigo, 1972a, 1972b), many stomatal pores remain open on nonwaxed fruit. Some edible coatings, especially resin-based ones, covering the entire fruit especially block stomatal pores and thus restrict gas exchange between inside the fruit and surrounding environment. In addition, like with apples, commercial washing of citrus greatly removes the natural wax on the fruit surface, making it susceptible to shriveling and decay and making it necessary to apply wax coatings.

Current commercial citrus coatings are shellac, wood rosin, carnauba, polyethylene-based, or a mixture of the above waxes (Table 7.4). These coatings have different properties to satisfy various commodities and marketing needs. Shellac and wood rosin-based coatings offer the most shine but often cause extremely low O_2 and excessive buildup of internal CO_2, thus causing fermentation, off-flavor, and other disorders (Hagenmaier, 2001b, 2002; Hagenmaier and Baker, 1994a; Hagenmaier and Shaw, 2002). Carnauba-based coatings offer excellent water loss protection and a mild O_2 and CO_2 resistance. Recently, coatings formulated from shellac and carnauba mixtures are widely used commercially to combine properties of both materials. On the other hand, polyethylene waxes are used for low internal O_2/high internal CO_2 sensitive cultivars (Hagenmaier, 2002; Hagenmaier and Baker, 1994a; Hagenmaier and Shaw, 2002). Storage temperature greatly influences internal O_2 and CO_2 levels. For example, the internal CO_2 of 'White' grapefruit coated with a shellac-based wax was about two times higher at 21°C than at 1°C (Petracek et al., 1999). Similarly, the concentrations of CO_2 in citrus fruit coated with various coatings were two to three times higher at 18°C than at 5°C (Cuquerella et al., 1981). Steady-state internal O_2 and CO_2 levels may be attained within 5 to 6 hours after coating (Petracek et al., 1999), and the internal gas levels often change little once steady state is attained (Hasegawa and Iba, 1980). The degree of decrease of internal O_2 concentration is generally higher than the increase of internal CO_2, partially because coating films usually are more permeable to CO_2 than to O_2 (Mir and Beaudry, 2004; Petracek et al., 1999).

Citrus fruit is sensitive to anaerobic conditions. During fruit maturation, aerobic respiration decreases, and the supplementary anaerobic pathway increases even without application of coatings or storage in a low O_2 atmosphere (Bruemmer, 1989). Ethanol and acetaldehyde levels (indicating anaerobic respiration) have also been found to increase in citrus over the harvest season (Davis, 1970). Application of waxes can exacerbate this anaerobic condition (Nisperos-Carriedo et al., 1990), especially in tangerines, resulting in off-flavors due to increased internal ethanol, acetaldehyde, and CO_2, as well as decreased internal O_2 (Cohen et al., 1990; Vines et al., 1968). A similar situation occurred when citrus was subjected to CA storage with low O_2 and high CO_2, which caused increases in ethanol and acetaldehyde in fruit tissue, resulting in reduced flavor scores (Davis et al., 1973; Ke and Kader, 1990). Coating grapefruit with a polyethylene wax emulsion reduced the incidence of rind pitting when the fruit was stored at 0 to 10°C (Davis and Harding, 1960). However, Dou (2004) found that the high gas barrier shellac coating protected chilling injury of grapefruit better than polyethylene.

Semperfresh extended the storage life of citrus, but storage decay was a limiting factor. The coating is, however, compatible with most fungicides used on citrus fruit (Curtis, 1988). An indirect and little understood effect of these coatings is that coated fruit remain more turgid and juicy than uncoated fruit despite some loss in weight. Coated citrus fruit also had better flavor, which appeared to be related to acidity measurements (Curtis, 1988). In another study, TAL Pro-long coated 'Valencia' oranges resulted in higher flavor volatile levels in the juice and relatively low ethanol levels compared to oil- or wax-coated fruit (Nisperos-Carriedo et al., 1990). Because high shine is important for citrus, Semperfresh was applied in combination with shellac to citrus fruits. This treatment resulted in fruit with higher turgidity, less decay, and good flavor, although ethanol levels were increased. Whereas wax coatings have been reported to adversely affect flavor in citrus after long-term storage, fruit coated with Semperfresh exhibited enhanced flavor (Curtis, 1988).

'Persian' limes were treated with 1.5, 2, and 2.5% TAL Pro-long and stored at various temperature-RH combinations (8 to 25°C and 40% to 95% RH). The coating extended the shelf life of limes by controlling weight loss and degreening when stored at high RH and low temperatures (Motlagh and Quantick, 1988).

Yamauchi et al. (2003, 2008) found that the effect of degreening suppression of a sour citrus by sucrose fatty acid esters depended on the fatty acid portion in the coating. Sucrose laurate ester was more effective than any other sucrose fatty acid esters, such as myristate, palmitate, stearate, or laurate, because laurate ester inhibited activities of chlorophyllase and chlorophyll-degrading peroxidase.

Chitosan has also been used as a citrus coating. Ratanachinakorn et al. (2005) showed that application of chitosan decreased water loss of pummelos and extended their storage life.

7.5.3 Stone fruits

Waxing of peaches started on a commercial scale in the early 1960s. The principal benefit from waxing was to reduce water loss and to control decay through the use of fungicides carried in the wax. "Defuzzing" of peaches was necessary before waxing. Washed and defuzzed peaches lost weight (water) faster than unwashed controls, while waxed, defuzzed fruit lost slightly less weight than controls and had a more attractive appearance (Kraght, 1966). A beeswax-coconut oil emulsion extended shelf life and quality of peaches stored at room temperature (Erbil and Muftugil, 1986). In another study, Maftoonazad et al (2008) reported that peaches coated with sodium alginate and methyl cellulose had lower respiration rates and water loss, thus having 21 and 24 days shelf life, respectively, in comparison with 15 days in the noncoated control. When nectarines lose 4% to 5% of their fresh weight, shriveling becomes apparent. Waxing of fruit resulted in less weight loss and, therefore, reduced this defect (Mitchell et al., 1963). Sümnü and Bayindirli (1995b) found that

1% or 1.5% Semperfresh coatings of apricots decreased respiration rates, weight loss, color change, and losses of titratable acidity and soluble solids.

Sweet cherry fruit deteriorate rapidly after harvest and, in some cases, do not reach consumers at optimal quality after transport and marketing. The main causes of sweet cherry deterioration are weight loss, color changes, softening, surface pitting, stem browning, and loss of acidity and decay (Bernalte et al., 2003). Cherries are generally coated with vegetable oil emulsion commercially for decreasing water loss and stem browning. Cherries coated with Semperfresh showed reduced moisture loss and softening (Drake et al., 1988; Yaman and Bayoindirli, 2002). Unfortunately, none of the coated fruit showed reduced stem discoloration, which greatly influences consumer perception of cherry quality (Drake et al., 1988). Storability of 'Burlat' cherries was improved by edible composite coatings based on locust bean gum, shellac, and beeswax (Rojas-Argudo et al., 2005). The most hydrophobic solution produced better effects on the decrease of losses of weight and firmness; however, no benefits were achieved in color and titratable acidity retention (Rojas-Argudo et al., 2005). Sodium caseinate and milk protein concentrate–based coatings reduced water loss and had a beneficial effect on the sensory quality, especially the formulations containing glycerol and stearic-palmitic acid blend (Certel et al., 2004). Derivatives of fatty acids and polysaccharides decreased respiration rate and weight loss of sweet cherries (Alonso and Alique, 2004). Martinez-Romero et al. (2006) developed an *Aloe vera* gel coating that delayed water loss, color change, fruit softening, stem browning, and microbial growth.

Chitosan and hypobaric treatments, alone or combined, significantly reduced brown rot, grey mold, and total rots, the latter also including blue mold, *Alternaria, Rhizopus,* and green rots. A combined treatment with 1.0% chitosan and hypobaric 0.50 atm was the best in controlling decay (Romanazzi et al., 2003).

7.5.4 Grapes

An edible coating based on *Aloe vera* gel reduced weight loss, retarded color change, softening, ripening, rachis browning, and berry decay in 'Crimson Seedless' table grapes (Valverde et al., 2005). This coating was able to reduce the initial microbial counts for both mesophilic aerobic and yeast and molds, which significantly increased in uncoated berries during storage.

The effect of chitosan coating alone or in combination with application of ethanol on postharvest decay incidence caused by *Botrytis cinerea* in table grape was investigated by Romanazzi et al. (2007). In both 'Autumn Seedless' and 'Thompson Seedless' cultivars, the combination of 0.5% chitosan with 10% or 20% ethanol improved decay control with respect to single treatments, while combinations of 0.1% chitosan with 10% or 20% ethanol did not (Romanazzi et al., 2007). Chitosan and grapefruit seed extract (GSE) treatments, alone or combined, significantly reduced postharvest fungal rot of 'Red Globe' table grapes compared with control challenged with *Botrytis cinerea* (Xu et al., 2007). GSE also reduced weight loss, retarded color change, ripening, rachis browning, and dehydration, and combination of GSE and chitosan showed a synergistic effect in reducing postharvest fungal rot and maintaining fruit quality (Xu et al., 2007). Chitosan applied as a preharvest spray or postharvest coating reduced decay in table grapes and affected the content of total phenolic compounds and the activities of antioxidative enzymes (Meng et al., 2008). Chitosan coating also reduced water loss regardless of whether or not a preharvest spray was applied (Meng et al., 2008).

Coatings based on gelatin and native starches from sorghum and rice were applied to 'Crimson' grapes, and they extended the storage life without affecting flavor (Fakhouri et al., 2007). Sorghum starch coating also imparted shininess (Fakhouri et al., 2007). Postharvest application of wax containing fungicides reduced storage loss in 'Concord' grapes (Blanpied and Hickey, 1963).

7.5.5 Strawberries and other berries

Loss of quality in strawberries is mostly due to the relative high metabolic activity and sensitivity to fungal decay, mainly gray mold (*Botrytis cinerea*) (Hernández-Muñoz et al., 2006). They have a very short postharvest life, and losses can reach 40% during storage (Satin, 1996). Water loss causes shriveling and faster depletion of appearance and flavor (Nunes et al., 1998). With the use of cold storage and atmosphere modification, including with edible coatings, quality deterioration and spoilage could be minimized (Ueda and Bai, 1993). Modified atmosphere storage can, however, adversely affect strawberry color (Holcroft and Kader, 1999) and cause off-flavor development (Ke et al., 1991; Ueda and Bai, 1993).

In recent years, much attention has been paid to the potential of natural polymers such as polysaccharides and proteins in strawberry coatings (Del-Valle et al., 2005; Hernández-Muñoz et al., 2006; Xu et al., 2007). Several authors have reported beneficial effects of chitosan and its derivatives on strawberries (El Ghaouth et al., 1991a; Han et al., 2004). Although a postharvest dip with a calcium salt solution increased fruit firmness and delayed postharvest decay (Bitencourt de Souza et al., 1999; Garcia et al., 1996), adding calcium gluconate to a chitosan coating did not cause any additional benefit on strawberry quality attributes in comparison with chitosan coating alone (Hernández-Muñoz et al., 2006). However, Ribeiro et al. (2007) found that the addition of calcium chloride to starch, carrageenan, and chitosan coatings decreased microbial growth; the minimum rate of microbial growth was obtained with chitosan.

A prickly pear cactus–mucilage based edible coating led to increased firmness retention and shelf life of strawberries (Del-Valle et al., 2005). Starch-based coatings reduced weight loss, maintained firmness, and reduced decay in strawberries. The higher the amylase content ratio, the better the conservation effect (Garcia et al., 2000, 2001).

Tanada-Palmu and Grosso (2005) made coatings with wheat gluten and lipids (beeswax and fatty acids). All coatings containing gluten had beneficial effects on firmness retention, weight loss reduction, and visual quality and flavor. However, addition of lipids to the gluten in bilayer coatings resulted in unacceptable appearance and taste.

7.5.6 Kiwifruit

Surface of a kiwifruit is covered with short fuzzy hair, which makes application of a coating difficult. Nevertheless, Xu et al. (2001) coated kiwifruit with a composite coating containing 0.775% soybean protein isolate, 0.165% stearic acid, and 0.08%

pullulan. The coating retarded softening and extended shelf life for three times longer than the control.

Hardy kiwifruit (*Actinidia arguta*) is closely related to the kiwifruit (*A. deliciosa*) botanically. The fruit are generally green, without fuzz, and about the size of grapes. Coating with calcium caseinate, chitosan, PrimaFresh® 50-V (from Pace International, Seattle, Washington), and Semperfresh (from AgriCoat NatureSeal Ltd., Westport, Connecticut) provided an attractive sheen to the fruit surface and reduced water loss (Fisk et al., 2008).

7.5.7 Melons

Melons coated with Citruseal, a polyethylene-based wax, had less weight loss but more anaerobic tissue breakdown after 6 weeks at 8°C and 5 days at 15 to 25°C (Edwards and Blennerhassett, 1990). A carnauba-paraffin wax emulsion and a wax-sodium bisulfite mixture were used in an experiment to control decay of cantaloupes. The latter treatment was effective; however, it was unattractive due to a breakdown of the emulsion (Barger et al., 1948).

The bilayer coating of a chitosan and polyethylene wax microemulsion reduced water loss, the decrease of ascorbic acid, and increase of pH in 'Hami' melon (Cong et al., 2007). A combination of natamycin with the bilayer coating prevented the growth of *Alternaria alternata* and *Fusarium semitectum*, the major pathogens causing postharvest decay of 'Hami' melon (Cong et al., 2007). However, Batista et al. (2007) coated AF-682 melon with PVC [poly(vinyl chloride)] and a composite coating containing carnauba wax and cassava starch, and found that neither coating extended storage life, and both coatings caused fermentation and rotting (Batista et al., 2007). The quality of waxed melon is probably highly dependent on melon type, cultivar, physiological stage, and storage conditions.

Encouraging results were reported when polyethylene-based waxes, containing no or very low shellac, were applied to 'Galia'-type melon (Fallik et al., 2005). However, a polyethylene-shellac coating caused increased off-flavor in the melon fruit due to high-internal levels of CO_2, ethanol, acetaldehyde, and ethyl acetate. Untreated fruit, or fruit that were coated with beeswax, had the best taste and enhanced ethyl butanoate—(responsible for fruity-pleasant notes), butyl acetate, and 2-methylpropyl acetate. Nevertheless, the fruit lost firmness quickly and suffered from high-decay incidence (Fallik et al., 2005).

7.5.8 Bananas

Wax-type coatings have been reported to prolong the shelf life of tropical fruits. A finely divided crystalline-paraffin wax emulsion (Lawson, 1960), paraffin wax (Blake, 1966), sisal-paraffin, sugarcane, and polyethylene wax emulsions (Dalal et al., 1971), all increased shelf life, decreased bruising, and improved the appearance of bananas, without affecting flavor. The polyethylene emulsion was less effective than the others, but all extended the storage life of bananas by delaying ripening. These results, however, were affected by harvest maturity in that bananas harvested prior to 100 days from emergence of the inflorescence showed no delayed ripening due to coating treatments (Dalal et al., 1971). In another study, coating green bananas with polyethylene wax emulsion (15% solids) extended their storage life by 1 to 2 weeks, added glossy shine, decreased O_2 and increased CO_2 in the internal atmosphere of the fruit, and decreased shrinkage, respiration rate, and emergence of senescent spotting (Ben-Yehoshua, 1966). It was suggested that the decrease in spotting was due to the decreased internal O_2, resulting in decreased activity for the enzyme PPO, which oxidizes peel compounds leading to production of brown pigments. A candelilla coating extended banana shelf life and slowed ripening as evidenced by a retarded decrease in starch degradation (usually 20% to 25% in green bananas decreasing to 2% in ripe fruit) and slowed color change (Siade and Pedraza, 1977).

Delayed ripening, along with increased internal CO_2 and decreased O_2 levels, was observed in coated bananas (Banks, 1983, 1984, 1985b). Sucrose esters of fatty acids were labeled with ^{14}C and included in fruit coatings which retarded ripening (Bhardwaj et al., 1984). The esters remained intact on surfaces of bananas, apples, and pears during storage at 17°C for 30 days. Only a small amount of ^{14}C was detected in pulpy tissue of the fruits. The mode of action of sucrose esters in retarding ripening was not dependent, therefore, on their migration into the pulpy tissue (Bhardwaj et al., 1984).

TAL Pro-long was also applied to banana fruit (Banks, 1985b). The fruit exhibited decreased O_2 levels, and the coating prevented the climacteric rise in ethylene production, if applied before initiation of ripening. If ripening had already been initiated, the coating stopped climacteric ethylene in mid rise. Coated fruit also showed delayed chlorophyll loss (loss of green color) and climacteric rise in respiration, but these effects did not occur as rapidly as those described for ethylene, and declined relative to ripening initiation. The effect of the coating on color and respiration, therefore, appeared to be independent of its effect on ethylene production.

The fruit surface was analyzed using light microscopy, scanning electron microscopy, chlorine diffusion, and dispersive X-ray analysis (with coating labeled with aurothioglucose) (Banks, 1984). It was observed that the coating caused stomatal blockage that physically impeded gaseous diffusion through stomata, the principal route of gas exchange across banana fruit skin.

Several polysaccharide-based coatings displayed retarded color development, reduced CO_2 evolution and weight loss, and maintained greater firmness. Chitosan-based coatings were much superior in prolonging the shelf life and quality of banana and mango compared with coatings consisted of modified starch or cellulose (Kittur et al., 2001).

7.5.9 Mangoes and other tropical fruits

Aqueous wax emulsions, consisting of plant (sisal, sugarcane, and carnauba) waxes and mineral petroleum (paraffin) with and without shellac and emulsifiers, increased storage life of mango, pineapple, banana, papaya, guava, and avocado (Dalal et al., 1971). Coating mango with a paraffin wax emulsion (7% solids) resulted in increased shelf life (Mathur and Srivastava, 1955), especially at reduced storage temperature (12.8°C) (Bose and Basu, 1954), while coating with refined mineral oil resulted in fruit injury (Mathur and Srivastava, 1955). The oil coating appeared to decrease respiration more than did the wax coating, resulting in anaerobic conditions severe enough to injure the fruit. Both coatings, however, decreased weight loss. Mango fruits also exhibited retarded ripening and therefore increased storage life when coated with 0.75% to 1% TAL Pro-long and stored at 25°C (Dhalla and Hanson, 1988). The authors also reported reduced weight loss in the coated fruit compared to uncoated controls, and increased ethanol formation in fruit pulp after 13 days with 1% TAL Pro-long. No adverse effect on sensory quality was detected when a 0.75% formulation was used, however. Coated mangoes showed a slower decrease in titratable acidity and ascorbic acid as well as a retarded softening and carotenogenesis (loss of green color).

More recent studies tested carnauba, shellac, zein, and cellulose-based polysaccharide coatings on mangoes under two storage conditions (Baldwin et al., 1999a; Hoa et al., 2002). In general, coatings delayed ripening as measured by various indicators (firmness, acidity), except the carnauba wax. Carnauba wax was the only coating that reduced water loss in both studies.

Application of gelatin-starch coatings delayed the ripening of avocado, as indicated by a better pulp firmness and retention of skin color, lower weight loss, and a delayed climacteric respiratory rise (Aguilar-Méndez et al., 2008). Avocadoes coated with a methyl cellulose–based coating had lower respiration rates, greener color, and higher firmness as associated with delayed ripening (Maftoonazad and Ramaswamy, 2005). Feygenberg et al. (2005) applied organic coatings, BeeCoat (a colloidal solution based on beeswax) and Decco Lustr (a carnauba-based wax), on avocadoes and mangoes and found that both coatings reduced water loss, shrinkage, chlorophyll breakdown, chilling injury symptoms, and decay development; however, carnauba wax resulted in better shine.

Room temperature storage life of guavas was extended 80% by coating with a 3% carnauba-paraffin wax emulsion (Srivastava et al., 1962). No effect on ripening was evident, however. Coating pineapples in the "shipping green" stage with a 4% wax emulsion did not interfere with color development but reduced acid concentration in the juice and fruit weight loss by 35% to 45%. Increasing the amount of wax, however, interfered with normal color development (Schappelle, 1941). Coating of fully husked coconuts with paraffin wax resulted in an increase of 8 to 10 months storage life before appreciable loss of nut water occurred. Shellac coatings, however, were not as successful (Muliyar and Marar, 1963).

Waxing of papaya fruit with various resin-, polyethylene-, and carnauba-based waxes resulted in some delay in certain ripening parameters, a decrease in weight loss, and, in some cases, some off-flavor (Paull and Chen, 1989). Early work with mineral oil on mango fruits proved unsuccessful because of subsequent anaerobic conditions due to reduced gas exchange (Mathur and Subramanyam, 1956).

Preclimacteric mamey sapote fruit coated with carnauba wax maintained good visual appearance and had low water loss. However, the wax treatment accelerated respiration rate, ethylene production, and fruit softening, and reduced storage life (Ergun et al., 2005).

Rapid postharvest browning of litchi fruit pericarp is the result of PPO activity, anthocyanin hydrolysis, and nonenzymatic polymerization of o-quinones into melanins (Zhang et al., 2005). Effects of chitosan coating on browning of litchi (*Litchi chinensis* Sonn.) fruit were investigated by several workers (Jiang et al., 2005; Joas et al., 2005; Zhang and Quantick, 1997). In these studies, coating was formed by dipping fruits in chitosan solution. Zhang and Quantick (1997) reported that chitosan coating, irrespective of concentration (1% and 2% dissolved in 2% glutamic acid), delayed changes in contents of anthocyanins, flavonoids, and total phenolics. It also delayed the increase in PPO activity and partially inhibited the increase in peroxidase activity. Jiang et al. (2005) also similarly observed that chitosan (2% in 5% acetic acid) coating delayed the decrease in anthocyanin content and the increase in PPO activity. Such effects of chitosan coatings were also observed with peeled litchi fruit (Dong et al., 2004), longan fruit (Jiang and Li, 2001), and fresh-cut Chinese water chestnut vegetable (Pen and Jiang, 2003). Dependence of browning rate of chitosan-coated litchi fruit on the initial pericarp water content (Caro and Joas, 2005), pericarp pH, and dehydration rate during storage (Joas et al., 2005) has been reported.

Breadfruit coated with Semperfresh F (AgriCoat NatureSeal Ltd., UK), Nutri-Save (Nova Chem Ltd, Nova Scotia, Canada), Sta-Fresh MP (JBT FoodTech, Lakeland, Florida), and chitosan had low respiration rate and ethylene production rate, and retarded softening; however, all coatings caused off-odor and flesh discoloration although internal oxygen concentrations were higher than 8% and CO_2 concentration mostly lower than 12% (Worrell et al., 2002).

7.5.10 Tomatoes and other fruit vegetables

Coatings similar to those applied to fruits were also applied to vegetables as early as the 1930s. By the 1950s water waxes consisting of water, vegetable waxes (usually carnauba, sometimes candelilla), paraffin wax, and emulsifying agents (soaps, detergents) were usually formulated to contain 6% to 10% wax solids (Hartman and Isenberg, 1956). A typical formula of this type included carnauba wax, oleic acid, triethanolamine, water, and paraffin wax. Wax emulsions decreased water loss, shriveling, and shrinkage in cucumbers, summer squash, pumpkins, eggplant, peppers, and tomatoes. A water emulsion of a wax mixture (carnauba-paraffin) sharply reduced weight loss and spoilage in coated cucumber, but increased low-temperature injury symptoms (pitting) in cold storage (Mack and Janer, 1942). For a time, the waxing of stem scars of tomatoes was attempted because the majority of tomato gas exchange occurs through this area (Brooks, 1938; Emmert and Southwick, 1954). Waxing the stem scars with mixtures of paraffin, beeswax, and mineral oil effectively delayed ripening of mature green tomatoes (Ayres et al., 1964; Hartman and Isenberg, 1956), but the process was labor intensive.

Another advantage of wax coatings was the fact that coatings acted as lubricants to reduce surface scarring and chafing, especially for tomatoes and peppers. Coating of peppers with a carnauba wax emulsion with either 3% or 6% solids prolonged storage, reduced shriveling and decay, enhanced shine, reduced rates of respiration, and resulted in better retention of ascorbic acid (vitamin C) (Habeebunnisa et al., 1963). TAL Pro-long was not effective in decreasing respiration rate and water loss in fruits of Solanaceae such as tomato and sweet pepper (Lowings and Cutts, 1982; Nisperos-Carriedo and Baldwin, 1988), nor was Semperfresh effective in retarding water loss of melons (Edwards and Blennerhassett, 1990).

Park et al. (1994) reported that applying zein coating on tomato surface delayed color change and weight loss and maintained firmness during storage. Zapata et al. (2008) applied zein and alginate coatings to tomato fruit and found that both coatings decreased respiration rate, ethylene production, losses of firmness and weight, and color change.

El Ghaouth et al. (1992) reported that chitosan coating (1% and 2% in 0.25 N HCl) reduced the respiration rate and ethylene production of tomato, with a greater effect observed at 2% than 1% chitosan. Chitosan-coated tomatoes were firmer, higher in titratable acidity, and less decayed, and they exhibited less red pigmentation than the control fruit after 4 weeks storage at 20°C. Similarly, Kim et al. (1999) observed that chitosan (318 kDa; 2% in 0.2-M acetic acid) coated tomatoes showed less weight loss and higher flesh firmness than the noncoated control during storage at room temperature for 18 days. Liu et al. (2007) reported that chitosan coating provided an effective control of blue mold and gray mold of tomato fruit stored at 25 and 2°C. Chitosan strongly inhibited spore germination, germ tube elongation, and mycelial growth of *B. cinerea* and *P. expansum in vitro*, and damaged the plasma membranes of spores of both pathogens. These findings suggest that the effects of chitosan on gray mold and blue mold in tomato fruit may be associated with a direct fungitoxic property against the pathogens, as well as the elicitation of biochemical defense responses in fruit (Liu et al., 2007). Similar results were confirmed by Badawy and Rabea (2009). An et al. (2008) used a silver nanoparticles-PVP coating to green asparagus and found that the coating decreased the growth of microorganisms, retarded tissue hardening, and significantly extended storage life.

Mineral oil–based coatings are widely used for fruit and vegetables. Because their application has an effect of dewatering the fruit surface, no drying is required before or after the coating process. Mineral oil–based coatings are effective in controlling water loss, improving appearance, and reducing abrasive injuries during handling and transportation (Grant and Burns, 1994; Hardenburg, 1967).

7.5.11 Root and tuber crops

Paraffin wax is frequently used to coat root crops such as rutabaga, turnips, and yucca (Grant and Burns, 1994). To adhere well to the surface, vegetables must be clean before coating application. Molten paraffin waxes are applied to root vegetable surfaces above the boiling point of water, any free water on the commodity will form "steam pockets" and will flake during subsequent handling, thus the product must be completely dry before application of paraffin (Grant and Burns, 1994). The effect of waxing on the storage life of cassava roots was first studied in the 1940s (Castagnino, 1943), and the technique was subsequently refined. Dipping the roots in a fungicidal wax followed by storage at ambient temperature substantially increased the storage potential (Subramanyam and Mathur, 1956). Waxing approximately halved the weight loss during the first 2 weeks of storage and extended the useful storage time period (i.e., <10% loss) from 2 to 10 days. An increase in respiration was also delayed. The principle benefit of waxing seems to be derived from reduced moisture loss rather than pathogen inhibition. Paraffin wax dips (90 to 95°C for 45 s) averted serious quality losses for 1 to 2 months (Anon., 1972), ensuring sufficient time for export or storage

(Burton, 1970; De Buckle et al., 1973; Zapata, 1978). Water-based carnauba and paraffin waxes are comparable in root quality maintenance (Sargent et al., 1995).

Turnips are waxed to improve appearance, as well as to reduce water loss and wilting (Franklin, 1961). Enrobing of turnip roots with a water-miscible, carnauba-based wax also temporarily darkened and intensified the purple color (Perkins-Veazie and Collins, 1991). Wax coatings decreased O_2 consumption in carrots with no effect on CO_2 production, but because the respiratory quotient of carrots (and most root crops) is less than one, the shift in respiration was unlikely to have harmful effects. Paraffin-waxed potatoes showed reduced sprouting, weight loss, chlorophyll (greening), and solanine synthesis (Wu and Salunkhe, 1972) without adversely affecting respiration. Dalal et al. (1971) reported that treatments with vegetable waxes and paraffin prolonged the shelf life of carrots, potatoes, tomatoes, muskmelon, and cucumbers. Turnips were waxed to reduce weight loss, and hot waxes (paraffin) worked better than cold (colloidal waxes in water suspension) (Franklin, 1961). Starch-coated radishes packaged in an antifog-type of film (OPP-Coex-film) kept soluble pectins as in the control, but glucosinolates, the health components of radish, were only maintained by the film packaging (Schreiner et al., 2003).

Casariego et al. (2008) determined the effects of the concentrations of glycerol and sorbitol (as hydrophilic plasticizers), Polyoxyethylene (20) sorbitan monooleate (as surfactant), and chitosan on the wettability of chitosan-based edible coatings in view of their application on tomato and carrot. The values of the polar and dispersive components of the superficial tension of the foods were determined to be 3.04 and 25.67 mN/m, respectively, for tomato, and 0.34 and 26.13 mN/m, respectively, for carrot, with the sum of the two components being the superficial tensions of tomato and carrot (28.71 and 26.48 mN/m, respectively). The best experimental values of wettability were obtained for the following coating composition: 1.5% (w/v) of chitosan and 0.1% (w/w) of Tween 80. The increase in the concentration of chitosan and glycerol or sorbitol as plasticizers decreased the values of wettability and adhesion coefficients.

7.5.12 Leafy vegetables

In spite of a large body of research existing on coating applications for fresh-cut vegetables, only a few reports were found for coating applications on intact leafy vegetables. The major functions of coatings or treatments for fresh-cut vegetables are antibrowning, sanitation, and firmness preservation, which will be dealt with in the next chapter. Little benefit was achieved by coating leafy vegetables (Platenius, 1939). Chitosan (43 kDa, DD=94%) coating was effective in controlling decay of lettuce; however, its applicability may be hampered due to a pronounced bitter taste developed after treatment (Devlieghere et al., 2004).

7.6 Conclusion

This chapter presented a detailed review of postharvest physiology and issues to take into account to optimize produce storage using coatings. Depending on ingredients, edible coatings can modify fruit internal atmosphere, prevent dehydration and shrinkage, provide a physical barrier against bruises and chaffing, and with the appropriate additives, also provide a chemical barrier to protect produce from decaying. Edible coatings provide a tool to improve appearance, maintain nutritional and flavor quality, and extend the shelf life of fruit and vegetables.

References

Aguilar-Méndez, M.A., E.S. Martín-Martínez, S.A. Tomás, A. Cruz-Orea, and M.R. Jaime-Fonseca. 2008. Gelatine starch films: Physicochemical properties and their application in extending the post-harvest shelf life of avocado *Persea americana. J Sci Food Agric.* 88: 185–193.

Akagi, A., and H.U. Stotz. 2007. Effects of pathogen polygalacturonase, ethylene, and firmness on interactions between pear fruits and *Botrytis cinerea. Plant Dis.* 91: 1337–1344.

Albrigo, L.G. 1972a. Ultrastructure of cuticular surfaces and stomata of developing leaves and fruit of the 'Valencia' orange. *J Amer Soc Hort Sci.* 97: 761–765.

Albrigo, L.G. 1972b. Variation in surface wax on oranges from selected groves in relation to fruit moisture loss. *Proc Fla State Hort Soc.* 85: 262–263.

Alonso, J., and R. Alique. 2004. Influence of edible coating on shelf life and quality of 'Picota' sweet cherries. *Eur Food Res Technol.* 218: 535–539.

Amarante, C., and N.H. Banks. 2001. Postharvest physiology and quality of coated fruits and vegetables. *Hort Rev.* 26: 161–237.

Amarante, C., N.H. Banks, and S. Ganesh. 2001a. Characterising ripening behaviour of coated pears in relation to fruit internal atmosphere. *Postharvest Biol Technol.* 23: 51–59.

Amarante, C., N.H. Banks, and S. Ganesh. 2001b. Effects of coating concentration, ripening stage, water status and fruit temperature on pear susceptibility to friction discolouration. *Postharvest Biol Technol.* 21: 283–290.

An, J., M. Zhang, S. Wang, and J. Tang. 2008. Physical, chemical and microbiological changes in stored green asparagus spears as affected by coating of silver nanoparticles-PVP. *LWT—Food Sci Technol.* 41: 1100–1107.

Anon. 1972. La yuca parafinada. *Rev Inst Invest Tecnol Bogota.* 14: 47–51.

Ashley, R.J. 1985. Permeability and plastics packaging, p. 269–303. In: J. Comyn (ed.). *Polymer Permeability.* Elsevier Applied Science Publishing, New York.

Ayranci, E., and S. Tunc. 2004. The effect of edible coatings on water and vitamin C loss of apricots (*Armeniaca vulgaris* L.) and green peppers (*Capsicum annuum* L.). *Food Chem.* 87: 339–342.

Ayres, J.C., A.A. Kraft, and L.C. Peirce. 1964. Delaying spoilage of tomatoes. *Food Technol.* 9: 100–103.

Badawy, M.E.I., and E.I. Rabea. 2009. Potential of the biopolymer chitosan with different molecular weights to control postharvest gray mold of tomato fruit. *Postharvest Biol Technol.* 51: 110–117.

Bai, J., V. Alleyne, R.D. Hagenmaier, J.P. Mattheis, and E.A. Baldwin. 2003a. Formulation of zein coatings for apples (*Malus domestica* Borkh). *Postharvest Biol Technol.* 28: 259–268.

Bai, J., E.A. Baldwin, and R.H. Hagenmaier. 2002a. Alternatives to shellac coatings provide comparable gloss, internal gas modification, and quality for 'Delicious' apple fruit. *HortScience.* 37: 559–563.

Bai, J., R.D. Hagenmaier, and E.A. Baldwin. 2002b. Volatile response of four apple varieties with different coatings during marketing at room temperature. *J Agric Food Chem.* 50: 7660–7668.

Bai, J., R.D. Hagenmaier, and E.A. Baldwin. 2003b. Coating selection for 'Delicious' and other apples. *Postharvest Biol Technol.* 28: 381–390.

Bai, J., E.A. Mielke, P.M. Chen, R.A. Spotts, M. Serdani, J.D. Hansen, and L.G. Neven. 2006. Effect of high-pressure hot-water washing treatment on fruit quality, insects, and disease in apples and pears: Part I. System description and the effect on fruit quality of 'D'Anjou' pears. *Postharvest Biol Technol.* 40: 207–215.

Bai, J., R.K. Prange, and P.A. Toivonen. 2009a. Pome fruits, p. 267–285. In: E. Yahia (ed.). *Modified and Controlled Atmospheres for the Storage, Transportation, and Packaging of Horticultural Commodities.* CRC Press, Boca Raton, FL.

Bai, J., P. Wu, J. Manthey, K. Goodner, and E. Baldwin. 2009b. Effect of harvest maturity on quality of fresh-cut pear salad. *Postharvest Biol Technol.* 51: 250–256.

Bajwa, B.E., and F.M. Anjum. 2007. Improving storage performance of *Citrus reticulata* Blanco mandarins by controlling some physiological disorders. *Int J Food Sci Technol.* 42: 495–501.

Baldwin, E.A. 1994. Edible coatings for fresh fruits and vegetables: Past, present, and future, p. 25–64. In: J.M. Krochta, E.A. Baldwin, and M.O. Nisperos-Carriedo (eds.). *Edible Coating and Films to Improve Food Quality.* Technomic, Lancaster, PA.

Baldwin, E.A. 1999. Coatings in food preservation, p. 577–610. In: S. Rahman (ed.). *Handbook of Food Preservation.* CRC Press, Boca Raton, FL.

Baldwin, E.A. 2004. Ethylene and postharvest commodities. *HortScience.* 39: 1538–1540.

Baldwin, E.A., J.K. Burns, W. Kazokas, J.K. Brecht, R.D. Hagenmaier, R.J. Bender, and E. Pesis. 1999a. Effect of two edible coatings with different permeability characteristics on mango (*Mangifera indica* L.) ripening during storage. *Postharvest Biol Technol.* 17: 215–226.

Baldwin, E.A., T.M.M. Malundo, R. Bender, and J.K. Brecht. 1999b. Interactive effects of harvest maturity, controlled atmosphere and surface coating on mango (*Mangifera indica* L.) flavor quality. *HortScience.* 34: 514.

Baldwin, E.A., M. Nisperos-Carriedo, P.E. Shaw, and J.K. Burns. 1995. Effect of coatings and prolonged storage conditions on fresh orange flavor volatiles, degrees brix, and ascorbic acid levels. *J Agric Food Chem.* 43: 1321–1331.

Baldwin, E.A., M.O. Nisperos, R.D. Hagenmaier, and R.A. Baker. 1997. Use of lipids in coatings for food products. *Food Technol.* 51: 56–64.

Baldwin, E.A., and B. Wood. 2006. Use of edible coating to preserve pecans at room temperature. *HortScience.* 41: 188–192.

Banks, N.H. 1983. Evaluation of methods for determining internal gases in banana fruit. *J Exp Bot.* 34: 871–879.

Banks, N.H. 1984. Some effects of TAL Pro-long coating on ripening bananas. *J Exp Bot.* 35: 127–137.

Banks, N.H. 1985a. Internal atmosphere modification in Pro-long coated apples. *Acta Hort.* 157: 105–112.

Banks, N.H. 1985b. Responses of banana fruit to Pro-long coating at different times relative to the initiation of ripening. *Sci Hort.* 26: 149–157.

Barger, W.R., J.S. Wiant, W.T. Pentzer, A.L. Ryall, and D.H. Dewey. 1948. A comparison of fungicidal treatments for the control of decay in California cantaloupes. *Phytopathology.* 38: 1019–1024.

Batista, P.F., A.E.O. Santos, M.M.M. Pires, B.F. Dantas, A.R. Peixoto, and C.A. Aragão. 2007. Use of plastic and edible films in the postharvest conservation of yellow melon. Hort Brasileirs. 25: 572–576.

Bauchot, A.D., and P. John. 1996. Scald development and the levels of alpha-farnesene and conjugated triene hydroperoxides in apple peel after treatment with sucrose ester-based coatings in combination with food-approved antioxidants. *Postharvest Biol Technol.* 7: 41–49.

Bauchot, A.D., P. John, Y. Soria, and I. Recasens. 1995. Sucrose ester-based coatings formulated with food-compatible antioxidants in the prevention of superficial scald in stored apples. *J Amer Soc Hort Sci.* 120: 491–496.

Beaudry, R.M. 1999. Effect of O_2 and CO_2 partial pressure on selected phenomena affecting fruit and vegetable quality. *Postharvest Biol Technol.* 15: 293–303.

Beaudry, R.M. 2000. Responses of horticultural commodities to low oxygen: Limits to the expanded use of modified atmosphere packaging. *HortTechnology.* 10: 491–500.

Beaudry, R.M., A.C. Cameron, A. Shirazi, and D.L. Dostal-Lange. 1992. Modified-atmosphere packaging of blueberry fruit: Effect of temperature on package O_2 and CO_2. *J Amer Soc Hort Sci.* 117: 436–441.

Ben-Yehoshua, S. 1966. Some effects of plastic skin coating on banana fruit. *Trop Agric—Trinidad.* 43: 219–232.

Bernalte, M.J., E. Sabio, M.T. Hernández, and C. Gervasini. 2003. Influence of storage delay on quality of 'Van' sweet cherry. *Postharvest Biol Technol.* 28: 303–312.

Bhardwaj, C.L., H.F. Jones, and I.H. Smith. 1984. A study of the migration of externally applied sucrose esters of fatty acids through the skins of banana, apple and pear fruits. *J Sci Food Agric.* 35: 322–331.

Bitencourt de Souza, A.L., S. Quintao, S.D.P., M.I.F. Chitarra, and A.B. Chitarra. 1999. Post-harvest application of CaCl2 in strawberry fruit (*Fragaria ananassa* Dutch cv. Sequoia): evaluation of fruit quality and post-harvest life. *Ciencia e agrotec., Lavras.* 23: 841–848.

Blake, J.R. 1966. Some effect of paraffin wax emulsions on bananas. *Queensland J Agric Anim Science.* 23: 49–56.

Blanpied, G.D., and K.D. Hickey. 1963. 'Concord' grape storage trials for control of *Botrytis cinerea* and *Penicillium* sp. *Plant Dis. Rep.* 47: 986–992.

Bose, A.N., and G. Basu. 1954. Studies on the use of coating for extension of storage life of fresh Fajli mango. *J Food Sci.* 19: 424–428.

Brancoli, N., and G.V. Barbosa-Canovas. 2000. Browning of apple slices treated with polysaccharide films, p. 225–232. In: G.V.a.G.G.W. Barbosa-Canovas (ed.). *Innovations in Food Processing.* Technomic, Lancaster, PA.

Brooks, C. 1938. Some effects of waxing tomatoes. *Amer Soc Hort Sci Proc.* 35: 720.

Bruemmer, J.H. 1989. Terminal oxidase activity during ripening of Hamlin orange. *Phytochemistry.* 28: 2901–2902.

Burton, C.L. 1970. Diseases of tropical vegetables in the Chicago market. *Trop Agric—Trinidad.* 47: 303–313.

Cameron, A.C., R.M. Beaudry, N.H. Banks, and M.V. Yelanich. 1994. Modified-atmosphere packaging of blueberry fruit: Modeling respiration and package oxygen partial pressures as a function of temperature. *J Amer Soc Hort Sci.* 119: 534–539.

Cameron, A.C., and S.F. Yang. 1982. A simple method for the determination of resistance to gas diffusion in plant organs. *Plant Physiol.* 70: 21–23.

Caro, Y., and J. Joas. 2005. Postharvest control of litchi pericarp browning (cv. Kwai Mi) by combined treatments of chitosan and organic acids: II. Effect of the initial water content of pericarp. *Postharvest Biol Technol.* 38: 137–144.

Casariego, A., B.W.S. Souza, A.A. Vicente, J.A. Teixeira, L. Cruz, and R. Díaz. 2008. Chitosan coating surface properties as affected by plasticizer, surfactant and polymer concentrations in relation to the surface properties of tomato and carrot. *Food Hydrocolloid.* 22: 1452–1459.

Castagnino, G.A. 1943. Conservacion de la raiz de mandioca. *El Campo (BuenosAires).* 27: 23.

Certel, M., M.K. Uslu, and F. Ozdemir. 2004. Effects of sodium caseinate and milk protein concentrate-based edible coatings on the postharvest quality of 'Bing' cherries. *J Sci Food Agric.* 84: 1229–1234.

CFR. 1990. Oxidized Polyethylene. In: Code of Federal Regulations (Ed., 21 CFR 172.260, U.S. GPO, Washington, DC.
Chien, P.-J., F. Sheu, and H.-R. Lin. 2007a. Quality assessment of low molecular weight chitosan coating on sliced red pitayas. *J Food Eng*. 79: 736–740.
Chien, P.-J., F. Sheu, and F.-H. Yang. 2007b. Effects of edible chitosan coating on quality and shelf life of sliced mango fruit. *J Food Eng*. 78: 225–229.
Chiumarelli, M., and M.D. Ferreira. 2006. Qualidade pos-colheita de tomates 'Debora' com utilizacao de diferentes coberturas comestiveis e temperaturas de armazenamento. *Hort Brasileira*. 24: 381–385.
Chlebowska-Śmigiel, A., M. Gniewosz, and E. Świńczak. 2007. An attempt to apply a pullulan and pullulan-protein coatings to prolong apples shelf-life stability. *Acta Sci Pol Technol Alimen*. 6: 49–56.
Cho, S.Y., J.-W. Park, H.P. Batt, and R.L. Thomas. 2007. Edible films made from membrane processed soy protein concentrates. *LWT-Food Sci Technol*. 40:418–423.
Cho, S.Y. and C. Rhee. 2004. Mechanical properties and water vapor permeability of edible films made from fractionated soy proteins with ultrafiltration. *LWT-Food Sci Technol*. 37:833–839.
Choi, K.H., G.H. Lee, Y.J. Han, and J.M. Bunn. 1995. Tomato maturity evaluation using color image analysis. *Trans ASAE*. 38: 171–176.
Choi, W.Y., H.J. Park, D.J. Ahn, J. Lee, and C.Y. Lee. 2002. Wettability of chitosan coating solution on 'Fuji' apples skin. *J Food Sci*. 67: 2668–2672.
Chung, D., M.L. Chikindas, and K.L. Yam. 2001. Inhibition of saccharomyces cerevisiae by slow release of propyl paraben from a polymer coating. *J Food Protect*. 64: 1420–1424.
Cisneros-Zevallos, L., and J.M. Krochta. 2003. Whey protein coatings for fresh fruits and relative humidity effects. *J Food Sci*. 68: 176–181.
Claypool, L.L. 1939. The waxing of deciduous fruits. *Pro Amer Soc Hort Sci*. 37: 443–447.
Cohen, E., Y. Shalom, and I. Rosenberg. 1990. Postharvest ethanol buildup and off-flavor in Murcott tangerine fruits. *J Amer Soc Hort Sci*. 115: 775–778.
Conforti, F.D., and J.A. Totty. 2007. Effect of three lipid/hydrocolloid coatings on shelf life stability of 'Golden Delicious' apples. *Int J Food Sci Technol*. 42: 1101–1106.
Cong, F., Y. Zhang, and W. Dong. 2007. Use of surface coatings with natamycin to improve the storability of 'Hami' melon at ambient temperature. *Postharvest Biol Technol*. 46: 71–75.
Cuppett, S.L. 1994. Edible coatings as carriers of food additives, fungicides, and natural antagonists, p. 121–137. In: J.M. Krochta, E.A. Baldwin, and M.O. Nisperos-Carriedo (eds.). *Edible Coating and Films to Improve Food Quality*. Technomic, Lancaster, PA.
Cuquerella, J., J.M. Martizez-Javega, and M. Jimenez-Cuesta. 1981. Some physiological effects of different wax treatments on Spanish citrus fruit during cold storage. *Proc Int Soc Citricult*. 2: 734–737.
Curtis, G.J. 1988. Some experiments with edible coatings on the long-term storage of citrus fruits. *Proc 6th Int Citrus Cong*. 3: 1514–1520.
Dalal, V.B., W.E. Eipeson, and N.S. Singh. 1971. Wax emulsion for fresh fruits and vegetables to extend their storage life. *India Food Packer*. 25: 9–15.
Dangaran, K.L., P. Cooke, and P.M. Tomasula. 2006. The effect of protein particle size reduction on the physicl properties of CO_2-precipitated casein films. *J Food Sci*. 71:E196–E201.
Davis, P.L. 1970. Relation of ethanol content of citrus fruits to maturity and to storage condition. *Proc Fla State Hort Soc*. 83: 294–298.
Davis, P.L., and P.C. Harding. 1960. The reduction of rind breakdown of 'Marsh' grapefruit by polyethylene emulsion treatments. *J Amer Soc Hort Sci*. 75: 271–274.

Davis, P.L., B. Roe, and J.H. Bruemmer. 1973. Biochemical changes in citrus fruits during controlled-atmosphere storage. *J Food Sci.* 38: 225–229.

De Buckle, S., S. Teresa, H. Castelblanco, L.E. Zapata, M.F. Bocanegra, and L.E. Rodriguez. 1973. Preservation of fresh cassava by the method of waxing. *Rev Inst Invest Tecnol Bogota* 15: 33–47.

Del-Valle, V., P. Hernández-Muñoz, A. Guarda, and M.J. Galotto. 2005. Development of a cactus-mucilage edible coating (*Opuntia ficus-indica*) and its application to extend strawberry (*Fragaria ananassa*) shelf-life. *Food Chem.* 91: 751–756.

Devlieghere, F., A. Vermeulen, and J. Debevere. 2004. Chitosan: antimicrobial activity, interactions with food components and applicability as a coating on fruit and vegetables. *Food Microbiol.* 21: 703–714.

Dhalla, R., and S.W. Hanson. 1988. Effect of permeable coatings on the storage life of fruits. *Int J Food Sci Technol.* 23: 107–112.

Diab, T., C.G. Biliaderis, D. Gerasopoulos, and E. Sfakiotakis. 2001. Physicochemical properties and application of pullulan edible films and coatings in fruit preservation. *J Sci Food Agric.* 81: 988–1000.

Dong, H., L. Cheng, J. Tan, K. Zheng, and Y. Jiang. 2004. Effects of chitosan coating on quality and shelf life of peeled litchi fruit. *J Food Eng.* 64: 355–358.

Dong, T., R. Xia, M. Wang, Z. Xiao, and P. Liu. 2008. Changes in dietary fibre, polygalacturonase, cellulase of 'Navel' orange (*Citrus sinensis* (L.) Osbeck 'Cara Cara') fruits under different storage conditions. *Sci Hort.* 116: 414–420.

Dou, H. 2004. Effect of coating application on chilling injury of grapefruit cultivars. *HortScience.* 39: 558–561.

Drake, S.R., J.K. Fellman, and J.W. Nelson. 1987. Postharvest use of sucrose polyesters for extending the shelf-life of stored 'Golden Delicious' apples. *J Food Sci.* 52: 1283–1285.

Drake, S.R., E.M. Kupferman, and J.K. Fellman. 1988. 'Bing' sweet cherry (*Prunus avium* L.) quality as influenced by wax coatings and storage temperature. *J Food Sci.* 53: 124–126.

Ducamp-Collin, M.-N., H. Ramarson, M. Lebrun, G. Self, and M. Reynes. 2008. Effect of citric acid and chitosan on maintaining red colouration of litchi fruit pericarp. *Postharvest Biol Technol.* 49: 241–246.

Durango, A.M., N.F.F. Soares, and N.J. Andrade. 2006. Microbiological evaluation of an edible antimicrobial coating on minimally processed carrots. *Food Control.* 17: 336–341.

Eastman Chemical Inc. 1984. A guide to the use of epolene waxes under United States FDA food additive regulations. Publication No. F-243D.

Eastman Chemical Inc. 1990. Emulsification of epolene waxes. Publication No. F302.

Eckert, J.W., and G.E. Brown. 1986. Postharvest citrus diseases and their control, p. 315–360. In: S.N.a.W.G. W.F. Wardowski (ed.). *Fresh Citrus Fruit.* Van Nostrand Reinhold, New York.

Edwards, M., and R. Blennerhassett. 1990. The use of postharvest treatments to extend storage life and to control postharvest wastage of honeydew melons (*Cucumis melo* L. var. inodorus Naud.) in cool storage. *Australia J Exp Agric.* 30: 693–697.

Edwards, M., and R. Blennerhassett. 1994. Evaluation of wax to extend the postharvest storage life of honeydew melons (*Cucumis melo* L. var. inodorus Naud.). *Australia J Exp Agric.* 34: 427–429.

El Ghaouth, A., J. Arul, R. Ponnampalam, and M. Boulet. 1991a. Chitosan coating effect on storability and quality of fresh strawberries. *J Food Sci.* 56: 1618–1620.

El Ghaouth, A., J. Arul, R. Ponnampalam, and M. Boulet. 1991b. Use of chitosan coating to reduce water loss and maintain quality of cucumber and bell pepper fruits. *J Food Process Preserv.* 15: 359–368.

El Ghaouth, A., R. Ponnampalam, F. Castaigne, and J. Arul. 1992. Chitosan coating to extend the storage life of tomatoes. *HortScience*. 27: 1016–1018.

Elson, C.M., E.R. Hayes, and P.D. Lidster. 1985. Development of the differentially permeable fruit coating Nutri-Save for the modified atmosphere storage of fruit. *Proc 4th Nat CA Res Conf: Controlled Atmospheres for Storages and Transport of Perishable Agricultural Commodities.* Raleigh, NC. pp. 248–262.

Emmert, F., and F. Southwick. 1954. The effect of maturity, apple emanations, waxing, and growth regulators on the respiration and red color development of tomato fruits. *Amer Soc Hort Sci Proc*. 63: 393–401.

Erbil, H.Y., and N. Muftugil. 1986. Lengthening the postharvest life of peaches by coating with hydrophobic emulsions. *J Food Process Preserv*. 10: 269–279.

Ergun, M., S. Sargent, A. Fox, J. Crane, and D. Huber. 2005. Ripening and quality responses of mamey sapote fruit to postharvest wax and 1-methylcyclopropene treatments. *Postharvest Biol Technol*. 36: 127–134.

Eswarandam, S., N.S. Hettiarachchy, and J.F. Meulleneteffect. 2006. Effect of malic and lactic acid incorporated soy protein coatings on the sensory attributes of whole apple and fresh-cut cantaloupe. *J Food Sci*. 71: 307–313.

Fakhouri, F.M., L.C.B. Fontes, P.V.D.M. Goncalves, C.R. Milanez, C.J. Steel, and F.P. Collares-Queiroz. 2007. Films and edible coatings based on native starches and gelatin in the conservation and sensory acceptance of 'Crimson' grapes. *Ciencia e Tecnolgia de Alimentos*. 17: 369–375.

Falcão-Rodrigues, M.M., M. Moldão-Martins, and M.L. Beirão-da-Costa. 2007. DSC as a tool to assess physiological evolution of apples preserved by edibles coatings. *Food Chem*. 102: 475–480.

Fallik, E., Y. Shalom, S. Alkalai-Tuvia, O. Larkov, E. Brandeis, and U. Ravid. 2005. External, internal and sensory traits in 'Galia-type' melon treated with different waxes. *Postharvest Biol Technol*. 36: 69–75.

Farber, J.N., L.J. Harris, M.E. Parish, L.R. Beuchat, T.V. Suslow, J.R. Gorney, E.H. Garrett, and F.F. Busta. 2003. Microbiological safety of controlled and modified atmosphere packaging of fresh and fresh-cut produce. *Compr Rev Food Sci Food Safety*. 2: 142–160.

Farooqi, W.A., and E.G. Hall. 1973. Effect of wax coating containing diphenylamine on apples and pears during storage and ripening. *Australia J Exp Agric*. 13: 200–204.

Feygenberg, O., V. Hershkovitz, R. Ben-Arie, E. Jacob, S. Pesis, and T. Nikitenko. 2005. Postharvest use of organic coating for maintaining bio-organic avocado and mango quality. *Acta Hort*. 682: 507–512.

Fields, P.G., and N.D.G. White. 2002. Alternatives to methyl bromide treatments for stored-product and quarantine insects. *Ann Rev Entomol*. 47: 331–359.

Fischer, R.L., and A.B. Bennett. 1991. Role of cell wall hydrolases in fruit ripening. *Ann Rev Plant Physiol Plant Molecul Biol*. 42: 675–703.

Fisk, C.L., A.M. Silver, B.C. Strik, and Y. Zhao. 2008. Postharvest quality of hardy kiwifruit (*Actinidia arguta* 'Ananasnaya') associated with packaging and storage conditions. *Postharvest Biol Technol*. 47: 338–345.

Fornes, F., V. Almela, M. Abad, and M. Agustí. 2005. Low concentrations of chitosan coating reduce water spot incidence and delay peel pigmentation of 'Clementine' mandarin fruit. *J Sci Food Agric*. 85: 1105–1112.

Franklin, E.W. 1961. The waxing of turnips for the retail market. Canada Department of Agriculture, Publication No. 1120, 3 pp.

Franssen, L.R., T.R. Rumsey, and J.M. Krochta. 2004. Whey protein film composition effects on potassium sorbate and natamycin diffusion. *J Food Sci*. 69: C347–C350.

Garcia, J.M., S. Herrera, and A. Morilla. 1996. Effects of postharvest dips in calcium chloride on strawberry. *J AgricFood Chem.* 44: 30–33.

Garcia, M.A., M.N. Martino, and N.E. Zaritzky. 2000. Lipid addition to improve barrier properties of edible starch-based films and coating. *Food Sci.* 65: 941–947.

García, M.A., M.N. Martino, and N.E. Zaritzky. 2001. Composite starch-based coatings applied to strawberries (*Fragaria ananassa*). *Nahrung/Food*. 45: 267–272.

García, M.A., A. Pinotti, M.N. Martino, and N.E. Zaritzky. 2004. Characterization of composite hydrocolloid films. *Carbohydr Polym.* 56:339–345.

Gennadios, A., T.H. McHugh, C.L. Weller, and J.M. Krochta. 1994. Edible coatings and films based on proteins, p. 201–278. In: J.M. Krochta, E.A. Baldwin, and M.O. Nisperos-Carriedo (eds.). *Edible Coating and Films to Improve Food Quality*. Technomic, Lancaster, PA.

Gennadios, A., and C.L. Weller. 1990. Edible films and coatings from wheat and corn proteins. *Food Technol.* 44: 63–69.

Gennadios, A. and C.L. Weller. 1991. Edible films and coatings from soymilk and soy protein. *Cereal Food World.* 36:1004–1009.

Gontard, N., S. Guilbert, and J.-L. Cuq. 1992. Edible wheat gluten films: influence of the main process variables on film properties using response surface methodology. *J Food Sci.* 57:190–195.

Gontard, N., R. Thibault, B. Cuq, and S. Guilbert. 1996. Influence of relative humidity and film composition on oxygen and carbon dioxide permeabilities of edible films. *J. Agric Food Chem.* 44:1064–1069.

Gonzalez, J. 1997. Wax treatments meeting probit 9 requirements for controlling *Brevipalpus chilensis* in cherimoyas and citrus fruit. *Proc. Ann. Int. Res. Conf. Methyl Bromide Alternat. Emission Reduct.* p. 68.61–68.62.

Grant, L.A., and J. Burns. 1994. Application of coating, p. 189–200. In: J.M. Krochta, E.A. Baldwin, and M.O. Nisperos-Carriedo (eds.). *Edible Coating and Films to Improve Food Quality*. Technomic, Lancaster, PA.

Guilbert, S. 2000. Edible films and coatings and biodegradable packaging. *Bull Int. Dairy Feder.* 346:10–16.

Guilbert, S., N. Gontard, and L.G.M. Gorris. 1996. Prolongation of the shelf-life of perishable food products using biodegradable films and coatings. *LWT-Food Sci. Technol.* 29:10–17.

Habeebunnisa, M., C. Pushpa, and J. Srivastava. 1963. Studies on the effect of protective coating on the refrigerated and common storage of bell peppers (*Capsicum frutescense*). *Food Sci (Mysores).* 12: 192–196.

Hagenmaier, R.D. 1998. Selection of citrus 'wax' coatings on criteria other than short-term gloss. *Packinghouse Newsletter.* 182.

Hagenmaier, R.D. 2000. Evaluation of a polyethylene-candelilla coating for 'Valencia' oranges. *Postharvest Biol Technol.* 19: 147–154.

Hagenmaier, R.D. 2001a. Ethanol content of 'Murcott' tangerines harvested at different times and treated with coatings of different O_2 permeability. *Proc Fla State Hort Soc.* 114: 170–173.

Hagenmaier, R.D. 2001b. Ethanol content of 'Murcott' tangerines harvested at different times and treated with coatings of different O_2 permeability. *Proc Fla State Hort Soc.* 114: 170–173.

Hagenmaier, R.D. 2002. The flavor of mandarin hybrids with different coatings. *Postharvest Biol Technol.* 24: 79–87.

Hagenmaier, R.D., and R.A. Baker. 1993a. Citrus fruit with single or layered coatings compared with packinghouse-coated fruit. *Proc Fla State Hort Soc.* 106: 238–240.

Hagenmaier, R.D., and R.A. Baker. 1993b. Cleaning method affects shrinkage rate of citrus fruit. *HortScience*. 28: 824–825.

Hagenmaier, R.D., and R.A. Baker. 1993c. Reduction in gas exchange of citrus fruit by wax coatings. *J Agric Food Chem*. 41: 283–287.

Hagenmaier, R.D., and R.A. Baker. 1994a. Internal gases, ethanol content and gloss of citrus fruit coated with polyethylene wax, carnauba wax, shellac or resin at different application levels. *Proc Fla State Hort Soc*. 107: 261–265.

Hagenmaier, R.D., and R.A. Baker. 1994b. Wax microemulsions and emulsions as citrus coatings. *J Agric Food Chem*. 42: 899–902.

Hagenmaier, R.D., and R.A. Baker. 1995. Layered coatings to control weight loss and preserve gloss of citrus fruit. *HortScience*. 30: 296–298.

Hagenmaier, R.D., and R.A. Baker. 1996. Edible coatings from candelilla wax microemulsions. *J Food Sci*. 61: 562–565.

Hagenmaier, R.D., and R.A. Baker. 1997. Edible coatings from morpholine-free wax microemulsions. *J Agric Food Chem*. 45: 349–352.

Hagenmaier, R.D., and K. Goodner. 2002. Storage of 'Marsh' grapefruit and 'Valencia' oranges with different coatings. *Proc Fla State Hort Soc*. 115: 303–308.

Hagenmaier, R.D., and K. Grohmann. 1999. Polyvinyl acetate as a high-gloss edible coating. *J Food Sci*. 64: 1064–1067.

Hagenmaier, R.D., and P.E. Shaw. 1990. Moisture permeability of edible films made with fatty acid and hydroxypropyl methyl cellulose. *J Agric Food Chem*. 38: 1799–1803.

Hagenmaier, R.D., and P.E. Shaw. 1991a. Permeability of coatings made with emulsified polyethylene wax. *J Agric Food Chem*. 39: 1705–1708.

Hagenmaier, R.D., and P.E. Shaw. 1991b. Permeability of shellac coatings to gases and water vapor. *J Agric Food Chem*. 39: 825–829.

Hagenmaier, R.D., and P.E. Shaw. 1992. Gas permeability of fruit coating waxes. *J Amer Soc Hort Sci*. 117: 105–109.

Hagenmaier, R.D., and P.E. Shaw. 2002. Changes in volatile components of stored tangerines and other specialty citrus fruits with different coatings. *J Food Sci*. 67: 1742–1745.

Hall, D.J. 1981a. Innovations in citrus waxing—An overview. *Proc Fla State Hort Soc*. 94: 258–263.

Hall, D.J. 1981b. Innovations in citrus waxing: An overview. *Proc Fla State Hort Soc*. 94: 258–263.

Hall, D.J. 1989. Postharvest treatment of Florida fresh market tomatoes with fungicidal wax to reduce decay. *Proc Fla State Hort Soc*. 102: 365–367.

Hallman, G.J., and J.F. Foos. 1996. Coating combined with dimethoate as a quarantine treatment against fruit flies (Diptera: Tephritidae). *Postharvest Biol Technol*. 7: 177–181.

Hallman, G.J., R.G. McGuire, E.A. Baldwin, and C.A. Campbell. 1995. Mortality of feral Caribbean fruit fly (Diptera: Tephritidae) immatures in coated guavas. *J Econom Entomol*. 88: 1353–1355.

Hallman, G.J., M.O. Nisperos-Carriedo, E.A. Baldwin, and C.A. Campbell. 1994. Mortality of Caribbean fruit fly (Diptera: Tephritidae) immatures in coated fruits. J Econom Entomol. 87: 752–757.

Han, C., Y. Zhao, S.W. Leonard, and M.G. Traber. 2004. Edible coatings to improve storability and enhance nutritional value of fresh and frozen strawberries (*Fragaria ananassa*) and raspberries (*Rubus ideaus*). *Postharvest Biol Technol*. 33: 67–78.

Hardenburg, R.E. 1967. Wax and related coatings for horticultural products: A bibliography. Agriculture Research Bulletin 57-15, U.S. Department of Agriculture, Washington, DC.

Hartman, J., and F. Isenberg. 1956. Waxing vegetables. New York Agric. Exten. Ser. Bull. No. 965, pp. 3–14.

Hasegawa, Y., and Y. Iba. 1980. The effects of coating with wax on citrus fruit. Ministry of Agriculture Japan. Series B, 7, Bull Fruit Tree Research Station, pp. 85–97.

Hernández-Muñoz, P., E. Almenar, M.J. Ocio, and R. Gavara. 2006. Effect of calcium dips and chitosan coatings on postharvest life of strawberries (*Fragaria ananassa*). *Postharvest Biol Technol.* 39: 247–253.

Hernández-Muñoz, P., R. Villalobos, and A. Chiralt. 2004. Effect of thermal treatments on functional properties of edible films made from wheat gluten fractions. *Food Hydrocolloids.* 18:647–654.

Heyes, J.A., V.M. Burton, and L.A. de Vre. 1998. Cellular physiology of textural change in harvested asparagus. *Acta Hort.* 464: 455–460.

Hitz, C.W., and I.C. Haut. 1938. Effect of certain waxing treatments at time of harvest upon the subsequent storage quality of 'Grimes Golden' and 'Golden Delicious' apples. *Proc Amer Soc Hort Sci.* 36: 440–447.

Hitz, C.W., and I.C. Haut. 1942. Effects of waxing and pre-storage treatments upon prolonging the edible and storage qualities of apples. University of Maryland Agricultural Experiment Station Technical Bulletin No. A14, 44 pp.

Hoa, T.T., M.-N. Ducamp, M. Lebrun, and E.A. Baldwin. 2002. Effect of different coating treatments on the quality of mango fruit. *J Food Qual.* 25: 471–486.

Holcroft, D.M., and A.A. Kader. 1999. Controlled atmosphere-induced changes in pH and organic acid metabolism may affect color of stored strawberry fruit. *Postharvest Biol Technol.* 17: 19–32.

Holcroft, D.M., and E.J. Mitcham. 1996. Postharvest physiology and handling of litchi (*Litchi chinensis* Sonn.). *Postharvest Biol Technol.* 9: 265–281.

Hwang, Y., Y. Kim, and J. Lee. 1998. Effect of postharvest application of chitosan and wax, and ethylene scrubbing on the quality changes in stored 'Tsugaru' apples. *J Korea Soc Hort Sci.* 39: 579–582.

Jagannath, J.H., C. Nanjappa, D. Das Gupta, and A.S. Bawa. 2006. Studies on the stability of an edible film and its use for the preservation of carrot (*Daucus carota*). *Int J Food Sci Technol.* 41: 498–506.

Jiang, Y., J. Li, and W. Jiang. 2005. Effects of chitosan coating on shelf life of cold-stored litchi fruit at ambient temperature. *LWT—Food Sci Technol.* 38: 757–761.

Jiang, Y., and Y. Li. 2001. Effects of chitosan coating on postharvest life and quality of longan fruit. *Food Chem.* 73: 139–143.

Jiménez-Cuesta, M., J. Cuquerella, and M.-J. JM. 1981. Determination of a color index for citrus degreening. *Proc Int Soc Citricult.* 2: 750–753.

Jin, T., D. Sun, J.Y. Su, H. Zhang, and H.J. Sue. 2008. Antimicrobial efficacy of zinc oxide quantum dots against *Listeria monocytogenes, Salmonella, Enteritidis,* and *Escherichia coli* O157:H7. *J Food Sci.* 74: M46–M62.

Joas, J., Y. Caro, M. Ducamp, and M. Reynes. 2005. Postharvest control of pericarp browning of litchi fruit (*Litchi chinensis* Sonn cv Kwa'i Mi) by treatment with CTS and organic acids. I. Effect of pH and pericarp dehydration. *Postharvest Biol Technol.* 38: 128–136.

Ju, Z., and E.A. Curry. 2000a. Stripped corn oil controls scald and maintains volatile production potential in 'Golden Supreme' and 'Delicious' apples. *J AgricFood Chem.* 48: 2173–2177.

Ju, Z., and E.A. Curry. 2000b. Stripped corn oil emulsion alters ripening, reduces superficial scald, and reduces core flush in 'Granny Smith' apples and decay in 'D'Anjou' pears. *Postharvest Biol Technol.* 20: 185–193.

Ju, Z., Y. Duan, and Z. Ju. 2000. Plant oil emulsion modifies internal atmosphere, delays fruit ripening, and inhibits internal browning in Chinese pears. *Postharvest Biol Technol.* 20: 243–250.

Jung, D.M., J.S. de Ropp, and S.E. Ebeler. 2000. Study of interactions between food phenolics and aromatic flavors using one- and two-dimensional 1H NMR spectroscopy. *J Agric Food Chem.* 48: 407–412.

Kamper, S.L., and O.R. Fennema. 1985. Use of an edible film to maintain water vapor gradients in foods. *J Food Sci.* 50: 382–384.

Kaplan, H.J. 1986. Washing, waxing and color-adding, p. 379–395. In: W.F. Wardowski, S. Nagy, and W. Grierson (eds.). *Fresh Citrus Fruits.* AVI, New York.

Ke, D., L. Goldstein, M. O'Mahony, and A.A. Kader. 1991. Effects of short term exposure to low O_2 and high CO_2 atmospheres on quality attributes of strawberries. *J Food Sci.* 56: 50–54.

Ke, D., and A.A. Kader. 1990. Tolerance of 'Valencia' oranges to controlled atmospheres, as determined by physiological responses and quality attributes. *J Amer Soc Hort Sci.* 115: 779–783.

Kerbel, E., F.G. Mitchell, A.A. Kader, and G. Meyer. 1989. Effect of 'Semperfresh' coating on postharvest life, internal atmosphere modification and quality maintenance of 'Granny Smith' apples. *Proc. 5th Intl. Cont. Atm. Res. Conf.,* Wenatchee, WA. p. 14–16.

Kerstiens, G. 1996. Cuticular water permeability and its physiological significance. *J Exp Bot.* 47: 1813–1832.

Kester, J.J., and O.R. Fennema. 1986. Edible films and coatings: A review. *Food Technol.* 40: 47–59.

Kevers, C., M. Falkowski, J. Tabart, J.-O. Defraigne, J. Dommes, and J.L. Pincemail. 2007. Evolution of antioxidant capacity during storage of selected fruits and vegetables. *J Agric Food Chem.* 55: 8596–8603.

Khwaldia, K., C. Perez, S. Banon, S. Desobry, and J. Hardy. 2004. Milk proteins for edible films and coatings. *Crit Rev Food Sci Nutr.* 44:239–251.

Kim, H., B. Son, S. Park, and K. Lee. 1999. A study on the properties and utilization of chitosan coating. 2. Changes in the quality of tomatoes by chitosan coating. *J Korea Fish Soc.* 32: 568–572.

Kittur, F.S., N. Saroja, Habibunnisa, and R.N. Tharanathan. 2001. Polysaccharide-based composite coating formulations for shelf-life extension of fresh banana and mango. *Eur Food Res Technol.* 213: 306–311.

Kraght, A.J. 1966. Waxing peaches with the consumer in mind. *Produce Marketing.* 9: 20–21.

Krochta, J.M., and C. De Mulder-Johnston. 1997. Edible and biodegradable polymer films: Challenges and opportunities. *Food Technol.* 51: 61–74.

Kupferman, E. 1998. Postharvest applied chemicals to pears: A survey of pear packers in Washington, Oregon and California. *Tree Fruit Postharv J.* 9: 3–24.

Lau, O.L., and M. Meheriuk. 1994. The effect of edible coatings on storage quality of 'McIntosh' 'Delicious' and 'Spartan' apples. *Canadian J Plant Sci.* 74: 847–852.

Lawson, J.A. 1960. Banana packing and waxing. *West Australia Dept Agric J (Ser. 4).* 1: 41–45.

Le Tien, C.L., C. Vachon, M.-A. Mateescu, and M. Lacroix. 2001. Milk protein coating prevent oxidative browning of apples and potatoes. *J Food Sci.* 66: 512–516.

Lendzian, K.J. 1982. Gas permeability of plant cuticles. *Planta.* 155: 310–315.

Li, H., and T. Yu. 2001. Effect of chitosan on incidence of brown rot, quality and physiological attributes of postharvest peach fruit. *J Sci Food Agric.* 81: 269–274.

Link, S.O., S.R. Drake, and M.E. Thiede. 2004. Prediction of apple firmness from mass loss and shrinkage. *J Food Qual.* 27: 13–26.

Liu, L., J.F. Kerry, and J.P. Kerry. 2006. Effect of food ingredients and selected lipids on the physical properties of extruded edible films/casings. *Int J Food Sci Technol.* 41:295–302.

Liu, J., S. Tian, X. Meng, and Y. Xu. 2007. Effects of chitosan on control of postharvest diseases and physiological responses of tomato fruit. *Postharvest Biol Technol.* 44: 300–306.

Lowings, P., and D. Cutts. 1982. The preservation of fresh fruits and vegetables. *Proc Inst Food Sci Technol Ann Symp*, Nottingham, UK. p. 52.

Luchsinger, L., and C. Walsh. 1993. Changes in ethylene rate and ground color in peaches (cv. Red Haven and Marqueen) and nectarines (cv. Fantasia) during maturation and ripening. *Acta Hort.* 343: 70–72.

Mack, W.B., and J.R. Janer. 1942. Effects of waxing on certain physiological processes of cucumbers under different storage conditions. *J Food Sci.* 7: 38–47.

Maftoonazad, N., and H. Ramaswamy. 2005. Postharvest shelf-life extension of avocados using methyl cellulose-based coating. *LWT—Food Sci Technol.* 38: 617–624.

Maftoonazad, N., H.S. Ramaswamy, and M. Marcotte. 2008. Shelf-life extension of peaches through sodium alginate and methyl cellulose edible coatings. *Int J Food Sci Technol.* 43: 951–957.

Maguire, K.M., N.H. Banks, A. Lang, and I.L. Gordon. 2000. Harvest date, cultivar, orchard, and tree effects on water vapor permeance in apples. *J Amer Soc Hort Sci.* 125: 100–104.

Maguire, K.M., N.H. Banks, and L. Opara. 2001. Factors affecting weight loss of apples. *Hort Rev.* 25: 197.

Maharaj, R., and C.K. Sankat. 1990. The shelf-life of breadfruit stored under ambient and refrigerated conditions. *Acta Hort.* 269: 411–424.

Martínez-Romero, D., N. Alburquerque, J.M. Valverde, F. Guillén, S. Castillo, D. Valero, and M. Serrano. 2006. Postharvest sweet cherry quality and safety maintenance by Aloe vera treatment: A new edible coating. *Postharvest Biol Technol.* 39: 93–100.

Martin-Polo, M., C. Mauguin, and A. Voilley. 1992. Hydrophobic films and their efficiency against moisture transfer. 1. Influence of the film preparation technique. *J Agric Food Chem.* 40:407–412.

Mate, J.I., E.N. Frankel, and J.M. Krochta. 1996. Whey protein isolate edible coatings: Effect on the rancidity process of dry roasted peanuts. *J Agric Food Chem.* 44:1736–1740.

Mathur, P.B., and H.C. Srivastava. 1955. Effect of skin, coatings on the storage behaviour of mangoes. *J Food Sci.* 20: 559–566.

Mathur, P.B., and H. Subramanyam. 1956. Effect of a fungicidal wax coating on the storage behavior of mangoes. *J Sci Food Agric.* 7: 673–676.

McGuire, R.G., and E.A. Baldwin. 1994. Composition of cellulose coatings affect populations of yeasts in the liquid formulation and on coated grapefruits. *Proc Fla State Hort Soc.* 107: 293–297.

McGuire, R.G., and G.J. Hallman. 1995. Coating guavas with cellulose or carnauba-based emulsions interferes with postharvest ripening. *HortScience.* 30: 294–295.

McHugh, T.H., C.C. Huxsoll, and J.M. Krochta. 1996. Permeability properties of fruit puree edible films. *J Food Sci.* 61: 88–91.

McHugh, T. and J. Krochta. 1994. Water vapor permeability properties of edible whey protein-lipid emulsion films. *J Amer Oil Chem Soc.* 71:307–312.

McHugh, T.H., and E. Senesi. 2000. Apple wraps: A novel method to improve the quality and extend the shelf life of fresh-cut apples. *J Food Sci.* 65: 480–485.

Meheriuk, M., and O.L. Lau. 1988. Effect of two polymeric coatings on fruit quality of 'Bartlett' and 'D'Anjou' pears. *J Amer Soc Hort Sci.* 113: 222–226.

Mei, Y. and Y. Zhao. 2003. Barrier and mechanical properties of milk protein-based edible films containing nutraceuticals. *J Agric Food Chem.* 51:1914–1918.

Mei, Y., Y. Zhao, J. Yang, and H.C. Furr. 2002. Using edible coating to enhance nutritional and sensory qualities of baby carrots. *J Food Sci.* 67: 1964–1968.

Meng, X., B. Li, J. Liu, and S. Tian. 2008. Physiological responses and quality attributes of table grape fruit to chitosan preharvest spray and postharvest coating during storage. *Food Chem.* 106: 501–508.

Mercado-Silva, E., P. Benito-Bautista, and M. de los Angeles García-Velasco. 1998. Fruit development, harvest index and ripening changes of guavas produced in central Mexico. *Postharvest Biol Technol.* 13: 143–150.

Min, S., and J.M. Krochta. 2007. Edible coatings containing bioactive antimicrobial agents, p. 29–52. In: J.H. Han (ed.). *Packaging for Nonthermal Processing of Food.* Blackwell/IFT Press. Ames, IA.

Mir, N., and R.M. Beaudry. 2004. Modified atmosphere packaging. In: K. Gross, and C.Y. Wang (eds.). *The Commercial Storage of Fruits, Vegetables, and Florist and Nursery Stocks.* www.ba.ars.usda.gov/hb66/015map.pdf.

Mir, N., M. Wendorf, R. Perez, and R.M. Beaudry. 1998. Chlorophyll fluorescence in relation to superficial scald development in apple. *J Amer Soc Hort Sci.* 123: 887–892.

Mitchell, F. 1987. Preparing peaches and nectarines for export marketing. *The Orchardist of New Zealand.* 60: 150–152.

Mitchell, F.G., J.H. Larue, J.P. Gentry, and M.H. Gerdts. 1963. Packing nectarine to reduce shrivel. *Calif Agric.* 17: 10–11.

Motlagh, F.H., and P.C. Quantick. 1988. Effect of permeable coatings on the storage life of fruits. *Int J Food Sci Technol.* 23: 99–105.

Muliyar, M., and M. Marar. 1963. Studies on the keeping quality of ripe coconuts in storage. *India Coconut J.* 17: 13–18.

Muramatsu, N., K. Kiyohide, and O. Tatsushi. 1996. Relationship between texture and cell wall polysaccharides of fruit flesh in various species of citrus. *HortScience.* 31: 114–116.

Navarro-Tarazaga, M.L., M.A. Del Rio, J.M. Krochta, and M.B. Pérez-Gago. 2008a. Fatty acid effect on hydroxypropyl methylcellulose-beeswax edible film properties and postharvest quality of coated 'Ortanique' mandarins. *J Agric Food Chem.* 56: 10689–10696.

Navarro-Tarazaga, M.L., R. Sothornvit, and M.B. Pérez-Gago. 2008b. Effect of plasticizer type and amount on hydroxypropyl methylcellulose-beeswax edible film properties and postharvest quality of coated plums (Cv. Angeleno). *J Agric Food Chem.* 56: 9502–9509.

Ngarmsak, M., P. Delaquis, P. Toivonen, T. Ngarmsak, B. Ooraikul, and G. Mazza. 2006. Antimicrobial activity of vanillin against spoilage microorganisms in stored fresh-cut mangoes. *J Food Protect.* 69: 1724–1727.

Nisperos-Carriedo, M.O., and E.A. Baldwin. 1988. Effect of two types of edible films on tomato fruit ripening. *Proc Fla State Hort Soc.* 101: 217–220.

Nisperos-Carriedo, M.O., and E.A. Baldwin. 1994. Method of increasing the stability of fruits, vegetables, or fungi and composition thereof. U.S. Patent 5,376,391.

Nisperos-Carriedo, M.O., P.E. Shaw, and E.A. Baldwin. 1990. Changes in volatile flavor components of pineapple orange juice as influenced by the application of lipid and composite films. *J Agric Food Chem.* 38: 1382–1387.

No, H.K., S.P. Meyers, W. Prinyawiwatkul, and Z. Xu. 2007. Applications of chitosan for improvement of quality and shelf life of foods: A review. *J Food Sci.* 72: R87–R101.

Nunes, M.C.N., J.K. Brecht, A.M.M.B. Morais, and S.A. Sargent. 1998. Controlling temperature and water loss to maintain ascorbic acid levels in strawberries during postharvest handling. *J Food Sci.* 63: 1033–1036.

Olivas, G.I., J.E. Davila-Avina, N.A. Salas-Salazar, and F.J. Molina. 2008. Use of edible coatings to preserve the quality of fruits and vegetables during storage. *Stewart Postharvest Rev.* 4: 1–10.

Olivas, G.I., D.S. Mattinson, and G.V. Barbosa-Cánovas. 2007. Alginate coatings for preservation of minimally processed 'Gala' apples. *Postharvest Biol Technol.* 45: 89–96.

Olivas, G.I., J.J. Rodriguez, and G.V. Barbosa-Canovas. 2003. Edible coatings composed of methylcellulose, stearic acid, and additives to preserve quality of pear wedges. *J Food Process Preserv.* 27: 299–320.

OMRI. 2008. OMRI products list, web edition. http://www.omri.org/sites/default/files/opl-pdf/complete_company.pdf.

Özden, Ç., and L. Bayindirli. 2002. Effects of combinational use of controlled atmosphere, cold storage and edible coating applications on shelf life and quality attributes of green peppers. *Eur Food Res Technol.* 214: 320–326.

Pao, S., C.L. Davis, D.F. Kelsey, and P.D. Petracek. 1999. Sanitizing effects of fruit waxes at high pH and temperature on orange surfaces inoculated with *Escherichia coli*. *J Food Sci.* 64: 359–362.

Papineau, A.M., D.G. Hoover, K. Knorr, and D.F. Farkas. 1991. Antimicrobial effect of water-soluble chitosans with high hydrostatic pressure. *Food Biotechnol.* 5: 45–57.

Park, H.J. 1999. Development of advanced edible coatings for fruits. *Trend Food Sci Technol.* 10: 254–260.

Park, H.J., and M.S. Chinnan. 1995. Gas and water vapor barrier properties of edible films from protein and cellulosic materials. *J Food Eng.* 25: 497–507.

Park, H.J., M.S. Chinnan, and R.L. Shewfelt. 1994. Edible coating effects on storage life and quality of tomatoes. *J Food Sci.* 59: 568–570.

Park, S.I., B.I. Lee, S.T. Jung, and H.J. Park. 2001. Biopolymer composite films based on k-carrageenan and chitosan. *Mater Res Bull.* 36:511–519.

Paull, R., and N. Chen. 1989. Waxing and plastic wraps influence water loss from papaya fruit during storage and ripening. *J Amer Soc Hort Sci.* 114: 937–942.

Pen, L.T., and Y.M. Jiang. 2003. Effects of chitosan coating on shelf life and quality of fresh-cut Chinese water chestnut. *Lebensmittel-Wissenschaft und -Technologie.* 36: 359–364.

Pérez-Gago, M.B., C. Rojas, and M.A. Del Rio. 2002. Effect of lipid type and amount of edible hydroxypropyl composite methylcellulose-lipid composite coatings used to project postharvest quality of mandarins cv. fortune. *J Food Sci.* 67: 2903–2910.

Pérez-Gago, M.B., C. Rojas, and M.A. del Río. 2003a. Effect of hydroxypropyl methylcellulose-lipid edible composite coatings on plum (cv. Autumn giant) quality during storage. *J Food Sci.* 68: 879–883.

Pérez-Gago, M.B., M. Serra, M. Alonso, M. Mateos, and M.A. del Rio. 2003b. Effect of solid content and lipid content of whey protein isolate-beeswax edible coatings on color change of fresh-cut apples. *J Food Sci.* 68: 2186–2191.

Pérez-Gago, M.B., M. Serra, M. Alonso, M. Mateos, and M.A. del Río. 2005. Effect of whey protein- and hydroxypropyl methylcellulose-based edible composite coatings on color change of fresh-cut apples. *Postharvest Biol Technol.* 36: 77–85.

Pérez-Gago, M.B., M. Serra, and M.A. del Río. 2006. Color change of fresh-cut apples coated with whey protein concentrate-based edible coatings. *Postharvest Biol Technol.* 39: 84–92.

Perkins-Veazie, P.M., and J.K. Collins. 1991. Color changes in waxed turnips storage. *J Food Qual.* 14: 313–319.

Petracek, P.D., R.D. Hagenmaier, and H. Dou. 1999. Waxing effects on citrus fruit physiology, p. 71–92. In: M. Schirra (ed.). *Advances in Postharvest Diseases and Disorders Control of Citrus Fruit.* Research Signpost, India.

Petracek, P.D., L. Montalvo, H. Dou, and C. Davis. 1998. Postharvest pitting of 'Fallglo' tangerine. *J Amer Soc Hort Sci.* 123: 130–135.

Pieniazek, S.A. 1944. Physical characteristics of the skin in relation to apple fruit transpiration. *Plant Physiol.* 19: 529–536.

Pinotti, A., M.A. García, M.N. Martino, and N.E. Zaritzky. 2007. Study on microstructure and physical properties of composite films based on chitosan and methylcellulose. *Food Hydrocolloids.* 21:66–72.

Platenius, H. 1939. Was emulsions for vegetables, p. 43 pp. Cornell University Agriculture Experiment Station Bulletin No. 723.

Plotto, A., and B. Baker. 2004. Review of wax-based and other coatings for fruits and vegetables. *IFOAM.* 23 pp.

Potjewijd, R., M.O. Nisperos, J.K. Burns, M. Parish, and E.A. Baldwin. 1995. Cellulose-based coatings as carriers for *Candida guillermondii* and *debaryomyces* sp. in reducing decay of oranges. *HortScience.* 30: 1417–1421.

Quezada Gallo, J.A., A. Gramin, C. Pattyn, M.R. Díaz Amaro, F. Debeaufort, and A. Voilley. 2005. Biopolymers used as edible coating to limit water transfer, colour degradation and aroma compound 2-pentanone lost in Mexican fruits. *Acta Hort.* 682: 1709–1716.

Ratanachinakorn, B., W. Kumsiri, Y. Buchsapawanich, and J. Singto. 2005. Effect of chitosan on the keeping quality of pummelos. *Acta Hort.* 682:1769–1772

Ribeiro, C., A.A. Vicente, J.A. Teixeira, and C. Miranda. 2007. Optimization of edible coating composition to retard strawberry fruit senescence. *Postharvest Biol Technol.* 44: 63–70.

Rojas-Argudo, C., M.B. Perez-Gago, and M.A. del Rio. 2005. Postharvest quality of coated cherries cv. 'Burlat' as affected by coating composition and solids content. *Food Sci Technol Int.* 11: 417–424.

Rojas-Graü, M.A., R.J. Avena-Bustillos, M. Friedman, P.R. Henika, O. Martin-Belloso, and T.H. McHugh. 2006. Mechanical, barrier, and antimicrobial properties of apple puree edible films containing plant essential oils. *J Agric Food Chem.* 54: 9262–9267.

Rojas-Grau, M.A., R.M. Raybaudi-Massilia, R.C. Soliva-Fortuny, R.J. Avena-Bustillos, T.H. McHugh, and O. Martin-Belloso. 2007. Apple puree-alginate edible coating as carrier of antimicrobial agents to prolong shelf-life of fresh-cut apples. *Postharvest Biol Technol.* 45: 254–264.

Romanazzi, G., O.A. Karabulut, and J.L. Smilanick. 2007. Combination of chitosan and ethanol to control postharvest gray mold of table grapes. *Postharvest Biol Technol.* 45: 134–140.

Romanazzi, G., F. Nigro, and A. Ippolito. 2003. Short hypobaric treatments potentiate the effect of chitosan in reducing storage decay of sweet cherries. *Postharvest Biol Technol.* 29: 73–80.

Romanazzi, G., F. Nigro, A. Ippolito, D. DiVenere, and M. Salerno. 2002. Effects of pre and postharvest chitosan treatments to control storage grey mold of table grapes. *J Food Sci.* 67: 1862–1867.

Sagoo, S., R. Board, and S. Roller. 2002. Chitosan inhibits growth of spoilage micro-organisms in chilled pork products. *Food Microbiol.* 19: 175–182.

Saltveit, M.E. 2004. Respiratory metabolism. In: K. Gross, C.Y. Wang, and M. Saltveit (eds.), *The Commercial Storage of Fruits, Vegetables, and Florist and Nursery Stocks*, www.ba.ars.usda.gov/hb66/019respiration.pdf.

Sargent, S.A., T.B.S. Correa, and A.G. Soares. 1995. Application of postharvest coatings to fresh cassava roots (*Manihot esculenta*, Crantz) for reduction of vascular streaking, p. 331–338. In: L. Kushwaha, R. Serwatowski, and R. Brook (eds.). *Proc. Harv. Postharv. Technol. Fresh Fruits and Veg.*, ASAE, Washington, DC.

Satin, M. 1996. The prevention of food losses after harvesting, pp. 81–94. *Food Irradiation* (2nd ed). Technomic, Lancaster, PA.

Schappelle, N.A. 1941. A physiological study on the effects of waxing pineapples of different stages of maturity. Puerto Rico University Station Research Bulletin. 3: 30–32.

Schreiner, M., S. Huyskens-Keil, A. Krumbein, H. Prono-Widayat, and P. Lüdders. 2003. Effect of film packaging and surface coating on primary and secondary plant compounds in fruit and vegetable products. *J Food Eng.* 56: 237–240.

Shewfelt, R.L. 1993. Measuring quality and maturity, p. 99–124. In: R.L. Shewfelt and S.E. Prussia (eds.). *Post Harvest Handling: A Systems Approach*. Academic Press, New York.

Shon, J., and Z.U. Haque. 2007. Efficacy of sour whey as a shelf-life enhancer: Use in antioxidative edible coating of cut vegetables and fruit. *J Food Qual.* 30: 581–593.

Siade, G., and E. Pedraza. 1977. Extension of storage life of banana (Giant Cavendish) using natural wax canelilla. *Acta Hort.* 62: 327–335.

Singh, K. 1971. Storage behaviour of sweet oranges and mandarins. *Tech Bull ICAR.* 35:106.

Smith, S.M., and J.R. Stow. 1984. The potential of a sucrose ester coating material for improving the storage and shelf-life qualities of 'Cox's Orange Pippin' apples. *Ann Appl Biol.* 104: 383–391.

Smock, R.M. 1935. Certain effects of wax treatments on various varieties of apples and pears. *Proc Amer Soc Hort Sci.* 33: 284–289.

Sothornvit, R., and N. Pitak. 2007. Oxygen permeability and mechanical properties of banana films. *Food Res Int.* 40: 365–370.

Sothornvit, R., and P. Rodsamran. 2008. Effect of a mango film on quality of whole and minimally processed mangoes. *Postharvest Biol Technol.* 47: 407–415.

Srivastava, H.C., N.S. Kapur, V.B. Dalal, H. Subramanyam, S. D'Souza, and K.S. Rao. 1962. Storage behavior of skin coated guavas (*Psidium guajava*) under modified atmosphere. *Food Sci.* 11: 284–289.

Steele, N.M., M.C. McCann, and K. Roberts. 1997. Pectin modification in cell walls of ripening tomatoes occurs in distinct domains. *Plant Physiol.* 114: 373–381.

Subramanyam, H., and P.B. Mathur. 1956. Effect of a fungicidal wax coating on the storage behaviour of tapioca roots. *Bull Cent Food Technol Res Inst.* Mysore. 5: 110–111.

Sümnü, G., and L. Bayindirli. 1995a. Effects of coatings on fruit quality of 'Amasya' apples. *Lebensmittel-Wissenschaft und -Technologie.* 28: 501–505.

Sümnü, G., and L. Bayindirli. 1995b. Effects of sucrose polyester coating on fruit quality of apricots (*Prunus armeniaca* (L)). *J Sci Food Agric.* 67: 537–540.

Tanada-Palmu, P.S., and C.R.F. Grosso. 2005. Effect of edible wheat gluten-based films and coatings on refrigerated strawberry (*Fragaria ananassa*) quality. *Postharvest Biol Technol.* 36: 199–208.

Tapia, M.S., M.A. Rojas-Grau, F.J. Rodriguez, A. Carmona, and O. Martin-Belloso. 2007. Alginate and gellan-based edible films for probiotic coatings on fresh-cut fruits. *J Food Sci.* 72: 190–196.

Tasdelen, Ö., and L. Bayindirli. 1998. Controlled atmosphere and edible coating effects on storage life and quality of tomatoes. *J Food Process Preserv.* 22: 303–320.

Turrell, F.M., and L.J. Klotz. 1940. Density of stomata and oil glands and incidence of water spot in the rind of Washington navel orange. *Bot Gaz.* 101: 862–870.

U.S. Environmental Protection Agency. 2006. Drinking water standards and health advisories.

Ueda, Y., and J. Bai. 1993. Effect of short term exposure of elevated CO_2 on flesh firmness and ester production of strawberry. *J Japan Soc Hort Sci.* 62: 457–464.

Valverde, J.M., D. Valero, D. Martinez-Romero, F. Guillen, S. Castillo, and M. Serrano. 2005. Novel edible coating based on Aloe vera gel to maintain table grape quality and safety. *J Agric Food Chem.* 53: 7807–7813.

Vargas, M., A. Albors, A. Chiralt, and C. González-Martínez. 2006. Quality of cold-stored strawberries as affected by chitosan-oleic acid edible coatings. *Postharvest Biol Technol.* 41: 164–171.

Vargas, M., A. Chiralt, A. Albors, and C. González-Martínez. 2009. Effect of chitosan-based edible coatings applied by vacuum impregnation on quality preservation of fresh-cut carrot. *Postharvest Biol Technol.* 51: 263–271.

Vargas, M., C. Pastor, A. Chiralt, D.J. McClements, and C. Gonzalez-Martinez. 2008. Recent advances in edible coatings for fresh and minimally processed fruits. *Crit Rev Food Sci Nutr.* 48: 496–511.

Vines, H.M., W. Grierson, and G.J. Edwards. 1968. Respiration, internal atmosphere, and ethylene evolution of citrus fruit. *J Amer Soc Hort Sci.* 92: 227–234.

Vojdani, F., and J.A. Torres. 1990. Potassium sorbate permeability of methylcellulose and hydroxypropyl methylcellulose coatings: Effect of fatty acids. *J Food Sci.* 55: 841–846.

Wakabayashi, K. 2000. Changes in cell wall polysaccharides during fruit ripening. *Journal of Plant Research.* 113: 231–237.

Waks, J., M. Schiffmann-Nadel, E. Lomaniec, and E. Chalutz. 1985. Relation between fruit waxing and development of rots in citrus fruit during storage. *Plant Dis.* 69: 869–870.

Waldron, K.W., M.L. Parker, and A.C. Smith. 2003. Plant cell walls and food quality. *Compr Rev Food Sci and Food Safety.* 4: 101–119.

Waldron, K.W., and R.R. Selvendran. 1992. Cell wall changes in immature asparagus stem tissue after excision. *Phytochemistry.* 31: 1931–1940.

Weiss, J., P. Takhistov, and D.J. McClements. 2006. Functional materials in food nanotechnology. *J Food Sci.* 71: R107–R116.

Worrell, D.B., C.M. Sean Carrington, and D.J. Huber. 2002. The use of low temperature and coatings to maintain storage quality of breadfruit, *Artocarpus altilis* (Parks.) Fosb. *Postharvest Biol Technol.* 25: 33–40.

Wu, M.T., and D.K. Salunkhe. 1972. Control of chlorophyll and solanine synthesis and sprouting of potato tubers by hot paraffin wax. *J Food Sci.* 37: 629–630.

Xu, S., X. Chen, and D. Sun. 2001. Preservation of kiwifruit coated with an edible film at ambient temperature. *J Food Eng.* 50: 211–216.

Xu, W.-T., K.-L. Huang, F. Guo, W. Qu, J.-J. Yang, Z.-H. Liang, and Y.-B. Luo. 2007. Postharvest grapefruit seed extract and chitosan treatments of table grapes to control *Botrytis cinerea*. *Postharvest Biol Technol.* 46: 86–94.

Yaman, Ö., and L. Bayoindirli. 2002. Effects of an edible coating and cold storage on shelf-life and quality of cherries. *Lebensmittel-Wissenschaft und -Technologie.* 35: 146–150.

Yamauchi, N., K. Eguchi, Y. Tokuhara, Y. Yamashita, A. Oshima, M. Shigyo, and K. Sugimoto. 2003. Control of degreening by sucrose laurate ester and heat treatment of citrus nagato-yuzukichi fruit. *J Hort Sci Biotechnol.* 78: 563–567.

Yamauchi, N., Y. Tokuhara, Y. Ohyama, and M. Shigyo. 2008. Inhibitory effect of sucrose laurate ester on degreening in citrus nagato-yuzukichi fruit during storage. *Postharvest Biol Technol.* 47: 333–337.

Young, D.H., H. Kohle, and H. Kauss. 1982. Effect of chitosan on membrane permeability of suspension-cultured glycine max and phaseolus vulgaris cells. *Plant Physiol.* 70: 1449–1454.

Zapata, L.E. 1978. Preservation of fresh cassava, Proceedings of the Third Advanced Seminar of Bogota Food Technology. Bogata, D.E. Colombia.

Zapata, P.J., F. Guillén, D. Martínez-Romero, S. Castillo, D. Valero, and M. Serrano. 2008. Use of alginate or zein as edible coatings to delay postharvest ripening process and to maintain tomato (*Solanum lycopersicon* Mill) quality. *J Sci Food Agric.* 88: 1287–1293.

Zhang, D., and P.C. Quantick. 1997. Effects of chitosan coating on enzymatic browning and decay during postharvest storage of litchi (*Litchi chinensis* Sonn.) fruit. *Postharvest Biol Technol.* 12: 195–202.

Zhang, Z., X. Pang, D. Xuewu, Z. Ji, and Y. Jiang. 2005. Role of peroxidase in anthocyanin degradation in litchi fruit pericarp. *Food Chem.* 90: 47–52.

Zhuang, R.Y., and Y.W. Huang. 2003. Influence of HPMC edible coating on fresh-keeping and storability of tomato. *J Zhejiang Univ Sci.* 4: 109–113.

8 Coatings for minimally processed fruits and vegetables

Sharon Dea, Christian Ghidelli, Maria B. Pérez-Gago, Anne Plotto

Contents

8.1	Introduction	244
8.2	Physiology of minimally processed fruit and vegetables	244
	8.2.1 Ethylene production as a response to stress	245
	8.2.2 Increased respiration as a response to stress	246
	8.2.3 Enzyme activity	246
	8.2.4 Nutritional and flavor changes	247
	8.2.5 Color changes/browning	249
	8.2.6 Texture and softening	250
	8.2.7 Surface microbial flora	251
8.3	Techniques and approach to increase shelf life of minimally processed products	252
	8.3.1 Quality of the raw product	253
	8.3.2 Sanitation treatments prior processing	253
	8.3.3 Processing practices	254
	8.3.4 Postprocessing treatments	255
	8.3.4.1 Temperature	255
	8.3.4.2 Chemical treatments	255
	8.3.4.3 Physical treatments	256
8.4	Edible coatings for minimally processed products	258
	8.4.1 Films versus emulsion coatings	259
	8.4.2 Polysaccharides	263
	8.4.2.1 Chitosan	263
	8.4.2.2 Alginates	264
	8.4.2.3 Gellan	265
	8.4.2.4 Carrageenan	265
	8.4.2.5 Xanthan gum	265
	8.4.2.6 Cellulose and cellulose derivatives	266
	8.4.2.7 Pectin	267
	8.4.2.8 Starch and starch derivatives	267
	8.4.3 Proteins	268
	8.4.3.1 Whey protein and casein	268
	8.4.3.2 Soy protein	269

8.4.4	Lipids	270
8.4.5	Composite and bilayer coatings	271
8.4.6	Coating optimization	272
8.4.7	Future needs and trends of minimally processed fruits and vegetables	273
References		276

8.1 Introduction

Before dealing with coatings, this chapter first addresses the basic physiology and quality issues of fresh-cut fruit and vegetables, the better to appreciate the coatings devised to improve quality of these products.

First available to restaurants, hospitals, and other food service operators, minimally processed fruits and vegetables, also known as fresh-cut fruit and vegetables, are now readily available in supermarkets and sold in numerous package sizes, including individual (i.e., single-serving) portions. In the United States, the fresh-cut industry is the fastest growing sector in the fresh produce business (Premier et al., 2007).

Minimal processing is defined to include all operations such as washing, sorting, trimming, peeling, slicing, dicing, chopping, and shredding that would not extensively affect the fresh-like quality of the produce (Shewfelt, 1987). The product remains biologically and physiologically active, in that the tissues are living and respiring, with a shifting of cellular processes and interactions in response to the tissue damage inflicted by the operations. Therefore, minimal processing increases the degree of perishability of the processed materials, thus challenging their marketability (Gorny et al., 1999).

Quality of fresh-cut items is determined by a consistent and fresh appearance, acceptable texture, characteristic flavor, and sufficient shelf life to survive the distribution system. Those characteristics can be preserved to a certain extent by understanding the physiology of the fresh-cut fruit or vegetable and applying different techniques. Such techniques would be to choose the proper cultivar and ripeness stage, to use adequate sanitation procedures and processing techniques, to maintain low temperature from harvest to retail (cold chain), and to apply physical and chemical treatments such as modified atmosphere packaging (MAP) or dips and edible coatings.

8.2 Physiology of minimally processed fruit and vegetables

Fresh-cut products are wounded tissues that are prone to rapid deterioration; therefore, their physiology differs from that of intact fruits and vegetables (Brecht, 1995; Toivonen and Brummell, 2008). Within seconds after removal of the protective skin (pericarp) or cutting the product into pieces, wound signals of different natures (i.e., electrical, chemical, and hormonal) are sent through the tissues and initiate defense responses that promote wound healing, guard against bacterial attack, and generally protect cells from further stress. Additionally, as a result of peeling and cutting, the subcellular compartmentalization at the cut surface is disrupted, mixing substrates and enzymes, and initiates reactions that do not normally occur in the intact fruit or vegetable. These physiological changes cause increased respiration rate and synthesis of wound-induced ethylene as well as increased ion leakage, loss of components, alteration in flux potential, and loss of turgor. These changes are also accompanied by loss of firmness and flavor, discoloration of cut surfaces, possible decrease in vitamins, and increase in water activity (a_w) at the cut surface which accelerates water loss (Beaulieu and Gorny, 2004; González-Aguilar et al., 2007b). In addition, cut fruit surfaces provide a favorable environment for microbial growth, due to increased moisture, sugars, and analytes leaking from open cells.

8.2.1 Ethylene production as a response to stress

One of the most common responses to wounding in plant tissue is an increase in both respiration rate and ethylene production (Saltveit, 1997; Escalona et al., 2003). Wound-induced ethylene is caused by the increased formation of 1-aminocyclopropane-1-carboxylic acid (ACC) and the subsequent conversion of ACC to ethylene (Boller and Kende, 1980; Yu and Yang, 1980). An increased rate of wound-induced ethylene may cause physiological disorders and consequently affect the quality attributes of the product. Physiological changes induced by elevated ethylene concentration include increased cell permeability, loss of compartmentation, increased senescence and respiratory activity, and increased activity of enzymes (Hyodo et al., 1983). Some specific examples include accumulation of phenolic compounds in carrots, sprouting in potatoes, lignification in asparagus, brown spot (russet spotting) in lettuce, and general softening (Reid, 1992).

The increased rate of ethylene production, in response to cutting, have been shown in kiwifruit (Watada et al., 1990; Agar et al., 1999), apple (Hu et al., 2007b), papaya (Paull and Chen, 1997), strawberry (Rosen and Kader, 1989), endive (Salman et al., 2009), tomato (Lee et al., 1970; Mencarelli et al., 1989; Abeles et al., 1992; Brecht, 1995; Artés et al., 1999), and squash (Abeles et al., 1992; Hu et al., 2007a). However, Gorny et al. (2000) reported no change in ethylene production following the slicing of pears. Also, preparation of fresh-cut mango cubes resulted in ethylene production rates 1.5 times lower than that of the whole fruit (Chatanawarangoon, 2000). In fact, it was found that the peel was the major contributor to ethylene production for mango.

When ethylene is induced by wounding, the increased rate varies depending on the type of commodity, cultivar, ripeness stage, and storage temperature. The rate of ethylene production stimulated by stress typically occurs with an initial short lag period, followed by a progressive increase, reaching a peak before subsiding to a stable level. Hyodo et al. (1983) reported that cut winter squash showed an increased rate of ethylene production after a 3-hour lag, and the rate reached its peak in 30 hours, followed by a decline to a low level, 40 hours after incubation at 24°C. Although the rate increase is usually in the range of 5- to 20-fold, more than 100-fold increase in response to wounding has been reported in some cases (Hyodo and Nishino, 1981; Hoffman and Yang, 1982; Hyodo et al., 1983). Moreover, the wound response is usually greater in fruit at the preclimacteric and climacteric stages than in postclimacteric fruit (Toivonen and DeEll, 2002).

Storage temperature has a significant effect on wound-induced ethylene production as well. It has been shown that storage of cantaloupe pieces at 0 to 2.5°C will almost completely suppress wound-induced ethylene as compared to higher storage temperatures (Toivonen and DeEll, 2002). Artés-Hernández et al. (2007) observed, in fresh-cut lemons, that the production rate of ethylene at 0°C was four times higher than that of whole fruit, while at 10°C, the production was up to 10 times higher.

8.2.2 Increased respiration as a response to stress

The effects of processing on tissue metabolism can be observed very rapidly, often within minutes to a few hours after cutting (Toivonen and Brummell, 2008). The initiation of respiration in response to wounding is delayed compared to that found for wound-induced ethylene (Brecht, 1995). In general, high respiration rates, measured by the CO_2 produced in fresh-cut products, are directly associated with a rapid increase in the tissue metabolism and, consequently, with accelerated loss of acids, sugars, and other components that determine flavor quality and nutritive value (Cantwell and Suslow, 2002). Biochemical changes triggered by cellular response include stimulation of degradation of carbohydrates, activation of glycolysis and the pentose phosphate pathway, activation in mitochondrial activity, and increase in the synthesis of proteins and enzyme activities (Uritani and Asahi, 1980). These activations and accelerations in cellular processes serve to provide energy and precursors for the biosynthesis of secondary metabolites that are important to wound healing. Wound-induced respiration has been associated with enhanced synthesis of enzymes involved in the respiratory pathway and to a transitory increase in aerobic respiration in fresh-cut carrots (Surjadinata and Cisneros-Zevallos, 2003).

Increase of respiration has been observed in many fresh-cut products including banana (Palmer and McGlasson, 1969), tomato (MacLeod et al., 1976), cantaloupe (McGlasson and Pratt, 1964), celery (Gómez and Artés, 2005), salad mixture (lettuce, celery, carrot, radish, onion, endive) (Priepke et al., 1976), pear, strawberry (Rosen and Kader, 1989), and lettuce (Cantwell and Suslow, 2002). In these studies, the typical increase in respiration is in the range of 2- to 10-fold. However, Watada et al. (1996) reported that for some fresh-cut products, such as zucchini, muskmelons, honeydews, and crenshaws, the respiration rate was found similar or lower than that of the intact product at 0, 5, and 10°C, but dramatically higher at 20°C, probably due to rapid physiological deterioration and microbial growth.

8.2.3 Enzyme activity

Mechanical injury from cutting of a fruit or vegetable destroys the integrity of cell tissues and the compartmentation of endogenous enzymes and substrates. Some of these enzyme-catalyzed changes are well known: enzymatic browning and cell wall degradation due to enzymatic hydrolysis of the cell wall pectin substances.

The amount of phenolic compounds is increased through wound induction of phenylalanine ammonia lyase (PAL), the committed enzyme in phenolic biosynthesis; these phenolics can be oxidized by polyphenol oxidase (PPO) and peroxidase (POD) to quinones that ultimately polymerize to produce the browning appearance common to wounded lettuce (Degl'Innocenti et al., 2005). Such increase in enzymatic activity as a result of the fresh-cutting process has been reported for fresh-cut potato strips (Cantos et al., 2002), broccoli florets (Gong and Mattheis, 2003), jicama cylinders (Aquino-Bolaños et al., 2000), carrots (Goldberg et al., 1985), and lettuce segments (Hisaminato et al., 2001; Murata et al., 2004).

Generally, the activity of lipolytic enzymes, including phospholipase and lipoxygenase (LOX), increases during senescence, and they have been implicated as one of the major causes of tissue breakdown in a number of vegetables (Galliard et al., 1976; Wardale and Galliard, 1977). The phospholipase activity releases unsaturated fatty acids from the membrane that can serve as substrates for LOX. In response to physical wounding, LOX may act positively, through its role in the production of defense-related signaling molecules, or negatively, through participation in autocatalytic peroxidation reactions. LOX hydroperoxides can contribute to tissue damage through inactivation of protein synthesis and deterioration of cellular membranes (Karakurt and Huber, 2003). LOX catalyzes formation of fatty acid radicals that can react with cell components, leading to further breakdown. Bleaching of plant pigments, loss of β-carotene and chlorophyll a, has been shown to occur during LOX-mediated reactions (Schieberle et al., 1981; Klein et al., 1984).

In addition to the above examples, numerous less visible reactions occur unnoticed, but can cause drastic changes to the flavor, texture, and palatability of the product. Enzyme reactions initiate and catalyze biological pathways that may otherwise be inactive, generating undesirable products. These reactions frequently catalyze and activate further enzyme reactions, resulting in a cascade of events (Schwimmer, 1983).

Stress due to wounding or infection often induces plant tissues to accumulate unusual metabolites, such as glycoalkaloids in damaged potato tubers, and polyphenolic compounds in injured sweet potatoes (Haard and Cody, 1978; Grisebach, 1987). Moreover, chlorogenic acid, isochlorogenic acid, caffeic acid, and methyl caffeate are among the common products formed via the phenylpropanoid pathway. Increased production of polyphenols due to wounding was also accompanied by lignin synthesis. Moreover, suberization, common in injured tomato, potato, and bean pod, involves the synthesis of suberin polymers consisting of ω-hydroxy and dicarboxylic acids of long chain fatty acids and alcohols as major components (Dean and Kolattukudy, 1976; Sukumaran et al., 1990).

8.2.4 Nutritional and flavor changes

The initial nutritional value of a fresh-cut product can only be as good as its whole counterpart. Thus, any preharvest or postharvest event that affects the quality of the whole fruit or vegetable can jeopardize the final flavor and nutritional value of the fresh-cut product.

Antioxidants, such as ascorbic acid (AA), lycopene, β-carotene, and phenolics, are of great interest regarding the nutritional content of fruit and vegetables. If included in the diet, these compounds are known to prevent oxidation caused by reactive oxygen species that lead to damage the cells and DNA, and cause some degenerative diseases (Hu and Jiang, 2007).

The compound AA is a key marker for determining the extent of oxidation in fresh-cut fruits and vegetables and is easily oxidized during fresh-cut processing. Changes in AA content in fresh-cut produce are a result of both biosynthesis and degradation reactions during storage (Hu and Jiang, 2007). Increased ascorbate content with time was observed for fresh-cut slices of mangoes kept at 5 and 13°C, even though the levels never reached that of whole fruit (Tovar et al., 2001). Gil et al. (2006) reported an increase in AA content in fresh-cut slices and whole strawberries stored at 5°C, while a loss of AA was observed in pineapple pieces, kiwifruit slices, and in cantaloupe cubes. Increases in AA were attributed to de-novo synthesis from monosaccharides (e.g., D-glucose) (Tolbert and Ward, 1982; Loewus and Loewus, 1987; Liao and Seib, 1988). One study demonstrated initiation of AA-synthesis enzymes upon potato tuber injury, suggesting that the same metabolic process may also occur in other wounded tissue (Ôba et al., 1994). Furthermore, considering that the AA concentration is often reported on a fresh weight basis, an increase in AA may be a result of water loss during storage rather than an actual increase in AA (Nunes et al., 1998).

Carotenoid degradation is accelerated with exposure to unfavorable conditions such as low pH, oxygen, or light exposure (Wright and Kader, 1997a), conditions created during fresh-cut processing. Ethylene production induced by wounding hastens tissue senescence, including fatty acid oxidation by LOX, which in turn contributes to carotenoid co-oxidation (Wright and Kader, 1997a, 1997b; Brecht et al., 2004). A reduction of 25% of the initial total carotenoid content of fresh-cut mango was observed 9 days after slicing and storage at 5°C (Gil et al., 2006). Odriozola-Serrano et al. (2008) reported that four of six cultivars of fresh-cut tomatoes retained their initial lycopene content for a period of 21 days at 4°C. For the remaining two cultivars, one kept the initial lycopene content for up to 14 days to then decrease for the rest of the storage period, while the other had its lycopene content depleted slightly and continuously throughout storage.

Reyes et al. (2007) demonstrated that the increase in antioxidant capacity after wounding depends on the type of fruit or vegetable tissue. They measured changes in antioxidant capacity, total soluble phenolics, AA, total carotenoids, and total anthocyanins after wounding in zucchini, white and red cabbage, iceberg lettuce, celery, carrot, parsnips, red radish, sweet potato, and potato. The phenolic changes ranged from 26% decrease to an increase up to 191%, while antioxidant capacity changes ranged from a 51% decrease to an increase up to 442%. Reduced ascorbic acid decreased up to 82%, whereas the changes in anthocyanins and carotenoids were less evident.

Nevertheless, even if some nutritional loss is expected during the shelf life of a fresh-cut product, it has been shown that for several fresh-cut fruits (i.e., strawberry, persimmon, peach, papaya, mango, strawberry, pineapple, kiwifruit, cantaloupe, and watermelon), the visual quality was appreciably reduced before any significant nutrient decrease has occurred (Wright and Kader, 1997a, 1997b; Lamikanra and Richard, 2002; Rivera-López et al., 2005; Gil et al., 2006).

Cut fruit products rapidly lose their typical flavor, even when stored under refrigerated conditions. It is well known that they can develop staleness or loss of freshness within a day of refrigerated storage (Lamikanra and Richard, 2002). Moreover, physical stress that inevitably occurs during fresh-cut processing results in enzymes coming in contact with substrates, and contributes to changes in flavor. Such changes are mainly due to the loss of the principal flavor-related volatiles and the synthesis of stress-related off-flavor volatiles such as ethanol (Lamikanra et al., 2002; Hodges and Toivonen, 2008).

During storage of fresh-cut cantaloupe, the breakdown of esters is an early and important reaction step that, by providing precursors for the synthesis of secondary aroma volatile compounds, leads to loss of freshness (Lamikanra et al., 2002; Lamikanra and Richard, 2002; Beaulieu, 2005). Sothornvit and Rodsamran (2008) showed that longer storage time and higher temperature significantly damaged fresh-cut mango flavor by favoring the development of off-flavor associated with fermentative metabolites such as ethanol and acetaldehyde, which sensory panelists identified as the main attribute affecting flavor.

In general, biochemical parameters associated with sugars and acids, such as pH, titratable acidity, soluble solids content, and organic and amino acids are important indicators of the overall flavor of fruit and vegetables. However, for fresh-cuts, these parameters are not recommended to be used as quality indicators because they are not significantly affected by storage (Lamikanra and Richard, 2002). For example, the pH, titratable acidity, Brix, organic acids, sugars, and amino acids measured in cut cantaloupe after 2 weeks of storage at 4°C were not significantly different from the amounts present in the freshly cut fruit (Lamikanra et al., 2000). Similar observations have also been reported for fresh-cut mango (Tovar et al., 2001; Donadon et al., 2004; Ngarmsak et al., 2005; Gil et al., 2006) and lemons (Artés-Hernández et al., 2007).

Taste aroma quality are important attributes for consumers and therefore should be carefully evaluated when determining the shelf life of fresh-cut products, because an acceptable postcutting visual appraisal does not necessarily imply that a product has satisfactory flavor quality (Beaulieu and Gorny, 2004). It is difficult to establish overall shelf-life limits for fresh-cut fruit, taking flavor quality into consideration, because of the effects of initial fruit variability, the different postcutting treatments and effect of packaging (Beaulieu and Gorny, 2004). Moreover, the optimum visual appearance of fresh-cut fruits accepted by the retailers and the consumers is often attained when the fruit are processed immature or unripe, thus compromising taste and aroma (Beaulieu and Gorny, 2004).

8.2.5 Color changes/browning

The visual quality of fresh-cut fruits and vegetables can be assessed using several attributes, including overall appearance, absence of defects, shape and size, glossiness, and most importantly, color. In fact, appearance is a primary quality attribute that greatly influences the consumer purchase decision. From initial maturity to wound-related effects and microbial colonization, many factors have major effects on the appearance of fresh-cut products (Beaulieu and Gorny, 2004; Toivonen and Brummell, 2008).

Enzymatic browning is one of the most limiting factors on the shelf life of fresh-cut products, and consequences of enzymatic browning are not restricted to discoloration; undesirable flavors and nutrients loss may result. Enzymatic browning is a complex process that can be subdivided in two parts. The first part is mediated by PPO, resulting in the formation of o-quinones (slightly colored), which through nonenzymatic reactions, lead to the formation of complex brown pigments. o-Quinones are highly reactive and can rapidly undergo oxidation and polymerization. Usually brown pigments are formed, but, reddish-brown, blue-gray, and even black discolorations can be synthesized (Garcia and Barrett, 2002).

In green vegetables, the senescence process usually leads to a yellow coloration of the tissues, normally considered the major consequence of chlorophyll degradation (Toivonen and Brummell, 2008). Also, in some minimally processed green vegetables, the synthesis of pheophytin, an olive-colored pigment, appears when the chlorophyll loses its bond with the magnesium atom and substitutes it with a hydrogen atom (Artés et al., 2007). The maintenance of a low temperature and a high relative humidity, combined with atmospheres lowered in O_2 and moderately rich in CO_2, are shown to be the main advisable techniques to delay this disorder (Artés et al., 2007).

Carrots may develop "white blush," also known as "white bloom," a discoloration defect that results in the formation of a white layer of material on the surface of peeled carrots, giving poor appearance to the product (Garcia and Barrett, 2002). This superficial whitish layer has been associated with the synthesis of lignin (Lavelli et al., 2006), natural healing of the tissue, although it has also been related to dehydration of cells that are damaged or removed from the tissue (Tatsumi et al., 1993; Avena-Bustillos et al., 1994).

Besides, browning on the leaf edge of nonphotosynthetic tissues has been associated with CO_2 injury when present in concentrations higher than 2 kPa during cold storage (Artés et al., 2007). The "russet spotting" characterized by the presence of pink to brown stains in the mid-rib of the leaf is quite frequent and has been related to an ethylene concentration levels higher to 0.1 ppm during cold storage (Artés et al., 2007). Onions, garlic, and leeks can develop pink, red, green, blue-green, or blue discolorations as a consequence of cell disruption (Toivonen and Brummell, 2008).

8.2.6 Texture and softening

Wounding hastens senescence and induces tissue softening, which is considered a major shelf-life limitation for fresh-cut produce (Soliva-Fortuny and Martin-Belloso, 2003; Beaulieu and Gorny, 2004). Many of the textural changes occurring on fresh-cut fruit are a continuation of the normal ripening events that lead to softening (Toivonen and Brummell, 2008). In whole fruit, cell walls undergo a natural degradation during fruit ripening, reducing cell wall firmness and intercellular adhesion. Softening is attributed to changes in turgor pressure and in the structure and composition of cell walls, such as disassembly of the pectin matrix, mediated at least in part by the sequential action of pectin methyl esterase (PME) and polygalacturonase (PG) enzymes (Beaulieu and Gorny, 2004; Pinheiro and Almeida, 2008). In addition, softening may be attributable to the accumulation of osmotic solutes in the intercellular space and partly to postharvest water loss from ripening fruit (Toivonen and Brummell, 2008). In climacteric fruit, wound-induced ethylene would have the same effect as treating tissue with exogenous ethylene, causing a hastening of ripening and softening.

Compared with fruit, vegetables generally have a much greater proportion of cells with thickened secondary walls and, consequently, are much firmer and less susceptible to softening. The loss of textural quality is related to aging processes and senescence, water loss, reduced turgor, and wounding effects, including the leakage of osmotic solutes. Wilting is the major cause of loss of visual appearance and texture in delicate leafy produce such as lettuce and spinach.

Flesh translucency, characterized by the alteration of flesh texture to become dark and glassy, or an overripe appearance, seriously limit the use of fruit by the fresh-cut fruit industries (Lana et al., 2006). Translucency is caused by the filling of cellular free space with liquid, giving to the tissue a transparent appearance. Fresh-cut cucumber melon, papaya, pears, tomatoes, and watermelon are susceptible to this disorder (Artés et al., 2007).

8.2.7 Surface microbial flora

Fresh-cut produce are very susceptible to microbial spoilage due to their high water activity (a_w), the presence of nutrients at the cut surface, and the absence of preservative processes known to delay undesirable biological and biochemical changes, such as bleaching, freezing, or sterilization.

Raw fruit and vegetables have a naturally occurring microflora that is affected by several external factors, namely, product origin, agricultural production practices, harvesting and processing techniques, initial quality and maturity, transportation mode, storage temperature, and the use of controlled (CA) and modified (MA) atmosphere (Ngarmsak et al., 2006). The microbial load can further be enhanced by the different processing methods, such as handling, cutting, shredding, slicing, and grating (Abadias et al., 2008). In addition, during distribution and storage, temperature fluctuations and the high humidity present in packages provide a favorable environment and incubation time for proliferation of spoilage organisms and microorganisms of public health significance (Heard, 2002; Fan and Song, 2008).

The major microbial concerns related to fresh-cut produce are mesophilic and psychrotrophic microorganisms affecting product shelf life, and human pathogens. Most microbes on fresh fruits and vegetables are bacteria, and 80% to 90% of bacteria are Gram-negative rods, predominantly *Pseudomonas*, *Enterobacter*, or *Erwinia* species (Zhang, 2007). Lactic acid bacteria such as *Leuconostoc mesenteroides* and *Lactobacilus* spp., and several species of yeast and molds are also commonly found (Brackett, 1987; Fan and Song, 2008). Those microorganisms may degrade the sensory quality by affecting the appearance, cause off-odor/off-flavor, and to a lesser extent, may cause texture loss.

The detection of off-odors or obvious visual defects on fresh-cut vegetables is often accompanied by a bacterial count exceeding 8 log colony forming units (cfu)/g or a yeast count exceeding 5 log cfu/g (Ragaert et al., 2007). For instance, unacceptable changes of appearance during storage of minimally processed artichoke at 4°C appeared at day 15, when the psychrotrophic microbial count reached 8.8 log cfu/g (Giménez et al., 2003). Li et al. (2001) also found that the visual quality of minimally processed iceberg lettuce became unacceptable at day 14 during storage at 5°C, which corresponded to aerobic psychrotrophic and yeast counts of 8.8 and 6.4 log cfu/g, respectively.

Although there have been a number of reports about microbiological contamination involving whole fresh produce and fresh-cut vegetables (Abadias et al., 2008), there is still little information about microbial contamination of fresh-cut fruits. For most fruit, due to their low pH, the natural microflora is restricted to acid-tolerant microorganisms, such as fungi and lactic acid bacteria. Fruits may also be a vehicle for non-acid-tolerant microorganisms, although these may not grow (Brackett, 1987). Beaulieu and Gorny (2004) indicated that aerobic plate count, total plate count, and, more significantly, yeast and mold count correlated closely with the shelf life of fresh-cut fruits.

In fresh-cut fruit products, microorganism minimum detection levels based on visual observation of spoilage may vary depending on the microorganism and type of product. For example, in mango cubes, mesophilic and psychrotrophic aerobic and lactic bacterial counts detection level was reached at 2.4 log CFU/g, while for yeast the detection level was 3 log CFU/g (Poubol and Izumi, 2005a, 2005b). For fresh-cut melon, spoilage became detectable by consumers when yeast counts reached a level above 5 log CFU/g and aerobic psychrophilic counts reached 8 log (CFU/g) (Oms-Oliu et al., 2008a). In melon cubes, the cause of off-odor was associated with yeast and mold counts above 7 log CFU/g (Bai et al., 2003).

The incidence of food-borne outbreaks caused by contaminated fresh fruit and vegetables has increased in recent years. The pathogens most frequently associated with produce-related outbreaks include bacteria (*Salmonella*, *Escherichia coli*), viruses (Norwalk-like, hepatitis A), and parasites (*Cryotosporidium*, *Cyclospora*), with *Salmonella* and *E. coli* O157:H7 being the leading causes of produce-related outbreaks in the United States (Abadias et al., 2008). *Listeria monocytogenes* is also considered a potential vehicle of food-borne outbreaks caused by the consumption of contaminated minimally processed fresh vegetables (Ryser and Marth, 1991). Previous works have shown that *L. monocytogenes* can grow or survive at refrigeration temperatures on many raw or processed vegetables, such as cabbage and shredded cabbage (Beuchat et al., 1986; Kallander et al., 1991), iceberg lettuce (Steinbruegge et al., 1988; Beuchat and Brackett, 1990), asparagus, broccoli, and cauliflower (Berrang et al., 1989).

Moreover, while the acidic pH of most fruits prevents the development of most pathogens, they are not totally without risk (Brackett, 1987). Human pathogens may gain entry to fruit products when animal fertilizers or contaminated irrigation water or water used for rinsing come into contact with fruits during production and processing (Brackett, 1987; Bordini et al., 2007).

8.3 Techniques and approach to increase shelf life of minimally processed products

Cultural practices, harvest maturity, postharvest handling, ripeness stage, storage conditions (i.e., temperature, humidity, atmosphere), and storage duration are all factors affecting the wound response in fresh-cut tissue (Portela and Cantwell, 2001; Cantwell and Suslow, 2002; Beaulieu and Gorny, 2004), and therefore affect the shelf-life quality of the minimally processed product. This section will review the intrinsic and external factors and techniques that affect the overall product quality and can help optimize its shelf life.

8.3.1 Quality of the raw product

The choice of the cultivar (genotype) is of prime importance to assure the optimal quality of a fresh-cut product. Cultivars often differ in organoleptic, compositional, and nutritional qualities and consequently behave differently when processed into fresh-cut products. For example, it was shown that the shelf life of 14 cultivars of peaches and 8 cultivars of nectarines varied between 2 and 12 days at 0°C, and their positive response to CA and to an antibrowning treatment varied greatly (Gorny et al., 1999). Similarly, Beaulieu (2005) noticed differences in the volatile and quality attributes in six cantaloupe cultivars, and a range of quality attributes for different cultivars of fresh-cut mango cubes was reported by Rattanapanone et al. (2001) and Poubol and Izumi (2005a, 2005b).

It is well known that fruit physiological and metabolic activities change with the ripeness stage (Allong et al., 2001). In general, when selecting less mature fruit for processing, a longer shelf life is expected due to better firmness retention and decreased changes in appearance compared with processed ripe fruit. However, when using unripe fruit for processing, the amount and composition of volatiles present or released by the end product, and consequently the flavor, will not be satisfactory, and the final fresh-cut product will lack good sensory quality (Gorny et al., 2000; Beaulieu and Lea, 2003; Beaulieu and Gorny, 2004). A mature fruit on the other hand will have superior eating quality but shorter shelf life (Watada and Qi, 1999). Because some vegetables, which are the actual "fruit" of the plant (squash, bell pepper, cucumber), do not usually ripen once harvested, and other vegetables, which include non-fruit parts of the plant such as roots, stems, and flowers (potato, carrot, asparagus, broccoli), are better tasting when harvested immature, an optimal harvest maturity must be carefully targeted for each commodity. Thus, determining the optimal ripeness stage and harvest maturity that combine acceptable shelf life and eating quality is the key to successful commercialization of fresh-cut fruits and vegetables.

Biotechnology can help change the characteristics of fresh-cut fruits and vegetables in order to improve their shelf-life quality (Zhang, 2007). For example, changes in flavor, starch, vitamin and anthocyanin contents, enzymatic activity, or simply characteristics for superior processing or improved visual quality could be engineered. The Flavr Savr tomato developed in the early 1990s by Calgene, Inc., a biotechnology company with headquarters in Davis, California, is a good example of genetically modified product that may benefit the fresh-cut industry. This tomato has a gene that slows the natural softening process that accompanies ripening, thus the biotechnologically modified tomatoes can remain on the vine longer and soften more slowly, resulting in more flavor and color.

8.3.2 Sanitation treatments prior processing

Sanitizers are commonly used in fresh-cut processing operations to prevent contamination of food products by maintaining low levels of microorganisms in the environment. The utilization of decontamination methods to prolong the shelf life of fresh-cut produce should reduce the risk of food-borne infections and intoxications, decrease the microbial spoilage, preserve the fresh attributes and the nutritional quality, and leave the product free of unacceptable toxic residues or by-products (Gómez-López et al., 2009). Sustainable sanitation techniques for keeping quality and safety of fresh-cut products have been recently reviewed by Artés et al. (2009).

Washing with chlorinated water is a widely employed sanitation procedure accomplished by immersing the product in solutions containing between 50 and 200 µL/L free chlorine (HOCl) during less than 5 minutes (Ngarmsak et al., 2006; Rico et al., 2007). Even though the application of chlorine is not considered very effective in reducing microbial levels in contaminated tissue, chlorine reduces microbial loads in the water and prevents cross-contamination. In addition, chlorine rinse acts directly on the tissue by inhibiting browning reactions while it also helps remove cellular contents present on the cut surfaces of fruits and vegetables that may promote browning (Brecht et al., 1993). However, there is some controversy about using chlorine as an antimicrobial agent due to the possible formation of carcinogenic chlorinated compounds in the rinsing water, namely, chloramines and trihalomethanes (Rico et al., 2007). Chlorine dioxide (ClO_2), a strong oxidizing and sanitizing agent, has a broad and high biocidal effectiveness and is also less affected than chlorine by pH and organic matter (Zhang, 2007). Hydrogen peroxide (H_2O_2), acidified sodium chloride, peroxyacetic acid, and organic acids have also been used to sanitize fresh-cut fruits and vegetables (Brecht et al., 2004; Narciso and Plotto, 2005; Ngarmsak et al., 2006; Rico et al., 2007). Peroxyacetic acid (PAA) is currently the most commonly used sanitizer in commercial fresh-cut processing facilities.

8.3.3 Processing practices

The sharpness of the cutting blade used for processing greatly affects the quality attributes of fresh-cut products (Hodges and Toivonen, 2008). That is, as a consequence of membrane rupture, a blunt blade will cause accumulation of liquid in the intercellular spaces, which can, in turn, reduce gas diffusion and induce anaerobic respiration, producing off-odor due to ethanol synthesis, whereas a sharp blade minimizes tissue damage and associated wound stress responses such as increased respiration and ethylene production (Portela and Cantwell, 2001; González-Aguilar et al., 2007a; Oms-Oliu et al., 2008a).

The cutting shape may also influence the metabolism of fresh-cut tissue. For example, when stored at 5°C or 10°C, slices from fresh-cut papaya had better soluble solid content retention, lower weight loss, and better overall quality index than cubes from the same papaya fruit (Riviera-Lopez et al., 2005). Shredded root of radish had a higher respiration rate and lower content of soluble solids and AA than sliced radishes (Aguila et al., 2006). Moreover, trapezoidal cuts were shown to extend melon shelf life compared to slices or cylinder cuts (Aguyao et al., 2004). The surface area of the cut tissue is often the reason for such differences.

Washing after cutting may improve firmness retention of fresh-cut fruit by removing from the cut surfaces solutes and stress-related signaling compounds such as acetaldehyde and phenolics (Toivonen and Brummell, 2008). Moreover, washing increases the activities of catalase, peroxidase, and superoxide dismutase enzymes, which are involved in scavenging oxygen free radicals that contribute to membrane injury.

8.3.4 Postprocessing treatments

8.3.4.1 Temperature

Storage temperature has received the widest attention among the various postharvest environmental factors, as it plays an important role in the postharvest physiology and quality of horticultural crops. In addition to providing effective action against fruit decay, low temperature reduces the respiration rate (RR) (Q_{10}), which provides the energy to drive the reactions occurring during ripening (Kays, 1991; Mohammed and Brecht, 2002). In addition, low temperature slows the quantitative and qualitative changes in the normal complement of enzymes that bring about the characteristic synthetic and degradative changes associated with ripening, such as softening, color changes, and flavor and compositional changes, thus increasing fruit postharvest life (Kays, 1991; Ponce de León et al., 1997).

Proper temperature management during postharvest handling, processing, and distribution is the most important external factor that must be controlled to preserve the quality and safety of fresh-cut fruits and vegetables (Zhang, 2007). Temperature has a direct relationship with the shelf life of fresh-cut products. That is, the lower the temperature, the longer the shelf life of the fresh-cut fruit or vegetable. For instance, the marketable period of fresh-cut mango cubes was 3 to 5 days at 10°C, but could be extended 5 to 8 days at 5°C (Rattanapanone et al., 2001). Moreover, temperature was the main factor affecting postcutting life of fresh-cut pineapple, which ranged from 4 days at 10°C to over 14 days at 2.2 and 0°C (Marrero and Kader, 2006).

Because fresh-cut products are stored or displayed at the retail store for only a short period of time, and because they are extremely perishable compared with the whole fruits or vegetables, exposure to a temperature that causes a slight amount of chilling injury (CI) is preferred over a temperature that causes more rapid deterioration due to ripening and senescence (Watada and Qi, 1999). A significant number of fresh-cut fruits do not seem to be as chilling sensitive as the corresponding intact fruit. Furthermore, CI symptoms are often only manifested when fruit are transferred to nonchilling temperatures and may never become visible if the product is maintained exclusively at chilling temperatures. In fact, some authors have suggested that fresh-cut products may be subject to CI despite little visual manifestation of injury. For example, higher respiration rates of fresh-cut products compared with the corresponding whole fruit may in some cases, and to some extent, be an indicator of CI (Brecht et al., 2004). Moreover, poor flavor retention in fresh-cut products, especially fruits, due to the inhibition of aroma volatile production, is a widely recognized problem and may be caused by CI (Beaulieu and Gorny, 2004).

8.3.4.2 Chemical treatments

Several chemical and physical treatments may be applied in synergy with proper temperature management and handling practices to extend the shelf life and maintain the product quality. For example, dips in organic acids solutions (e.g., lactic acid, citric acid, acetic acid, tartaric acid) have been described as having a strong antimicrobial action against psychrophilic and mesophilic microorganisms in fresh-cut fruits and vegetables (Rico et al., 2007). The powerful antimicrobial action of organic acids is attributed to their capacity to reduce external pH, disrupt membrane transport or permeability, cause anion accumulation, and reduce the internal cellular pH by dissociation of hydrogen ions from the acid fraction (Rico et al., 2007).

Reducing agents, most commonly AA or its isomer erythorbic acid, isoascorbate or sodium erythorbate, are some of the most commonly used agents to reduce or eliminate cut surface discoloration, which is mainly attributed to enzymatic browning (Beaulieu and Gorny, 2004; Brecht et al., 2004). The use of isoascorbic acid has been shown to be more effective than AA or N-acetyl-l-cysteine, another reducing agent, in preventing tissue softening, surface browning, and decay on fresh-cut pineapple slices, extending shelf life to 14 days compared to a shelf life of 9 days for the nontreated slices (González-Aguilar et al., 2004).

Calcium and its salts have been used to decrease tissue softening of a great variety of fresh-cut fruits (Soliva-Fortuny and Martin-Belloso, 2003; Toivonen and Brummell, 2008; Aguayo et al., 2008). These compounds help maintain cell wall integrity by interacting with pectin to form calcium pectate, and help reduce tissue softening by cross-linking with cell wall and middle lamella pectins (Luna-Guzman et al., 1999; Rico et al., 2007). A combination of calcium chloride, AA, and citric acid significantly reduced color deterioration and loss of firmness, without affecting sensory characteristics, of fresh-cut mangoes stored at 5°C (González-Aguilar et al., 2007a).

The use of 1-MCP was proven effective to slow the changes associated with loss of quality and extend the shelf life of fresh-cut product (Vilas-Boas and Kader, 2007). The action of 1-MCP is mediated through the inhibition of ethylene perception of plant tissues by interacting with the receptor and competing with ethylene for binding sites (Watkins, 2006). Therefore, the effectiveness of inhibition of ripening or senescence of fruit and vegetables is a function of the concentration of 1-MCP applied, up to saturation of the binding sites (Watkins, 2006). Response of fresh-cut products to 1-MCP treatment depends on the dose applied, the type of crop, the maturity or the ripeness stage, the exposure time, and the temperature (Blankenship and Dole, 2003). 1-MCP can be applied as a pre- or postcutting treatment.

Slices made from 1-MCP-treated papayas had double the shelf life compared to slices made with untreated papayas (Ergun et al., 2006). Arias et al. (2009) reported that slices made from 1-MCP-treated 'Blanquilla' pears (treated just after harvest) were firmer and had improved color compared to nontreated fruit. In pineapple, fresh-cut slices made from 1-MCP-treated fruit, had lower respiration rates, browning, and hydrolysis of ascorbic acid compared to nontreated fruit (Budu and Joyce, 2003). Exposure of fresh-cut apples slices to 1-MCP (after processing) decreased ethylene production, respiration, softening, color change, and synthesis of aroma compounds (Perera et al., 2003; Calderón-López et al., 2005).

8.3.4.3 Physical treatments

Along with good temperature management, the use of ozone, radiation, ultraviolet (UV) light, heat treatment, and modified atmosphere packaging (MAP), are physical treatments that can be applied to fresh-cut products in order to extend their shelf life.

Ozone (O_3) is a strong antimicrobial agent with high reactivity and penetrability, and spontaneously decomposes to O_2 in air, or to $O_2 + H_2O$ in water (Rico et al., 2007). A low dose of gamma irradiation is also very effective in reducing bacterial, parasitic, and protozoan pathogens in raw food (Rico et al., 2007). A maximum irradiation level of 1 kGy was approved by the U.S. FDA for use on fruits and vegetables and could be used without damage to the cut product (Fan and Sokorai, 2008). Another effective antimicrobial agent is UV light, referred to as UV-A, UV-B, or UV-C according to the wavelength from shortest to longest, respectively. UV light damages the DNA of microorganisms as well as acting indirectly against spoilage pathogens due to the induction of resistance mechanisms in different fruits and vegetables (Boynton, 2004). UV-C irradiation was reported to improve the total antioxidant capacity of fresh-cut mango (González-Aguilar et al., 2007a) and improve the phenolic profile of honey pineapple, banana, and guava (Alothman et al., 2009).

Mild heat treatments have been shown to potentially benefit the texture of fresh-cut fruit and vegetables. The treatment strengthens the cell wall due to the activation of the enzymes PME which promotes de-esterification of the pectin molecule, thus increasing the number of calcium binding sites. For example, in minimally processed celery, treatment by immersion in hot water allowed a better retention of the original color and the total chlorophyll content (Viña et al., 2007). In fresh-cut peach, heating the fruit at 50°C for 10 minutes, 4 hours before cutting, effectively controlled browning and retained firmness during storage (Koukounaras et al., 2008). A hot air treatment (48°C for 3 hours), applied to fresh-cut broccoli, was successful in delaying yellowing, maintaining chlorophyll content, and retaining tissue integrity compared to controls (Lemoine et al., 2009). Similar results have been observed for fresh-cut apples (Kim et al., 1993; Barrancos et al., 2003). However, low heat (38°C for 12 or 24 hours) of whole mangoes accelerated softening of fresh-cut slices (Plotto et al., 2003). Such treatments are difficult to take into commercial applications because of the multiple combinations of treatment time and temperature that need to be adjusted for each commodity, cultivar, and ripeness stage.

A low O_2 or elevated CO_2 atmosphere, plus saturated or near-saturated humidity environments generated in MAP have been successfully used to extend fresh-cut produce shelf life. In fact, MAP systems in association with low temperature are extensively used commercially to extend shelf life of fresh-cut products by reducing respiration rate, cell wall degradation, water loss, phenolic oxidation, microbial growth, and ethylene biosynthesis and action (Gorny, 2003). Due to the active metabolism of fresh-cut fruits, MAP alters the atmosphere composition surrounding the product, affecting the concentrations of O_2, CO_2, water vapor, and other volatiles compounds that impact the physiology and overall quality of the product (Forney, 2007).

MAP systems are designed to maintain a respiring product in a favorable atmosphere, usually incorporating reduced O_2 and elevated CO_2. The MA is created and maintained through the interplay of product respiration and gas permeation through the package (Yam and Lee, 1995; Mir and Beaudry, 2004). It can be created either passively or actively.

A passive MAP system is generated by allowing the desired atmosphere to develop naturally because of the product respiration and the diffusion of gases through the selected film or perforations (Moleyar and Narasiam 1994; Yam and

Lee, 1995). In passive MAP systems, the development of a desirable atmosphere composition may take a considerable amount of time (up to several days), depending on the product respiration rate and the void volume within the package (Rodov et al., 2007). A similar situation can occur with coatings although with much less control. On the other hand, in an active MAP system, the atmosphere is created rapidly by flushing the headspace of the package with a desired gas mixture, usually consisting of N_2 mixed with O_2 and CO_2 concentrations that are near the anticipated equilibrium concentrations of those gases. In both cases, once the MA is established, the dynamic equilibrium of respiration and permeation maintain the appropriate atmosphere. Bai et al. (2003) reported that fresh-cut honeydew cubes packed in active MAP (5 kPa O_2 + 5 kPa CO_2) had better color retention, reduced respiration rate, and microbial population and a longer shelf life than those stored in passive MAP (Bai et al., 2003).

The tolerance to and physiological effects of elevated CO_2 are highly variable and depend on the commodity, maturity or ripeness stage, and storage temperature. In addition, elevated CO_2 may alter the response of the product to reduced O_2 concentrations, because with an increase in CO_2 concentration, the tolerance limits to reduced O_2 decrease (Watkins, 2000; Kader, 2002; Kader and Saltveit, 2003). In some cases, elevated CO_2 may cause discoloration and softening, and induce fermentative metabolism (Mir and Beaudry, 2004). Nevertheless, elevated CO_2 concentration (>8 to 10 kPa) may inhibit ethylene action and can effectively inhibit microorganism growth.

Holding fresh-cut mango slices in 10 kPa O_2 and CO_2 at 5°C retarded browning and softening (Limbanyen et al., 1998). MAP led to lower microbial infection, less translucency, and better color retention in fresh-cut cantaloupe cubes stored in active MAP (4 kPa O_2 plus 10 kPa CO_2) (Bai et al., 2001). Moreover, MAP-treated fresh-cut kohlrabi, stored in 6 kPa O_2 plus 13 kPa CO_2 and 13 kPa O_2 plus 8 to 9 kPa CO_2, had reduced microbial population growth and better color retention than air stored material (Escalona et al., 2003).

One hazard that may occur with the use of a nonoptimal MAP system for a given fresh-cut commodity is the generation of anaerobic conditions or accumulation of high CO_2 levels in the package. This can be caused by the utilization of inappropriate film or gas flushing protocols, variation in respiration rates from different cultivars or varieties, seasonal variation, and storage duration of the product prior to processing (Kim et al., 2005a, 2005b; Hodges and Toivonen, 2008). Such conditions can cause the product to pass from aerobic to anaerobic respiration, which causes the production of ethanol and acetaldehyde leading to off-flavors and odors (Saltveit, 2003; Beaulieu, 2006).

8.4 Edible coatings for minimally processed products

Edible coatings improve the quality and extend the shelf life of lightly processed fruit and vegetables by acting as a barrier to water loss and gas exchange, creating a micromodified atmosphere around the product. In addition, edible coatings can serve as carriers for other generally recognized as safe (GRAS) compounds, such as preservatives and other functional ingredients from natural sources (Baldwin et al., 1995; Olivas and Barbosa-Cánovas, 2005; Vargas et al., 2008). For example, the addition of a texture enhancer, such as calcium chloride, in an edible coating formulation may enhance fruit quality during storage by maintaining firmness (Olivas and Barbosa-Canovas, 2005). Furthermore, calcium in the form of calcium ascorbate provides a dual function of cross-linking (from Ca++) and preventing the cut surface from browning (from ascorbate) (Wong et al., 1994). The incorporation of natural antioxidants, such as AA, citric acid, cysteine, and antimicrobials such as lactic acid, peracetic acid, can help in reducing enzymatic browning and controlling microbial growth of fresh-cut products. Furthermore, edible coatings contribute to the reduction of synthetic packaging waste (Vargas et al., 2008).

Due to the particular physical and chemical properties of fresh-cut produce, edible coatings must be designed and formulated to suit their specific physiology. Some coatings may not adhere well to fresh-cut surface, while others may offer good adherence but may be a poor barrier to moisture or not resist water vapor diffusion (Garcia and Barrett, 2002). On a general basis, edible coatings used with fresh-cut products must be transparent, tasteless, and odorless, in addition to containing safe and food-grade substances. They must have an appropriate water vapor permeability (WVP), solute permeability, and selective permeability to gases and volatile compounds. Further, the cost of technology and raw materials from which coatings are made has to be relatively low (Vargas et al., 2008).

The following sections will describe the nature and the properties of the main components found in edible coatings for fresh-cut fruit (i.e., proteins, polysaccharides, lipids, and resins), and how their combination can improve the shelf life of minimally processed fruits and vegetables. Table 8.1 summarizes the use of coating materials and additives for fresh-cut fruits and vegetables, with corresponding references. More information on edible coatings for minimally processed fruit and vegetables can be found in the following reviews: Baldwin et al., 1995; Olivas and Barbosa-Cánovas, 2005; Lin and Zhao, 2007; Bourtoom, 2008; Vargas et al., 2008; Rojas-Graü et al., 2009.

8.4.1 Films versus emulsion coatings

Minimally processed products can be coated with stand-alone films preformed by casting or extrusion processes, or by formation of the film layer directly on the surface of the product by dipping or spraying the coating solution. Stand-alone edible films are usually used to cover, wrap, or separate food components from each other and from the environment. Many of the films studied are actually dry films of layered structures, and most films are not suitable for foods with high surface water activity because they swell, dissolve, or disintegrate on contact with water (Guilbert, 1986). However, adding a hydrophobic portion may contribute to decreasing water permeability. In this sense, McHugh et al. (1996) developed stand-alone edible films from fruit and vegetable purees that presented good mechanical and barrier properties to be used to wrap fresh-cut fruits. A film made from apple puree, beeswax, pectin, glycerol, ascorbic acid, and citric acid was effective at reducing moisture loss and browning in fresh-cut apples (McHugh and Senesi, 2000). Another film made from pure mango puree reduced weight loss of whole mango fruit and extended the shelf life of fresh-cut mango slices by 2 to 3 days (Sothornvit and Rodsamran, 2008).

Table 8.1 Edible coatings for minimally processed fruits

Commodity	Coating material	Additives and plasticizer	References
Apple	Alginate; apple puree	N-acetylcysteine, $CaCl_2$, oregano, lemongrass, vanillin, Gly	Rojas-Graü et al., 2007
	Alginate, gellan; sunflower oil	N-acetylcysteine, $CaCl_2$,	Rojas-Graü et al., 2008
	Alginate	N-acetylcysteine, glutathione, cinnamon, clove, lemongrass, citral, cinnamaldehyde, eugenol, calcium lactate, malic acid, Gly	Raybaudi-Massilia et al., 2008b
	Apple puree, BW, vegetal oils	AA, citric acid, Gly	McHugh and Senesi, 2000
	Alginate, AMG, linoleic acid	$CaCl_2$	Olivas et al., 2007
	Carrageenan, CMC, WPC	$CaCl_2$, Gly, PEG	Lee et al., 2003
	CMC, CC, WPI	$CaCl_2$, Gly	Le Tien et al., 2001
	Nature Seal™, CMC, SPC	AA, PS, SB, soy oil, $CaCl_2$, Gly, PEG	Baldwin et al., 1996
	SWP, SPI, CC	Sorbitol	Shon and Haque, 2007
	WPI, BW	Gly	Pérez-Gago et al., 2003
	WPI, WPC, HPMC, BW, CW	Gly	Pérez-Gago et al., 2005a
	WPC, BW	AA, Cys, Gly	Pérez-Gago et al., 2006
	Candelilla wax	Aloe vera juice, ellagic acid, gallic acid	Saucedo-Pompa et al., 2007
Avocado	Candelilla wax	Aloe vera juice, ellagic acid, gallic acid	Saucedo-Pompa et al., 2007
Banana	Carrageenan	AA, Cys, $CaCl_2$, Gly, PEG	Bico et al., 2009
	Candelilla wax	Aloe vera juice, ellagic acid, gallic acid	Saucedo-Pompa et al., 2007
Carrot	Alginate	Citric acid, $CaCl_2$	Amanatidou et al., 2000
	Chitosan, yam starch	Gly	Durango et al., 2006
	Chitosan, yam starch	Gly	Simões et al., 2009
	Chitosan, MC, oleic acid	—	Vargas et al., 2009
	Xanthan gum	Gluconal cal, vitamin E	Mei et al., 2002

	Cellulose-based		Peiyin and Barth, 1998
	Pectin, CMC, CC, WPI	Cinnamaldehyde, Gly	Caillet et al., 2006
	HPMC, sucrose ester		Villalobos-Carvajal et al., 2009
	CC, WPI	Gly	Lafortune et al., 2005
	SWP, SPI, CC	Sorbitol	Shon and Haque, 2007
Celery	CC, AMG		Avena-Bustillos et al., 1997
Eggplant	SPI, BW	AA, Cys, Gly	Ghidelli et al., 2010b
Grapefruit	Wax microemulsion		Baker and Hagenmaier, 1997
Lettuce	Alginate	$CaCl_2$	Tay et al., 2004
Litchi	Chitosan		Dong et al., 2003
Mango	Chitosan		Chien et al., 2007
	CMC, chitosan, dextrin, stearic acid	AA, citric acid, calcium lactate	Ducamp-Collin et al., 2009
	CMC, maltodextrin	Calcium ascorbate, N-acetylcysteine	Plotto et al., 2004
Melon	Alginate, gellan, pectin, sunflower oil	$CaCl_2$, Gly,	Oms-Oliu et al., 2008a
	Alginate	Cinnamon, palmarosa, lemongrass, eugenol, geraniol, citral, malic acid, calcium lactate, Gly	Raybaudi-Massilia et al., 2008
	SPI	malic acid, lactic acid, Gly	Eswaranandam et al., 2006
Mushroom	Chitosan		Eissa, 2007
Onion	SWP, SPI, CC	Sorbitol	Shon and Haque, 2007
Papaya	Alginate, gellan	AA, $CaCl_2$, Gly	Tapia et al., 2007
Pear	Alginate, gellan, pectin	N-acetylcysteine, glutathione, $CaCl_2$, Gly	Oms-Oliu et al., 2008b
	MC, stearic acid	AA, PS, $CaCl_2$, Gly, PEG	Olivas et al., 2003
Persimmon	WPI, BW	Sodium ascorbate, Gly	Pérez-Gago et al., 2005b
	SPI	citric acid, $CaCl_2$, Gly	Ghidelli et al., 2010a

(*Continued*)

Table 8.1 Edible coatings for minimally processed fruits (Continued)

Commodity	Coating material	Additives and plasticizer	References
Potato	Nature Seal™, CMC, SPC	AA, PS, SB, soy oil, $CaCl_2$, Gly, PEG	Baldwin et al., 1996
	CMC, CC, WP	$CaCl_2$, Gly	Le Tien et al., 2001
	SWP, SPI, CC	Sorbitol	Shon and Haque, 2007
Strawberry	Chitosan	—	Campaniello et al., 2008
Water chestnut	Chitosan	—	Pen and Jiang, 2003

Notes: BW, beeswax; AMG, acetylated monoglyceride; CMC, carboxymethyl cellulose; WPC, whey protein concentrate; CC, calcium caseinate; SPC, soy protein concentrate; SWP, sour whey powder; WPI, whey protein isolate; HPMC, hydroxypropyl methylcellulose; CW, carnauba wax; MC, methylcellulose; SPI, soy protein isolate; $CaCl_2$, calcium chloride; Gly, glycerol; AA, ascorbic acid; PEG, polyethylene glycol; Cys, cysteine; PS, potassium sorbate; SB, sodium benzoate.

Coating formulations applied directly to food products usually, but not necessarily, consist of emulsions that contain immiscible components, with one component dispersed as fine droplets (dispersed phase) in the other (continuous phase). This way, gas exchange, adherence, and moisture barrier properties of the composite coating are improved, with results very beneficial to fresh-cut fruits and vegetables (Baldwin et al., 1995). Emulsion coatings require the consideration of stability, which is related to a balance between attractive and repulsive forces, including van der Waals, electrostatic, steric, and hydration forces.

8.4.2 Polysaccharides

Polysaccharides (chitosan, alginate, carageenan, cellulose, starch, pectin, gums) are widely used as edible coatings for fresh-cut fruits (Krochta and de Mulder-Johnston, 1997). These coatings present generally good gas—mostly oxygen—barriers due to their hydrogen-bonded network structure (McHugh et al., 1994), and adhere well to the hydrophilic cut surfaces of fruits and vegetables. However, they are poor water barriers (Baldwin et al., 1995), which may increase product desiccation and weight loss.

8.4.2.1 Chitosan

Chitosan [2-amino-2 deoxy-β-D-glucan] is a cationic polymer obtained from the deacetylation of chitin originating from the exoskeleton of crustaceans (crab, shrimp, and crayfishes), and from the cell wall constituents of fungi and insects (Andrady and Xu, 1997; Hirano, 1999). Chitosans are described in terms of degree of deacetylation and average molecular weight. Extraction methods make the differences in the degree of deacetylation, the distribution of acetyl groups, the chain length, and the conformational structure of chitin and the chitosan molecule (Tsai et al., 2002). Chitosans have been widely studied because of their antimicrobial properties as well as their cationic character and their film-forming properties (Muzzarelli, 1996; Geraldine et al., 2008). They are also used by the food industry as clarifying agents, antioxidants, and as enzymatic browning inhibitors (Devlieghere et al., 2004). The amino groups in the chitosan molecule provide positive charges that make possible modifications to the molecule by covalent bonding with anions, such as those from fatty acids and other proteins under the right pH (Janjarasskul and Krochta, 2010). It is also believed that the cationic property of chitosan is partially responsible for its antimicrobial properties, by binding to the negatively charged microbial cell membranes (Young et al., 1982; Papineau et al., 1991). The positive charges of the chitosan molecule also provide some antioxidant activity, as well as the ability to carry and slow-release functional ingredients (Coma et al., 2002).

Chitosan edible coatings were shown to effectively maintain quality and extend the shelf-life of many fresh-cut products such as carrot (Vargas et al., 2009), Chinese water chestnut (Pen and Jiang, 2003), litchi (Dong et al., 2004), mango (Chien et al., 2007; Ducamp-Collin et al., 2009), mushrooms (Eissa, 2007), and strawberries (Campaniello et al., 2008). Combination of chitosan with other polysaccharides has also shown to improve its functional properties. For example, an edible yam starch and chitosan coating was successful in controlling the microbial growth, preventing surface whitening, and maintaining the sensory quality of minimally processed carrots (Durango et al., 2006; Simões et al., 2009). In another study, chitosan and cellulose were made into a stand-alone film, with polyethylene glycol added as a plasticizer, which controlled microbial growth on fresh cut melon and pineapple as compared to uncoated fruit, or fruit wrapped in a commercial stretch film (Sangsuwan et al., 2008).

8.4.2.2 Alginates

Alginates are the salts of alginic acid, a linear copolymer of D-mannuronic and L-guluronic acid monomers, extracted from brown seaweeds of the *Phaeophyceae* class (Sime, 1990; Cha and Chinnan, 2004; Vargas et al., 2008). Alginates form films or gels by reacting with divalent and trivalent cations (e.g., calcium, magnesium, and ferric ions) that are added as texturing and gelling agents (Cha and Chinnan, 2004). Using calcium as the gelling agent with alginates has shown to provide additional properties, such as maintaining firmness of fresh-cut apples (Olivas et al., 2007; Rojas-Graü et al., 2007, 2008; Raybaudi-Massilia et al., 2008b), carrots (Amanatidou et al., 2000), lettuce (Tay and Perera, 2004), melon (Oms-Oliu et al., 2008a; Raybaudi-Massilia et al., 2008a), papaya (Tapia et al., 2008), and pears (Oms-Oliu et al., 2008b).

Alginates produce uniform, transparent, and water-soluble films that give the fruit a bright, translucent, and fresh-like appearance (Olivas et al., 2007). These films are also effective gas barriers and were shown to decrease the respiration rate and the ethylene production by creating a modified atmosphere around the cut pieces of apple (Raybaudi-Massilia et al., 2008b; Rojas-Graü et al., 2007), melon (Raybaudi-Massilia et al., 2008a), and pears (Oms-Oliu et al., 2008b).

Like other hydrophilic polysaccharides, alginate coatings require the addition of plasticizers to increase coating flexibility and decrease brittleness by reducing the internal hydrogen bonds between the polymer chains and increasing molecular spaces. In addition, due to their hydrophilic nature, the incorporation of a lipidic substance to the alginate coating mix may be necessary to improve water vapor barrier properties. Tapia et al. (2008) studied the effect of the addition of glycerol, ascorbic acid, and sunflower oil to alginate-based coatings in order to improve their moisture barrier on fresh-cut papaya. The coatings improved the firmness of the fresh-cut product during the period studied, and the addition of 0.025% (w/w) sunflower oil resulted in a 16% increase in the water vapor resistance of the coated samples.

Several studies (Rojas-Graü et al., 2007, 2008; Raybaudi-Massilia et al., 2008a, 2008b; Oms-Oliu et al., 2008b; Tapia et al., 2008) successfully used alginate edible coatings with antibrowning agents, such as ascorbic acid, N-acetyl-l-cysteine, and glutathione, to control enzymatic browning and extend shelf life of fresh-cut commodities. Lemongrass and cinnamon (0.7%), citral (0.5%), or cinnamaldehyde (0.5%) essential oils added to an alginate-based coating (alginate, malic acid, N-acetyl-l-cysteine, glutathione, and calcium lactate) effectively prevented microbiological growth on fresh-cut apples (Raybaudi-Massilia et al., 2008b). The addition of palmarosa oil (0.3%) to a similar alginate-based coating inhibited the native flora and reduced *Salmonella enteritidis* population while maintaining the fresh-cut melon quality parameters (Raybaudi-Massilia et al., 2008a).

8.4.2.3 Gellan

Gellan is a polysaccharide of microbial origin and is secreted by the bacterium *Sphingomonas elodea* (formerly referred to as *Pseudomonas elodea*). Like alginate, gellan coatings are made by adding calcium ions (calcium chloride) to cross-link the carbohydrate polymer (Oms-Oliu et al., 2008b).

In a comparative work between gellan and alginate, the gellan films showed better vapor barrier properties than alginate stand-alone films, and water solubility values found for gellan films at 25°C were significantly lower (0.47 to 0.59 g soluble solids/g total solids) than the values found for the alginate film (0.74 to 0.79) (Tapia et al., 2007). This implies that gellan films, which are more difficult to dissolve in water, present a slightly higher hydrophobicity that could be desirable for their use on fresh-cut fruits. Also in this study, fresh-cut apples and papaya cylinders were successfully coated with these films, and the addition of sunflower oil (0.025 mL/100 mL film-forming solution) to the gellan coating applied to fresh-cut apples improved its water barrier property, thus reducing moisture loss and maintaining texture.

Gellan-based edible films and coatings are good carriers for antioxidant agents such as glutathione, *N*-acetyl-l-cysteine, and ascorbic and citric acids (Rojas-Graü et al., 2008). For example, incorporation of *N*-acetyl-l-cysteine and glutathione to a gellan-based coating effectively controlled enzymatic browning, reduced ethylene production, and maintained the desirable quality characteristics of fresh-cut apples (Rojas-Graü et al., 2008) and pears (Oms-Oliu et al., 2008b). Moreover, this coating maintained the vitamin C content and antioxidant potential in pears and in fresh-cut melon (Oms-Oliu et al., 2008a).

8.4.2.4 Carrageenan

Extracted from several red seaweeds, mainly *Chondrus crispus,* carrageenan is a complex mixture of at least five different water-soluble galactose polymers (Karbowiak et al., 2007; Vargas et al., 2008). Carrageenans form gels in the presence of monovalent or divalent cations during moderate drying, leading to a three-dimensional network formed by polysaccharide double helices which becomes a solid film after solvent evaporation (Karbowiak et al., 2007). Among the different types of polymers (κ, ι, and λ carrageenan), ι-carrageenan is preferred for coatings as it makes clear and elastic gels (Karbowiak et al., 2007).

Application of carrageenan edible coatings with an antibrowning agent was a good method to prevent weight and firmness losses, to reduce respiration rate, to maintain eating quality, and to maintain microbial growth within acceptable limits during storage of fresh-cut banana (Bico et al., 2009) and fresh-cut apples (Lee et al., 2003).

8.4.2.5 Xanthan gum

Xanthan gum is the product of glucose fermentation by the *Xanthomonas campestri* bacterium. It is an anionic polymer with cellulosic backbone substituted on alternate glucose residues with a trisaccharide side chain (Mei et al., 2002). A xanthan gum coating was used as carrier of calcium and vitamin E on fresh-cut baby carrots (Mei et al., 2002). These formulations were efficient at delaying the white surface discoloration of carrots, probably by acting as surface moisturizer, and they did not affect the aroma, flavor, sweetness, crispness, and carotene levels of the baby carrots.

8.4.2.6 Cellulose and cellulose derivatives

Cellulose, the structural material of plant cell walls, is composed of linear chains of (1→4)-β-d-glucopyranosyl units. Because it is naturally insoluble in water, chemical modification of cellulose by etherification is necessary to make it usable. Water-soluble cellulose derivatives are methylcellulose (MC), hydroxypropylcellulose (HPC), hydroxypropylmethylcellulose (HPMC), and ionic carboxymethylcellulose (or sodium carboxymethylcellulose CMC). These cellulose derivatives tend to have excellent film-forming property, and films are generally odorless, tasteless, transparent, flexible, and of moderate strength. Regarding their physical properties, cellulose derivative films provide some barrier to oil and fats and have moderate barrier properties to moisture and oxygen (Gennnadios et al., 1997; Krochta and Mulder-Johnson, 1997; Turhan and Sahbaz, 2004; Maftoonazad and Ramaswamy, 2005). MC is the most hydrophobic of the cellulose derivatives (Kester and Fennema, 1986), even though the WVP of cellulose stand-alone films is still relatively high. MC-based coatings were used on fresh-cut apples (Vargas et al., 2009) and pear wedges (Olivas et al., 2003). In both studies, the application of MC edible coating did not reduced weight loss, probably due to the high relative humidity conditions. However, the incorporation of stearic acid, a fatty acid (hydrophobic molecule), to MC coatings reduced the weight loss of pear wedges compared to MC alone.

Scarce reports of application of HPMC-based coatings to fresh-cut fruit are available. Villalobos-Carvajal et al. (2009) studied the barrier properties of HPMC edible coatings containing surfactant mixtures prepared in aqueous or hydroalcoholic solution, applied to fresh-cut carrots. HPMC edible coating prepared in a hydroalcoholic solution provided greater resistance to water vapor migration than those prepared in an aqueous solution. In addition, coatings prepared in an alcoholic media induced lesser color changes (whitening) than those prepared in an aqueous solvent, probably because the film formed was thinner (Villalobos-Carvajal et al., 2009). In fresh-cut apples coated with HPMC containing beeswax or carnauba wax formulations, there was no significant reduction in either weight loss or browning (Pérez-Gago et al., 2005b).

CMC, or cellulose gum, is available for food applications in a variety of types based on degree of substitution, particle size, viscosity, and hydration characteristics (Nisperos-Carriedo, 1994). On fresh-cut fruit, the use of CMC-based edible coatings with an antioxidant agent such as N-acetyl-l-cysteine or ascorbic or citric acid, gave positive results in controlling enzymatic browning, improving visual appearance, and maintaining fruit aroma and flavor in fresh-cut mangoes (Plotto et al., 2004; Ducamp et al., 2009). Combinations of CMC with other hydrocolloids, such as whey or milk protein, showed good synergy, and were useful in reducing respiration rate, delaying browning, maintaining firmness, and increasing the antioxidative power of coated fresh-cut apples (Lee et al., 2003) and potatoes (Le Tien et al., 2001). It was hypothesized that the carboxyl groups in CMC may effectively trap peroxide radicals (Le Tien et al., 2001). Similar effectiveness was demonstrated by Baldwin et al. (1996) with minimally processed potato and apple, where the antibrowning activity of CMC-based edible coating with ascorbic acid was more effective than an ascorbic acid aqueous solution alone. The latter results suggest that ascorbic acid undergoes less degradation within the cellulose matrix, thus enhancing its antibrowning activity. Moreover, application of CMC retarded water loss, and the addition of potassium sorbate and citric acid showed a synergistic effect for microbial control (Baldwin et al., 1996). Finally, two commercial cellulose-based edible coatings of varying pH (2.7 and 4.6) significantly retarded surface whitening, preserved carotene content, retarded peroxidase activity, and maintained fresh appearance of lightly processed carrots (Li and Barth, 1998).

8.4.2.7 Pectin

Pectin is a soluble plant fiber derived from plant cell walls, and the major commercial source for food applications is from citrus peel and apple pomace. Pectin is divided into types according to the degree of esterification (DE) that is an indication of the content and branching of methyl esters on the polymer chain composed of (1→4)-α-d-galactopyranosyluronic acid units. The DE imparts solubility and gellation properties. Pectins with a DE of 50% are divided into low- and high-methoxyl pectins. Low-methoxyl pectin can form gels under a narrow pH range and with the presence of soluble solids, while high-methoxyl pectin can form gels in the presence of calcium ion (Nisperos-Carriedo, 1994). Pectin-based edible coatings with added sunflower oil contributed to reduced moisture and firmness loss and inhibition of ethylene synthesis in fresh-cut melon and pears. Further addition of N-acetyl-l-cysteine and glutathione as antioxidants to pectin was effective in avoiding browning in fresh-cut pears. The addition of these antioxidants preserved vitamin C and phenolic content in fresh-cut pears, which was attributed to a reduction of O_2 diffusion. Also, fresh-cut melon coated with pectin maintained their quality attributes better compared to samples coated with gellan or alginate after 1 week storage, without affecting their taste (Oms-Oliu et al., 2008a,b).

8.4.2.8 Starch and starch derivatives

Starch is a polymeric carbohydrate composed of anhydroglucose units [(1→4)-α-d-glucopyranosyl monomers]. Most starches contain two types of glucose polymers: a linear chain molecule termed *amylose* and a branched polymer of glucose termed *amylopectin*. Starch edible coatings exhibit different properties than other polysaccharides, which is attributed to the amylose content (Lawton, 1996).

Among the polysaccharides available commercially for production of edible coatings, starch is the natural biopolymer most commonly used because it is abundant, relatively low cost, and has a wide range of functionalities (Mali et al., 2002). Mali et al. (2002) showed that yam (*Dioscorea* sp.) starch, with about 30% amylase, presented good film-forming properties; therefore, it was a good source for the production of edible coatings. Yam starch was used on minimally processed carrots (Durango et al., 2006). When combined with chitosan, it effectively controlled microbiological growth and enhanced visual quality and phenolic content of carrot sticks (Simões et al., 2009).

Dextrin, derived from starch with smaller molecular size, could be used in edible coating providing a better water vapor resistance than starch coatings (Lin and Zhao, 2007). Ducamp-Collin et al. (2009) reported the effectiveness of dextrin potato starch with calcium lactate and ascorbic acid to reduce respiration rate in fresh cut mangoes. Meanwhile, combination of maltodextrin with CMC did not improve mango flavor or texture in minimally processed mangoes (Plotto et al., 2004).

8.4.3 Proteins

Proteins used in edible coatings destined for fresh-cut fruits and vegetables are usually whey protein and casein extracted from milk, or soy proteins (Baldwin and Baker, 2002). Due to their chemical nature, proteins can impart a range of physical and chemical properties to coatings. The type, sequence, and amount of amino acids will determine their molecular size, shape (globular, random coil, or helix conformation), and charges depending on the pH. Therefore, coatings and films made with proteins will vary in their flexibility (rigid versus flexible), thermal stability, and barrier properties (Vargas et al., 2008). Proteins, like polysaccharides, are highly polar polymers and are capable of forming strong films, with low permeability to O_2 and CO_2 but poor water barrier properties (Kester and Fennema, 1986; McHugh and Krochta, 1994; Baldwin and Baker, 2002). Plasticizers can improve the flexibility and strength of those films (Brault et al., 1997; Sothornvit and Krochta, 2001); however, their application implies a decrease in film and coating moisture barrier. In spite of the inherent hydrophilicity of proteins, protein-based coatings made with high levels of hydrophobic amino acids, like those found in soy protein and casein, present greater moisture barrier properties than proteins with less hydrophobic amino acids, especially if they are combined with lipids.

8.4.3.1 Whey protein and casein

Milk proteins include approximately 80% of casein and 20% of whey protein (Gennadios et al., 1994) and are attractive for coating manufacturers due to their numerous functional properties (Chen, 2002; Krochta, 1997, 2002; Vargas et al., 2008). During manufacturing, casein is separated from whey protein by adjusting milk to the isoelectric pH of casein, then centrifuging the casein precipitate (Gennadios et al., 1994). The acid casein can be converted to soluble caseinates through neutralization with an alkali, obtaining sodium, calcium, magnesium, potassium, and ammonium caseinates. Caseins have low levels of cysteine, and as a result, have an open random coil shape. Because of that nature, they can easily form films from aqueous solutions without further treatment. Casein-based edible coatings are transparent, flavorless, flexible, and have high nutritional qualities and excellent sensory and gas barrier properties (Lin and Zhao, 2007; Vargas et al., 2008). Different casein products may result in films of different permeabilities and mechanical properties, for example, films made with calcium caseinate were stronger and more flexible and presented lower WVP than films made with sodium caseinate (Brault et al., 1997).

Liquid whey is a by-product of cheese processing and is commercially purified to produce whey protein concentrate (WPC) with 25% to 80% protein content, or whey protein isolate (WPI) with protein content above 80%, prepared from WPC by adding an ion-exchange step (Gennadios et al., 1994; Krochta, 2002). Unlike caseins, whey proteins require heat denaturation for film formation (Gennadios et al., 1994; McHugh et al., 1994). Whey protein produces transparent, flavorless, and flexible water-based edible coatings.

Casein and WPI coatings efficiently delayed browning of apple and potato slices by acting as oxygen barriers. Moreover, they were shown to have a high antioxidant activity due to the presence of amino acids and the simple sugar lactose known for its free radical quenching effect (Le Tien et al., 2001).

The same effectiveness in controlling enzymatic browning was reported in the application of both WPI and WPC edible coatings with the addition of lipid compounds (beeswax or carnauba wax) on fresh-cut apples. This was attributed to a possible antioxidant effect of cysteine, presented in high levels in whey proteins, or the oxygen barrier provided by the coatings (Pérez-Gago et al., 2003, 2005a). The addition of lipids to these whey protein coatings did not significantly reduce moisture loss. However, combination of casein-coating with a hydrophobic compound that acted as a plasticizer (acetylated monoglyceride) contributed to significantly increase water vapor resistance, thus reducing moisture loss, and respiration rate of celery sticks (Avena-Bustillos et al., 1997). Browning of fresh-cut apples and persimmon was further retarded when whey protein coating was combined with antibrowning agents. Ascorbic acid (Pérez-Gago et al., 2006) or cysteine (Pérez-Gago et al., 2005b) reduced enzymatic browning more effectively when incorporated into the whey protein coatings than when applied alone. The polymer coating might offer a protective effect to degradation of the antioxidants, such as observed earlier in a CMC coating (Baldwin et al., 1996).

Application of WPI and casein edible coating with irradiation treatment (1 kGy) and packed under air was used to prevent the whitening of baby carrots, maintaining firmness and quality during a 21-day storage period (Lafortune et al., 2005; Caillet et al., 2006).

8.4.3.2 Soy protein

Soy protein (SP) is the major plant origin protein studied as coating material for minimally processed products. SP coatings can be prepared from soy protein isolates (SPI) or concentrates (SPC), containing about 90% and 70% protein, respectively, with a plasticizer, typically glycerol or sorbitol, to improve flexibility (Gennadios et al., 1994). Like with other carbohydrate and protein coatings, SP coatings exhibit poor moisture resistance and water vapor barrier properties due to the inherent hydrophilicity of protein and plasticizers added (Rhim et al., 2000).

Few studies reported the application of soy protein on fresh-cut fruits and vegetables. SPI coating effectiveness was shown to control browning in potato slices and reduce moisture loss in carrots and apple slices (Shon and Haque, 2007). A SPI-cysteine-based edible coating was successful at maintaining the quality of fresh-cut eggplants and minimally processed persimmon (Ghidelli et al., 2010a, 2010b). The incorporation of cysteine as an antibrowning agent further delayed the enzymatic browning and prevented softening of fresh-cut eggplant tissue (Ghidelli et al., 2010b).

In addition, SPI coating combined with low O_2 and high CO_2 modified atmosphere showed a positive synergic effect in controlling tissue browning and maintained the general visual quality of fresh-cut persimmon up to 8 to 10 days at 5°C (Ghidelli et al., 2010a).

The influence of the SPI application alone or with additives (malic or lactic acid) on the sensory quality of fresh-cut cantaloupe was studied by Eswaranandam et al. (2006). SPI alone or with lactic acid improved the sweetness of fresh-cut cantaloupe but did not have any added effect on the taste, overall appearance, enzymatic browning, or moisture loss compared to controls.

8.4.4 Lipids

Lipids used in edible coatings for fresh-cut fruits include beeswax, candelilla and carnauba wax, tryglicerides, acetylated monoglycerides, fatty acids, fatty alcohols, and surfactants such as sucrose esters. Lipids have long been used to protect horticultural crops from dehydration, to slow senescence, and to improve surface appearance (Kester and Fennema, 1986; Hagenmaier and Baker, 1994; Hernandez, 1994; Morillon et al., 2002). However, because of their hydrophobic properties, lipids tend to form thicker and brittle coatings, and they do not adhere well to the moist surface of fresh-cut tissue. Consequently, for fresh-cut produce, the best use of lipids in a coating is to combine them with hydrophilic film-forming agents such as polysaccharides or proteins.

In stand-alone films, many studies describe the effect of lipid type on WVP, showing an effect of lipid polarity created by chemical groups (e.g., carboxylic, alcohol), aliphatic chain length, and degree of unsaturation (Morillon et al., 2002). The study by McHugh and Senesi (2000), where different types of lipids were added to an apple puree–based edible film, is a good example of property modification due to the lipidic phase. Waxes, high molecular weight fatty acids, and fatty alcohols showed good adhesion to the casting surface. Vegetable oil significantly reduced WVP, and the puree film with vegetable oil did not need additional plasticizers. Among the fatty acids tested, oleic acid exhibited the best water barrier properties. Moreover, increasing concentrations of lipids resulted in significant decrease in WVP for coatings with oleic acid, palmitic acid, and beeswax. These apple-based wraps significantly reduced moisture loss and browning in fresh-cut apples, and color was preserved for 12 days at 5°C.

The use of candelilla wax by itself improved the shelf life of fresh-cut avocados, bananas, and apples (Saucedo-Pompa et al., 2007). Moreover, the addition of *Aloe vera* juice or ellagic acid, an antioxidant compound, reduced weight loss, retarded browning, and retained firmness of fresh-cut fruits.

The addition of lipids to protein or polysaccharide coatings helped improve the coating moisture barrier. Stearic acid added to a MC coating significantly reduced fresh-cut pear weight loss (Olivas et al., 2003). However, incorporation of beeswax and carnauba wax in coatings composed of whey or soy proteins or HPMC did not significantly reduce moisture loss of fresh-cut apples and eggplants (Pérez-Gago et al., 2003, 2005a; Ghidelli et al., 2010b). Nevertheless, these combinations improved visual appearance by reducing enzymatic browning of these fresh-cut produces.

Due to their hydrophobic nature, lipid compounds do not adhere well to the high moisture surface of fresh-cut products. Because wax coatings can withstand high a_w with little loss of integrity, Baker and Hagenmaier (1997) developed wax microemulsion coatings to inhibit fluid leakage from grapefruit segments. Generally, wax microemulsion coatings provided an acceptable means with which to control fluid leakage from grapefruit segments; coatings made with either polyethylene wax or carnauba wax with C12–C18 fatty acids were the most effective coatings, without compromising appearance or general acceptability.

Acetylated monoglyceride is a vegetable oil derivative which is solid at room temperature. Therefore, its application as a coating for fruits or vegetables requires an emulsifier such as calcium or sodium caseinate. Optimization of caseinate-acetylated monoglyceride coating produced a 75% reduction in moisture loss and minimized wound response in celery sticks by reducing respiration (Avena-Bustillos et al., 1997).

8.4.5 Composite and bilayer coatings

As reviewed in the preceding section, each coating material has its own physicochemical properties: hydrocolloids (proteins and carbohydrates) tend to form hydrophilic networks with good barriers to oxygen and carbon dioxide but poor water permeability, and lipids form hydrophobic coatings with good water barrier properties, but they do not adhere well to fresh-cut tissue. Thus, when combined in proper proportions, they complement each other and form successful edible coatings for fresh-cut fruits.

An edible composite coatings can be formed as a bilayer or as an emulsion. In bilayer edible coatings, the polysaccharide or protein solution is applied on the fruit surface, and after drying of the coating, a second layer with the lipid is applied. In emulsion composite edible coatings, the lipid is dispersed and entrapped in the hydrophilic phase forming a homogeneous emulsion that is directly applied on the fruit surface (Krochta, 1997). Therefore, emulsion composite coatings are more convenient to apply than bilayer coatings, because they only require one application and drying step. They also have better adherence to a larger number of surfaces due to the presence of both polar and nonpolar components, and they exhibit good mechanical resistance provided by the continuous polymer matrix (Quezada-Gallo et al., 2000; Pérez-Gago and Krochta, 2001). Therefore, only very few studies can be found in the literature where fresh-cut fruits and vegetables are coated with a bilayer composite coating. Wong et al. (1994) reported a reduction of water loss and of internal oxygen concentration of fresh-cut apple cylinders coated with a bilayer coating composed of a first layer of polysaccharides (pectin, carrageenan, alginate, and microcrystalline cellulose), followed by a layer containing acetylated monoglyceride. Recently, application of composite emulsion coatings on fresh-cut products has been reported by several authors and was discussed in previous sections (Avena-Bustillos et al., 1997; Baker and Hagenmaier, 1997; Olivas et al., 2003; Pérez-Gago et al., 2003, 2005a, 2005b, 2006; Villalobos-Carvajal et al., 2009; Ghidelli et al., 2010b).

8.4.6 Coating optimization

The success of an edible coating is based on the physicochemical and barrier properties of its components (proteins, polysaccharides, lipids). Thus, determining the proper composition and proportions of the components is of prime importance in order to extend the shelf life and enhance the quality of fresh-cut fruit and vegetables. Considerable work can be found in the literature regarding the effect of the different components on the barrier and mechanical properties of stand-alone films (Guilbert, 1986; Kester and Fennema, 1986; Krochta and De Mulder-Johnston, 1997). For example, the ratio of polymer to plasticizer has phenomenal effects on stand-alone film strength and elasticity. Brault et al. (1997) showed that glycerol had a double function by acting as a plasticizer in a calcium caseinate film, thereby improving film viscoelasticity, and also enhancing formation of cross-links within the caseinate chains, resulting in a stronger film.

Similarly, many factors, such as coating composition and formulation solid content, proven to affect the performance of edible coatings on the postharvest life of whole fruit, have not been studied in detail on fresh-cut products. The investigation of these factors prior to the incorporation of additives such as antioxidants or antimicrobials is very important to optimize the coating performance on fresh-cut produce. In this sense, Pérez-Gago et al. (2003) found that the solid content of the formulation and lipid content of WPI-beeswax edible coatings without incorporation of antioxidants also had an effect in the degree of browning of fresh-cut apples. As beeswax and solid content increased, the browning index of cut apples decreased (Figures 8.1a and 8.2a). However, high beeswax or solid content imparted a whitish appearance to the coated apples that was considered as unacceptable by the sensory panel. The optimum solid content of the emulsion and beeswax content in order to reduce browning were 16% and 20% (dry basis), respectively (Figure 8.1b, 8.2b). In a more recent study, Pérez-Gago et al. (2005a) showed that the selection of the hydrophilic component is important in the formulation of coatings for fresh-cut products. Results showed that whey protein–based coatings without incorporation of antioxidants were more effective in reducing enzymatic browning of 'Golden Delicious' apples than HPMC-based coatings, probably due to the antioxidant effect of some amino acids such as cysteine, or the higher oxygen barrier that the protein exerts. However, no differences on visual appearance were found between the uses of WPI having 98% protein or WPC with 65% protein content. When different lipids (carnauba wax or beeswax) were incorporated into the formulations, results indicated that beeswax was more effective at reducing browning as measured with the colorimeter, but visual differences between waxes were less evident at the end of the storage time (Figures 8.3a, b).

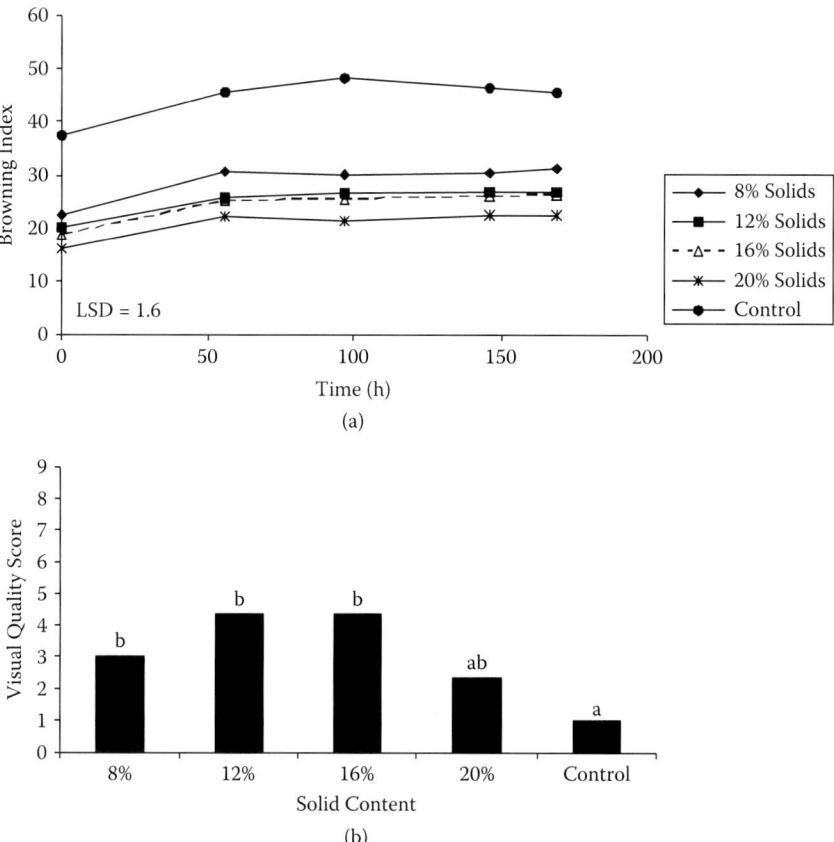

Figure 8.1 Effects of solid content in whey protein isolate–beeswax coating on (a) browning index and (b) visual quality (1 = poor quality, 9 = excellent) of fresh-cut apples during storage at 5°C. (Adapted from Pérez-Gago, M.B., Serra, M., Alonso, M., Mateos, M., and del Río, M.A., *J. Food Sci.*, 68(7), 2186–2191, 2003.)

8.4.7 Future needs and trends of minimally processed fruits and vegetables

It is widely known that edible coatings provide significant benefits in extending shelf life and enhancing quality and microbial safety of fresh-cut fruits and vegetables. Nevertheless, the use of edible coatings on a wide range of fresh-cut products and on a commercial scale is still limited by several factors. Many available edible coating formulations are characterized by high hydrophilicity, which does not provide a satisfactory moisture barrier to the fresh-cut products with their high water activity. Higher moisture on the cut side of the flesh or presence of natural hydrophobic waxy layer on the peel side of the fruit can also create problems with coating adhesion and durability. Therefore, more studies are required to develop new formulations with higher moisture barrier and surface adhesion, as well as to understand the functionality and interactions among the different components.

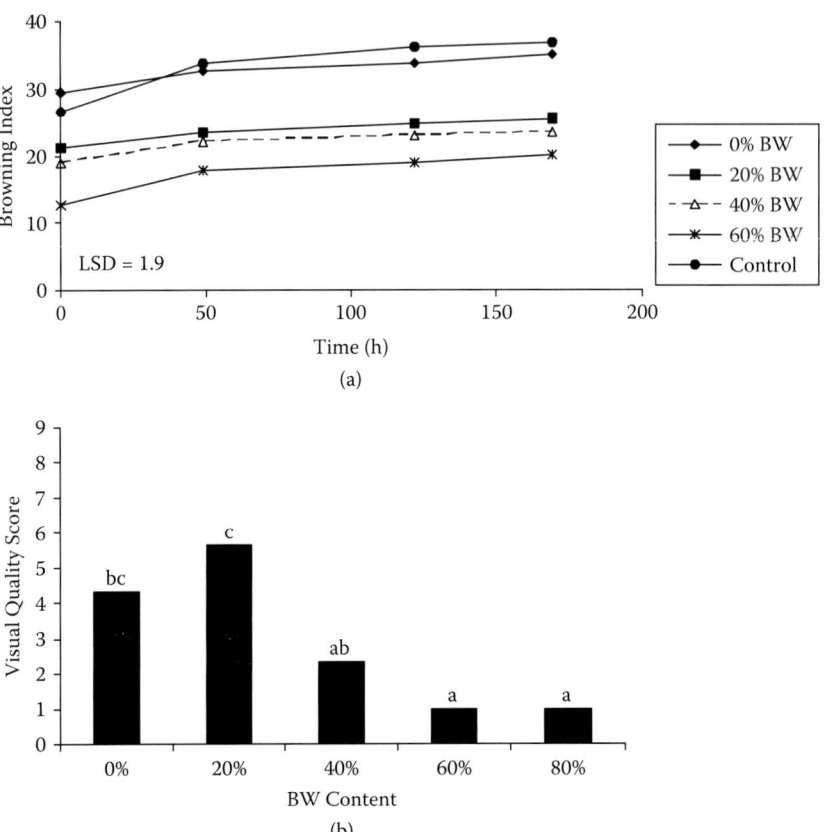

Figure 8.2 Effects of beeswax content in whey isolate-beeswax coating on (a) browning index and (b) visual quality (1 = poor quality, 9 = excellent) of fresh-cut apples during storage at 5°C. (Adapted from Pérez-Gago, M.B., Serra, M., Alonso, M., Mateos, M., and del Río, M.A., *J. Food Sci.*, 68(7), 2186–2191, 2003.).

It is important to investigate sensory quality of coating materials and coated products, including appearance, color, aroma, taste, and texture, because they are important factors that influence commercial success of fresh-cut products. Application of edible coatings, alone or with additives such as antibrowning or antimicrobial agents, could give the product an unattractive surface appearance or develop exogenous flavors affecting consumer repeat purchase.

The development of new technologies that allow more control of coating properties and functionalities are being investigated. Among them, a new technique, called micro- and nanoencapsulation—as opposed to macroencapsulation—consists in incorporating functional ingredients and antimicrobial compounds into edible coatings. This technology could pack solid, liquid, or gaseous substances in miniature (micro- or nanoscale) sealed capsules that can release their content at controlled rates under specific conditions (e.g., changes of pH, temperature, irradiation, osmotic shock). This technology shows the important advantage to protect encapsulated

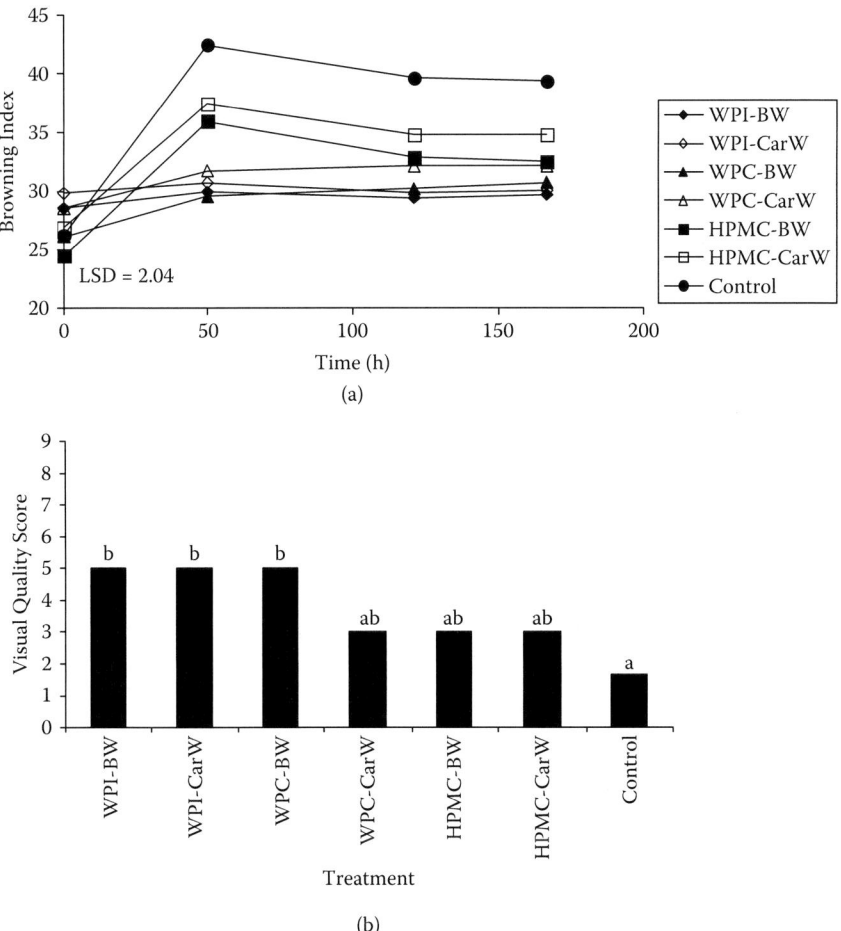

Figure 8.3 (a) Browning index and (b) visual quality (1 = poor quality, 9 = excellent) of fresh-cut apples coated with whey protein isolate (WPI), whey protein concentrate (WPC), hydroxypropyl methylcellulose (HPMC), beeswax (BW), or carnauba wax (CarW), during storage at 5°C. (Adapted from Pérez-Gago, M.B., Serra, M., Alonso, M., Mateos, M., and del Río, M.A., *Postharv. Biol. Technol.*, 36, 77–85, 2005.)

ingredients from moisture, heat, or other extreme storage conditions, as they are perishable in oxygen, light, or lipid oxidation. Most of the research so far has focused on nanoencapsulation of silver or zinc particles for microbial control on fruit surfaces (Rhim et al., 2006; An et al., 2008; Jin et al., 2008; Fayaz et al., 2009). In addition to increased functionality, incorporation of nanoparticles from clay derivatives advantageously modified physical properties (tensile strength and WVP) of a chitosan-based film (Rhim et al., 2006). One aspect that will need to be addressed with the development of nanotechnology in foods is their regulatory status. There is currently no regulation concerning nanoparticles, and it has been speculated that some harmless ingredients in their natural (i.e., "macro") form may become harmful under nanoparticle form (Sozer and Kokini, 2008).

Finally, another trend in the development of new coatings is incorporation of healthful additives, including probiotics (Rojas-Graü et al., 2009). Whether adding simple vitamins—ascorbic acid (Tapia et al., 2008), calcium, or vitamin E (Han et al., 2004; Mei et al., 2002)—or live probiotics, *Bifidobacterium lactis* (Tapia et al., 2007), the possibilities are endless, and each food developer can exert his or her creativity.

References

Abadias, M., Usall, J., Anguera, M., Solsona, C., Viñas, I., 2008. Microbiological quality of fresh, minimally-processed fruit and vegetables, and sprouts from retail establishments. *Int. J. Food Microbiol.* 123, 121–129.

Abeles, F.B., Morgan, P.W., Saltveit, M.E., 1992. *Ethylene in plant biology*, 2nd ed. Academic Press, San Diego, CA. 414 pp.

Agar, I.T., Massantini, R., Hess-Pierce, B., Kader, A.A., 1999. Postharvest CO_2 and ethylene production and quality maintenance of fresh-cut kiwifruit slices. *J. Food Sci.* 64, 433–440.

Aguayo, E., Escalona, V., Artés, F., 2004. Metabolic behavior and quality changes of whole and fresh processed melon. *J. Food Sci.* 69, 149–155.

Aguayo, E., Escalona, V.H., Artés, F.A., 2008. Effect of hot water treatment and various calcium salts on quality of fresh-cut 'Amarillo' melon. *Postharv. Biol. Technol.* 47, 397–406.

Aguila, J.S. del., Sasaki, F.F., Heiffig, L.S., Ortega, E.M.M., Jacomino, A.P., Kluge, R.A., 2006. Fresh-cut radish using different cut types and storage temperatures. *Postharv. Biol. Technol.* 40, 149–154.

Allong, R., Wickham, L.D., Mohammed, M., 2001. Effect of slicing on the rate of respiration, ethylene production and ripening of mango fruit. *J. Food Quality*. 24, 405–419.

Alothman, M., Bhat, R., Karim, A.A., 2009. UV radiation-induced changes of antioxidant capacity of fresh-cut tropical fruits. *Innovative Food Sci. Emerg. Technol.* 10, 512–516.

Amanatidou, A., Slump, R.A., Gorris, L.G.M., Smid, E.J., 2000. High oxygen and high carbon dioxide modified atmosphere for shelf-life extension of minimally processed carrots. *J. Food Sci.* 65(1), 61–66.

An, J., Zhang, M., Wang, S., Tang, J., 2008. Physical, chemical and microbiological changes in stored green asparagus spears as affected by coating of silver nanoparticles-PVP. *LWT—Food Sci. Technol.* 41, 1100–1107.

Andrady, A.L., Xu, P., 1997. Elastic behaviour of chitosans films. *J. Polymer Sci.* 35, 517–521.

Aquino-Bolaños, E.N., Cantwell, M.I., Peiser, G., Mercado-Silva, E., 2000. Changes in the quality of fresh-cut jicama in relation to storage temperatures and controlled atmospheres. *J. Food Sci.* 65, 1238–1243.

Arias, E., López-Buesa, P., Oria, R., 2009. Extension of fresh-cut 'Blanquilla' pear (*Pyrus communis* L.) shelf-life by 1-MCP treatment after harvest. *Postharv. Biol. Technol.* 54, 53–58.

Artés, F., Conesa, M.A., Hernández, S., Gil, M.I., 1999. Keeping quality of fresh-cut tomato. *Postharv. Biol. Technol.* 17, 153–162.

Artés, F., Gómez, P.A., Artés-Hernández, F., 2007. Physical, physiological and microbial deterioration of minimally fresh processed fruits and vegetables. *Food Sci. Tech. Int.* 13, 179–190.

Artés, F., Gómez, P., Aguayo, E., Escalona, V., Artés-Hernández, F., 2009. Sustainable sanitation techniques for keeping quality and safety of fresh-cut plant commodities. *Postharv. Biol. Technol.* 51, 287–296.

Artés-Hernández, F., Rivera-Cabrera, F., Kader, A.A., 2007. Quality retention and potential shelf-life of fresh-cut lemons as affected by cut type and temperature. *Postharv. Biol. Technol.* 43, 245–254.

Avena-Bustillos, R.J., Cisneros-Zevallos, L.A., Krochta, J.M., Saltveit, M.E., 1994. Application of casein-lipid edible film emulsions to reduce white blush on minimally processed carrots. *Postharv. Biol. Technol.* 4, 319–329.

Avena-Bustillos, R.J., Krochta, J.M., Saltveit, M.E., 1997. Water vapor resistance of Red Delicious apples and celery sticks coated with edible caseinate-acetylated monoglyceride films. *J. Food. Sci.* 62(2), 351–354.

Bai, J., Saftner, R.A., Watada, A.E., Lee, Y.S., 2001. Modified atmosphere maintains quality of fresh-cut cantaloupe (*Cucumis melo* L.). *J. Food Sci.* 66, 1207–1211.

Bai, J., Saftner, R.A., Watada, A.E. 2003. Characteristics of fresh-cut honeydew (*Cucumis×melo* L.) available to processors in winter and summer and its quality maintenance by modified atmosphere packaging. *Postharv. Biol. Technol.* 28, 349–359.

Baker, R.A., Hagenmaier, R.D., 1997. Reduction of fluid loss from grapefruit segments with wax microemulsion coatings. *J. Food. Sci.* 62(4), 789–792.

Baldwin, E.A., Baker, R.A., 2002. Use of proteins in edible coatings for whole and minimally processed fruits and vegetables. In: A. Gennadios (Ed.). *Protein-based films and coatings*. Boca Raton, FL: CRC Press, pp. 501–515.

Baldwin, E.A., Nisperos-Carriedo, M.O., Baker, R.A., 1995. Use of edible coatings to preserve quality of lightly (and slightly) processed products. *Crit. Rev. Food. Sci. Nutr.* 35, 509–524.

Baldwin, E.A, Nisperos, M.O., Chen, X., Hagenmaier, R.D., 1996. Improving storage life of cut apple and potato with edible coating. *Postharv. Biol. Technol.* 9, 151–163.

Barrancos, S., Abreu, M., Goncalves, E.M., Beirão-da Costa, S., Beirão-da-Costa, M.L., Moldão-Martins, M., 2003. The effect of heat pre-treatment on quality and shelf-life of fresh-cut apples. *Acta Hort.* 599, 595–601.

Berrang, M.E., Brackett, R.E., Beuchat, L.R., 1989. Growth of *Listeria monocytogenes* on fresh vegetables stored under controlled atmosphere. *J. Food Protect.* 52, 702–705.

Beaulieu, J.C., Lea, J.M., 2003. Volatile and quality changes in fresh-cut mangos prepared from firm-ripe and soft-ripe fruit, stored in clamshell containers and passive MAP. *Postharv. Biol. Technol.* 30, 15–28.

Beaulieu, J.C., and Gorny, J.R., 2004. Fresh-cut fruits. In: K.C. Gross, C.Y. Wang, M. Saltveit (Eds.). *The commercial storage of fruits, vegetables and florist and nursery stocks*. USDA Handbook 66.

Beaulieu, J.C., 2005. Within-season volatile and quality differences in stored fresh-cut cantaloupe cultivars. *J. Agric. Food Chem.* 53, 8679–8687.

Beaulieu, J.C., 2006. Volatile changes in cantaloupe during growth, maturation, and in stored fresh-cuts prepared from fruit harvested at various maturities. *J. Am. Soc. Hort. Sci.* 131, 127–139.

Beuchat, L.R., Brackett. R.E., 1990. Survival and growth of *Listeria monocycogenes* on lettuce as influenced by shredding, chlorine treatment, modified atmosphere packaging and temperature. *J. Food Sci.* 55, 755–758, 870.

Beuchat, L.R., Brackett, R.E., Hao, D.Y.Y., Conner, D.E., 1986. Growth and thermal inactivation of *Listeria monocytogenes* in cabbage and cabbage juice. *Can. J. Microbial.* 32, 791–795.

Bico, S.L.S., Raposo, M.F.J., Morais, R.M.S.C., Morais, A.M.M.B., 2009. Combined effect of chemical dip and/or carragenna coating and/or controlled atmosphere on quality of fresh-cut banana. *Food Control.* 20, 508–514.

Blankenship, S.M., Dole, J.M., 2003. 1-methylcyclopropene: a review. *Postharv. Biol. Technol.* 28, 1–25.

Boller, T., Kende, H., 1980. Regulation of wound ethylene synthesis in plants. *Nature.* 286, 259–260.

Bordini, M.E.B., Ristori, C.A., Jakabi, M., Gelli, D.S., 2007. Incidence, internalization and behavior of Salmonella in mangoes, var. Tommy Atkins. *Food Control.* 18, 1002–1007.

Bourtoom, T., 2008. Edible films and coatings: characteristics and properties. *Int. Food Res. J.* 15(3), 237–248.

Boynton, B.B., 2004. Determination of the effects of modified atmosphere packaging and irradiation on sensory characteristics, microbiology, texture and color of fresh-cut cantaloupe using modeling for packaging design. PhD Dissertation. University of Florida.

Brackett, R.E., 1987. Microbiological consequences of minimally processed fruits and vegetables. *J. Food Quality.* 10, 195–206.

Brault, D., D'Aprano, G., Lacroix, M., 1997. Formation of free-standing sterilized edible films from irradiated caseinates. *J. Ag. Food Chem.* 45, 2964–2969.

Brecht, J.K., 1995. Physiology of lightly processed fruits and vegetables. *HortScience.* 30, 18–22.

Brecht, J.K., Sabaa-Srur, A.U.O., Sargent, S.A., Bender, R.J., 1993. Hypochlorite inhibition of enzymic browning of cut vegetables and fruits. *Acta Hort.* 343, 342–344.

Brecht, J.K., Saltveit, M.E., Talcott, S.T., Schneider, K.R., Felkey, K., Bartz, J.A., 2004. Fresh-cut vegetables and fruit. *Hort. Rev.* 30, 185–251.

Budu, A.S., Joyce, D.C., 2003. Effect of 1-methylcyclopropene on the quality of minimally processed pineapple fruit. *Aust. J. Exp. Agric.* 43, 177–184.

Caillet, S., Millette, M., Salmiéri, S., Lacroix, M., 2006. Combined effects of antimicrobial coating, modified atmosphere packaging, and gamma irradiation on *Listeria innocua* present in ready-to-use carrots (*Daucus carota*). *J. Food Prot.* 69(1), 80–85.

Calderón-López, B., Bartsch, J.A., Lee, C.Y., Watkins, C.B., 2005. Cultivar effects on quality of fresh cut apple slices from 1-methylcyclopropene (1-MCP) treated apple fruit. *J. Food Sci.* 70, S221–S227.

Campaniello, D., Bevilacqua, A., Sinigaglia, M., Corbo, M.R., 2008. Chitosan: antimicrobial activity and potential applications for preserving minimally processed strawberries. *Food Microbiol.* 25, 992–1000.

Cantos, E., Tudela, J.A., Gil, M.I., Espin, J.C., 2002. Phenolic compounds and related enzymes are not rate-limiting in browning development of fresh-cut potatoes. *J. Agric. Food Chem.* 50, 3015–3023.

Cantwell, M.I., Suslow, T.V., 2002. Postharvest handling systems: fresh-cut fruits and vegetables. In: A.A. Kader (Ed.). *Postharvest technology of horticultural crops*, 3rd ed. Publication 3311. University of California, Berkeley, pp. 445–464.

Cha, D.S., Chinnan, M.S., 2004. Bipolymer-based antimicrobial packaging. A review. *Cri. Rev. Food Sci. Nutr.* 44(4), 223–237.

Chantanawarangoon, S., 2000. Quality maintenance of fresh-cut mango cubes. Masters thesis. University of California–Davis.

Chen, H. 2002. Formation and properties of casein films and coatings. In: A. Gennadios (Ed.). *Protein-based films and coatings*. Boca Raton, FL: CRC Press, pp. 181–211.

Chien, P.J., Sheu, F., Yang, F.H., 2007. Effects of edible chitosan coating on quality and shelf life of sliced mango fruit. *J. Food Eng.* 78, 225–229.

Coma, V., Martial-Gros, A., Garreau, S., Copinet, A., Salin, F., Deschamps, A., 2002. Edible antimicrobial films based on chitosan matrix. *J. Food Sci.* 67, 1162–1169.

Dean, B.R., Kolattukudy PE., 1976. Synthesis of suberin during wound-healing in jade leaves, tomato fruit, and bean pods. *Plant Physiol.* 58(3), 411–416.

Degl'Innocenti, E., Guidi, L., Pardossi, A., Togoni, F., 2005. Biochemical study of leaf browning in minimally processed leaves of lettuce (*Lactuca sativa* L. Var. Acephala). K. *Agric. Food Chem.* 53, 9980–9984.

Devlieghere, F., Vermeulen, A., Debevere, J., 2004. Chitosan: antimicrobial activity, interactions with food components and applicability as a coating on fruit and vegetables. *Food Microbiol.* 21, 703–714.

Donadon, J.R., Durigan, J.F., de Almeida Teixeira, G.H., Lima, M.A., Sarzi, B., 2004. Production and preservation of fresh-cut 'Tommy Atkins' mango chunks. *Acta Hort.* 645, 257–260.

Dong, H., Cheng, L., Tan, J., Zheng, K., Jiang, Y., 2004. Effects of chitosan coating on quality and shelf life of peeled lichi fruti. *J. Food Eng.* 64, 355–358.

Ducamp-Collin, M.-M., Reynes, M., Lebrun, M., Freire, Jr. M., 2009. Fresh-cut mango fruits: evaluation of edible coatings. *Acta Hort.* 820, 761–767.

Durango, A.M., Soares, N.F.F., Andrade, N.J., 2006. Microbiological evaluation of an edible antimicrobial coating on minimally processed carrots. *Food Control.* 17, 336–341.

Eissa, H.A.A., 2007. Effect of chitosan coating on shelf life and quality of fresh-cut mushroom. *J. Food Quality.* 30, 623–645.

Ergun, M., Huber, D.J., Jeong, J., Bartz, J.A., 2006. Extended shelf-life and quality of fresh-cut papaya derived from ripe fruit treated with the ethylene antagonist 1-methylcyclopropene. *J. Am. Soc. Hort. Sci.* 131, 97–103.

Escalona, V.H., Aguayo, E., Artés, F., 2003. Quality and physiological changes of fresh-cut kohlrabi. *HortScience.* 38, 1148–1152.

Eswaranandam, S., Hettiarachchy, N.S., Meullenet, J.F., 2006. Effect of malic and lactic acid incorporated soy protein coatings on the sensory attributes of whole apple and fresh-cut cantaloupe. *J. Food Sci.* 71(3), S307–S313.

Fan, L., Song, J., 2008. Microbial quality assessment methods for fresh-cut fruits and vegetables. *Stewart Postharv. Rev.* 4, 1–9.

Fan, X., Sokorai, K.J., 2008. Retention of quality and nutritional value of thirteen fresh-cut vegetables treated with low dose radiation. *J. Food Sci.* 73(7), S367–S372.

Fayaz, A.M., Balaji, K., Girilal, M., Kalaichelvan, P.T., Venkatesan, R., 2009. Mycobased synthesis of silver nanoparticles and their incorporation into sodium alginate films for vegetable and fruit preservation. *J. Agric. Food Chem.* 57, 6246–6252.

Forney, C.F., 2007. New innovations in the packaging of fresh-cut produce. *Acta Hort.* 746, 53–60.

Galliard, T., Matthew, A., Fishwick, M., Wright, A., 1976. The enzymic degradation of lipids resulting from physical disruption of cucumber (*Cucumis sativis*) fruit. *Phytochemistry.* 15, 1647–1650.

Garcia, E., Barrett, D.M., 2002. Preservative treatments for fresh-cut fruits and vegetables. In: O. Lamikanra (Ed.). *Fresh-cut fruits and vegetables: science, technology and market.* CRC Press, Boca Raton, FL, pp. 267–304.

Gennadios, A., Hanna, M.A., Kurth, B., 1997. Application of edible coatings on meats, poultry, and seafood: a review. *Lebensm.Wissen. Technol.* 30, 337–350.

Gennadios, A., McHugh, T.H., Weller, G.L., Krochta, J.M., 1994. Edibles coatings and film based on proteins. In: J.M. Krochta, E.A. Baldwin, M.O. Nisperos-Carriedo (Eds.). *Edible coatings and films to improve food quality.* Lancaster, PA: Technomic, pp. 201–277.

Geraldine, R.M., de Fátima Ferreira Soares N., Alvarenga Botrel, D., de Almeida Gonçalves, L., 2008. Characterization and effect of edible coatings on minimally processed garlic quality. *Carbohydrate Polym.* 72, 403–409.

Ghidelli, C., Mateos, M., Sanchís, E., Rojas-Argudo, C., del Río, M.A., Pérez-Gago, M.B., 2010a. Effect of edible coating and modified atmosphere packaging on enzymatic browning of fresh-cut persimmons cv. Rojo brillante. *Acta Hort.* (ISHS) 876, 341–348

Ghidelli, C., Mateos, M., Rojas-Argudo, C., Sanchís, E., del Río, M.A., Pérez-Gago, M.B., 2010b. Application of soy protein-beeswax edible coating with antioxidant on reducing enzymatic browning of fresh-cut eggplants. *Acta Hort.* (ISHS) 877, 591–596.

Gil, M.I., Aguayo, E., Kader, A.A., 2006. Quality changes and nutrient retention in fresh-cut versus whole fruits during storage. *J. Agric. Food Chem.* 54, 4284–4296.

Giménez, M., Olarte, C., Sanz, S., Lomas, C., Echávarri, J.F., Ayala, F., 2003. Relation between spoilage and microbiological quality in minimally processed artichoke packaged with different films. *Food Microbiol.* 20, 231–242.

Goldberg, R., Le, T., Catesson, A.-M., 1985. Localization and properties of cell wall enzyme activities related to the final stages of lignin biosynthesis. *J. Exp. Bot.* 36, 503–510.

Gómez, P.A., Artés, F., 2005. Improved keeping quality of minimally fresh processed celery sticks by modified atmosphere packaging. *LWT—Food Sci. Technol. Intl.* 38, 323–329.

Gómez-López, V.M., Rajkovic, A., Ragaert, P., Smigic, N., Devlieghere, F., 2009. Chlorine dioxide for minimally processed produce preservation: a review. *Trends Food Sci. Technol.* 20, 17–26.

Gong, Y., Mattheis, J.P., 2003. Effect of ethylene and 1-methylcyclopropene on chlorophyll catabolism of broccoli florets. *Plant Growth Regul.* 40, 33–38.

González-Aguilar, G.A., Ruiz-Cruz, S., Cruz-Valenzuela, R., Rodriguez-Felix, A., Wang, C.Y., 2004. Physiological and quality changes of fresh-cut pineapple treated with antibrowning agents. *LWT—Food Sci. Technol.* 37, 369–376.

González-Aguilar, G.A., Celis, J., Sotelo-Mondo, R.R., Rosa, L.A., Rodrigo-Garcia, J., Alvarez-Parrilla, E., 2007a. Physiological and biochemical changes of different fresh-cut mango cultivars stored at 5°C. *Intl. J. Food Sci. Technol.* 43, 91–101.

González-Aguilar, G.A., Villegas-Ochoa, M.A., Martinez-Tellez, M.A., Gardea, A.A., Ayala-Zavala, J.F., 2007b. Improving antioxidant capacity of fresh-cut mangoes treated with UV-C. *J. Food Sci.* 72, S197–S202.

Gorny, J.R., 2003. A summary of CA and MA requirements and recommendations for fresh-cut (minimally processed) fruits and vegetables. *Acta Hort.* 600, 609–614.

Gorny, J.R., Hess-Pierce, B., Kader, A.A., 1999. Postharvest physiology and quality maintenance of fresh-cut nectarines and peaches. *Acta Hort.* 485, 173–179.

Gorny, J.R., Cifuentes, R.A., Hess-Pierce, B., Kader, A.A., 2000. Quality changes in fresh-cut pear slices as affected by cultivar, ripeness stage, fruit size, and storage regime. *J. Food Sci.* 65, 541–544.

Grisebach, H. 1987. New insights into the ecological role of secondary plant metabolites. *Comments Agric. Food Chem.* 1, 27–45.

Guilbert, S., 1986. Technology and application of edible protective films. In: M. Mathlouthi (Ed.). *Food packaging and preservation: theory and practice.* London, Elsevier Applied Science, London, UK, pp. 371–394.

Haard, N.F., Cody, M., 1978. Stress metabolites in postharvest fruits and vegetables-role of ethylene. In: H.O. Hultin, M. Milner (Eds.). *Postharv. Biol. Biotechnol.* Westport, CT: Food Nutr. Press, pp. 111–135.

Hagenmaier, R.D., Baker, R.D., 1994. Wax microemulsion and emulsions as citrus coatings. *J. Agric. Food Chem.* 42, 899–902.

Han, C., Zhao, Y., Leonard, S.W., Traber, M.G. 2004. Edible coatings to improve storability and enhance nutritional value of fresh and frozen strawberries (*Fragaria*×*ananassa*) and raspberries (*Rubus ideaus*). *Postharv. Biol. Technol.* 33, 67–78.

Heard, G.M., 2002. Microbiology of fresh-cut produce. In: O. Lamikanra (Ed.). *Fresh-cut fruits and vegetables: science, technology and market.* CRC Press, Boca Raton, FL, pp. 187–248.

Hernandez, E., 1994. Edible coatings from lipids and resins. In: J.M. Krochta, E.A. Baldwin, M.O. Nisperos-Carriedo (Eds.). *Edible coatings and films to improve food quality.* Lancaster, PA: Technomic, pp. 279–303.

Hirano, S., 1999. Chitin and chitosan as novel biotechnological materials. *Polym. Int.* 48, 732–734.

Hisaminato, H., Murata, M., Homma, S., 2001. Relationship between the enzymatic browning and phenylalanine ammonia-lyase activity of cut lettuce, and the prevention of browning by inhibitors of polyphenol biosynthesis. *Biosci. Biotech. Biochem.* 65, 1016–1021.

Hodges, D.M., Toivonen, P.M.A., 2008. Quality of fresh-cut fruits and vegetables as affected by exposure to abiotic stress. *Postharv. Biol. Technol.* 48, 155–162.

Hoffman, N.E., Yang, S.F., 1982. Enhancement of wound-induced ethylene synthesis by ethylene in preclimacteric cantaloupe. *Plant Physiol.* 69, 317–322.

Hu, W., Jiang, Y., 2007. Quality attributes and control of fresh-cut produce. *Stewart Postharv. Rev.* 3, 1–9.

Hu, W.Z., Jiang, A.L., Qi, H.P., Pang, K., 2007a. Changes in wound-induced ethylene production and ACC oxidase ion fresh-cut squash. *Acta Hort.* 746, 357–362.

Hu, W.Z., Pang, K., Jiang, A.L. Tian, M.X., 2007b. Changes in ethylene production, respiration and polyphenol oxidase of fresh-cut apple. *Acta Hort.* 746, 369–374.

Hyodo, H., Nishino T., 1981. Wound-induced ethylene formation in albedo tissue of citrus fruit. *Plant Physiol.* 67, 421–423.

Hyodo, H., Tanaka K., Watanabe, K., 1983. Wound-induced ethylene production and l-aminocyclopropane-l-carboxylic acid synthase in mesocarp tissue of winter squash fruit. *Plant Cell Physiol.* 24, 963–969.

Janjarasskul, T., Krochta, J.M., 2010. Edible packaging materials. *Annu. Rev. Food Sci. Technol.* 1, 415–448.

Jin, T., Sun, D., Su, J.Y., Zhang, H., Sue, H.J., 2008. Antimicrobial efficacy of zinc oxide quantum dots against *Listeria monocytogenes*, *Salmonella Enteritidis*, and *Escherichia coli* O157:H7. *J. Food Sci.* 74, M46–M52.

Kader, A.A., 2002. Modified atmospheres during transport and storage, In: A.A. Kader (Ed.). *Postharvest technology of horticultural crops,* 3rd ed. Publication 3311. University of California, Berkeley, pp. 135–144.

Kader, A.A., Saltveit, M.E., 2003. Respiration and gas exchange. In: J.A. Bartz, J.K. Brecht (Eds.). *Postharvest physiology and pathology of vegetables.* Marcel Dekker, New York, pp. 7–29.

Kallander, K.D., Hitchins, A.D., Lancette, G.A., Schmieg, J.A., Garcia, G.R., Solomon H.M., Sofos, J.N., 1991. Fate of *Listeria monocytogenes* in shredded cabbage stored at 5 and 25°C under a modified atmosphere. *J. Food Protect.* 54, 302–304.

Karakurt, Y., Huber, D.J., 2003. Activities of several membrane and cell-wall hydrolyses, ethylene biosynthetic enzymes, and cell wall polyuronide degradation during low-temperature storage of intact and fresh-cut papaya (*Carica papaya*) fruit. *Postharv. Biol. Technol.* 28, 219–229.

Karbowiak, T., Debeaufort, F., Volley, A., 2007. Influence of thermal process on structure and functional properties of emulsion-based edible films. *Food Hydrocolloid.* 21, 879–888.

Kays, S.J., 1991. Postharvest physiology of perishable plant products. Van Nostrand Reinhold, New York. 532 pp.

Kester, J.J., Fennema, O.R., 1986. Edible films and coatings: a review. *Food Technol.* 40(12), 47–59.

Kim, D.M., Smith, N.L., Lee, C.Y., 1993. Apple cultivar variations in response to heat treatment and minimal processing. *J. Food Sci.* 58, 1111–1124.

Kim, J.G., Luo, Y., Saftner, R.A., Gross, K.C., 2005a. Delayed modified atmosphere packaging of fresh-cut romaine lettuce: effects on quality maintenance and shelf-life. *J. Am. Soc. Hort. Sci.* 130, 116–123.

Kim, J.G., Luo, Y., Tao, Y., Saftner, R.A., Gross, K.C., 2005b. Effect of initial oxygen concentration and film oxygen transmission rate on the quality of fresh-cut romaine lettuce. *J. Sci. Food Agric.* 85, 1622–1630.

Klein, B.P., Grossman, S., King, D., Cohen, B.S., Pinsky, A., 1984. Pigment bleaching, carbonyl production and antioxidant effects during the anaerobic lipoxygenase reaction. *Biochem. Biophys. Acta.* 793, 72–79.

Koukounaras, A., Diamantidis, G., Sfakiotakis, E., 2008. The effect of heat treatment on quality retention of fresh-cut peach. *Postharv. Biol. Technol.* 48, 30–36.

Krochta, J.M., 1997. Edible protein films and coatings. In: S. Damodaran, A. Paraf (Eds.). *Food proteins and their applications*. Marcel Dekker, New York, pp. 529–550.

Krochta, J.M., 2002. Proteins as raw materials for films and coatings: definition, current status, and opportunities. In: A. Gennadios (Ed.). *Protein-based films and coatings*. CRC Press, Boca Raton, FL, pp. 1–41.

Krochta, J.M., de Mulder-Johnston, C., 1997. Edible and biodegradable polymer films: challenges and opportunities. *Food Technol.* 51(2), 61–74.

Lafortune, R., Caillet, S., Lacroix, M., 2005. Combined effects of coating, modified atmosphere packaging, and gamma irradiation on quality maintenance of ready-to-use carrots (*Daucus carota*). *J. Food Prot.* 68(2), 353–359.

Lamikanra, O., Richard, O.A., 2002. Effect of storage on some volatile aroma compounds in fresh-cut cantaloupe melon. *Agric. Food Chem.* 50, 4043–4047.

Lamikanra, O., Chen, J.C., Banks, D., Hunter, P. A., 2000. Biochemical and microbial changes during the storage of minimally processed cantaloupe. *J. Agric. Food Chem.* 48, 5955–5961.

Lamikanra, O., Richard, O.A., Parker, A., 2002. Ultraviolet induced stress response in fresh cut cantaloupe. *Phytochemistry.* 60, 27–32.

Lana, M.M., Tijskens, L.M.M., van Kooten, O., 2006. Modelling RGB colour aspects and translucency of fresh-cut tomatoes. *Postharv. Biol. Technol.* 40, 15–25.

Lavelli, V., Pagliarini, E., Ambrosoli, R., Minati, J.L., Zanoni, B., 2006. Physicochemical, microbial, and sensory parameters as indices to evaluate the quality of minimally-processed carrots. *Postharv. Biol. Technol.* 40, 34–40.

Lawton, J.W., 1996. Effect of starch type on the properties of starch containing films. *Carbohydrate Polym.* 29, 203–208.

Lee, J.Y., Park, H.J., Lee, C.Y., Choi, W.Y., 2003. Extending shelf-life of minimally processed apples with edible coatings and antibrowning agents. *LWT—Food Sci.Technol.* 36, 323–329.

Lee, T.H., McGlasson, W.B., Edwards, R.A., 1970. Physiology of disks of irradiated tomato fruit. I. Influence of cutting and infiltration on respiration, ethylene production and ripening. *Rad. Bot.* 10, 521–529.

Lemoine, M.L., Civello, P., Chaves, A., Martínez, A., 2009. Hot air treatment delays senescence and maintains quality of fresh-cut broccoli florets during refrigerated storage. *LWT—Food Sci. Technol.* 42, 1076–1081.

Le Tien, C., Vachon, C., Mateescu, M.A., Lacroix, M., 2001. Milk protein coatings prevent oxidative browning of apples and potatoes. *J. Food. Sci.* 66(4), 512–516.

Li, P., Barth, M.M., 1998. Impact of edible coatings on nutritional and physiological changes in lightly-processed carrots. *Postharv. Biol. Technol.* 14, 51–60.

Li, Y., Brackett, R.E., Shewfelt, R.L., Beuchat, L.R., 2001. Changes in appearance and natural microflora on iceberg lettuce treated in warm, chlorinated water and then stored at refrigeration temperature. *Food Microbiol.* 18, 299–308.

Liao, M.L., Seib, P.A., 1988. Chemistry of l-ascorbic acid related to foods. *Food Chem.* 30, 289–312.

Limbanyen, A., Brecht, J.K., Sargent, S.A., Bartz, J.A., 1998. Fresh-cut mango fruit slices. *HortScience.* 33, 457.

Lin, D., Zhao, Y., 2007. Innovation in development and application of edible coatings for fresh and minimally processed fruits and vegetables. *Comp. Rev. Food Sci. Food Safety.* 6(3), 60–75.

Loewus F.A., Loewus M.W., 1987. Biosynthesis and metabolism of ascorbic acid in plants. *CRC Crit. Rev. Plant Sci.* 5, 101–119.

Luna-Guzmán, I., Cantwell, M., Barrett, D.M., 1999. Fresh-cut cantaloupe: effects of $CaCl_2$ dips and heat treatments on firmness and metabolic activity. *Postharv. Biol. Technol.* 17, 201–213.

MacLeod, R.F., Kader, A.A., Morris, L.L., 1976. Stimulation of ethylene and CO_2 production of mature-green tomatoes by impact bruising. *Hort. Sci.* 11, 604–606.

Maftoonaazad, N., Ramaswamy,H.S., 2005. Postharvest shelf-life extension of avocados using methyl cellulose-based coating. *Lebens. Wissen. Technol.* 38, 617–624.

Mali, S., Grossmann, M.V.E., Garcia, M.A., Martino, M.N., Zaritzky, N.E., 2002. Microstructural characterization of yam starch films. *Carbohydrate Polym.* 50, 379–386.

Marrero, A., Kader, A.A., 2006. Optimal temperature and modified atmosphere for keeping quality of fresh-cut pineapples. *Postharv. Biol. Technol.* 39, 163–168.

McGlasson, W.B., Pratt, H.K., 1964. Effects of wounding on respiration and ethylene production by cantaloupe fruit tissue. *Plant Physiol.* 39, 128–132.

McHugh, T.H., Krochta, J.M., 1994. Permeability properties of edible films. In: J.M. Krochta, E.A. Baldwin, M.O. Nisperos-Carriedo (Eds.). *Edible coatings and films to improve food quality*. Technomic, Lancaster, PA, pp. 139–187.

McHugh, T.H., Aujard, J.F., Krotcha, J.M., 1994. Plasticized whey protein edible films: water vapor permeability properties. *J. Food Sci.* 59, 416–419, 423.

McHugh, T.H., Huxsoll, C.C., Krochta, J.M., 1996. Permeability properties of fruit puree edible films. *J. Food Sci.* 61, 88–91.

McHugh, T.H., Senesi, E., 2000. Apple wraps: a novel method to improve the quality and extend the shelf life of fresh-cut apples. *J. Food Sci.* 65(3), 480–485.

Mei, Y., Zhao, Y., Yang, J., Furr, H.C., 2002. Using edible coating to enhance nutritional and sensory qualities of baby carrots. *J. Food. Sci.* 67(5), 1964–1968.

Mencarelli, F., Saltveit, M.E., Jr, Massantini, R., 1989. Lightly processed foods: ripening of tomato fruit slices. *Acta Hort.* 244, 193–200.

Mir, N., Beaudry, R.M., 2004. Modified atmosphere packaging. In: K.C. Gross, C.Y. Wang, M. Saltveit (Eds.). *The commercial storage of fruits, vegetables, and florist and nursery stocks*. USDA Handbook 66.

Mohammed, M., Brecht, J.K., 2002. Reduction of chilling injury in 'Tommy Atkins' mangoes during ripening. *Sci. Hort.* 95, 297–308.

Moleyar, V., Narasimham, P. 1994. Modified atmosphere packaging of vegetables: an appraisal. *J. Food Sci. Technol.* 31, 267–278.

Morillon, V., Debeaufort, F., Blond, G., Capelle, M., Voilley, A., 2002. Factors affecting the moisture permeability of lipid based edible films: a review. *Crit. Rev. Sci. Nutr.* 42, 67–89.

Murata, M., Tanaka, E., Minoura, E., Homma, S., 2004. Quality of cut lettuce treated by heat shock: prevention of enzymatic browning, repression of phenylalanine ammonialyase activity, and improvement on sensory evaluation during storage. *Biosci. Biotech. Biochem.* 68, 501–507.

Muzzarelli, R.A.A., 1996. Chitosan-based dietary foods. *Carbohydrate Polym.* 29, 309–316.

Narciso, J., Plotto, A., 2005. A comparison of sanitation systems for fresh-cut mango. *HortTechnol.* 15, 837–842.

Ngarmsak, M., Ngarmsak, T., Delaquis, P. J., Toivonen, P. M., 2005. Effect of sanitation treatments on the microbiology of fresh-cut Thai mango. *Proceeding of the APEC Symposium on Assuring Quality and Safety of Fresh Produce.* Pp. 347–354.

Ngarmsak, M., Delaquis, P., Toivonen, P., Ngarmsak, T., Ooraikul, B., Mazza, G., 2006. Microbiology of fresh-cut mangoes prepared from fruit sanitized in hot chlorinated water. *Food Sci. Technol. Int.* 12, 95–103.

Nisperos-Carriedo, M.O., 1994. Edible coatings and films based on polysaccharides. In: J.M. Krochta, E.A. Baldwin, M. Nisperos-Carriedo (Eds.). *Edible coatings and films to improve food quality.* Technomic, Lancaster, PA, pp. 305–335.

Nunes, M.C.N., Brecht, J.K., Morais, A.M.M.B., Sargent, S.A., 1998. Controlling temperature and water loss to maintain ascorbic acid levels in strawberries during postharvest handling. *J. Food Sci.* 63, 1033–1036.

Ôba, K., Fukui, M., Imai, Y., Iriyama, S., Nogami, K., 1994. L-galactono-γ-lactone dehydrogenase, partial characterization, induction of activity and role in the synthesis of ascorbic acid in wounded white potato tuber tissue. *Plant Cell Physiol.* 35, 473–478.

Odriozola-Serrano, I., Soliva-Fortuny, R., Martín-Belloso, O., 2008. Effect of minimal processing on bioactive compounds and color attributes of fresh-cut tomatoes. *LWT—Food Sci. Technol.* 41, 217–226.

Olivas, G.I., Barbosa-Cánovas, G.V., 2005. Edible coatings for fresh-cut fruits. *Critical Rev. Food Sci. Nutr.* 45, 657–670.

Olivas, G.I., Rodriguez, J.J., Barbosa-Cánovas, G.V., 2003. Edible coatings composed of methylcellulose, stearic acid, and additives to preserve quality of pear wedges. *J. Food Process. Preserv.* 27, 299–320.

Olivas, G.I., Mattinson, D.S., Barbosa-Cánovas, G.V., 2007. Alginate coatings for preservation of minimally processed 'Gala' apples. *Postharv. Biol. Technol.* 45, 89–96.

Oms-Oliu, G., Soliva-Fortuny, R., Martín-Belloso, O., 2008a. Using polysaccharide-based edible coatings to enhance quality and antioxidant properties of fresh-cut melon. *LWT—Food Sci.Technol.* 41, 1862–1870.

Oms-Oliu, G., Soliva-Fortuny R., Martín Belloso, O., 2008b. Edible coatings with antibrowning agents to maintain sensory quality and antioxidant properties of fresh-cut pears. *Postharv. Biol. Technol.* 50, 87–94.

Palmer, J.K., McGlasson, W.B., 1969. Respiration and ripening of banana fruit slices. *Aust. J. Bio. Sci.* 22, 87–99.

Papineau, A.M., Hoover, D.G., Knorr, K., Farkas, D.F., 1991. Antimicrobial effect of water-soluble chitosans with high hydrostatic pressure. *Food Biotechnol.* 5, 45–57.

Paull, R.E., Chen, W., 1997. Minimal processing of papaya (*Carica papaya* L.) and the physiology of halved fruit. *Postharv. Biol. Technol.* 12, 93–99.

Pen, L.T., Jiang, Y.M., 2003. Effects of chitosan coating on shelf life and quality of fresh-cut Chinese water chestnut. *LWT—Food Sci.Technol.* 36, 359–364.

Perera, C.O., Balchin, L., Baldwin, E., Stanley, R., Tian, M., 2003. Effect of 1-MCP on the quality of fresh-cut apple slices. *J. Food Sci.* 68, 1910–1914.

Pérez-Gago, M.B., Krochta, J.M., 2001. Lipid particle size effect on water vapor permeability and mechanical properties of whey protein-beeswax emulsion films. *J. Agric. Food Chem.* 49, 996–1002.

Pérez-Gago, M.B., Serra, M., Alonso, M., Mateos, M., del Río, M.A., 2003. Effect of solid content and lipid content of whey protein isolate-beeswax edible coatings on color change of fresh-cut apples. *J. Food Sci.* 68(7), 2186–2191.

Pérez-Gago, M.B., Serra, M., Alonso, M., Mateos, M., del Río, M.A., 2005a. Effect of whey protein-and hydroxypropyl methylcelluylose-based edible composite coatings on color change of fresh-cut apples. *Postharv. Biol. Technol.* 36, 77–85.

Pérez-Gago, M.B., Serra, M., del Río, M.A., 2005b. Effect of whey protein-beeswax edible composite coating on color change of fresh-cut persimmons cv. 'Rojo Brillante'. *Acta Hort.* 682, 1917–1924.

Pérez-Gago, M.B., Serra, M., del Río, M.A., 2006. Color change of fresh-cut apples coated with whey protein concentrate-based edible coatings. *Postharv. Biol. Technol.* 39, 84–92.

Pinheiro, S.C.F., Almeida, D.P.F., 2008. Modulation of tomato pericarp firmness through pH and calcium: implications for the texture of fresh-cut fruit. *Postharv. Biol. Technol.* 47, 119–125.

Plotto, A., Bai, J., Baldwin, E.A., Brecht, J.K., 2003. Effect of pretreatment of intact 'Kent' and 'Tommy Atkins' mangoes with ethanol vapor, heat or 1-methylcyclopropene *n* quality and shelf life of fresh-cut slices. *Proc. Fla. State Hort. Soc.*, 116, 394–400.

Plotto, A., Goodner, K.L., Bai, J., Rattanapanone, N., Baldwin A.E., 2004. Effect of polysaccharide coatings on quality of fresh-cut mangoes (*Mangifera indica*). *Proc. Fla. State Hort. Soc.*, 117, 382–388.

Ponce de León, L., Muñoz, C., Pérez, L., Díaz de León, F., Kerbel., C., Pérez Flores, L., Esparza, S., Bósquez, E., Trinidad, M., 1997. Hot-water quarantine treatment and water-cooling of 'Haden' mangoes. *Acta Hort.* 455, 786–796.

Portela, S.I., Cantwell, M.I., 2001. Cutting blade sharpness affects appearance and other quality attributes of fresh-cut cantaloupe melon. *J. Food Sci.* 66, 1265–1270.

Poubol, J., Izumi, H., 2005a. Physiology and microbiological quality of fresh-cut mango cubes as affected by high-O_2 controlled atmospheres. *J. Food Sci.* 70, M286–M291.

Poubol, J., Izumi, H., 2005b. Shelf-life and microbial quality of fresh-cut mango cubes stored in high CO_2 atmospheres. *J. Food Sci.* 70, M69–M74.

Premier, R., Jaeger, J., Tomkins, B., 2007. Microbial quality considerations for the ready-to-eat fresh produce industry. *Acta Hort.* 746, 25–32.

Priepke, P.E., Wei, L.S., Nelson, A.I., 1976. Refrigerated storage of prepackaged salad vegetables. *J. Food Sci.* 41, 379–382.

Quezada-Gallo, J.A., Debeaufort, F., Callegarin, F., Voilley, A., 2000. Lipid hydrophobicity, physical state and distribution effects on the properties of emulsion-based edible films. *J. Membrane Sci.* 180, 37–46.

Ragaert, P., Deliedhere, F., Debevere, J., 2007. Role of microbial and physiological spoilage mechanisms during storage of minimally processed vegetables. *Postharv. Biol. Technol.* 44, 185–194.

Rattanapanone, N., Lee, Y., Wu, T., Watada, A.E., 2001. Quality and microbial changes of fresh-cut mango cubes held in controlled atmosphere. *HortScience.* 36, 1091–1095.

Raybaudi-Massilia, R.M., Mosqueda-Melgar, J., Martín-Belloso, O., 2008a. Edible alginate-based coating as carrier of antimicrobials to improve shelf-life and safety of fresh-cut melon. *Inter. J. Food Microbiol.* 121, 313–327.

Raybaudi-Massilia, R.M., Rojas-Graü, M.A., Mosqueda-Melgar, J., Martín-Belloso, O., 2008b. Comparative study on essential oil incorporated into an alginate-based edible coating to assure the safety and quality of fresh-cut Fuji apples. *J. Food Protec.* 71(6), 1150–1161.

Reid, M.S., 1992. Maturation and maturity indices. In: A.A. Kader (Ed.). *Postharvest technology for horticultural crops.* University of California Division of Agriculture and Natural Resources Publication 3311, pp. 21–28.

Reyes, L.F., Villarreal, J.E., Cisneros-Zevallos, L., 2007. The increase in antioxidant capacity after wounding depends on the type of fruit or vegetable tissue. *Food Chem.* 101, 1254–1262.

Rhim, J.W., Gennadios, A., Handa, A., Weller, C.L., Hanna, M.A., 2000. Solubility, tensile, and color properties of modified soy protein isolate films. *J. Agric. Food Chem.* 48, 4937–4941.

Rhim, J.-W., Hong, S.-I., Park, H.-M., Ng, P.K.W., 2006. Preparation and characterization of chitosan-based nanocomposite films with antimicrobial activity. *J. Agric. Food Chem.* 54, 5814–5822.

Rico, D., Martín-Diana, A.B., Barat, J.M., Barry-Ryan, C., 2007. Extending and measuring the quality of fresh-cut fruit and vegetables: a review. *Trends Food Sci. Technol.* 18, 373–386.

Rivera-López, J., Vázquez-Ortiz, F.A., Ayala-Zavala, F., Sotelo-Mundo, R.R., González-Aguilar, G.A., 2005. Cutting shape and storage temperature affect overall quality of fresh-cut papaya cv. 'Maradol'. *J. Food Sci.* 70, 482–489.

Rodov, V., Horev, B., Goldman, G., Vinokur, Y., Fishman, S., 2007. Model-driven development of microperforated active modified-atmosphere packaging for fresh-cut produce. *Acta Hort.* 746, 83–88.

Rojas-Graü, M.A., Raybaudi-Massilia, R.M., Soliva-Fortuny, R., Avena-Bustillos, R.J., McHugh, T.H., Martín-Belloso O., 2007. Apple puree-alginate edible coating as carrier of antimicrobial agents to prolong shelf life of fresh-cut apples. *Postharv. Biol. Technol.* 45, 254–264.

Rojas-Graü, M.A., Tapia, M.S., Martín-Belloso, O., 2008. Using polysaccharide-based edible coatings to maintain quality of fresh-cut Fuji apples. *LWT—Food Sci.Technol.* 41, 139-147.

Rojas-Graü, M.A., Soliva-Fortuny, R., Martín-Belloso O., 2009. Edible coatings to incorporate active ingredients to fresh-cut fruits: a review. *Trends Food Sci. Technol.* 20, 438–447.

Rosen, J.C., Kader, A.A., 1989. Postharvest physiology and quality maintenance of sliced pear and strawberry fruits. *J. Food Sci.* 54, 656–659.

Ryser, E.T., Marth, E.H., 1991. *Listeria, listeriosis and food safety.* Marcel Dekker, New York, pp. 632.

Salman, A., Filgueiras, E., Cristescu, S., Lopez-Lauri, F., Harren, F., Sallanon, H., 2009. Inhibition of wound-induced ethylene does not prevent red discoloration in fresh-cut endive (*Cichorium intybus* L.). *Eu. Food Res. Technol.* 228, 651–657.

Saltveit, M.E., 1997. Physical and physiological changes in minimally processed fruits and vegetables. In: F.A. Tomás-Barberán, R.J. Robins (Eds.). *Phytochemistry of fruits and vegetables.* Oxford University Press, New York, pp. 205–220.

Saltveit, M.E., 2003. Is it possible to find an optimal controlled atmosphere? *Postharv. Biol. Technol.* 27, 3–13.

Sangsuwan, J., Rattanapanone, N., Rachtanapun, P., 2008. Effect of chitosan/methyl cellulose films on microbial and quality characteristics of fresh-cut cantaloupe and pineapple. *Postharv. Biol. Technol.* 49, 403–410.

Saucedo-Pompa, S., Jasso-Cantu, D., Ventura-Sobrevilla, J., Sáenz-Galindo, A., Rodríguez-Herrera, R., Aguilar, C.N., 2007. Effect of candelilla wax with natural antioxidants on the shelf life quality of fresh-cut fruits. *J. Food. Qual.* 30, 823–836.

Schieberle, P., Grosch, W., Kexel, H., Schmidt, H.L., 1981. A study of oxygen isotope scrambling in the enzymic and non-enzymic oxidation of linoleic acid. *Biochem. Biophys. Acta.* 666, 322–326.

Schwimmer, S., 1983. Biochemistry in the food industry: molecular pabulistics and thanatobolism; *TIBS.* 8, 306–310.

Shewfelt, R.L., 1987. Quality of minimally processed fruits and vegetables. *J. Food Qual.*, 10, 143–156.

Shon, J., Haque, Z.U., 2007. Efficacy of sour whey as a shelf-life enhancer: use in antioxidative edible coatings of cut vegetable and fruit. *J. Food Qual.* 30, 581–593.

Sime, W.J., 1990. Alginates. In: P. Harris (Ed.). *Food gels.* Elsevier Applied Science, London, pp. 53–58.

Simões, A.D.N., Tudela, J.A., Allende, A., Puschmann, R., Gil, M.I., 2009. Edible coatings containing chitosan and moderate modified atmosphere maintain quality and enhance phytochemicals of carrot sticks. *Postharv. Biol. Technol.* 51, 364–370.

Soliva-Fortuny, R.C., Martín-Belloso, O., 2003. New advances in extending the shelf-life of fresh-cut fruits: a review. *Trends Food Sci. Technol.* 14, 341–353.

Sothornvit, R., Krochta, J.M., 2001. Plasticizer effect on mechanical properties of beta-lactoglobulin films. *J. Food. Eng.* 50, 149–155.

Sothornvit, R., Rodsamran, P., 2008. Effect of a mango film on quality of whole and minimally processed mangoes. *Postharv. Biol. Technol.* 47, 407–415.

Sozer, N., and Kokini, J.L., 2008. Nanotechnology and its applications in the food sector. *Trends Biotechnol.* 27(2), 82–89.

Steinbruegge, E.G., Maxcy, R.B., Liewen, M.B., 1988. Fate of *Listeria monocytogenes* on ready to serve lettuce. *J. Food Protect.* 51, 596–599.

Sukumaran, N.P., Jassal, S., Verma, S.C., 1990. Quantitative determination of suberin deposition during wound healing in potatoes (*Solanum tuberosum* L.). *J. Sci. Food Agric.* 51, 271–274.

Surjadinata, B.B., Cisneros-Zevallos, L., 2003. Modeling wound-induced respiration of fresh-cut carrots (*Daucus Carota* L.). *J. Food Sci.* 68, 2735–2740.

Tapia, M.S., Rojas-Graü, M.A., Rodríguez F.J., Ramírez, J., Carmona, A., Martín-Belloso, O., 2007. Alginate- and gellan-based edible films for probiotic coatings on fresh-cut fruits. *J. Food Sci.*, 72, E190–E196.

Tapia, M.S., Rojas-Graü, M.A., Carmona, A., Rodríguez F.J., Soliva-Fortuny, R., Martín Belloso, O., 2008. Use of alginate- and gellan-based coatings for improving barrier, texture and nutritional properties of fresh-cut papaya. *Food Hydrocolloids.* 22, 1493–1503.

Tatsumi, Y., Watada, A.E., Ling, P.P., 1993. Sodium chloride treatment or waterjet slicing effects on white tissue development of carrot sticks. *J. Food Sci.* 56, 1390–1392.

Tay, S.L., Perera, C.O., 2004. Effect of 1-methylcyclopropene treatment and edible coatings on the quality of minimally processed lettuce. *J. Food Sci.* 69(2), 131–135.

Toivonen, P.M.A., DeEll, J.R., 2002, Physiology of fresh-cut fruits and vegetables. In: O. Lamikanra (Ed.). *Fresh-cut fruits and vegetables: science, technology and market.* CRC Press, Boca Raton, FL, pp. 91–124.

Toivonen, P.M.A., Brummell, D.A., 2008. Biochemical bases of appearance and texture changes in fresh-cut fruit and vegetables. *Postharv. Biol. Technol.* 48, 1–14.

Tolbert, B.M., Ward, J.B., 1982. Dehydroascorbic acid. In: P.A. Seib, B.M. Tolbert (Eds.). *Ascorbic acid: chemistry, metabolism and uses. Advances in chemistry,* Vol. 200, American Chemical Society, Washington, DC, pp. 101–123.

Tovar, B., Garcia, H.S., Mata, M., 2001. Physiology of pre-cut mango. I. ACC and ACC oxidase activity of slices subjected to osmotic dehydration. *Food Res. Intl.* 34, 207–215.

Tsai, G.J., Su, W.H., Chen, H.C., Pan, C.L., 2002. Antimicrobial activity of shrimp chitin and chitosan from different treatments and applications of fish preservation. *Fish. Sci.* 68, 170–177.

Turhan, K.N., Sahbaz, F., 2004. Water vapor permeability, tensile properties and solutions of methylcellulose-based edible films. *J. Food Eng.* 61, 459–466.

Uritani, I., Asahi T., 1980. Respiration and related metabolic activity in wounded and infected tissues. In: P.K. Stumpf, E.E. Conn (Eds.). *The biochemistry of plants*, MIL. Academic Press, Orlando, FL, pp. 463–485.

Vargas, M., Pastor, C., Chiralt, A., McClements, D.J., González-Martínez, C., 2008. Recent advances in edible coatings for fresh and minimally processed fruits. *Crit. Rev. Food Sci. Nutr.* 48, 496–511.

Vargas, M., Chiralt, A., Albors, A., González-Martínez, C., 2009. Effect of chitosan-based edible coatings applied by vacuum impregnation on quality preservation of fresh-cut carrot. *Postharv. Biol. Technol.* 51, 263–271.

Vilas-Boas, E.V. de B, Kader, A.A., 2007. Effect of 1-methylcyclopropene (1-MCP) on softening of fresh-cut kiwifruit, mango and persimmon slices. *Postharv. Biol. Technol.* 43, 238–244.

Villalobos-Carvajal, R., Hernández-Muñoz, P., Albors, A., Chiralt, A., 2009. Barrier and optical properties of edible hydroxypropypyl methylcellulose coatings containing surfactants applied to fresh-cut carrot slices. *Food Hydrocolloids.* 23, 526–535.

Viña, S.Z., López Osornio, M.M., Chaves, A.R., 2007. Quality changes in fresh-cut celery as affected by heat treatment and storage. *J. Sci. Food Agric.* 87, 1400–1407.

Wardale, D.A., Galliard, T., 1977. Further studies on the subcellular localization of lipid-degrading enzymes. *Phytochemistry.* 16, 333–338.

Watada, A.E., Abe, K., Yamauchi, N., 1990. Physiological activities of partially processed fruits and vegetables. *Food Technol.* 20, 116, 118, 120–122.

Watada, A.E., Ko, N.P., Minott, D.A., 1996. Factors affecting quality of fresh-cut horticultural products. *Postharv. Biol. Technol.* 9, 115–125.

Watada, A.E., Qi, L., 1999. Quality of fresh-cut produce. *Postharv. Biol. Technol.* 15, 201–205.

Watkins, C.B., 2000. Responses of horticultural commodities to high carbon dioxide as related to modified atmosphere packaging. *HortTechnol.* 10, 501–506.

Watkins, C.B., 2006. The use of 1-methylcyclopropene (1-MCP) on fruits and vegetables-Research review paper. *Biotechnol. Adv.* 24, 389–409.

Wong, D.W.S., Camirand, W.M., Pavlath, A.E., 1994. Development of edible coatings for minimally processed fruits and vegetables. In: J.M. Krochta, E.A. Baldwin, M. Nisperos-Carriedo (Eds.). *Edible coatings and films to improve food quality.* Technomic, Lancaster, PA, pp. 65–88.

Wong, D.W.S., Tillin, S.J., Hudson, J.S., Pavlath, A.E., 1994. Gas exchange in cut apples with bilayer coatings. *J. Agric. Food Chem.* 42, 2278–2285.

Wright, K.P., Kader, A.A., 1997a. Effect of controlled-atmosphere storage on the quality and carotenoid content of sliced persimmons and peaches. *Postharv. Biol. Technol.* 10, 89–97.

Wright, K.P., Kader, A.A., 1997b. Effect of slicing and controlled-atmosphere storage on the ascorbate content and quality of strawberries and persimmons. *Postharv. Biol. Technol.* 10, 39–48.

Yam, K.L., Lee, D.S., 1995. Design of modified atmosphere packaging for fresh produce. In: M.L. Rooney (Ed.). *Active food packaging*. Chapman & Hall, Glasgow, pp. 55–73.
Young, D.H., Kohle, H., Kauss, H., 1982. Effect of chitosan on membrane permeability of suspension-cultured glycine max and phaseolus vulgaris cells. *Plant Physiol*. 70, 1449–1454.
Yu, Y-B., Yang. S.F., 1980. Biosynthesis of wound ethylene. *Plant Physiol*. 66, 281–285.
Zhang, X., 2007. New approaches on improving the quality and safety of fresh cut fruit and vegetables. *Acta Hort*. 746, 97–102.

9 Applications of edible films and coatings to processed foods

Tara H. McHugh and Roberto de Jesús Avena-Bustillos

Contents

9.1	Introduction	291
9.2	Meat films and coatings	291
9.3	Cereal coatings	297
9.4	Raisin and nut coatings	298
9.5	Confectionary coatings	300
9.6	Strips and pouches	303
9.7	Coatings for fresh fruit and vegetables	305
	9.7.1 NatureSeal®	305
	9.7.2 Nutri-Save™	307
	9.7.3 Pro-long™	308
	9.7.4 Semperfresh™	309
	9.7.5 Shellac	312
	9.7.6 Waxes	312
References		313

9.1 Introduction

Edible coatings have been successfully applied in processed foods such as meat, cereals, confectionaries, dried fruits, nuts, and fresh and fresh-cut fruits and vegetables (Donhowe and Fennema, 1994; Baldwin and Baker, 2002; Cagri et al., 2004). These coatings are used to improve the quality and shelf life of foods. Furthermore, different food ingredients, derived from meats, cereals, nuts, fruits, and vegetables, are being used to produce edible films for strips and pouches. These films act as novel packaging systems and control the release of active compounds such as antioxidants, flavors, and antimicrobial agents (Rojas-Graü et al., 2006, 2007). A comprehensive update of experimental and commercial edible coating and film applications is provided in this chapter.

9.2 Meat films and coatings

Collagen is a complex, fibrous, structural animal protein found in skin, connective tissue, and tendons. Natural collagen casings from beef, pork, or lamb intestines for comminuted meat are one of the earliest uses of protein films, and edible packaging materials in general. Subsequently, artificial collagen casings made from extruded collagen fibers supplanted the natural casings. Fabricated collagen casings allowed meat processors to manufacture portion-controlled, value-added products from meat trimmings on high-speed automated equipment (Gennadios, 2004).

Collagen casings and film wraps for meat products are produced by extruding a viscous (4% to 10% solids) aqueous suspension of purified acidified collagen into a neutralizing coagulation bath, followed by washing, plasticizing, and drying. Thermoplastic extrusion is a potential alternative to form protein casings and films, avoiding the need to add and then remove solvent by drying. Research suggests that some proteins display thermoplastic behavior (Krochta, 2002). However, inducing protein thermoplastic behavior generally has not been much explored or exploited for edible film production. Successful, efficient production of protein edible films using conventional extrusion equipment would certainly improve commercialization potential (Krochta, 2002).

Extruded collagen films are generally transparent with slight opacity and texture and adhere well to damp food products, such as fresh meat. Such films also are flexible when nonhydrated and become increasingly flexible with increased hydration. Collagen shrinks and can be sealed upon heating. It also is nearly sterile because it is extruded under high heat and pressure before passing through an acid solution. Also, collagen films are quite tasteless and even seem to acquire the flavor of the coated food. Collagen films are described as ideal based on sensory characteristics (Conca, 2002).

Besides collagen casings, collagen edible films have been used commercially since the 1980s as overwraps for boneless meat products that are heat-processed in nettings. Such films, marketed by Globe Packaging Co. (Carlstadt, New Jersey) under the trade name Coffi™, aid removal of nonedible netting after heat processing and improve product appearance (Gennadios, 2004). Coffi film is applied in sheet or roll form on meat products using pneumatic commercial devices for vacuum filling and net shirring (Bisson and Weisenfels, 1992).

As mentioned, collagen films are used commercially as overwraps for boneless meats. In addition, Conca (2002) showed that extruded collagen films were as effective as plastic bags in maintaining the quality of beef during frozen storage. Conca (2002) compared noncoated and collagen-coated meat samples pan-fried from the frozen state without removing the collagen coating. Mean sensory ratings for all samples were within acceptable ranges. In general, higher ratings were given to coated meat for odor, flavor, texture, and overall quality. The collagen film tended to hold in meat juices during cooking, thereby making the coated samples more flavorful, tender, and juicy. There was little or no difference in elasticity between noncoated and collagen-coated beef samples during storage, although differences were observed among withdrawal periods for other cuts of meat.

These and other storage study results have shown that collagen packaging maintains meat quality at frozen temperatures, comparable to that of plastic bags (Conca, 2002). Also, films or coatings from other types of proteins (i.e., corn zein, wheat gluten, soy protein, and egg albumen) carrying antioxidants have shown promise for reducing the rate of lipid oxidation in meat products (Herald et al., 1996; Wu et al., 2000; Armitage et al., 2002).

An economical and environmentally viable film consisting of pea starch, glycerol, and grape seed extract, and its effect on meat preservation was studied by

Corrales et al. (2009). This investigation involved the evaluation of pea starch-based edible films containing 1% grape seed extract to control *Brochothrix thermosphacta* growth on surface inoculated pork loins. Flavonoids and phenolics from the grape seed extract migrated further inside the meat and the surface phenolics concentration codiluted with contact time. The incorporation of grape seed extract into pea starch films possessed bacteriostatic effects against meat microbial surface load during at least the first 4 days of incubation. The migration rate of phenolic compounds into the meat surface may play an important role in the growth inhibition of *B. thermosphacta* and the antimicrobial effect was related to the higher phenolic acid migration at the initial incubation stages. The migration was affected due to the physical entrapment of grape seed extract phenolic compounds in the amylose helix, and consequently, their release rate was reduced. Hydroxyl groups of phenolics, mainly responsible for extract antibacterial and antioxidant properties, may interact with hydroxl groups in starch films resulting in reductions of antibacterial properties. Grape seed extract films presented remarkable advantages for marketability as they inhibit the growth of undesirable pathogens in meat, improving meat quality and extending its shelf life. They are totally biodegradable and edible reducing environmental constraints occasioned by polypropylene-based packaging. However, the control of particle release from pea starch films requires further research to maximize efficiency (Han and Krochta, 2007).

Starch-based materials are usually marketed for coating drug tablets. Starch and dextrin formulations are also marketed as protective coatings, integrity maintainers, appearance enhancers, and seasoning adhesives for meats (Krochta, 2002). Meat carcasses and meat pieces can be protected by a calcium alginate film, which both reduces water loss and improves the bacteriological quality. The same system may be applied to poultry and to hamburger-like products (Onsøyen, 2001). NewGem™ Foods Glaze Sheets made with fruit and vegetable-based edible films is currently used to save liquid glaze material for ham glazing (Table 9.1).

The use of edible coatings as a substitute for plastic bags used to ship frozen meats, allowing for direct cooking without coating removal, is a potential solution to the problem of dumping or storing food-contaminated plastic waste. Despite benefits, very few edible films are used due to performance and functional limitations of edible films in comparison to traditional polymer films. Inherently, edible packaging materials are themselves susceptible to biodegradation over time, similar to the foods that they are meant to protect.

Chocolate and shellac-based coatings are used within the military ration system (Conca, 2002). However, increased environmental awareness has led to new policies restricting waste disposal. As a result, growing interest in development of biodegradable packaging and edible films has emerged (Conca, 2002). During the 1970s, the U.S. Army explored the restructuring process of flaked and formed meats. Various edible coatings were tested for their ability to prevent moisture loss and deleterious oxidative changes occurring in frozen meats during lengthy storage periods required by the military. Most of the studied coatings were either fat or carbohydrate based. However, protein-based coatings such as gelatin/propylene glycol (Gelcote, Atlantic Gelatin, Woburn, Massachusetts) and soy protein isolate also were tested. Meat portions were dipped into the coating medium, allowed to air dry, packed between sheets of parchment paper, and

Table 9.1 Some commercial edible coatings and films

Commercial name	Composition	Uses	Company	Web site
NatureSeal®	Ascorbic acid, calcium chloride, hydroxypropyl methylcellullose	Browning inhibition; maintain taste, texture, and color of fresh-cut fruits and vegetables	Mantrose-Haeuser, Co., Inc.	www.natureseal.com
Semperfresh™	Sucrose esters of short-chain unsaturated fatty acid and sodium salts of carboxymethyl cellulose	Coating of whole pears and cherries to control weight loss and excess respiration, retain moisture, and preserve natural color of fruit	Agricoat Industries Ltd.	www.paceint.com
Pro-long™	Sucrose polyesters of fatty acids and sodium salts of carboxy-methyl cellulose	Coating for fresh fruits and vegetables	Courtaulds Ltd., Derby, United Kingdom	
Mantrocel®	Hydroxypropyl methylcellulose	Film coating for tablets and capsules, binder, filler, matrix, stabilizer	Mantrose-Haeuser, Co., Inc.	www.mantrose.com
Crystalac®	Shellac	Confectionary glaze	Mantrose-Haeuser, Co., Inc.	www.mantrose.com
Crystalac® Z2	Zein	Confectionary glaze	Mantrose-Haeuser, Co., Inc.	www.mantrose.com
Origami® Wraps	Fruit and vegetable-based films with bilayer protein films	Wrapping, pouches, sachets	NewGem™ Foods, LLC	www.origamifoods.com

Applications of edible films and coatings to processed foods 295

Product	Composition	Application	Company	Website
NewGem™ Foods Glaze Sheets	Fruit and vegetable-based films with spices and bilayer protein films	Ham glaze	NewGem™ Foods, LLC	www.origamifoods.com
Flavored film strips	Films made from hydrocolloids, plasticizers, active compounds, flavors, and colorants	Breath fresheners, oral hygiene; sugar-free candy, caffeine/energy, and vitamin/nutrient strips	Watson Inc.	www.watson-inc.com/film_technology.php
Listerine® PocketPak® Mint Breath Strips	Films made from hydrocolloids, plasticizers, active compounds, flavors, and colorants	Breath fresheners, oral hygiene	Johnson & Johnson	www.listerine.com/product-pocket-paks.jsp
Chloraseptic® Strips	Hydroxypropyl methylcellulose, starch, flavors, colorant, and active compounds	Fruit flavor strips for kids' sore throat relief	Prestige Brands, Inc.	www.chloraseptic.com
Coffi™	Collagen film nettings	Wraps for heat-processed boneless meat products	Globe Packaging Co.	http://globecasing.com/coffifilm
Durkex 500	Hydrogenated and partially hydrogenated vegetable oil blend	Protective coating and gloss enhancer for dried fruits, nuts, cereal, and snack products	Loders Croklaan	www.croklaan.com

blast frozen at −29°C before being stored at −3°C. After 24 hours, the coated meat portions were visually inspected, grilled, and evaluated for off-odors and flavors. Materials that produced even, translucent coatings without off-odors or flavors were stored at −3°C for 8 weeks and at −12°C for 5 months. After storage, coated meat samples were again inspected for visual appearance, odor, and flavor (Conca, 2002).

Consumer acceptance is ultimately essential for the commercialization of edible packaging materials. Therefore, it is imperative that sensory evaluation is integrated into edible packaging research to detect the introduction of any adverse sensory effects by the tested films and coatings. However, few studies report the sensory properties of foods treated with edible coatings (Herald et al., 1996). Du et al. (2009) evaluated the effect of adding carvacrol (the active ingredient of oregano essential oil) and cinnamaldehyde (the active ingredient of cinnamon oil) to apple- and tomato-based film-forming solutions on sensory properties of wrapped cooked chicken. Preference tests indicated that baked chicken wrapped with tomato and apple films containing 0.5% carvacrol or cinnamaldehyde were equally preferred over chicken wrapped with tomato or apple films without the plant antimicrobials. The consumers preferred carvacrol-containing tomato film chicken wraps over the corresponding apple film wraps. These findings suggest that films and coatings containing antibacterial essential oils can be used to protect raw chicken pieces against bacterial contamination without adversely affecting sensory preferences of cooked wrapped chicken pieces.

Protein-based coatings have been investigated for their potential to reduce oil absorption by coated foods during frying and, secondarily, to retain natural juices and flavors, enhance texture and appearance, and reduce moisture loss. Spraying aqueous gelatin solutions on battered and breaded meats reduced oil absorption upon frying due to the grease resistance of gelatin films formed around meat products (Olson and Zoss, 1985). However, protein films and coatings applied on food products that carry proteolytic enzymes, such as meat products, may degrade prematurely, thus losing their utility.

Collagen and gelatin coatings have been used on meats and sausages to reduce gas or water vapor permeability (Hood, 1987). Gelatin is a good gas barrier, but it is highly hydrophilic. Cross-linking of gelatin membranes reduces their solubility in water. Another approach to increasing moisture resistance is the use of laminated membranes (Arvanitoyannis, 2002). Todd (1982) developed a pliable sheet wrap to give an edible coating a fried appearance, taste, and texture upon baking of meat and other foodstuffs. The composition was made up of a mixture of a fat, water and a particulate farinaceous material suspended in a pliable form by a gelatin matrix.

Snyder and Kwon (1987) reported the use of yuba films (resulting from the dried supernatant layer after cooling of soymilk) shaped into sheets, sticks, and flakes for further use as wrappers for meats and vegetables in cooking. Enrobing buffalo meat with an egg white and starch emulsion coating improved binding properties and quality of cooked meat, reducing crumbling and breakage that cause loss of cooking performance, economic value, and consumer acceptability (Eyas Ahamed et al., 2007a). Also enrobing improved the shelf life and acceptability of buffalo meat cutlets, increasing shelf life to 90 days under frozen storage (Eyas Ahamed et al., 2007b).

Methyl cellulose and hydroxypropyl methylcellulose edible films were applied to marinated whole chicken strips prior to breading, after breading, or were incorporated in the breading mix. Films applied to chicken strips prior to breading had fried crusts with higher fat and lower moisture levels. Frying oil degradation was reduced by 50%, and fat absorption was lowered when hydroxypropyl methylcellulose edible films were applied to chicken strips prior to breading. It is postulated that the edible films hindered migration of moisture and acetic acid into the frying oil, and this activity was responsible for reduced free fatty acid generation in those oils used to fry the coated products. It is further suggested that extension of the life of frying oils may be accomplished through appropriate selection and application of edible films to prefried products (Holownia et al., 2000).

9.3 Cereal coatings

Gorton (1993) discussed technology developments for the application of coatings and toppings to baked foods and snacks. Specific examples included improvements in salt delivery systems for the salt topping of pretzels; rotary systems for depositing dry coatings (e.g., sesame and poppy seeds, bran flakes, cinnamon sugar, chopped nuts, etc.) onto bread, buns, or confectionery; and equipment for application of toppings at the end of the production process (e.g., sugar onto doughnuts). A new system, based on electrostatic principles, is designed to spray liquid coatings (oils, release agents, and flavorings) onto snack foods and bakery products to provide uniform and controllable coatings. Another coating technology for snack foods employs flexible nylon fibers set in inverted-pitch spirals that rotate in opposite directions through a long, narrow trough; snacks and flavorings are carried forward by the spiral brushes. The latter has been used successfully for snack crackers, nuts, extruded snacks, and crisps.

Coating of breakfast cereals for nutritional or flavoring purposes includes two coating phases:

1. First coating phase, controlling nutritional fortification, texture, and shelf life, covering conveyor belt application (spray system, product bed); liquid spray systems (pressure, interference, or force, combination systems); blending drum application (spray system, degree and duration of blending, coating drum, flights, drum slope, parameter adjustments)
2. Second coating phase, controlling sugar or flavor coatings, covering conveyor belt processing (restrictive-orifice spray system, adjustable-orifice spray system, spinning disc system, belt cleaning); coating-blending drum application (spray system, recirculating mode, dead-end mode) and dry flavor bit application (e.g., nuts, dried fruit, coconut bits); and formulations (coatings incorporating sugar, syrup, edible fat, and flavoring materials) (Burns and Fast, 1991)

Sugar coating of ready-to-eat breakfast cereals is done commercially by different methods. A low-sugar-content ready-to-eat breakfast cereal was produced by coating cereal pieces with oil, then coating with a sweetener powder containing high-conversion maltodextrin or low-conversion corn syrup and a high-intensity sweetener (Green and Nowakowski, 2005). Solis-Morales et al. (2009) used a fluidized bed processor, built with a top-spraying nozzle, as an alternative to a tumbling vessel for coating puffed wheat particulates with a sweet chocolate cover. The fluidized bed technique was compared with the tumbling method in which syrup was applied by spraying, as well as with a commercial sample. No significant difference was perceived in color, but the fluidized bed treated sample was considered crisper with better chocolate flavor than the commercial sample. In terms of attrition, the fluidized bed sample lost about 1% weight, while the tumbled-coated sample lost around 5% weight, and the commercial sample lost nearly 10% weight.

Cereal products are coated for a variety of purposes. Leusner (2009) patented a method for fiber fortification of breakfast cereals by coating with fiber to increase their fiber content to approximately 50% (by wt.) without affecting their flavor. Furthermore, Petersen (2008) enriched breakfast cereals with probiotics by coating with an encapsulated probiotic preparation using lipid as an encapsulating agent. Hoitink et al. (2004) described a cereal product that changes or loses its color upon addition of an aqueous liquid. The color change was intended to make the breakfast cereal more appealing for both children and adults. The cereal may be prepared by applying different coatings on a classic extruded cereals base, whereby an outer coating provides a certain color and is essentially washed off upon addition of milk. Rice fortified with vitamins and minerals has been coated with zein/stearic acid/wood resin mixtures to prevent vitamin and mineral losses during washing in cold water (Padua and Wang, 2002).

Several coating formulas were developed to meet the needs of various dry, bite-sized cube categories such as entrees, sandwiches, desserts, and cereals for space flights (Conca, 2002). Food for space flights is required to withstand harsh conditions, temperatures of 43°C, 100% relative humidity (RH), 100% oxygen atmosphere, cabin pressures of 5.3 psi, launch acceleration, vibration, and acoustic noise without formation of crumbs, dust, or fragmentation. Some problems discovered during the early space flights resulted in routine coating of bite-sized food cubes to prevent crumb formation and to reduce cube greasiness or stickiness (Conca, 2002). Aqueous gelatin coatings were applied to dessert cubes (e.g., date and pineapple fruitcake, graham cracker, and ice cream) and apricot, orange/lemon, and strawberry cereal cubes. Some of the food cubes were soaked in gelatin solutions, freeze dried, and then coated along the edges with acetylated monoglyceride. The fruitcake cubes were coated in gelatin and then wrapped in an edible starch paper to prevent stickiness. Combinations of zein, acetylated monoglycerides, citric acid, butylated hydroxyanisole (BHA), and butylated hydroxytoluene (BHT) were used to coat chocolate, coconut, apricot, pineapple, strawberry, and peanut dessert cubes. Special coating combinations of encapsulated protein, fat, and carbohydrate not only prevented crumbling and stickiness, but also preserved freshness. These coatings, which were developed by Pillsbury (Minneapolis, Minnesota), were also used for other dessert items such as brownies and gingerbread (Hollender, 1965).

9.4 Raisin and nut coatings

Raisins are commonly mixed with crispy breakfast cereals. Raisins need to be coated in order to reduce moisture migration and toughness due to the low water activity conditions inside cereal packaging. Guzman and Hegarty (2000) developed a method for improving softness of raisins based on hydration of a batch of raisins; application of a glycerol coating to the hydrated raisins; forming a 6-foot-high column of glycerol-coated raisins in a heated vessel, with an excess of glycerol; holding the coated raisins in the heated vessel for approximately 16 hours; and periodically mixing the raisins with the excess glycerol during the 16-hour holding period. Shellac-based coatings are used for macroencapsulation of components found in raisin–nut mixes to prevent oxidation of the high-fat nuts and moisture transfer between components of different water activities (Conca, 2002).

Other successful applications of coatings to raisins and nuts include the following. Starch and dextrin formulations are marketed as protective coatings, integrity maintainers, appearance enhancers, and seasoning adhesives for nuts, snacks, and cereals (Krochta, 2002). Caseinate-based and whey protein–based coatings have been applied on raisins and peanuts to provide a barrier to oxygen and moisture transfer for extending shelf life of the products (Chen, 1995, 2002; Maté and Krochta, 1996). Casein coatings did not show any significant effect on reducing moisture loss of coated raisins during storage (Watters and Brekke, 1961). In other studies, egg albumen and soy protein coatings significantly reduced moisture loss from coated raisins (Bolin, 1976).

Trezza and Krochta (2002) reviewed published research on the use of protein-based coatings for nuts stating that protein films can function as oxygen, moisture, and lipid barriers on nuts. Corn zein is one of a few proteins, such as collagen and gelatin, used commercially as an edible coating. Zein coatings are used as oxygen, lipid, and moisture barriers for nuts, candies, confectionery products, and other foods (Padua and Wang, 2002). A quick-drying alcohol-based zein (Optazein, Opta Food Ingredients, Inc., Cambridge, Massachusetts) was used as a coating to improve quality of nuts in a "raisin nut mix," an item found in military Cold Weather Rations. It contained 13% solids, 80% ethanol, and 5% glycerol (added to aid in spreading and to impart flexibility). This product was packaged in the military trilaminate pouch. The study showed that zein-coated samples had higher sensory ratings and held up better than wheat gluten–coated samples. Wheat gluten–coated and zein-coated samples stored at 38°C were dropped from the study after 12 and 18 months of storage, respectively, due to low sensory ratings. Zein-coated samples received a mean rating of 6.2 for appearance, odor, and texture, and of 5.5 for flavor and overall quality after 24 months at 27°C (Conca, 2002).

Whey protein coatings substantially reduced oxygen uptake and rancidity of roasted peanuts (Maté et al., 1996) and walnuts (Maté and Krochta, 1997) depending on levels of plasticizer in coating solution, thickness of dry coating, and conditions of storage. Lee et al. (2002a, 2002b) reported favorable sensory properties for whey protein–coated peanuts and chocolate-covered almonds coated with whey protein/lipid formulations.

Whey protein–based edible coatings effectively reduced lipid oxidation during the storage of dry roasted peanuts (Maté and Krochta, 1996; Maté et al., 1996), and walnuts (Mate and Krochta, 1997). However, like any polymer material, whey protein coatings allow some permeation of oxygen that can eventually produce rancid foods. It is hypothesized that the oxidation of foods by the permeation of oxygen can be markedly retarded by incorporating antioxidants into whey protein films and coatings that diffuse into the food at an appropriate rate to scavenge the free radicals created in the food during the first stage of oxidation.

The early exploration of casein films revealed several useful concepts, including extending the shelf life of coated bakery products and chocolate candies, reducing moisture loss from raisins to breakfast cereals. Some of these applications employed casein-lipid composite formulations to improve resistance to moisture loss (Chen, 2002).

Edible films or coatings may also be used to carry or support flavors at the product surface. These flavors will then be promptly released upon consumption of the product. In fact, a few products that use this flavoring concept have already been commercialized. For instance, there are roasted peanuts with a curry-flavored coating that instantaneously dissolves in the mouth providing the perception of the Indian spice. Another example involves application of multiple sugar coatings on sweets with each coating containing different tastes and flavors. Gum arabic or another hydrocolloid layer separates the coatings to prevent aroma migration. For such an application, diffusivity of volatiles should be very low, and the volatiles should have high affinity for the highly soluble coating (Debeaufort et al., 2002).

Processes for coating nuts and peanuts have been described (Krochta et al., 2005; Cleophas, 2006; Rapp et al., 2006; Mie et al., 2008) including pan coating and coating nuts with a film-forming coating to delay the development of rancidity. The film-forming coating is applied in a solution that contains an amount of surfactant greater than that which reduces the surface energy of the solution to its lowest level.

9.5 Confectionary coatings

Coating confectionary surfaces enhances flavor impact or nutritional value, provides qualitative points of difference, and protects product characteristics. Confectionary coating is usually done by panning. Groves (1977) discussed techniques of hard panning, soft panning, and chocolate coating and some panning equipment. Hard panning results from deposition of small sugar crystals on the centers which are held in a revolving drum. Crystals form by evaporation of a concentrated (60% to 65%) sucrose solution using external heat application or warm air. Smooth finishes result from less concentrated syrups, and a final polish may be applied using carnauba-or rice-bran wax. Soft panning employs alternate applications of syrup and powdered sugar. Soft-panned items require a 1- to 2-day drying period before being surface finished. Chocolate coating involves application of melted chocolate at 98 to 100°F to the centers, which sets by being cooled. Both soft panning and chocolate coating build up the thickness of the coating quite rapidly. Chocolate coating may be polished with gum arabic or corn syrup, and a final gloss applied with an alcoholic shellac solution. Hildebrandt (2009) also discussed developments in panned chocolates, including trends toward healthier centers (e.g., nuts or fruits) and use of exotic flavors.

Lees (2000) reviewed numerous faults that may occur during production or storage of panned confectionery with reference to common faults in soft panned goods (goods adhering together, poor buildup of cooling layers, surface spotting or pitting, lack of firmness in the primary coat, coarse grainy coating, uneven shapes, poor dispersion of flavorings/colorants, poor quality glazing, discoloration). Lees also studied dependence of stability of panned confectionery on development of a firm and even shell with low grain size and layers of sugar syrup at varying concentrations; important factors for successful panning (development of successful precoating, even-sized centers, careful treatment of nuts); causes of product clumping during pan rolling (addition of coating syrup too quickly, excessive syrup doses, interruption of pan movement during panning, incorrect sugar content in charge syrup, addition of syrup layers to damp precoating, incorrect drying conditions); causes of cracked coatings (too low temperature of coating, use of too large a charge of syrup, uneven coating thickness, sharply edged centers); causes of spotting on the product surface (use of oil-based flavorings, migration of oil from nut-based centers); causes of surface speckling (inadequate filtration of dust from air, damp air, rubbing of product on unlined metal trays during storage, inadequate cleaning of glazing pans); glazing (types of glaze and glazing substitutes, importance of adequate ventilation, low temperature air delivery, correct spray rate, and adequate drying between coating applications); importance of sealing chocolate-coated confectionery before polishing; and causes of surface pinholes (addition of syrup layers too quickly, use of too high concentrations of syrup).

Cooke (2001) discussed and compared different types of chocolate panning equipment, including rotating pans, pocket pans, and automatic computer-controlled pans. Equipment and processes used for panning at a U.S. company that manufactures chocolate-coated products including nuts and fruits were also described, together with the advantages of pocket panners over conventional pans in belt panning systems.

Chocolate, gum balls, sprinkles, cake decorations, panned candies, sugar-coated licorice, and jelly bean coatings are commonly based on shellac. Shellac is a natural polymer obtained by refining the secretion of *Kerria lacca*, a parasitic insect found on tress in Southeast Asia. It has been used alone in hard confectionary and pharmaceutical coatings (Hagenmaier and Shaw, 1991; Phan The et al., 2008). This odorless and tasteless confectionary coating provides several benefits such as providing scuff resistance, acting as a fat and moisture barrier, enhancing and protecting a high-gloss shine. Shellac-based glazes that enhance and protect confectionary gloss and provide brilliant shine with less tack are commercialized by Mantrose-Haeuser Co., Inc. (Westport, Connecticut) under the Crystalac® and Certified® brands. Shellac is mostly used for drugs, chocolate, and hard and soft sugar shelled coatings, but it is also used for coating a variety of food products such as imitation and enriched rice, moisture-resistant sugar, apples, pears, mandarins, cakes, doughnuts, icings, biscuits, edible ink base, seeds, nuts and specialty pharmaceutical glazes.

Chocolate coating involves application of melted chocolate at 36.7 to 37.8°C to the centers, which sets by being cooled, building up the thickness of the coating quite rapidly. Chocolate coating may be polished with gum arabic or corn syrup, and a final gloss applied with an alcoholic shellac solution (Groves, 1977).

Shellac coating of confectionery products is typically done by panning. Adele (2005) discussed principles of and developments in automated panning equipment for sugar confectionery. Aspects considered included pan drum design; air circulation and conditioning in pan coating systems; syrup dosing, distribution, and spraying systems; flavoring injection; chocolate dosing and distribution systems; use of glazes and shellac coatings; addition of dry ingredients (e.g., cocoa powder and glucose); unloading of automatic pans; control systems; stages involved in automated chocolate panning (gumming, smoothing, varnishing); automated panning of hard-coated sugar and sugar-free confectionery (gumming, engrossing, smoothing, coloring, waxing); developments in hard-panning using polyols for sugar-free coatings; and developments in soft panning for jelly-centered confectionery. Dreier (1991) discussed application of flavors, seasonings, chocolate, or vitamins by coating or enrobing. Examples included sugar-frosted cereals, cheese-seasoned crackers and snacks, candy- or shellac-coated chocolate morsels, and smear-resistant particulates in ready-to-serve cake frostings. Aspects considered included types of equipment (drum coaters, batch or continuous; coating applicators, spray nozzles, multiport applicators, dribble tubes, and pneumatic applicators; vented pans; candy pans; fluid and spouted bed coating devices; and spray dryers, drum or pan coaters (drum coaters coat larger, irregularly shaped materials); seasoning expanded snacks; and factors affecting coating applications.

Shobu et al. (2004) described a shellac-based coating, foods and drugs coated with this composition, and methods for their manufacture. The patent also includes a glaze produced from the coating, oil-based confectionery coated with the glaze, and a method for glazing of the confectionery. The coating includes shellac, a basic amino acid, or a basic phosphate. The glaze is produced by dissolving the coating in water and adding a thickener or a sugar.

Asatake and Yamaguchi (2007) developed a method to manufacture sugar-coated tablets including sugar confectionery or medicines. The coating provides a hard, crunchy texture to the product. The method involves coating the tablet with a fat and oil layer, covering this layer with liquid shellac which is subsequently dried, and coating the dried shellac with sugar solution which is then dried.

Mantrose-Haeuser Co., Inc., developed a line of wax products for use in nuts and confectionaries. Licabee is a liquid wax used to enhance gloss, decrease polishing time, and reduce white spotting in hard and soft sugar shelled coatings. Powdered carnauba wax is used as a polishing agent on hard and soft sugar shelled confections. Aqueous gum-based products (Certiseal®) are used as precoating agents to seal the centers of nuts and dried fruits to improve adhesion or to prevent fat migration, and also as polishing agents to smooth the surface of chocolate confections and provide gloss and a solvent barrier solution. Furthermore, Mantrose-Haeuser Co., Inc., commercializes a custom line of formulated oil and natural wax products under the brand name CertiCoat® used as polishing agents for jellies and gummies to improve gloss, antistick properties, and moisture retention.

Zein-based coating formulations have long been commercially available as finishing agents imparting surface gloss on confectionery products (Boutin, 1997). Zein coatings are used as oxygen, lipid, and moisture barriers for candies and confectionery products (Padua and Wang, 2002).

Edible films or coatings may be used to carry or support flavors at the product surface. These flavors will then be promptly released upon consumption of the product. In fact, a few products that use this flavoring concept have already been commercialized. An example involves application of multiple sugar coatings on sweets with each coating containing different tastes and flavors. Gum arabic or another hydrocolloid layer separates the coatings to prevent aroma migration. For such an application, diffusivity of volatiles should be very low, and the volatiles should have high affinity for the highly soluble coating (Debeaufort et al., 2002).

Singh et al. (2010) investigated the combined treatment of edible coatings and osmotic dehydration on pineapple samples. Pineapple samples were coated with 0.5% to 5% (w/v) sodium alginate solution by dipping for 60 seconds and 120 seconds and then dried at 50°C for 10 and 40 minutes. Coating may provide a solution to the problem of solid gain without affecting water removal during osmotic dehydration. The application of coating before osmotic dehydration will help in retaining nutrients and flavor more than those without coating. The combination of coating and osmotic dehydration may provide a practical method for the preservation of pineapple with better quality characteristics. The osmotically dehydrated coated fruits after rehydration can be added in ice cream, yogurt, and confectionary products.

Chocolate coating is used for nuts, wafers, cookies, and fruits. Morando (2009) developed a method for producing confectionery involving freezing fresh fruit; coating the frozen fruit with sugar solids sufficient to reach a critical mass; placing the coated fruit onto a wafer; and sealing the coated fruit and wafer in a confectionery layer such as chocolate. The wafer may be formed from chocolate or yogurt. Wu (2008) described a method to prepare a cereal-based snack food, which allows coating of the food with a thermally sensitive ingredient to improve its appearance. The method involves coating a formed cereal product with a thermally sensitive edible material, such as chocolate or cinnamon powder; applying a different, thermal-resistant coating to the resulting coated cereal product; and heating the double-coated product at 35 to 350°C for between 30 seconds and 10 hours.

Akutagawa (2007) developed a decorative sugar confectionery. It includes a base material having an oil-containing surface (e.g., chocolate), an aqueous coating layer on a surface of the base material which may be formed by silk screen printing, and an image formed on this coating using aqueous edible ink and ink-jet printing. Gautier (1990) and Wolff (1969) developed methods for chocolate coating of the interior of edible cups and ice cream cones, and wafers, respectively.

9.6 Strips and pouches

MacQuarrie (2006) developed orally disintegrating or dissolving edible strips (such as breath fresheners) for use as a matrix for retaining and delivering nutrients, flavors, and medicinal compounds made from new liquid film casting compositions including a major proportion of high bloom gelatin (at least 50% by wt.). The particularly low melting range for hydrated gelatin produces films that leave virtually no residue upon dissolving in the mouth and can be used in the form of thicker films and strips than known edible films. MacQuarrie (2008) also developed orally disintegrating or dissolving edible strips that can be used as a matrix for retention and delivery of nutrients, flavors, and medicinal films. They are made from new liquid film casting compositions containing gelatin as their main ingredient.

Recently, MacQuarrie (2010) developed edible strips that also dissolve easily in the mouth without leaving residues and are a suitable matrix for the delivery of nutrients, flavorings, or bioactive compounds. They are prepared from new liquid film casting compositions including predominantly low melting point hydrated gelatin.

They can be used to form films and strips that are thicker than any presently known edible films.

Hoffman et al. (2008) developed an edible film containing pullulan and one or more sweeteners at a concentration of 10% (dry solids basis). High-intensity sweeteners, such as sucralose, aspartame, acesulfame K, and brazzein, can be used. The film can be stored in the form of strips in a dispenser.

Quinn (2007) discussed the use of edible film strips to carry nutraceuticals. Edible strips are also made to simulate bacon. Edible crisp chips are flavored and colored to simulate bacon and may be in the form of flakes or strips. They are made from a flour-based dough and may be incorporated into other foods (Cooper and Melnick, 1971).

Nori (red algae, *Porphyra* spp.) is a staple edible seaweed in Japan (Ono et al., 1993). Nori strips are commonly used for sushi wrapping. NewGem™ Foods, LLC (Stockton, California) uses a technology developed jointly at Western Regional Research Center, U.S. Department of Agriculture, Agricultural Research Service (USDA-ARS) (Albany, California) to produce edible films (Origami® Wraps) based on fruit and vegetables purees and concentrates. These colorful and tasty films are currently used as a nutritious and attractive alternative for sushi wrapping.

Takeda (2009) patented a method of producing colored jelly products, including kuzukiri (an arrowroot jelly product) and strips of bean jelly, by addition of natural dye or edible food dye. Pollard (2009) described a method for forming a food product. It involves taking a stainless steel mold divided into multiple, equally spaced strips, and placing it onto a sheet of edible material. Each strip is then filled with a food material so that adjacent strips contain different food materials. The mold is lifted away from the edible sheet to leave the strips of food thereon, held together by the sheet material. The food product strips are then cut into shapes (e.g., squares, triangles, circles, or stars) to produce the finished food product. The mold may have 15 to 20 strips, each approximately 1 cm wide, 1 cm deep, and ≤20 cm long. The edible sheet may be a thin pastry sheet or a sheet of rice paper. The food materials placed on the sheet may include vegetables, fruits protein (e.g., meat), rice, and pasta. The method described provides healthy food products in colorful and attractive shapes.

Edible pouches and sachets are made from heat-sealable films and can be used as separate compartments of food ingredients and active compounds with antibacterial or antioxidant properties. Edible pouches can also be used to divide and store premeasured dry ingredients to be added into large production operations, thus avoiding operator errors in measurements and simplifying production. An early application of such edible pouches was to make a cereal package including a moisture-proof container enclosing servings of ready-to-eat breakfast cereal in edible, milk-soluble pouches (Ashley, 1973). Wissgott and Verschueren (1997) reported that soluble foods may be packaged in a sachet made from edible material which is soluble in hot water (e.g., a film made from carboxymethylcellulose, a plasticizer, and gelatin).

Quality changes in cheese slices individually packed in pouches made from four kinds of corn zein films during storage were investigated. Physicochemical properties and microbiological quality of the cheese slices were monitored over a 4-week period of storage at 5°C. Zein pouch materials also included different combinations of polyethylene glycol, glycerol, oleic acid, and zein-coated high-amylose corn starch film. Cheese quality was better maintained after storage in zein pouches with oleic acid added. Quality of cheese slices individually packed in this inner edible pouch and repacked in a plastic outer pouch was not significantly different from that of cheese slices individually packed in plastic inner pouches and repacked in a plastic outer pouch (Ryu et al., 2005).

An edible oxygen barrier film pouch was also fabricated from a heat-sealable corn zein layer laminated on soy protein isolate film and used to package olive oil condiments for use with instant noodles. The mechanical, barrier, and physical properties of this bilayer film were then investigated, and the oxidative stability of olive oil in the pouches was measured during storage under dry and intermediate relative humidity conditions. When compared to the soy protein isolate film, lamination with an additional layer of corn zein film led to increased tensile strength and water barrier properties, while it had a lower elongation at break and decreased oxygen barrier properties. Nevertheless, the oxygen permeability of the bilayer film was lower than that of nylon-metalocene catalyzed linear low-density polyethylene film that is the material usually used for such condiments. The corn zein/soy protein isolate bilayer films generated were heat sealable at 120 to 130°C and produced a seal strength greater than 300 N/m. The higher oxygen barrier property of the bilayer films resulted in reduced oxidative rancidity of olive oil packaged in the corn zein/soy protein isolate film when compared to olive oil packaged in polyethylene-based films (Cho et al., 2010).

Fruit- and vegetable-based edible films made by Origami Foods LLC (Stockton, California) using a technology developed jointly at Western Regional Research Center, USDA-ARS (Albany, California) are also used to make edible pouches to be filled with granola, chocolate chips, peanut butter, or other snack foods.

9.7 Coatings for fresh fruit and vegetables

In addition to coatings for processed food, the authors of this chapter would like to present a discussion of a few edible coatings used to control moisture loss and respiration of fresh and fresh-cut fruits and vegetables, and to alter surface appearance. Some of the most successfully commercial coatings currently used by the food industry are listed in Table 9.1.

9.7.1 NatureSeal®

NatureSeal® was jointly developed under a patent owned by Mantrose-Haeuser Co. (Westport, Connecticut) and USDA-ARS (Winter Haven, Florida) and later improved products were developed and patented by NatureSeal, Inc., a subsidiary of Mantrose-Haeuser Co., (Westport, Connecticut) and Western Regional Research Center USDA-ARS (Albany, California). NatureSeal contains a powdered blend of vitamins and minerals. NatureSeal is proven to maintain the natural taste, texture, and color of fresh-cut produce for up to 21 days. The NatureSeal line of products is allergen-free, sulfite-free, and Kosher certified. Several formulations are certified for use on organic produce.

Cliffe-Byrnes and O'Beirne (2007) evaluated the effect of NatureSeal on carrot discs packaged in microperforated oriented polypropylene film compared to plain oriented polypropylene films. Microperforated oriented polypropylene film generated useful modified atmospheres, which resulted in retention of good texture, aroma, and flavor characteristics, but allowed surface whitening and a loss of headspace aroma volatiles to occur. By contrast, plain oriented polypropylene film generated anaerobic atmospheres that resulted in an overall poor sensory quality with tissue softening and excess surface moisture, loss of aroma acceptability, and production of ethanol. Use of NatureSeal coating had beneficial effects on visual quality. It significantly improved sensory scores, which appeared to be caused by a reduction in surface whitening and moisture loss. Increasing storage temperature from 4 to 8°C resulted in a shorter shelf life. Although modified atmosphere had the biggest influence on quality, NatureSeal coating and storage at 4°C helped extend shelf life.

Vasantha Rupasinghe et al. (2006) examined the antimicrobial effect of vanillin against four pathogenic or indicator organisms—*Escherichia coli*, *Pseudomonas aeruginosa*, *Enterobacter aerogenes*, and *Salmonella enterica* subsp. *enterica* serovar Newport—and four spoilage organisms—*Candida albicans*, *Lactobacillus casei*, *Penicillum expansum*, and *Saccharomyces cerevisiae*—that could be associated with contaminated fresh-cut produce. The minimal inhibitory concentration of vanillin was dependent upon the microorganism, and this ranged between 6 and 18 mM. When incorporated with the commercial antibrowning dipping solution NatureSeal, 12 mM vanillin inhibited the total aerobic microbial growth by 37% and 66% in fresh-cut apples, during storage at 4°C for 19 days. Vanillin (12 mM), however, did not control enzymatic browning or softening as observed with NatureSeal.

Feng et al. (2004) tested commercially available and experimental coatings and antioxidants for their ability to reduce mechanically induced peel browning of pears (cv. Bartlett). Pears were assessed for flesh firmness and external peel color and then treated with coatings and antioxidants (ethoxyquin, diphenylamine, 2-mercaptobenzothiazole, NatureSeal, Semperfresh™, cysteine, ascorbic acid, calcium lactate, sodium bisulfate, and a combination of 2% ascorbic acid, 1% calcium lactate, and 0.5% cysteine) and stored at 0°C for up to 9 days. Mechanical injury of the pears was induced by vibrating, rolling, scuffing, and handling to simulate damage caused by refrigerated truck transportation. The tested compounds reduced the peel browning index by 3% to 15%, compared to untreated controls. Ethoxyquin, diphenylamine, cysteine, and the combination of ascorbic acid, calcium lactate, and cysteine showed the greatest control of peel browning in pears. Neither NatureSeal nor Semperfresh provided control of peel browning in Bartlett pears. It was suggested that to reduce peel browning, pears should be treated and packed immediately before transport.

Abbott et al. (2004) evaluated the effects of a commercial dipping procedure with an in-house method for the control of browning in apple pieces based on eating quality of Goldrush, Granny Smith, Golden Delicious, and Fuji apple slices. Apple slices were prepared, dipped, and subjected to cold storage for varying amounts of time, prior to equilibration at room temperature and sensory analysis. No significant differences in sensory properties were observed between slices subjected to the two dipping treatments, although the texture of NatureSeal-treated slices was rated as slightly better than that of slices dipped in the in-house solution. It was concluded that apple quality was maintained to the same extent by the in-house procedure and the commercial NatureSeal coating.

Bhagwat et al., (2004) determined the influence of the commercial wash solution NatureSeal for Apples and three experimental wash solutions adjusted to pH 2, 2.5, and 3 on survival of five foodborne pathogens (*Escherichia coli* O157:H7, *Salmonella Typhimurium*, *Shigella flexneri*, *Vibrio cholerae,* and *Listeria monocytogenes*) inoculated onto fresh-cut apple slices (cv. Fuji and Granny Smith). Also, the efficacy of the wash solutions on the sensory properties of fresh-cut apple slices following storage (at 5°C for 6 days) was evaluated by instrumental and sensory analyses. All wash solutions resulted in similar instrumental firmness, cut surface color, firmness, and flavor sensory scores for both apple cultivars throughout the storage period. All three experimental wash solutions, prior to their use with apple slices, reduced *Salmonella Typhimurium* and *V. cholerae* survival by ≥5 logs. Furthermore, the experimental wash solution at the lowest pH also reduced *E. coli* O157:H7, *L. monocytogenes,* and *S. flexneri* survival by ≤5 logs, while NatureSeal for Apples was only inhibitory against *V. cholerae*. As indicated by a decrease in conductivity, increases in soluble solids content and osmolality, and changes in pH, the compositions of wash solutions changed during treatment of apple slices, resulting in a loss of antibacterial activity. It is thus recommended that wash solutions not be reused on multiple batches of sliced apples, and it is suggested that alternative washing strategies need to be developed for fresh-cut apple slices that maintain the antimicrobial properties of the wash solutions.

Four liquid coating compounds, NatureSeal 2020, Nu-Film 17, Nu-Coat-Flo, and chitosan, at concentrations of 100%, 10%, 10.3%, and 1%, respectively, and an aqueous polyamine treatment, 5 mM putrescine, were applied to freshly harvested plums (*Prunus salicina* cv. Blue goose) to study their effects on fruit storage and quality. After dipping for 5 minutes in solutions of these compounds, the fruits were air dried and stored in ventilated bags at 3°C or –2°C and 90% to 95% RH for 4 weeks. The treatments had a beneficial effect on fruit storage and quality. Fruits treated with NatureSeal 2020 or Nu-Film 17 were attractive, marketable, and of better quality than fruits subjected to the other treatments (Basiouny and Baldwin 1998).

Coating preclimacteric avocados (*Persea americana*) with NatureSeal delayed ripening at 20°C by about 2 days, even when coated fruits were treated with 100 ppm ethylene for up to 3 days; however, 4 days of ethylene treatment overcame the ripening delay. Coating applied to initiated fruits had no effect on subsequent ripening. Following commercial application to late harvested avocados (Brooks Late) and storage for 4 weeks at 5°C, the coated fruits showed a consistent tendency for better storage performance. Internal O_2 and CO_2 levels were 15.2% and 3.7%, respectively, in uncoated avocados, and 10.2% and 10.1%, respectively, in coated avocados, suggesting that a modified atmosphere effect on ripening may explain the action of the coating on avocado ripening (Bender et al., 1993).

9.7.2 Nutri-Save™

Nutri-Save™ coating was developed around 1985 by Nova Chem Ltd. (Armdale, Nova Scotia, Canada). The coating is a biodegradable, water-soluble polysaccharide derived from shellfish, of pH 7.5 and viscosity 100 to 300 cP (1% solution), which is compatible with benomyl. Application of a 1% to 2% solution left a surface film of 1 to 3 mµ. It was evaluated during the early 1990s as an experimental coating of Bartlett and d'Anjou pears, Van and Stella cherries, and Golden Delicious, McIntosh, Delicious, and Spartan apples (Meheriuk and Lau, 1988; Lau and Yastremski, 1991; Meheriuk et al., 1991; Lau and Meheriuk, 1994). These reports indicated that normal fruit ripening was adversely affected or not noticeably beneficial by the application of Nutri-Save coating. As this coating formulation contains carboxy-methyl chitosan, it is not allowed for commercial food application in the United States and is no longer on the market.

9.7.3 Pro-long™

The commercial fruit coating Pro-long™ or TAL Pro-long™ was developed in England (Courtaulds Ltd., Derby) in the early 1980s and is an aqueous dispersion of sucrose esters of fatty acids, sodium salts of carboxymethylcellulose and mono- and diglycerides. This coating treatment has been used in fresh fruits and vegetables as a permeable membrane coating and can be produced on fruit surfaces by dipping in an aqueous dispersion of Pro-long (Dhalla and Hanson, 1988).

Banks (1983, 1984a) applied TAL Pro-long on bananas and reported that it depressed the oxygen and elevated the ethylene content of the fruit. Banks (1984b) concluded that the effects of TAL Pro-long are due to physical blockage of stomata, impeding gaseous diffusion, and reduction of the permeability of the skin. Coating banana fruit with Pro-long 24 hours after initiation of ripening decreased their skin permeability to gases and depressed their O_2 content. Coating the fruit immediately before ripening initiation delayed the onset of rapid ethylene production, which normally begins as the fruit starts to ripen. Coating slightly suppressed rapid ethylene production when applied immediately after ripening initiation, and exerted a strong, temporary inhibition of ethylene production when applied 24 hours later. A climacteric rise in respiration was absent from fruit coated with Pro-long immediately before or after ripening initiation; delaying coating by a further 24 hours arrested development of the climacteric in mid-rise. In mid-climacteric fruit, inhibition of ethylene production by coating occurred more rapidly than inhibition of respiration. Coating the fruit delayed the chlorophyll loss that normally accompanies ripening, but the magnitude of this effect declined as the coating treatment was delayed relative to ripening initiation. The effects of coating on skin color change and on respiration appeared to be largely independent of its effect on the fruit's ethylene content (Banks, 1985b). Bananas coated with Pro-long or stored under controlled atmosphere conditions ripened more slowly than untreated fruit, and this is reflected in the lower activity of some enzymes during storage of coated and controlled atmosphere-stored fruit (Dillon et al., 1989).

The effects of TAL Pro-long on the shelf life and selected quality attributes of plantain were also investigated (Olorunda and Aworh, 1984). Relative to untreated fruits, yellow color development was retarded by 4 to 8 days, and changes in pH and total acidity were delayed in plantain dipped in 1.5% or 2.5% Pro-long solution when stored at 30°C or at 20 to 24°C. Pro-long coating had less effect on the pulp:peel ratio of plantain and on changes in moisture contents of pulp and peel during storage. After ripening, fruits dipped in 1.5% Pro-long solution produced an acceptable, though slightly inferior, fried plantain product relative to untreated fruits.

Coating and modified atmosphere treatments reduced light-induced greening of King Edward potatoes. Coating caused small internal atmosphere changes, with CO_2 content being negatively correlated with chlorophyll content. Larger changes achieved by modifying the external atmosphere of such tubers increased greening suppression. However, the magnitude of response to coating and to the low O_2 treatment varied with maturity and growing conditions. The success of coating for inhibition of greening in potatoes will depend upon the physiological variability of the tubers to which it is applied (Banks, 1985a).

Meheriuk and Lau (1988) coated Bartlett and d'Anjou pears with Pro-long, reporting firmer, higher acid levels and greener skin color than comparable control fruit at different storage durations in air at 0°C. Incidence of breakdown in Bartlett and scald in d'Anjou were significantly reduced by the coating materials. However, ripening was adversely affected by the coatings, particularly in Bartlett, and many fruit failed to ripen properly even after 10 days at 20°C. Bartlett pears also developed a blotchy appearance (areas of green and yellow) when held at the ripening temperature.

A poststorage application of TAL Pro-long reduced the softening of low-oxygen stored McIntosh and controlled-atmosphere stored Delicious apples during a 21-day shelf-life period at 15°C and 90% to 95% RH. The treatment did not affect fruit firmness in McIntosh apples but did retard the loss of ground color. No physiological disorder was found in any treated fruit (Chu, 1986).

Motlagh and Quantick (1988) treated Brazilian limes (*C. aurantifolia* cv. Persian) with 1.5%, 2%, and 2.5% Pro-long stored under temperate, tropical, and cold store ambient conditions. Results suggested that Pro-long usefully extends the shelf life of limes, especially at high RH, low temperature, or both, by controlling weight loss and degreening; Pro-long was at least as effective as gibberellic acid in delaying degreening under degreening temperate ambient conditions (15 to 20°C; 40% to 60% RH).

Mangoes were treated with 0.75% and 1.0% w/v aqueous suspensions of Pro-long and stored at 25 ± 2°C/85% to 95% RH. Treatment with 0.75% Pro-long significantly increased storage life, retarded ripening, and reduced weight loss, without adversely affecting sensory quality. Pro-long coating at 1% resulted in undesirable increases in ethanol formation in the pulp of some mangoes (Dhalla and Hanson, 1988).

Nisperos-Carriedo and Baldwin (1988) coated green tomatoes with TAL Pro-long. Coated fruits showed only a slight decrease in ripening. The fruits, however, exhibited a more uniform color. Nisperos-Carriedo et al. (1990) reported that beeswax emulsion and TAL Pro-long alone or in combination were the most effective coatings in retaining or increasing volatile components considered important to fresh orange flavor (acetaldehyde, ethyl acetate, ethyl butyrate, and methyl butyrate) in pineapple oranges during storage at 21°C.

9.7.4 Semperfresh™

Semperfresh™, a fruit coating developed by Semper Bio Technology Ltd. (Reading, United Kingdom) consists of a combination of sucrose esters with a palm oil–based shortening ingredient and a thickener/stabilizer based on plant cellulose.

Semperfresh is provided in a dry powder form that is mixed with water and used to either dip or spray the fruit immediately after harvesting. It forms a coating around the fruit which reduces inflow of oxygen by approximately 90%, while hardly affecting outflow of CO_2. Semperfresh has been approved for use by the major health bodies (including the World Health Organization [WHO], U.S. Food and Drug Administration [FDA], European Economic Community [EEC], and China's Ministry of Health), and can help fruit stay fresh for up to twice as long (Semper Bio Technology, 1990). Semperfresh is similar to Pro-long but has more short-chain unsaturated fatty acid esters.

Semperfresh treatment reduced Golden Delicious and McIntosh apple ripening rates. Coated apples had delayed color development (internal and external Hunter color reflectance measurements) during 4 months of storage at 5°C. Semperfresh increased fruit firmness of both varieties during storage. Total acidity, pH, and soluble solids were not affected by the Semperfresh coating. Flavor and textural changes were not detected when apples coated with 1.2% Semperfresh were compared to untreated apples after 2 months storage (Santerre et al., 1989).

The effect of Semperfresh coating on storage quality of Golden Delicious, Ida Red, and MacIntosh apples varied for each cultivar. Semperfresh retarded apple ripening as shown by persistence of green tissue colors, and increased tissue firmness and titratable acidity. Soluble solids were not significantly affected by Semperfresh treatments. Semperfresh treatment improved consumer acceptability ratings for Golden Delicious and McIntosh apples but had no significant improvement for Ida Red apples (Chai et al., 1991). Sumnu and Bayindirli (1995) reported that Semperfresh coating at levels of 5, 10, 15, and 20 g/L extended shelf life of Amaysa apples by 10%, 25%, 30%, and 35%, respectively. Also, Ozdemir et al. (1994) and Oezdemir (1995) concluded that Semperfresh coating did not prevent the weight loss that normally occurs during 6 months of cold storage (1°C, 85% to 90% RH) of Starking Delicious and Golden Delicious apples. However, coated Starking Delicious apples were firmer, higher in acidity, higher in total soluble solids contents, and greener in skin color than uncoated fruits. It is concluded that coating with Semperfresh can be beneficial for stored apples at 1% and 2% applications. However, these concentrations have a limited effect on extending the cold storage period and maintaining the quality of Golden Delicious apples (Oezdemir, 1995). Semperfresh reduced scald incidence and severity appreciably when Granny Smith apples were stored in controlled atmosphere (1.6% CO_2:1.4% O_2) for 43 weeks but not when apples were stored in air. Semperfresh failed to improve the performance of low concentrations of diphenylamine to reduce scalding (Chellew and Little, 1995). Limited control of scald by Semperfresh plus ascorbyl palmitate is partially related to modification of the internal atmosphere of Granny Smith apples (Bauchot et al., 1995).

Al-Zaemey et al. (1989) reported that Semperfresh-treated bananas had a significantly lower weight loss than untreated fruit during ripening at 90% RH. The effect was less for fruit stored at 70% RH. Chemically ripened banana coated with Semperfresh significantly reduced the rate of ripening and increased shelf life; the effect was more pronounced when the coating was applied before initiation of chemical ripening (Goburdhun, 1994).

Avena-Bustillos et al. (1994) compared the effects of covering zucchini (*Cucurbita pepo*; melopepo) with different edible emulsion coatings and the commercial coating Semperfresh. Zucchini was coated with 0.5%, 0.75%, or 1% aqueous solutions of Semperfresh. Semperfresh did not increase water vapor resistance of zucchini or affect internal CO_2 and ethylene concentrations.

Suemnue and Bayindirli (1994) determined the effects of coating Ankara pears with different concentrations of Semperfresh and Jonfresh™ (carnauba wax and shellac based) after storage for 5 months at 0°C. Coating levels of 1% and 1.5% were effective for extending storage life and delaying ripening. Both Semperfresh- and Jonfresh-treated pears had better color, firmness, ascorbic acid, titratable acidity, and soluble solids retention than control pears. Although Semperfresh was effective in reducing respiration and weight loss, Jonfresh was found to be more efficient. As there was no significant difference in any other quality factor between the coatings, it was concluded that cost may be the determining factor in coating selection.

Feng et al. (2004) reported that Semperfresh coating provided control of peel browning in Bartlett pears. It is suggested that to reduce peel browning, pears should be treated and packed immediately before transport. It was found that the triple combination of Semperfresh, ascorbic acid, and cold storage improved the microbiological quality and physicochemical properties of coated quinces, leading to an extension in shelf life when compared with untreated quinces (Yurdugul, 2005).

Hardy kiwifruit coated with Semperfresh provided an attractive sheen to the fruit surface and did not impair ripening. Consumer tests indicated that both coated and uncoated fruit were well liked and indicate that this edible coating may be an alternative to costly low-vent packaging for reducing moisture loss and extending storage life of fresh kiwifruit (Fisk et al., 2008).

Bayindirli et al. (1995) also compared Semperfresh and Jonfresh coatings on Satsuma mandarins at 0.6% and 0.8%, respectively. Both coatings delayed ripening of Satsuma mandarins and effectively retained ascorbic acid, soluble solids, titratable acidity, and reduced respiration rate. Jonfresh coating was determined as the most efficient coating for minimizing weight loss. Jonfresh coating was also rated superior to Semperfresh coatings for Amasya apples (Sumnu and Bayindirli, 1995). Freshly harvested Kinnow mandarins coated with a combination of bavistin and Semperfresh could be kept for up to 40 days in a cool chamber (14.5 to 18.7°C and 84% to 96% RH) versus 15 days at room temperature (Pal et al., 1997).

Fully matured tomato fruits (cv. Pusa Early Dwarf) were dipped in Semperfresh (0.2% to 0.4% solution for 10 minutes) and subsequently stored at 24 to 31.5°C, 77% to 90% RH. Coating increased the shelf life by delaying ripening and decreasing physiological weight loss. The total soluble solids to acid ratio decreased initially (at the sixth day) but increased at later storage times (Kabir et al., 1995). Use of Semperfresh in combination with cold storage was significantly effective in retaining higher contents of vitamin C and total chlorophyll in coated green slender peppers. However, coating showed no synergistic benefit for improving the quality of peppers under controlled atmosphere storage and ambient conditions (Ozden and Bayindirli, 2002).

Haden mangoes coated with Semperfresh at 8, 16, and 24 g/L and then stored at 13°C and 85% RH had no effect on decay development. Ascorbic acid decreased in all stored fruit, but this decrease was slower in coated fruit, and there were no significant differences between the different Semperfresh concentrations (Carrillo-Lopez et al., 2000). Kensington Pride hard mature green mangoes coated with aqueous Semperfresh (0.6%) slightly delayed fruit ripening but reduced fruit aroma volatile development during storage at 21 ± 1°C and 55.2 ± 1.1% RH until the eating soft stage (Dang et al., 2008).

Sweet cherries treated with 10 or 20 g/L Semperfresh increased the shelf life of cherries by 26% and 21% at 0 and 30°C, respectively. At both temperatures, Semperfresh delayed changes in the majority of quality parameters (titratable acidity, ascorbic acid content, weight loss, firmness, and skin color) but did not influence sugar and soluble solid contents (Yaman and Bayoindirli, 2002).

9.7.5 Shellac

Shellac is a natural polymer obtained by refining the secretion of *Kerria lacca*, a parasitic insect found on trees in Southeast Asia. It has been used alone in hard confectionary and pharmaceutical coatings (Hagenmaier and Shaw, 1991; Phan The et al., 2008). Water-soluble shellac coatings showed higher oxygen, carbon dioxide, and water vapor permeabilities than alcohol-soluble shellac coatings (Hagenmaier and Shaw, 1991). This is convenient for coating fruits and vegetables to allow respiration gas transfer.

9.7.6 Waxes

Waxes commercially used in edible coatings are long-chain fatty acids from beeswax, paraffin, and carnauba. Waxes are commonly used to coat fruits and vegetables to reduce moisture loss during storage and extend shelf life. Many fruits and vegetables make their own natural waxy coating to help retain moisture, because most produce is 80% to 95% water. In some cases, waxes are applied to produce to replace natural waxes lost during washing. Wax coatings can also help inhibit mold growth, protect against bruising, and enhance appearance.

USDA-ARS scientists at the South Atlantic Area Citrus and Subtropical Products Laboratory (Winter Haven, Florida) coated oranges, grapefruit, and tangerines with hydrocarbon-containing wax microemulsions (Hagenmaeir and Baker, 1993; Hagenmaeir and Shaw, 2002). The fruit experienced less weight loss and exhibited internal carbon dioxide values less than half of those of commercially coated fruit in Florida packinghouses. Valencia oranges that received two layers of wax coatings had a weight loss of only 20% to 30% of the washed control with less skin shriveling, giving the peel a smooth appearance, even after prolonged storage. High-gloss levels could be obtained by applying a high-gloss coating as a second coating. Wax-coated oranges had lower internal carbon dioxide levels, higher internal oxygen levels, and better water resistance than fruit coated with shellac or resins. Fruits coated with wax had greater amounts of internal oxygen. They tended to contain less ethanol than fruit with traditional high-gloss coatings and tended to develop less off-flavors.

Thus, edible coatings and films have been developed for many processed food products, in many forms and from diverse materials. Natural edible products have been processed into coatings as well and used on both processed and fresh food products to improve quality and shelf life.

References

Abbott, J.A., Saftner, R.A., Gross, K.C., Vinyard, B.T., Janick, J. 2004. Consumer evaluation and quality measurement of fresh-cut slices of 'Fuji', 'Golden Delicious', 'GoldRush', and 'Granny Smith' apples. *Postharvest Biology and Technology* 33:127–140.

Adele, F. 2005. Automated panning technology. *Manufacturing Confectioner* 85(11):31–37.

Akutagawa, T. 2007. Decorative confectionery and method for producing the same. United States Patent Application Publication 20070237869. Available at: www.freepatentsonline.com/20070237869.pdf. Accessed December 28, 2010.

Al-Zaemey, A.B.S., Falana, I.B., Thompson, A.K. 1989. Effects of permeable fruit coatings on the storage life of plantain and bananas. *Aspects of Applied Biology* 20:73–80.

Armitage, D.B., Hettiarachchy, N.S., Monsoor, M.A. 2002. Natural antioxidants as a component of an egg albumen film in the reduction of lipid oxidation in cooked and uncooked poultry. *Journal of Food Science* 67:631–634.

Arvanitoyannis, I. 2002. Formation and properties of collagen and gelatin films and coatings. In *Protein-based films and coatings*. Gennadios, A., (ed.), CRC Press, Boca Raton, FL, pp. 275–304.

Asatake, M., Yamaguchi, S. 2007. Sugar-coated tablet and method for producing the same. Japanese Patent Application.

Ashley, J.D. 1973. Edible cereal package. United States Patent 3778515.

Avena-Bustillos, R.J., Krochta, J.M., Saltveit, M.E., Rojas-Villegas, R. de J., Sauceda-Perez, J.A. 1994. Optimization of edible coating formulations on zucchini to reduce water loss. *Journal of Food Engineering* 21:197–214.

Baldwin, E.A., Baker, R.A. 2002. Use of proteins in edible coatings for whole and minimally processed fruit and vegetables. In *Protein-based films and coatings*, Gennadios, A., (ed.), CRC Press, Boca Raton, FL, 501–515.

Banks, N.H. 1983. Evaluation of methods for determining internal gases in banana fruit. *Journal of Experimental Botany* 34:871–879.

Banks, N.H. 1984a. Some effects of TAL Pro-long coating on ripening bananas. *Journal of Experimental Botany* 35:127–137.

Banks, N.H. 1984b. Studies of the banana fruit surface in relation to the effects of TAL Pro-long coating on gaseous exchange. *Scientia Horticulturae* 24:279–286.

Banks, N.H. 1985a. Coating and modified atmosphere effects on potato tuber greening. *Journal of Agricultural Science, UK* 105:59–62.

Banks, N.H. 1985b. Responses of banana fruit to Pro-long coating at different times relative to the initiation of ripening. *Scientia Horticulturae* 26:149–157.

Basiouny, F.M., Baldwin, E.A. 1998. The use of liquid coating compounds as postharvest application to plum fruits. *Proceedings of the Florida State Horticultural Society* 110, 219–222.

Bauchot, A.D., John, P., Soria, Y., Recasens, I. 1995. Carbon dioxide, oxygen, and ethylene changes in relation to the development of scald in Granny Smith apples after cold storage. *Journal of Agricultural and Food Chemistry* 43:3007–3011.

Bayindirli, L., Sumnu, G., Kamadan, K. 1995. Effects of Semperfresh and Jonfresh fruit coatings on poststorage quality of Satsuma mandarins. *Journal of Food Processing and Preservation* 19:399–407.

Bender, R.J., Brecht, J.K., Sargent, S.A., Navarro, J.C., Campbell, C.A. 1993. Ripening initiation and storage performance of avocados treated with an edible-film coating. *Acta Horticulturae* 343:184–186.

Bhagwat, A.A., Saftner, R.A., Abbott, J.A. 2004. Evaluation of wash treatments for survival of foodborne pathogens and maintenance of quality characteristics of fresh-cut apple slices. *Food Microbiology* 21:319–326.

Bisson, A., Weisenfels, M. 1992. Machines and equipment for processing of COFFI films. I. Manual equipment. *Fleisch* 46:439–440.

Bolin, H.R. 1976. Texture and crystallization control in raisins. *Journal of Food Science* 41:1316–1319.

Boutin, B. 1997. Finishing agents. *The Manufacturing Confectioner* 77(12):102–109.

Burns, R.E., Fast, R.B. 1990. Application of nutritional and flavoring/sweetening coatings. In *Breakfast cereals and how they are made*, Fast, R.B., and Caldwell, E.F., (eds.), American Association of Cereal Chemists, Inc., St. Paul, MN, 195–220.

Cagri, A., Ustunol, Z., Ryser, E.T. 2004. Antimicrobial edible films and coatings. *Journal of Food Protection* 67:833–848.

Carrillo-Lopez, A., Ramirez-Bustamante, F., Valdez-Torres, J.B., Rojas-Villegas, R., Yahia, E.M. 2000. Ripening and quality changes in mango fruit as affected by coating with an edible film. *Journal of Food Quality* 23:479–486.

Chai, Y.L., Ott, D.B., Cash, J.N. 1991. Shelf-life extension of Michigan apples using sucrose polyester. *Journal of Food Processing and Preservation* 15:197–214.

Chellew, J.P., Little, C.R. 1995. Alternative methods of scald control in Granny Smith apples. *Journal of Horticultural Science* 70:109–115.

Chen, H. 1995. Functional properties and applications of edible films made of milk proteins. *Journal of Dairy Science* 78:2563–2583.

Chen, H. 2002. Formation and properties of casein films and coatings. In *Protein-based films and coatings*, Gennadios, A., (ed.), CRC Press, Boca Raton, FL, 181–211.

Cho, S.Y., Lee, S.Y., Rhee, C. 2010. Edible oxygen barrier bilayer film pouches from corn zein and soy protein isolate for olive oil packaging. *Lebensmittel-Wissenschaft und -Technologie* 43:1234–1239.

Chu, C.L. 1986. Poststorage application of TAL Pro-long on apples from controlled atmosphere storage. *HortScience* 21:267–268.

Cleophas, F.M.J. 2006. Process for coating of nuts and peanuts. Netherlands Patent.

Cliffe-Byrnes, V., O'Beirne, D. 2007. The effects of modified atmospheres, edible coating and storage temperatures on the sensory quality of carrot discs. *International Journal of Food Science and Technology* 42:1338–1349.

Conca, K.R. 2002. Protein-based films and coatings for military packaging applications. In *Protein-based films and coatings*, Gennadios, A., (ed.), CRC Press, Boca Raton, FL, 551–577.

Cooke, J.I. 2001. Chocolate belt panning. *Manufacturing Confectioner* 81:63–66.

Cooper, I., Melnick, D. 1971. Bacon flavoured chips. United States Patent.

Corrales, M., Han, J.H., Tauscher, B. 2009. Antimicrobial properties of grape seed extracts and their effectiveness after incorporation into pea starch films. *International Journal of Food Science and Technology* 44:425–433.

Dang, K.T.H., Singh, Z., Swinny, E.E. 2008. Edible coatings influence fruit ripening, quality, and aroma biosynthesis in mango fruit. *Journal of Agricultural and Food Chemistry* 56:1361–1370.

Debeaufort, F., Quezada-Gallo, J.A., Voilley, A. 2002. Edible films and coatings as aroma barriers. In *Protein-based films and coatings,* Gennadios, A., (ed.), CRC Press, Boca Raton, FL, 579–600.

Dhalla, R., Hanson, S.W. 1988. Effect of permeable coatings on the storage life of fruits. II. Pro-long treatment of mangoes (*Mangifera indica* L. cv. Julie). *International Journal of Food Science and Technology* 23:107–112.

Dillon, M., Hodgson, F.J.A., Quantick, P.C., Taylor, D.J. 1989. Use of the APIZYM testing system to assess the state of ripeness of banana fruit. *Journal of Food Science* 54:1379–1380.

Donhowe, G., Fennema, O.R. 1994. Edible films and coatings: characteristics, formation, definitions, and testing methods. In *Edible coatings and films to improve food quality*, Krochta, J.M., Saldwin, E.A., and Nisperos-Carriedo, M.O. (eds.), Technomic, Lancaster, PA, 1–24.

Dreier, W. 1991. The nuts and bolts of coating and enrobing. *Prepared Foods* 160:47–48.

Du, W-X., Avena-Bustillos, R.J., Woods, R., McHugh, T.H., Levin, C.E., Mandrell, R., Mendel Friedman. 2009. Hedonic evaluation of cooked chicken wrapped with apple and tomato films formulated with cinnamaldehyde and carvacrol. Presented at the Institute of Food Technologists Meeting, Anaheim, CA. June 6–10, 2009. Session No. 158 (Sensory Evaluation Division).

Eyas Ahamed, M., Anjaneyulu, A.S.R., Sathu, T., Thomas, R., Kondaiah, N. 2007a. Effect of different binders on the quality of enrobed buffalo meat cutlets and their shelf life at refrigeration storage (4±1°C). *Meat Science* 75:451–459.

Eyas Ahamed, M., Anjaneyulu, A.S.R., Sathu, T., Thomas, R., Kondaiah N. 2007b. Effect of enrobing on the quality and shelf life of buffalo meat cutlets under frozen storage. *Journal of Muscle Foods* 18:19–34.

Feng, X., Biasi, B., Mitcham, E.J. 2004. Effects of various coatings and antioxidants on peel browning of Bartlett pears. *Journal of the Science of Food and Agriculture* 84:595–600.

Fisk, C.L., Silver, A.M., Strik, B.C., Zhao, Y. 2008. Postharvest quality of hardy kiwifruit (*Actinidia arguta* 'Ananasnaya') associated with packaging and storage conditions. *Postharvest Biology and Technology* 47:338–345.

Gautier, J.P. 1990. Process for coating the interior of edible cups, cones, etc. (e.g. for ice cream) with a confectionery material, e.g. chocolate. French Patent Application.

Gennadios, A. 2004. Edible films and coatings from proteins. In *Proteins in food processing*, Yada, R.Y. (ed.)., Woodhead and CRC Press, Boca Raton, FL, pp. 442–467.

Goburdhun, S. 1994. Chemical ripening of Dwarf Cavendish bananas (cv. Naine). I. Effects of ethrel and ethylene on ripening. II. Extension of shelf-life of ripened fruits. *Revue Agricole et Sucriere de l'Ile Maurice* 73(3):36–43.

Gorton, L. 1993. Coating and topping—applying the finishing touch. *Baking & Snack* 45–46, 48.

Green, D.R., Nowakowski, C. 2005. Low sugar presweetened coated cereals and method of preparation. Canadian Patents 2505186.

Groves, R.J. 1977. Pan coating—three basic varieties considered. *Candy and Snack Industry* 142:33–34, 36–38.

Guzman, M.A., Hegarty, E. 2000. Method of improving the softness of raisins. United States Patent.

Hagenmaier, R.D., Shaw, P.E. 1991. Permeability of shellac coatings to gases and water vapor. *Journal of Agricultural and Food Chemistry* 39:825–829.

Hagenmeir, R.D., Baker, R.A. 1993. Reduction in gas exchange of citrus fruit by wax coatings. *Journal of Agricultural and Food Chemistry* 41:283–287.

Hagenmeir, R.D., Shaw, P.E. 2002. Changes in volatile components of stored tangerines and other specialty citrus fruits with different coatings. *Journal of Food Science* 67:1742–1745.

Han, J.H., Krochta, J.M. 2007. Physical properties of whey protein coating solutions and films containing antioxidants. *Journal of Food Science* 72:E308–E314.

Herald, T.J., Hachmeister, K.A., Huang, S., Bowers, J.R. 1996. Corn zein packaging materials for cooked turkey. *Journal of Food Science* 61:415–417, 421.

Hildebrandt, S. 2009. Panning healthier pleasures. *Candy Industry* 174(5):36, 38.
Hoffman, A.J., Shen, S., Harrison, M.D. 2008. Pullulan film containing sweetener. United States Patent Application Publication.
Hollender, H.A. 1965. Technology of Space Foods. R&D Associates, Activities Report, 17:19–31.
Hoitink, R., Guex, C., Voisin, I. 2004. Breakfast cereals. European Patent Application.
Holownia, K.I., Chinnan, M.S., Erickson, M.C., Mallikarjunan, P. 2000. Quality evaluation of edible film-coated chicken strips and frying oils. *Journal of Food Science* 65:1087–1090.
Hood, L.L. 1987. Collagen in sausage casings. In *Advances in meat research*, Vol. 4, Pearson, A.M., Dutson, T.R., and Bailey, A.J., (eds.), Van Nostrand Reinhold, New York, 109–129.
Kabir, J., Ghosh, B., Dutta Ray, S.K., Mitra, S.K. 1995. Post harvest use of edible coatings on shelf-life of tomato. *Indian Food Packer* 49(1):25–28.
Krochta, J.M. 2002. Proteins as raw materials for films and coatings: Definitions, current status, and opportunities. In *Protein-based films and coatings*, Gennadios, A., (ed.), CRC Press, Boca Raton, FL, 1–41.
Krochta, J.M., Dangaran, K.L., Shih-Yu Lin. 2005. Methods and formulations for providing gloss coatings to foods and for protecting nuts from rancidity. United States Patent Application Publication.
Lau, O.L., Yastremski, R. 1991. Retention of quality of Golden Delicious apples by controlled- and modified-atmosphere storage. *HortScience* 26:564–566.
Lau, O.L., Meheriuk, M. 1994. The effect of edible coatings on storage quality of McIntosh, Delicious and Spartan apples. *Canadian Journal of Plant Science* 74:847–854.
Lee, S.-Y., Trezza, T.A., Guinard, J.X., Krochta, J.M. 2002a. Whey-protein-coated peanuts assessed by sensory evaluation and static headspace gas chromatography, *Journal of Food Science* 67:1212–1218.
Lee, S.-Y., Dangaran, K.L., Guinard, J.X., Krochta, J.M. 2002b. Consumer acceptance of whey-protein-coated peanuts as compared to shellac-coated chocolate, *Journal of Food Science* 67:2764–2769.
Lees, R. 2000. The art and skill of panning. III. *Confectionery Production*. 66:18–19.
Leusner, S. 2009. Fiber fortified cereals, cereal bars and snacks and methods for making. United States Patent Application Publication.
MacQuarrie, R. 2006. Edible dissolving gelatin strips. United States Patent Application Publication.
MacQuarrie, R. 2008. Edible dissolving gelatin strips. United States Patent Application Publication.
MacQuarrie, R. 2010. Edible dissolving gelatin strips. United States Patent.
Maté, J.I., Krochta, J.M. 1996. Whey protein coating effect on the oxygen uptake of dry roasted peanuts. *Journal of Food Science* 61:1202–1206, 1210.
Maté, J.I., Frankel, E.N., Krochta, J.M. 1996. Whey protein isolate edible coatings: Effect on the rancidity process of dry roasted peanuts. *Journal of Agricultural and Food Chemistry* 44:1736–1740.
Maté, J.I., Krochta, J.M. 1997. Whey protein and acetylated monoglyceride edible coatings: effect on the rancidity process of walnuts. *Journal of Agricultural and Food Chemistry* 45:2509–2513.
Meheriuk, M., Lau, O.L. 1988. Effect of two polymeric coatings on fruit quality of Bartlett and d'Anjou pears. *Journal of the American Society for Horticultural Science* 113:222–226.
Meheriuk, M., Neilsen, G.H., McKenzie, D.L. 1991. Incidence of rain splitting in sweet cherries treated with calcium or coating materials. *Canadian Journal of Plant Science* 71:231–234.
Mie, M., Takano, T., Yamamoto, M., Yoshinaka, K., Kitamura, Y. 2008. Method for coating nut. Japanese Patent Application.

Morando, D.E. 2009. Edible candy confection with improved shelf-life and method of making thereof. U.S. Patent 7,618,666.

Motlagh, F.H., Quantick, P.C. 1988. Effect of permeable coatings in the storage life of fruits. I. Pro-long treatment of limes (*Citrus aurantifolia* cv. Persian). *International Journal of Food Science and Technology* 23:99–105.

Nisperos-Carriedo, M.O., Baldwin, E.A. 1988. Effect of two types of edible films on tomato fruit ripening. *Proceedings of the Florida State Horticultural Society* 101:217–220.

Nisperos-Carriedo, M.O., Shaw, P.E., Baldwin, E.A. 1990. Changes in volatile flavor components of pineapple orange juice as influenced by the application of lipid and composite films. *Journal of Agricultural and Food Chemistry* 38:1382–1387.

Olorunda, A.O., Aworh, O.C. 1984. Effects of Tal Pro-long, a surface coating agent, on the shelf life and quality attributes of plantain. *Journal of the Science of Food and Agriculture* 35:573–578.

Olson, S., Zoss, R. April 16, 1985. U.S. patent 4,511,583.

Ono, M., Yanagisawa, Y., Kawai, M. 1993. An examination of instrumental textural evaluation of nori products by texturometer. *Journal of Japanese Society of Food Science and Technology [Nippon Shokuhin Kogyo Gakkaishi]* 40:129–132.

Onsøyen, E. 2001. Alginate. Production, composition, physicochemical properties, physiological effects, safety, and food applications. In *Handbook of dietary fiber*, Cho, Susan Sungsoo, Dreher, Mark L., (eds.), CRC Press, Boca Raton, FL, 659–674.

Ozden, C., Bayindirli, L. 2002. Effects of combinational use of controlled atmosphere, cold storage and edible coating applications on shelf life and quality attributes of green peppers. *European Food Research and Technology* 214:320–326.

Ozdemir, A.E., Kaska, N., Agar, I.T., Dundar, O. 1994. Effects of Semperfresh treatments on the post-harvest physiology of cold stored apples. I. Starking Delicious apples. *Turkish Journal of Agriculture and Forestry* 18:473–478.

Oezdemir, A.E. 1995. Effect of Semperfresh treatments on the post-harvest physiology of cold stored apples. II. Golden Delicious. *Turkish Journal of Agriculture and Forestry* 19:11–15.

Padua, G.W., Wang, Q. 2002. Formation and properties of corn zein films and coatings. In *Protein-based films and coatings*. Gennadios, A., (ed.), CRC Press, Boca Raton, FL, 43–67.

Pal, R.K., Roy, S.K., Srivastava, Sanjay. 1997. Storage performance of Kinnow mandarins in evaporative cool chamber and ambient conditions. *Journal of Food Science and Technology, India* 34:200–203.

Petersen, W. 2008. Cultures encapsulated with compound fat breakfast cereals coated with compound fat and methods of preparation. United States Patent Application Publication.

Phan The, D., Debeaufort, F., Luu, D., Voilley, A. 2008. Moisture barrier, wetting and mechanical properties of shellac/agar on shellac/cassava starch bilayer bio-membrane for food applications. *Journal of Membrane Science* 277–283.

Pollard, D. 2009. Food product formed from strips of food material. UK Patent Application.

Quinn, C. 2007. Film studies. *Food Manufacture* 82(9):42–43.

Rapp, K.M., Hasslinger, B., Kowalczyk, J. 2006. Pan coating process. United States Patent Application Publication.

Rojas-Graü, M.A., Avena-Bustillos, R.J., Friedman, M., Henika, P.R., Martín-Belloso, O., McHugh, T.H. 2006. Mechanical, barrier and antimicrobial properties of apple puree edible films containing plant essential oils. *Journal of Agricultural and Food Chemistry* 54:9262–9267.

Rojas-Graü, M.A., Raybaudi-Massilia, R.M., Avena-Bustillos, R.J., Martín-Belloso, O., McHugh, T.H. 2007. Apple puree-alginate edible coating as carrier of antimicrobial agents to prolong shelf life of fresh-cut apples. *Postharvest Biology and Technology* 45:254–264.

Ryu, S.Y., Koh, K.H., Son, S.M., Oh, M.S., Yoon, J.R., Lee, W.J., Kim, S.S. 2005. Physical and microbiological changes of sliced process cheese packaged in edible pouches during storage. *Food Science and Biotechnology* 14:694–697.

Santerre, C.R., Leach, T.F., Cash, J.N. 1989. The influence of the sucrose polyester, Semperfresh®, on the storage of Michigan grown McIntosh and Golden Delicious apples. *Journal of Food Processing and Preservation* 13:293–305.

Semper Bio Technology. 1990. Sleeping beauties: Semperfresh, a new fruit coating that can help fruit stay fresh for up to twice as long. *Nutrition and Food Science* 90(6):12–13.

Shobu, T., Igusa, K., Ogasawara, T. 2004. Aqueous shellac coating agent and production process therefor, and coated food and production process therefor, coated drug and production process therefor, glazing composition for oil-based confectionary, glazing process, and glazed oil-based confectionary using same. United States Patent Application Publication.

Singh, C., Sharma, H.K., Sarkar, B.C. 2010. Influence of process conditions on the mass transfer during osmotic dehydration of coated pineapple samples. *Journal of Food Processing and Preservation* 34:700–714.

Snyder, H.E., Kwon, T.W. 1987. *Soybean Utilization*, New York: Van Nostrand Reinhold.

Solis-Morales, D., Saenz-Hernandez, C.M., Ortega-Rivas, E. 2009. Attrition reduction and quality improvement of coated puffed wheat by fluidised bed technology. *Journal of Food Engineering* 93:236–241.

Suemnue, G., Bayindirli, L. 1994. Effects of Semperfresh® and Jonfresh® fruit coating on poststorage quality of Ankara pears. *Journal of Food Processing and Preservation* 18:189–199.

Sumnu, G., Bayindirli, L. 1995. Effects of coatings on fruit quality of Amasya apples. *Lebensmittel-Wissenschaft und -Technologie* 28:501–505.

Takeda, Y. 2009. Colored kuzukiri, colored strip of bean jelly, and colored malony®. Japanese Patent Application.

Todd, J.M. 1982. Wrap food coating mix and method of using. United States Patent.

Trezza, T.A., Krochta, J.M. 2002. Application of edible protein coatings to nuts and nut-containing food products. In *Protein-based films and coatings,* Gennadios, A., (ed.), CRC Press, Boca Raton, FL, 527–549.

Vasantha Rupasinghe, H.P., Boulter-Bitzer, J., Taehyun Ahn, Odumeru, J.A. 2006. Vanillin inhibits pathogenic and spoilage microorganisms in vitro and aerobic microbial growth in fresh-cut apples. *Food Research International* 39:575–580.

Watters, G.G., Brekke, J.E. 1961. Stabilized raisins for dry cereal products. *Food Technology* 15:236–238.

Wissgott, U., Verschueren, K. 1997. Edible, hot water soluble sachet. European Patent Application.

Wolff, W. 1969. Process for the manufacture of chocolate-coated wafers. West German Patent Application.

Wu, R.Y.A. 2008. Process for preparing a hand-held snack item, and the product thereof. United States Patent.

Wu, Y., Rhim, J.W., Weller, C.L., Hamouz, F., Cuppett, S., Schnepf, M. 2000. Moisture loss and lipid oxidation for precooked beef patties stored in edible coatings and films. *Journal of Food Science* 65:300–304.

Yaman, O., Bayoindirli, L. 2002. Effects of an edible coating and cold storage on shelf-life and quality of cherries. *Lebensmittel-Wissenschaft und -Technologie* 35:146–150.

Yurdugul, S. 2005. Preservation of quinces by the combination of an edible coating material, Semperfresh, ascorbic acid and cold storage. *European Food Research and Technology* 220:579–586.

10 Application of commercial coatings

Yanyun Zhao

Contents

10.1	Introduction	319
10.2	Commercial applications of edible coatings	320
	10.2.1 Fresh and fresh-cut fruits and vegetables	320
	10.2.2 Processed foods	321
	10.2.3 Confectioneries	321
	10.2.4 Nuts	322
10.3	Methods of coating application	322
	10.3.1 Dipping	322
	10.3.2 Dripping	323
	10.3.3 Foaming	323
	10.3.4 Fluidized-bed coating	324
	10.3.5 Panning	324
	10.3.6 Spraying	325
	10.3.7 Electrostatic coating	325
10.4	Keys to successful coating application	326
	10.4.1 Moisture and gas barrier properties	326
	10.4.2 Surface characteristics of coatings	327
	10.4.3 Sensory attributes	327
	10.4.4 Incorporation of functional substances to enhance functionality	328
10.5	Summary	329
References		329

10.1 Introduction

The practice of coating fruits and vegetables dates back to the twelfth and thirteenth centuries, where, in China, commodities were commonly dipped in hot waxes to encourage fermentation (Kaplan, 1986; Yehoshua, 1987). In the last several decades, coatings composed of lipid, polysaccharide, and protein, alone or in combination, have been applied on fresh fruits and vegetables to retard moisture loss, improve appearance by imparting shine to the surface, provide a carrier for fungicides or growth regulators, and create a barrier for gas exchange between the commodity and the external atmosphere (Grant and Burns, 1994). Coatings are also applied for processed foods, from meat, poultry, and seafood, to candies and nuts for many of the same reasons as they are applied on fruits and vegetables. Depending on the surface characteristics of food items and the purpose of coatings, different coating procedures, including dipping, brushing, spraying, and panning, may be applied for evenly spreading the coatings on the surface of the food.

Appropriately formulated edible coatings can be utilized for most foods to meet challenges associated with stable quality, market safety, nutritional value, and economic production cost. The potential benefits of using edible coatings on fresh and processed food products include the following (Lin and Zhao, 2007):

- Coatings can serve as a moisture barrier on the surface of fresh and minimally processed produce for helping to alleviate the problem of moisture loss during postharvest storage, which leads to weight loss and changes in texture, flavor, and appearance.
- Coatings also function as a gas barrier for controlling gas exchange between the fresh produce and its surrounding atmosphere, and thereby decrease respiration and delay deterioration, retard enzymatic oxidation, and protect against from browning discoloration and loss of texture during storage.
- The same gas barriers also prevent the loss of natural volatile flavor compounds and color components from fresh produce. These also prevent the acquisition of foreign odors.
- Coatings protect produce from physical damage caused by mechanical impact, pressure, vibrations, and other factors.
- Coatings function as carriers of active ingredients, such as antimicrobials, antioxidants, nutraceuticals, colors, flavors, and other additives used to improve quality (Rooney, 2005).

Successful application of edible coatings on food is dominated by several critical factors, including the type of coating material and its specific formulation, the method of application, and the surface characteristics of the food. The type of coating materials and their formulation for specific food applications have been discussed extensively in the previous chapters. This chapter provides an overview of the application unit operation, including methods and their controls for ensuring the successful implementation.

10.2 Commercial applications of edible coatings

Edible coatings have been commercially applied to a wide range of food items since the 1930s and 1940s, including fresh and processed fruits and vegetables; fresh, frozen, and processed meat; poultry; seafood products; nuts; bakery goods; and confectioneries with some common and unique purposes. Several book chapters have described their applications (Baker et al., 1994; Wong et al., 1994; Baldwin, 2007; Ustunol, 2009). The following section gives a brief summary of the functions of edible coatings in several common food categories.

10.2.1 Fresh and fresh-cut fruits and vegetables

Most fruits and vegetables possess a natural waxy layer on the surface, called a cuticle. This waxy layer generally has a low permeability to water vapor. Applying an external coating will enhance this natural barrier or replace it in cases where this layer has been partially removed or altered during postharvest handling or processing. Coatings provide a partial barrier to moisture and gas exchange, improve the mechanical handling property of the product through helping maintain structural integrity, retain volatile flavor compounds, and carry other functional food ingredients (Lin and Zhao, 2007).

Proteins, polysaccharides, lipids, and resins are the most common major ingredients of food coatings. The physical and chemical characteristics of these biopolymers greatly influence the functionality of resulting coatings (Sothornvit and Krochta, 2000). Selection of materials is generally based on their cost, approval, and acceptability as food ingredients, film-forming properties, water solubility (hydrophilic and hydrophobic nature), ease of coating formation, and sensory properties. Lin and Zhao (2007) provided a comprehensive review on coating materials feasible for fruit and vegetable applications. Coating applications for suppression of respiration, control of moisture loss, and other functions have also been reviewed in several book chapters and review articles (Baldwin, 2007; Lin and Zhao, 2007; Olivas and Barbosa-Canovas, 2009).

10.2.2 Processed foods

Edible coatings may be applied to processed foods for the prevention of moisture loss, transfer of moisture between components of different water activity in a heterogenous food system, formation of ice in frozen food, exposure to oxygen, or diffusion of CO_2 (Baker et al., 1994; Cutter and Sumner, 2002; Baldwin, 2007; Ustunol, 2009). Some protect meat and nut products from oxidative rancidity and fat absorption. Others reduce movement of cooking oils into food pieces during frying, or leakage of shortening from the fried products. Still others reduce product shrinkage during later cooking, contribute to improved texture, or improve appearance of the final products (Baker et al., 1994). Coatings also function as carriers for antimicrobials, antioxidants, nutrients, colors, and flavoring ingredients (Lin and Zhao, 2007). With regard to the frozen foods, edible coatings reduce freezer burn by limiting the rate of transfer of water vapor from frozen food to surrounding atmosphere, and also reduce dripping during thawing and cooking (Duan et al., 2009, 2010). While normally designed to be thin and unnoticeable, coatings may be a significant component of a food, or may contribute to its structural integrity. More details about the coating applications in processed foods can be found in several book chapters (Baker et al., 1994; Baldwin, 2007; Ustunol, 2009).

10.2.3 Confectioneries

Many candies, including candied fruit, are coated to preserve and protect their structure and contents from stickiness, clumping, and absorption of moisture; to increase palatability; to impart shine to the product; and, in the case of chocolate and other lipid-containing confectioneries, to prevent loss of oil (Baker et al., 1994). Chocolate may be considered as an edible coating when it is used to enrobe a candy center, dried fruit piece, or cookie. Confectioneries are usually coated in a pan, a bowl that is motor driven to provide tumbling action. Coating materials for confectioneries usually include chocolate and sugar syrup and are sometimes followed by an application of small particle-sized granular sugar. Polishing of the confectioneries is accomplished by coating or glazing the panned pieces with waxes, oils, or shellac dissolved in accepted (food-grade) solvents for ease of application and drying (Baker et al., 1994; Nieto, 2009). A more complete discussion of confectioneries coating has been provided by Nieto (2009).

10.2.4 Nuts

The primary use of edible coatings on nuts is to prevent or delay onset of rancidity. Other potential functions of coating nuts are to provide better adhesion of salt, sugars, flavorings, colorants, and antioxidants (Baker et al., 1994). In addition, when nuts are granulated for use in food products, such as cake mixes, a coating is essential to protect the nut meat from absorption of moisture or reaction with other ingredients (Baker et al., 1994). Historically, corn zein and shellac products are marketed for coating of nuts. These coatings are applied in an ethanol solvent, whose low surface energy allows good wettability and coverage of the hydrophobic nut surfaces. However, the ethanol solvent also results in safety and environmental problems (Lin and Krochta, 2005). Recently, whey proteins have shown potential for coating nuts by reducing rancidity and extending shelf life of roasted nuts, and also adding additional benefits of using a water solvent and providing a glossy surface (Trezza and Krochta, 2002). Baldwin and Wood (2006) showed that coating with carboxy methylcellulose (CMC) imparted shine to the coated nuts. Coated nuts scored lower ratings for off-flavor and higher ratings for overall flavor by sensory panels. Hexanal levels, an indicator of rancidity, were reduced in nuts coated with CMC, and CMC coatings with added α-tocopherol were most effective (Baldwin and Wood, 2006). Detailed review of different coating materials on nuts and nut products, as well as the methods for applying coatings on nuts has been provided by Baker et al. (1994) and Trezza and Krochta (2002).

10.3 Methods of coating application

Several coating application methods including dipping, foaming, spraying, and panning may be applied for commercial coatings. The selection of an appropriate method depends on the characteristics of the food, the coating materials, the intended effect of the coating, and the cost. The following sections give a brief review of these methods.

10.3.1 Dipping

Edible coatings can be applied by dipping products in coating solutions and then allowing excess coating to drain as it dries and solidifies (Grant and Burns, 1994). Dipping has been commonly used for coating fruits, vegetables, and meat products. The commodity is directly dipped into the composite coating formulations in aqueous solution, removed, and allowed to dry, whereby a thin membranous film is formed over the commodity surface. The first reported dipping application was by the Florida citrus industry (Platenius, 1939), where the fruits were submerged into a tank of emulsion coating. Fruit was then generally conveyed to a drier or was air-dried under ambient conditions. Sometimes a porous basket can be used to drain excess coating (Grant and Burns, 1994). In this case, the coating thickness is determined by the balance of forces at the stagnation point on the liquid surface. A faster withdrawal speed pulls more fluid up onto the surface of the substrate before it has time to flow back down into the solution. The thickness is primarily affected by viscosity and density of coating solutions, and surface tension of coated products.

According to Grant and Burns (1994), dip-coating is excellent for producing high-quality, uniform coatings but requires precise control and a clean environment. Dipping application is adequate usually for small quantities of commodities. For achieving a uniform surface coverage, complete wetting of the commodity surfaces is critical. Therefore, the applied coating may remain wet for several minutes until the solvent evaporates. Heated drying of the wet surface sometimes is necessary by a variety of means including conventional thermal and ultraviolet (UV) techniques depending on the coating solution formulation. The continuous dipping of produce into a largely static milieu may result in unacceptable buildup of decay organisms, soil, and trash in the dip tank. The dip tanks can be equipped with a porous basket that can be lifted to strain and remove debris. In addition, fruit entering the dip tank must be completely dry to avoid dilution of the resin solution or emulsion coatings. For these reasons, other coating methods are more desirable where considerable quantities of fruits and vegetable are to be coated. Moreover, the residue of the coating solution can be an issue as they are all organic compounds.

10.3.2 Dripping

Overhead drip emitters can also apply coatings to fruits and vegetables. Different sizes of emitters may be used to deliver a variety of droplet sizes. Usually mounted on a dual-bank manifold, emitters are spaced 1 inch apart along the manifold and width of the bed. Irrigation tubing can be used to supply the emitters. A metering pump is used to deliver the coating at pressures not greater than 40 psi (276 kPa). The commodities tumble over rotating brush beds that become saturated with coating. As the fruits tumble and rotate over the saturated brushes, they become uniformly covered with coating (Grant and Burns, 1994). Control systems described for spray applications are also feasible for drip-type application.

This coating application method is the most economic. In addition, it has the ability to deliver the coating either directly to the commodity surface or to the brushes. However, due to the relatively large droplet sizes, good uniform coverage can only be achieved when the commodity has adequate tumbling action over several brushes that are saturated with the coatings (Grant and Burns, 1994).

10.3.3 Foaming

Foam application is useful for some emulsion coatings. A foaming agent is added to the coating or compressed air is blown into the applicator tank. Extensive tumbling action is necessary to break the foam for uniform distribution. The agitated foam is applied to commodities moving by on rollers and cloth flaps or brushes that distribute the emulsion over the surface of the commodity (Grant and Burns, 1994; Cutter and Sumner, 2002; Baldwin, 2007). Excess coating is removed by squeegees positioned below the rollers and is sometimes recirculated. Coating by spraying is the method generally used in most cases. Due to high pressure (60 to 80 psi), less coating solution is required. Programmable spray systems are available for automation during such operations (Tharanathan, 2003). This type of emulsion contains little water and, therefore, dries quite quickly, but inadequate coverage is often a problem.

10.3.4 Fluidized-bed coating

Fluidized-bed coating is a technique that can be used to apply a very thin layer onto dry particles of very low density or small size (Debeaufort and Voilley, 2009). It was originally developed as a pharmaceutical coating technique but is now increasingly being applied in the food industry. Dewettinck and Huyghebaert (1999) wrote comprehensive review of fluidized-bed coating technique and its application in food industry. The fluidization occurs when a flow of fluid upwards through a bed of particles reaches sufficient velocity to support the particles without carrying them away in the fluid stream. The bed of particles then assumes the characteristics of a boiling liquid, and hence, the term *fluidization*. Fluidized-bed coating may be applied to enhance or tune the effect of functional ingredients and additives such as processing aids (leavening agents and enzymes), preservatives (acids and salts), fortifiers (vitamins and minerals), flavors, spices to increase shelf life and mask taste, and other additives for ease of handling, controlled release, improved aesthetics, taste, and color (Dewettinck and Huyghebaert, 1999). Bakery products are commonly coated using the fluidized-bed technique including a variety of leavening system ingredients, vitamin C, acetic acid, lactic acid, potassium sorbate, sorbic acid, calcium propionate, and salt (Dezarn, 1995; De Pauw et al., 1996). In the meat industry, several food acids have been fluid-bed encapsulated to develop color and flavor systems. They are also used to achieve a reproducible pH in cured meat products and to shorten their processing time. Fluid-bed encapsulated salt is used in meats to prevent development of rancidity, as well as premature set due to myofibrillar binding (Dezarn, 1995). The downside of this coating application technique is its relatively high cost (Dewettinck and Huyghebaert, 1999).

10.3.5 Panning

Panning is usually employed for coating candies, nuts, and some processed fruits that are characterized by a smooth, regular surface obtained by the polishing action in the pan. The technology involves a stainless steel pan that is enclosed and perforated along the side panels. The coating is delivered by a pump to spray guns mounted in various parts of the pan. The coating is atomized by the spray guns (Grant and Burns, 1994). Panning is a slow process, in which the pan speeds vary based on the size of the center. For examples, large-size nuts require speeds of 15 rpm and sugar grains (hundreds and thousands) require speeds of 30 to 35 rpm. Air at 35 to 36°C is blown into the pan to give rapid drying of sugar layers, and to remove dust and frictional heat. Fellows (2000) describes this in greater detail, including the three main types of pan-coated products: hard, soft, and chocolate-coated products. Nieto (2009) further illustrated their specific applications for coating confectioneries using this technique.

10.3.6 Spraying

When a thin and uniform coating is required for certain surfaces, spraying is useful (Cutter and Sumner, 2002). In fact, early coating procedures involved sprays, with further distribution over food surfaces via rollers or brushes, followed by tumbling to evenly spread the coating (Grant and Burns, 1994; Cutter and Sumner, 2002). This is the most popular method for coating whole fruits and vegetables, especially with the development of high-pressure spray applicators and air-atomizing systems. Spray applications are also suitable when applying films to a particular side or when a dual application must be used for cross-linking, as is practiced with alginate coatings (Siragusa and Dickson, 1993; Donhowe and Fennema, 1994). Just as with foams, heated air can be applied after spraying to speed up the drying process or improve uniform distribution on the surfaces. Grant and Burns (1994) gave a good review on the history and application of this coating technique on fruits and vegetables.

In addition, spray coating may be used in combination with pan, fluidized-bed, and other coating techniques to deposit either thin or thick layers of aqueous solution or suspensions and molten lipids or chocolate. The spray nozzle plays a critical role in the coating process. The pressure, fluid viscosity, temperature, and surface tension of coating liquid, and nozzle shape directly determine the spraying efficiency (Fellows, 2000; Debeaufort and Voilley, 2009).

10.3.7 Electrostatic coating

Electrostatic coating is a process that employs charged particles to more efficiently coat a surface. Powdered particles or atomized liquid is initially projected toward a conductive surface using normal spraying methods, and then accelerated toward the surface by a powerful electrostatic charge. This coating procedure is historically used for painting but has lately been adapted for food application with the considerations of improving coating efficiency and surface adhesion, reducing waste, increasing productivity, and achieving more consistent coverage on the surface (Amefia et al., 2006; Barringer, 2006).

Both powders and liquids may be emitted to coat the food surfaces. The powder coatings include coating of seasonings, anticaking agents, and antimolding agents on cheese (Amefia et al., 2006; Setyo and Barringer, 2007). However, the use of liquid electrostatic coating in food is very limited, partially due to the complexity of the food system. Many food systems contain natural buffering components such as proteins that tend to reduce the charged nature of a solution (Abu-Ali, 2004). In addition, conductivity, viscosity, and surface tension of food vary significantly in comparison with nonfood systems. Hence, the exact relationship between these factors and the performance of liquid electrostatic coating system for food applications is not well known. In spite of these difficulties, liquid electrostatic coating has shown great promise in some applications, including the impregnation of bread with edible vegetable oil and coating of confectionary and chocolate products (Anonymous, 1978).

10.4 Keys to successful coating application

The success of an edible coating for meeting the specific needs of food strongly depends on its barrier property to gases, especially oxygen and water vapor, its adhesion to the surface, uniformity of coverage of coating, and also sensory quality of the coated food product. These properties, in turn, depend on the coating composition, especially the polymers it contains, the surface properties of coating and the food product, the suitability of the method of application (already discussed), and the storage conditions. What follows is a brief discussion of these critical properties.

10.4.1 Moisture and gas barrier properties

Permeation, absorption, and diffusion of water vapor, oxygen, and carbon dioxide are among the most important functional properties of any edible coating applied to food. Lin and Zhao (2007) summarized these permeability values of some edible films. Most polysaccharide-based and protein-based materials are hydrophilic, and coatings based on these polymers are therefore not good barriers to water vapor unless hydrophobic compounds are included in the coating formulation. Efforts are required to develop new coating materials and coating formulations that are good moisture barriers and have acceptable surface adhesion, and to understand the functionality and interactions among different components in the edible coating formula. Studies to improve the functionality of existing coating materials are also important. The incorporation of hydrophobic ingredients, such as lipids and fatty acids, for improving moisture barrier while maintaining desirable resistance to vapor, gas, or solute and also have sensory properties are critical in future research. Hydrophobic coating materials, such as sucrose polyesters, provide relatively better moisture barriers than hydrophilic materials (Lin and Zhao, 2007). In order to maximize the benefit of edible coatings, coating materials and formulations must be carefully selected and designed.

Barrier properties to oxygen, carbon dioxide, ethylene, and volatiles determine how a coating influences respiration and ripening of fresh produce and also the exchange of food aroma and flavor compounds with the environment. All coatings modify the atmosphere inside fruits and vegetables, depending on how application of the coating influences skin permeance. Too much restriction of gas exchange can result in ethanol and alcoholic flavors as a result of anaerobic fermentation associated with too high carbon dioxide or too low oxygen concentrations (Ben-Yehoshua, 1969). Therefore, control of skin permeance is critical in modifying the internal environment of fresh produce for the purpose of preserving food. In addition, oxygen permeability of coatings is critical for lipid oxygen of high-fat products and, thus, quality and shelf life of coated products. Antioxidant compounds, such as vitamins C and E, as well as essential oils, are sometimes incorporated into the coating materials for controlling the oxidation of coated products.

Environmental temperature and relatively humidity also influence the moisture and gas barrier properties of edible coatings, especially for processed foods that may be subject to different temperature and humidity conditions throughout processing, storage, and retail display. Duan and Zhao (2010) reviewed the impact of environment temperature and relative humidity on the permeability of some coating materials when applied on frozen foods. Retaining consistent and stable temperature and humidity is essential to best benefit from the functions of coatings.

10.4.2 Surface characteristics of coatings

For truly taking advantage of edible coatings, the coating must adhere to the food surface during processing, storage, and transportation. Adhesions of most hydrophilic edible coatings on the hydrophobic whole fruit surface are inherently poor due to the different chemical nature of the two surfaces. For improving surface adhesion of hydrophilic coatings, surfactants are typically added into coating formulations to improve wettability and adhesion of the coatings (Choi et al., 2002; Lin and Krochta, 2005). When applying a coating onto the wet surfaces, such as fresh-cut fruit or vegetable and fresh seafood and meat cuts, it presents an even more considerable challenge, because coatings may be dissolved and absorbed by the wet surfaces instead of drying to form a smooth and unique layer.

Surface wettability is essential for good coating adhesion. A liquid would perfectly wet the solid when the surface tension of a solid is greater than or equal to that of the liquid, which can be achieved by the addition of appropriate surfactant in the coating solutions. Addition of Tween 80 as surfactant into chitosan coating solutions reduced the surface tension of the coating solution and enhanced its wettability on 'Fuji' apple skin, resulting from surfactant-driven autophilicity for improving the adhesive force between the coating solution and associative apple skin (Choi et al., 2002). Adhesion of the coatings on the food surfaces with different characteristics still needs to be studied and improved. This is typically important for some fresh and fresh-cut fruits and vegetables where a natural hydrophobic waxy layer or a high-moisture and wet surface is already present, respectively.

The methods of coating application directly affect adhesion of coatings. For example, the dipping method, when using water containing detergent, may wash out the natural waxy layer on the surface of some fruits and vegetables, thus degrading the functionality of the coatings. Moreover, it may dilute the coating solution and results in significant residual (waste) of the coating materials. Hence, other coating application techniques, such as spraying or dripping, are necessary to increase coating efficiency and durability.

10.4.3 Sensory attributes

Sensory qualities of foods are strongly associated with the applications of edible films and coatings. Meanwhile, many active compounds used in the manufacture of edible films and coatings, including edible polymers, plasticizers, and other active agents, may impact the sensory attributes of wrapped or coated products because most active agents have their own characteristic flavor and color, and the interactions among the compounds may generate unique flavors. Because functional edible films and coatings are edible portions of the packaged or coated foods, it is expected that no components of the edible films and coatings should interfere with the organoleptic characteristics of the food product. Generally, tasteless edible films and coatings are desirable to minimize taste interference. Fortunately, the concentration of most active compounds used in edible films and coatings is usually very low; hence, their taste effects may be negligible. When high concentrations of natural active agents are added to edible films and coatings, the film and coating layers may possess the strong flavor of the incorporated active agents. This phenomenon becomes more significant when plant and herb essential oils/extracts or phenolic flavors are added to edible film layers. Unfortunately, few studies have examined sensory qualities of edible films and coating layers, as well as wrapped or coated food products.

One potential adverse effect of edible coating usage is the development of undesirable sensory properties on the coated products. Nonuniform or sticky surfaces may result, making the product unattractive to consumers (Zhao and McDaniel, 2005). Some of the important sensory attributes that should be considered when developing an edible coating for fresh and minimally processed produce include the following (Zhao and McDaniel, 2005; Lin and Zhao, 2007):

- Appearance as a result of surface dehydration, whitening, waxiness, and discoloration (as due to enzymatic browning). Selective coating materials can reduce moisture loss, control surface dehydration and discoloration, delay the surface whitening, and enhance the glossiness of food surfaces.
- Texture as mostly represented by firmness and crispness. Edible coatings can improve texture by reducing water loss and preventing dehydration. In addition, edible coatings may improve the mechanical integrity or handling characteristics of food products.
- Flavor and other sensory attributes. As already discussed, edible coatings can retard ethylene production and delay the ripening process, thus preventing the development of off-flavor and off-taste during postharvest storage of the produce.

10.4.4 Incorporation of functional substances to enhance functionality

One of the unique functions of edible coatings is the capability to incorporate functional ingredients into the matrix to enhance its functionality. This may include the following (Lin and Zhao, 2007):

- Improving basic coating functionality, such as plasticizers for improving mechanical properties and emulsifiers for stabilizing composite coatings and improving coating adhesion
- Improving quality, stability, and safety of coated foods by incorporating antioxidants, antimicrobial agents, nutraceuticals, flavors, and color agents

A plasticizer, in most cases, is required for making edible coatings, especially for polysaccharide- and protein-based coatings because the structure of such coatings is often brittle and stiff due to extensive interactions between polymer molecules (Krochta, 2002). Glycerol, acetylated monoglyceride, polyethylene glycol, and sucrose are common plasticizers incorporated into the polymeric coating matrix for decreasing glass transition temperature of the polymers and increasing coating flexibility (Guilbert and Gontard, 1995). Plasticizers are usually hygroscopic and attract water molecules. Water can also function as plasticizer, but it is easily lost due to dehydration in an environment with low relative humidity (Guilbert and Gontard, 1995). In addition to improving mechanical properties, the plasticizer affects the resistance of coatings to the permeation of vapors and gases (Sothornvit and Krochta, 2000, 2001), where the hydrophilic plasticizers usually increase the water vapor permeability of the coatings.

Emulsifiers are surface-active agents of an amphiphilic nature and are able to reduce the surface tension of water–lipid or water–air interfaces. Emulsifiers are essential for the formation of protein or polysaccharide coatings containing lipid emulsion particles. Emulsifiers also modify surface energy to control adhesion and wettability of the coating surfaces (Krochta, 2002). Addition of an emulsifier into whey protein coatings increases the hydrophilicity and coatability of peanut surfaces, thus improving oxygen barrier of the coatings (Lin and Krochta, 2005).

Other functional ingredients, such as antioxidants, antimicrobials, nutraceuticals, flavor, and color agents can be carried by edible coatings and retained on the food surface for enhancing food quality, stability, and safety. Common antimicrobial agents used in food systems, such as benzoic acid, sodium benzoate, sorbic acid, potassium sorbate, and propionic acid, may be incorporated into coatings. Lin and Zhao (2007) and Duan and Zhao (2010) provided comprehensive review of these applications in fresh, minimally processed fruits and vegetables, as well as in frozen foods.

10.5 Summary

Edible coatings can play a significant role in the preservation of foods by reducing moisture and gas transfer, minimizing lipid oxidation, and maintaining structural integrity. They also have the capability of enhancing quality and functionalities of foods by incorporating active substances into the coatings. For the successful application of coatings, coating material and formulation with desired barrier and sensory properties, the technique of application, and the surface characteristics of the food are the keys. Continuous efforts in all of the above areas are necessary.

References

Abu-Ali, JM. 2004. Electrostatic atomization, nonelectrostatic coating and electrostatic powder coating. M.S. Thesis. The Ohio State University.
Amefia, A, Abu-Ali, JM, and Barringer, SA. 2006. Improved functionality of food additives with electrostatic coating. *Innov Food Sci Emerg Technol* 7: 176–181.
Anonymous. 1978. Spray smoking of bacon and poultry. *Int Flavor Food Additiv* 9:262–266.

Baker, RA, Baldwin, EA, and Nisperos-Carriedo, MO. 1994. Edible coatings and films for processed foods. In: Krochta, JM, Baldwin, EA, and Nisperos-Carriedo, M (Eds.). *Edible Coatings and Films to Improve Food Quality*. Technomic, Lancaster, PA. pp. 89–101.

Baldwin, EA. 2007. Surface treatments and edible coatings. In: Rahman, MS (Ed.). *Handbook of Food Preservation*, 2nd ed, CRC Press/Taylor & Francis, Boca Raton, FL. pp. 477–508.

Baldwin, EA, and Wood, BW. 2006. Use of edible coating to preserve pecans at room temperature. *HortScience*. 41:188–192.

Barringer, SA. 2006. Coating snack foods. In: Hui YH (Ed.). *Handbook of Food Technology and Food Engineering*. Marcel Dekker, New York. pp. 169-1–169-9.

Ben-Yehoshua, S. 1969. Gas exchange, transportation, and the commercial deterioration in storage of orange fruit. *J Am Soc Hort Sci* 94:524–528.

Choi, WY, Park, HJ, Ahn, DJ, Lee, J, and Lee, CY. 2002. Wettability of chitosan coating solution on 'Fuji' apple skin. *J Food Sci* 67:2668–2672.

Cutter, CN, and Sumner, SS. 2002. Application of edible coatings on muscle foods. In: Gennadios, A (Ed.). *Protein-Based Films and Coatings*. CRC Press, Boca Raton, FL. pp. 467–484.

Debeaufort, F, and Voilley, A. 2009. Lipid-based edible films and coatings. In: Embuscado, M, and Huber, KC (Eds.). *Edible Films and Coatings for Food Applications*. Springer, New York. pp. 135–168.

De Pauw, P, Dewettinck, K, Arnaut, F, and Huyghebaert, A. 1996. Microencapsulation improves the action of bakery ingredients. *Voedingsmiddelentechnologie* 29:38–40.

Dewettinck, K, and Huyghebaert, A. 1999. Fluidized bed coating in food technology. *Trend Food Sci Technol* 10:163–168.

Dezarn, TJ. 1995. Food ingredient encapsulation. In: Risch, SJ, Reineccius, GA (Eds.). *Encapsulation and Controlled Release of Food Ingredients*. American Chemical Society, Washington, DC. pp. 74–86.

Donhowe, IG, and Fennema, O. 1994. Application of coatings. In: Krochta, JM, Baldwin, EA, and Nisperos-Carriedo, M (Eds.). *Edible Coatings and Films to Improve Food Quality*. Technomic, Lancaster, PA. pp. 1–24.

Duan, J, Chrieton, G, and Zhao, Y. 2010. Effect of combined chitosan-krill oil coating and modified atmosphere packaging on the storability of cold-stored lingcod (*Ophiodon elongates*) fillets. *Food Chem*. 122:1035–1042.

Duan, J, Cherian, G, and Zhao, Y. 2009. Fish oil incorporated chitosan coatings for enhancing quality and nutraceutical benefit of ling cod fish. *Food Chem*. 119:524–532.

Duan, J, and Zhao, Y. 2011. Edible coatings and films and their applications on frozen foods. In: Sun, D (Ed.). *Handbook of Frozen Food Processing and Packaging*, 2nd Edition, CRC Press/Taylor & Francis. Boca Raton, FL. In press.

Fellows, P. 2000. Coating or enrobing. In Fellows, P (Ed.). *Food Processing Technology: Principles and Practice*. 3rd ed. Woodhead, Cambridge, UK. pp. 455–461.

Grant, L, and Bums, JK. 1994. Ch. 8. Application of coatings. In: Krochta, JM, Baldwin, EA, and Nisperos-Carriedo, M (Eds.). *Edible Coatings and Films to Improve Food Quality*. Technomic, Lancaster, PA. pp. 189–200.

Guilbert, S, and Gontard, N. 1995. Edible and biodegradable food packaging. In: Ackermann, P, Jägerstad, M, Ohlsson, T (Eds.). *Foods and Packaging Materials: Chemical Interactions*. The Royal Society of Chemistry, Cambridge, UK. pp. 159–168.

Kaplan, HJ. 1986. Washing, waxing, and color adding. In: Wardowdki, WF, Nagy, S, Grierson, W (Eds.). *Fresh Citrus Fruit*. AVI, Westport, CT. pp. 379–395.

Krochta, JM. 2002. Proteins as raw materials for films and coatings: definitions, current status, and opportunities. In: Gennadios, A (Ed.). *Protein-Based Films and Coatings*, CRC Press, Boca Raton, FL. pp. 1–41.

Lin, SYD, and Krochta, JM. 2005. Whey protein coating efficiency on surfactant-modified hydrophobic surfaces. *J Agric Food Chem* 53:5018–5023.

Lin, D, and Zhao, Y. 2007. Innovations in the development and application of edible coatings for fresh and minimally processed fruits and vegetables. *Crit Rev Food Sci Food Safety* 6:60–75.

Nieto, MB. 2009. Structure and function of polysaccharide gum-based edible films and coatings. In: Embuscado, ME, and Huber, KC (Eds.). *Edible Films and Coatings for Food Application*. Springer, New York. pp. 57–112.

Olivas, GI, and Barbosa-Canovas, G. 2009. Edible films and coatings for fruits and vegetables. In: Embuscado, ME, and Huber, KC (Eds.). *Edible Films and Coatings for Food Application*. Springer, New York. pp. 211–244.

Platenius, H. 1939. Wax emulsions for vegetables. New York Agriculture Experiment Station Bulletin No. 723.

Rooney, ML. 2005. Introduction to active food packaging technologies. In: Han, JH (Ed.). *Innovations in Food Packaging*. Elsevier Academic Press, San Diego, CA. pp. 63–79.

Setyo, D, and Barringer, SA. 2007. Effect of hydrogen ion concentration and electrostatic polarity on food powder coating transfer efficiency and adhesion. *J Food Sci* 72:E356–E361.

Siragusa, GR, and Dickson, JS. 1993. Inhibition of *Listeria monocytogenes*, *Salmonella Typhimurium* and *Escherichia coli* O157:H7 on beef muscle tissue by lactic or acetic acid contained in calcium alginate gels. *J Food Safety* 13:147–158.

Sothornvit, R, and Krochta, JM. 2000. Plasticizer effect on oxygen permeability of β-lactoglobulin films. *J Agric Food Chem* 48:6298–6302.

Sothornvit, R, and Krochta, JM. 2001. Plasticizer effect on mechanical properties of β-lactoglobulin films. *J Food Eng* 50:149–155.

Tharanathan, RN. 2003. Biodegradable films and composite coatings: past, present and future. *Trend Food Sci Technol* 14:71–78.

Trezza, TA, and Krochta, JM. 2002. Application of edible protein-based coatings to nuts and nut-containing foods. In: Gennadios, A (Ed.). *Protein-Based Films and Coatings*. CRC Press, Boca Raton, FL. pp. 527–551.

Ustunol, Z. 2009. Edible films and coatings for meat and poultry. In: Embuscado, ME, and Huber, KC (Eds.). *Edible Films and Coatings for Food Application*. Springer, New York. pp. 244–268.

Wong, DWS, Camirand, WM, and Pavlath, AE. 1994. Ch. 3. Development of edible coatings for minimally processed fruits and vegetables. In: Krochta, JM, Baldwin, EA, and Nisperos-Carriedo, M (Eds.). *Edible Coatings and Films to Improve Food Quality*. Technomic, Lancaster, PA. pp. 65–82.

Yehoshua, SB. 1987. Transpiration. In: Weichmann, J (Ed.). *Postharvest Physiology of Vegetables*. Marcel Dekker, New York. pp. 113–170.

Zhao, Y, and McDaniel, M. 2005. Sensory quality of foods associated with edible film and coating systems and shelf-life extension. In: Han, JH (Ed.). *Innovations in Food Packaging*. Elsevier Academic Press, San Diego, CA. pp. 434–453.

11 Encapsulation of flavors, nutraceuticals, and antibacterials

Stéphane Desobry and Frédéric Debeaufort

Contents

11.1	Introduction	334
11.2	Capsules matrices	335
	11.2.1 Carbohydrates	336
	11.2.2 Proteins	337
11.3	Encapsulation methods	337
	11.3.1 Spray-drying	337
	11.3.2 Freeze-drying	337
	11.3.3 Spray-cooling and spray-chilling	338
	11.3.4 Extrusion	338
	11.3.5 Coacervation	338
11.4	Controlled release	339
	11.4.1 Release controlled by diffusion	339
	11.4.2 Release controlled by matrix degradation	339
	11.4.3 Release controlled by swelling	339
	11.4.4 Release controlled by melting	339
11.5	Edible films for volatile molecules (flavors, essential oils)	340
	11.5.1 Flavor compounds and matrices involved in edible films and coatings	340
	11.5.2 Retention and release of volatile compounds in edible films and coatings	343
	11.5.3 Barrier performances of edible films and coatings to aroma compounds	348
	11.5.4 Protection of flavor compounds against chemical degradation by edible films and coatings	350
	11.5.5 Volatile compound effect on structural and physical-chemical properties of edible films and coatings	354
	11.5.5.1 Impact on film appearance and transparency	356
	11.5.5.2 Changes in film microstructure induced by volatile compound encapsulation	356
	11.5.5.3 Influence of volatile compound encapsulation on film mechanical properties	356
	11.5.5.4 Impact of flavor compound encapsulation on the barrier properties of film	358
11.6	Edible films and coatings for nonvolatile molecule encapsulation (peptides, polyunsaturated fatty acids, antioxidants, and antibacterials)	361

		11.6.1	Nutraceuticals	361

Contents (restructured):

    11.6.1 Nutraceuticals 361
    11.6.2 Antibacterials 362
    11.6.3 Edible materials for nutraceuticals or antibacterial molecule encapsulation 363
        11.6.3.1 HPMC 363
        11.6.3.2 PLA 364
    11.6.4 Plasticizers 364
    11.6.5 Molecular diffusion from film and coating to food surface 364
    11.6.6 Chemical structure of migrant and matrix state 365
11.7 Conclusion 366
References 366

11.1 Introduction

Edible coatings provide a physical barrier against mass transport from the environment to food and from food to the environment, as shown in the previous chapters, and if these barrier properties are important for food passive protection. Consumers ask nowadays for better food safety and for higher nutritional and flavor properties. In recent years, active packaging has been developed to extend food shelf life by increasing coatings' positive effect. For example, more activity can be provided to edible coatings by adding active compounds, such as flavors, antibacterials, or nutraceuticals. Active compounds can be incorporated directly in the edible polymer matrix or can be encapsulated to better protect their activity and properties. The Eastern countries are the most advanced in this particular domain, North America follows, and Europe is now developing more and more active material solutions.

Active compound stability in different foods has been of increasing interest due to its relationship with food quality and acceptability. Manufacturing and storage processes, packaging materials, and ingredients in foods often cause modifications in overall flavor and nutritional value by reducing an active compound's activity or intensity or by producing newly formed components. Many factors such as physicochemical properties, concentration, and interaction of active molecules with food components affect the resulting quality. As a result, it is beneficial to encapsulate active ingredients prior to use.

Encapsulation is the technique by which one material or mixture of materials is coated with or entrapped within another material or system. The coated material is called the active or core material, and the coating material is called the shell, wall material, carrier, matrix, or encapsulant. The development of microencapsulation products started in the 1950s in the research of pressure-sensitive coatings for the manufacture of carbonless copying paper (Green and Scheicher, 1955). Encapsulation technology is now well developed and accepted within the pharmaceutical, chemical, cosmetic, foods, and printing industries. In food products, fats and oils, aroma compounds and oleoresins, vitamins, minerals, colorants, and enzymes have been encapsulated, while in films of coatings oils, aroma compounds, antimicrobials, and enzymes have been encapsulated.

Encapsulation processes are receiving more and more interest in food application, and recently, evolution from micro- to nanoencapsulation has been observed

(Shimoni, 2009). The main interest in reducing the size of the capsules was to limit the impact of capsule addition on food physical properties. Laboratories and industrial development teams have shown the high efficiency of this size reduction (Imran et al., 2010a, 2010b). This evolution is nevertheless limited by suspicion of possible toxicity of some nanoparticles for the human body after inhalation, skin permeation, and ingestion (Bouwmeester et al., 2009). Nanoparticle absorption through cell membranes may oxidize cellular compounds or develop inflammation. This risk exists in both cases of free nanocapsules or nanocapsules incorporated in edible films or coatings. A lot of studies are required now to better understand the real effect of nanocapsules incorporated in foods. In particular, the effect of solid or "soft" nanoparticles should be compared more precisely, because some studies have reported a negligible effect of soft and organic compounds easily degradable by the cell compared to solid mineral particles.

These studies will make possible the development of clear legislation in the countries that use nanoparticles. Waiting for that, the industry is not very confident in this particular technology, and only a few applications of nanoparticle incorporation in an edible matrix exist.

In this chapter, after describing the encapsulation matrix and techniques, two parts will be presented to cover the main aspects of encapsulation in edible films and coating. The first concerns volatile compound encapsulation, and the second concerns nonvolatile molecules such as nutraceuticals and antimicrobials.

11.2 Capsules matrices

Encapsulation of biomolecules can be achieved using two main methods. The first consists in making capsules in which the compound is included as a core or entrapped in a polymeric matrix (Figure 11.1a, b).

These encapsulation methods are used for large spectra applications. The protective efficiency of the two systems is highly different, with much more efficient barrier action in the core system, compared to easier release, and lower cost in the matrix system.

The second method consists of developing films or coatings in which the biomolecules are directly included and entrapped just as a matrix but on a larger scale (Figure 11.1c). In this matrix film system, the film or coating contains aroma, nutraceuticals, and active compounds. The release is completely controlled by the molecular diffusivity.

To increase the potentiality of the films, more and more applications combine the two previous methods (Figure 11.1a, b, c) to obtain film/capsules systems as presented in Figure 11.1d. In this structured film or coating, the amount of volatile compound can be reduced thanks to the optimized preservation, low release from the capsules, and low transfer through the film. These systems allow negligible oxidation of the active compounds due to reduced oxygen diffusivity.

Matrix material has to be chosen to face the high number of performance requirements placed on microcapsules when a limited number of encapsulating materials and methods exist. Since the late 1980s, they have been prepared from a large range of materials including proteins, carbohydrates, lipids, or gums (Brazel, 1999). Each group of materials has certain advantages and disadvantages, and these materials are combined

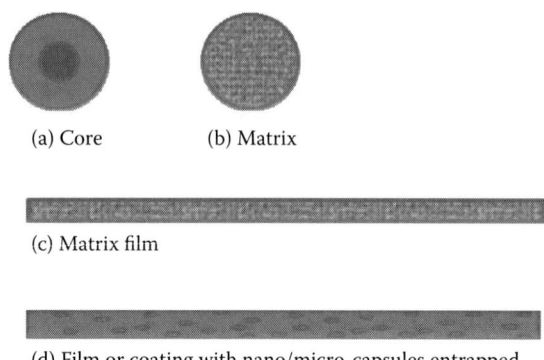

Figure 11.1 Core (a), matrix (b) in nano-/microcapsules systems, matrix film (c), and film or coating with nano-/microcapsules entrapped (d).

to produce high-quality systems. Functional characteristics of an effective wall material used to encapsulate active components must satisfy the following conditions:

- Be inert regarding the encapsulated molecules
- Allow a complete elimination of solvent used for matrix or capsule formation
- Bring maximum protection of the active ingredient against external factors
- Produce a stable system before solidification of capsules within a film
- Release the active compounds at the time and the place desired

Characteristics of major wall materials used for encapsulation are reported below.

11.2.1 Carbohydrates

Carbohydrates are used extensively in spray-dried encapsulations of food ingredients as the encapsulating support—that is, the wall material or carrier. The ability of carbohydrates to form a protective matrix, complemented by their diversity and low cost, make them the first choice for encapsulation. The main limit for their application is their high water sensitivity and fast hydration when in contact with water.

The preferred carbohydrate matrices are starch and starch-based ingredients (modified starches, maltodextrins, β-cyclodextrins). Maltodextrins possess matrix-forming properties important in a wall system. In selecting the wall materials for encapsulation, maltodextrin is a good compromise between cost and effectiveness, as it is bland in flavor, has low viscosity at high solid ratios, and is available in different average molecular weights. Gums and thickeners are generally bland or tasteless, but they can have a pronounced effect on the taste and flavor of foods. In general, hydrocolloids decrease sweetness, with much of the effect being attributed to viscosity and hindered diffusion. Gum arabic is also an excellent encapsulating material. Its solubility, low viscosity, emulsification characteristics, and good retention of active compounds make it very versatile for most encapsulation methods such as spray-drying. Its application is nevertheless limited due to its high cost compared to other carbohydrates, and its availability and cost are subject to high variation.

11.2.2 Proteins

Although food hydrocolloids are widely used as microencapsulants, food proteins (i.e., sodium caseinate, whey protein isolates, and soy protein isolates) have good ability to produce efficient encapsulation matrices. Whey protein isolates provide a good barrier against oxidation and an effective basis for microencapsulation by spray-drying. In combination with maltodextrins and corn syrup solids, whey proteins have been reported to be one of the most effective encapsulation materials during spray-drying. In such a system, whey proteins served as emulsifying and film-forming agents, while the carbohydrates acted as a matrix-forming material.

Protein-based materials such as polypeptone, soy protein, milk proteins, and gelatin derivatives are able to form stable emulsions with hydrophobic compounds. However, their solubilities in cold water, the potential to react with carbonyls, and their high cost limit potential applications.

11.3 Encapsulation methods

Encapsulation is accomplished by a large variety of methods (Madene et al., 2006). The two major industrial processes are spray-drying and extrusion, freeze-drying, however, coacervation, and adsorption techniques are also widely used in the case of heat-sensitive compounds.

11.3.1 Spray-drying

Spray-drying is the commercial process most widely used in large-scale production of encapsulated molecules. However, the merits of the process have ensured its dominance, including availability of equipment, low process cost, wide choice of carrier solids, good retention of volatiles, good stability of the finished product, and large-scale production in continuous mode. Production of encapsulated powders by spray-drying involves the formation of a stable emulsion in which the wall material acts as a stabilizer. When core materials of limited water solubility are encapsulated by spray-drying, the resulting capsules are of matrix-type structure. As such, the core has been shown to be organized in small wall-material-coated droplets embedded in the wall matrix.

The main disadvantage of spray-drying is that some low-boiling point aromatics or heat-sensible molecules can be lost, and some core material may also be on the surface of the capsule, where it is subject to oxidation. Another problem is that the product is a very fine powder, typically in the range of 10 to 100 µm in diameter, which needs further processing, to make agglomeras that are more readily soluble for liquid consumption.

11.3.2 Freeze-drying

The freeze-drying technique is one of the most useful processes for drying thermosensitive substances. This technique is based on low temperature dehydration under vacuum, avoiding water phase transition and oxidation. The dried mixture has a porous structure and must be ground, resulting in heterogeneous particles. This drying technique is less attractive than others, because its cost is up to 50 times higher than spray-drying. Freeze-drying, nevertheless, gives excellent preservation results and is adapted to high-value encapsulated molecules.

11.3.3 Spray-cooling and spray-chilling

The loss of heat-sensitive material during the spray-drying process has led to a number of alternative methods for dehydration of sprayed microcapsules. Spray-cooling and spray-chilling are similar to spray-drying, where the core material is dispersed in a liquefied coating or wall material and atomized at low temperature.

11.3.4 Extrusion

Encapsulation via extrusion is slightly more expensive compared to spray-drying. The principal advantage of the extrusion method is the high density of the produced matrix allowing a great stability of encapsulated materials against oxidation. In extrusion, there are two steps. First, the feed is extruded without cooking. Second, the raw ingredients are cooked by the combined action of heat, mechanical shearing, and pressure.

11.3.5 Coacervation

Coacervation consists in separating from the solution the colloidal particles that agglomerate into a separate liquid phase called coacervate. Coacervation can be simple or complex. Simple coacervation involves only one type of polymer with an addition of strongly hydrophilic agents to the colloidal solution. For complex coacervation, two or more types of polymers are used. Active molecules are entrapped in the matrix during coacervate formation by adjusting precisely the ratio between the matrix polymer and the entrapped molecule (Figure 11.2).

Figure 11.2 Gum arabic/ß-lactoglobulin coacervates (molecular model on the left; SEM (scanning electron microscopy) photograph on the right).

11.4 Controlled release

Controlled release is a method by which one or more active agents or ingredients are made available at a desired site and time and at a specific rate. For matrix systems encapsulating an active compound, release depends on diffusion process, particle type and geometry, and controlled degradation or dissolution of matrix material.

Compared to molecules added in food, advantages of controlled release are that the active ingredients are released at controlled rates over a prolonged period of time; loss of ingredients during processing, cooking, and even digestive molecular destruction can be strongly reduced; molecule bioavailability can be increased for nanocapsules.

11.4.1 Release controlled by diffusion

Diffusion is the main regulation phenomena for controlled release in stable matrix. The diffusion process is well known, thanks to thousands of experimental studies related to mass transfer through films, coatings, and capsules. The macromolecular network density of the matrix and the molecular weight of the encapsulated compounds are the key factors for release controlled by diffusion. The chemical potential difference at each side of the matrix is the major driving force influencing diffusion. The principal steps in the release of a compound from a matrix system are diffusion of the active agent to matrix surface; component partition between matrix and food; and finally, transport away from the matrix surface.

11.4.2 Release controlled by matrix degradation

The release of an active compound from a matrix-type delivery system may be controlled by a combination of diffusion and erosion. Heterogeneous erosion occurs when degradation is confined to a thin layer at the delivery system surface, whereas homogenous erosion is a result of degradation occurring at a uniform rate throughout the polymer matrix.

11.4.3 Release controlled by swelling

In swelling-controlled systems, the molecule dispersed in a polymeric matrix is unable to diffuse to any significant extent within the matrix. When the matrix polymer is placed in a thermodynamically compatible medium, the polymer swells owing to absorption of fluid from the medium. In the swollen matrix, the encapsulated molecule is able to diffuse due to larger intermolecular spaces.

11.4.4 Release controlled by melting

This mechanism of release involves the melting of the capsule wall to release the active material. It is readily accomplished in the food industry because there are numerous materials of low melting point which are approved for food use (lipids, modified lipids, or waxes). In such applications, coated particles are stored at temperatures well below coating melting point, and then heated above this temperature during preparation, cooking, and consumption.

11.5 Edible films for volatile molecules (flavors, essential oils)

Even if most research deals with interactions between packaging and volatile compounds in the case of beverages, similar problems occur for viscous liquids, gels, and solid foods. One possible way suggested to lower interactions between aroma compounds and plastics is to retain flavor molecules inside the food product by using an additional barrier. This could be an edible film or coating or a thin layer having a high selectivity against aroma transfer, and that can be eaten along with the protected food (Miller and Krochta, 1997; Debeaufort et al., 2002). This entails some kind of macroencapsulation of the product. The main application is using only barrier properties of the coating to retain aroma within the food; however, the film or coating could be used as a carrier or support for flavors at the surface of the product. So, edible films serve many purposes, including permitting the production of a dry free-flowing flavor (most flavors are liquids), protection of the flavoring from interaction with the food, or deleterious reactions such as oxidation, confinement during storage, and, finally, controlled release (Reineccius, 2009). The degree to which the edible film meets these requirements depends upon the process used to form the film around the flavoring and the film composition.

These volatile compounds will be released rapidly when the consumer tastes the product. Several products are already on the market using this technology for flavoring. For instance, there is a roasted peanut with a curry-flavored coating that is instantaneously dissolved in the mouth and gives immediately the perception of the Indian spice. Another example designed for children, is a multi-sugar-coated sweet in which each layer of the coating contains different tastes and flavors separated by arabic gum or hydrocolloid layers to prevent migrations of aroma compounds from one layer to another. For this application, diffusivity of volatile compounds should be very low and with a high affinity for the coating, which should be highly soluble in the mouth. As previously outlined, edible films and coatings can deliver and maintain desirable concentrations of color, flavor, spiciness, sweetness, saltiness, and so forth. Several commercial films, especially Japanese pullulan-based films, are available in a variety of colors, with spices and seasonings included (Guilbert and Gontard, 2005). For instance, Laohakunjit and Kerdchoechuen (2007) coated milled rice with sorbitol–rice starch coatings, containing 25% natural pandan leaf extract (*Pandanus amaryllifolius* Roxb.). The rice starch coating containing pandan extract allowed production of jasmine-flavored rice after cooking. Recently, Origami Foods (Pleasantown, California) commercialized vegetable and fruit edible films as alternatives to the seaweed sheets (nori) traditionally used for sushi and other Asian cuisine. They are made from broccoli, tomato, carrot, mango, apple, peach, pear, as well as a variety of other fruit and vegetable products, and they can contain spices, seasonings, colorants, flavors, vitamins, and other beneficial plant-derived compounds (Martin-Belloso et al, 2009).

11.5.1 Flavor compounds and matrices involved in edible films and coatings

The edible packaging materials are mainly composed of a film-forming substance that provides cohesiveness to the matrix (continuous network) or a barrier substance that lowers impermeability. These are usually polysaccharides, proteins, or lipids used alone or as mixtures. Only few substances have simultaneously good film-forming and barrier properties as, for example, wheat gluten–based films, which have satisfactory mechanical resistance and very low oxygen permeability. Moreover, the permeability of D-limonene (one of the main compounds of citrus flavor) in whey protein–based films is lower than in ethylene vinyl alcohol (EVOH) or poly-vinylidene chloride (PVDC) films (Fayoux et al., 1997a, 1997b; Miller et al., 1998). Because edible packaging containing volatile compounds is consumed along with the foods, their edibility and safety are essential. For this reason, flavor compounds can be used to obtain active edible packaging because they can act as antimicrobials, antioxidants, and flavoring agents. In particular, essential oils can be added to edible films and coatings to modify flavor, aroma, and odor, as well as to introduce antimicrobial properties (Sanchez-Gonzalez et al., 2010c).

Very little published data exist on the incorporation of plant essential oils into edible films and coatings. Essential oils are regarded as alternatives to chemical preservatives, and their use in foods meets the demands of consumers for minimally processed natural products, as reviewed by Burt (2004). Vanillin has been used recently as a bacteriostatic rather than a bactericidal agent in fresh-cut apples (Rupasinghe et al., 2006). Essential oils have also been evaluated for their ability to protect food against pathogenic bacteria in contaminated apple juice (Friedman et al., 2004; Raybaudi-Massilia et al., 2006) and other foods, and they are used as flavoring agents in baked goods, sweets, ice cream, beverages, and chewing gum (Fenaroli, 1995; Burt, 2004). Sanchez-Gonzalez et al. (2009, 2010a, 2010b) introduced tea tree and bergamot essential oils in chitosan or hydroxypropylmethylcellulose edible films at a range of 0 to 3% (w/w) in the film-forming suspension for antimicrobial properties. They displayed the antimicrobial efficiency at the higher concentration of bergamot on *Penicillium italicum* and at the lowest rate of tea tree oil on *Listeria monocytogenes*. However, no information was given on the concentration in the film after drying or on the sensory impact.

McHugh et al. (1996) developed the first edible films made from fruit purees which were shown to be a promising tool for improving quality and extending shelf life of minimally processed fruit. Rojas-Grau et al. (2006) recently investigated the effect of plant essential oils on antimicrobial and physical properties of apple puree edible films. Alginate-apple puree films, containing plant essential oils, were further explored as edible coatings by Rojas-Grau et al. (2007) with the aim of studying the effect of lemongrass, oregano oil, and vanillin on native psychrophilic aerobic bacteria, yeasts, molds, and inoculated *Listeria innocua* in fresh-cut 'Fuji' apples. Coatings with essential oils seemed to effectively inhibit the growth of *L. innocua* inoculated on apple pieces as well as psychrophilic aerobic bacteria, yeasts, and molds. In some cases, essential oils like thymol and carvacrol were added to a bio-based coating as antimicrobial agents and not as flavoring compounds (Ben Arfa et al., 2007; Del Nobile et al., 2008). Ponce et al. (2008) used oleoresins containing both volatile and nonvolatile compounds extracted from oregano, rosemary, olive, capsicum, garlic, onion, and cranberries in edible films based on sodium caseinate and carboxymethylcellulose and chitosan. These authors displayed that the use of chitosan enriched with rosemary and olive did not introduce deleterious effects on the sensorial acceptability of squash. Chitosan enriched with rosemary and olive improved the antioxidant protection of the minimally processed squash, offering a great advantage in the prevention of browning reactions that typically result in quality loss in fruits and vegetables. These coatings provide both antioxidant and antimicrobial properties of coatings at 1% oleoresin content without too much sensory disturbance.

Gelatin- and chitosan-based edible films incorporated with clove essential oil were tested for antimicrobial activity against six selected microorganisms: *Pseudomonas fluorescens, Shewanella putrefaciens, Photobacterium phosphoreum, Listeria innocua, Escherichia coli,* and *Lactobacillus acidophilus.* The clove-containing film inhibited all these microorganisms irrespective of the film matrix (Gómez-Estaca et al., 2010). The effect on the microorganisms during this period was in accordance with biochemical indexes of quality, indicating the viability of these films for fish preservation.

These authors also tested on 18 bacterial strains other essential oils: fennel (*Foeniculum vulgare* Miller), cypress (*Cupressus sempervirens* L.), lavender (*Lavandula angustifolia*), thyme (*Thymus vulgaris* L.), herb-of-the-cross (*Verbena officinalis* L.), pine (*Pinus sylvestris*), and rosemary (*Rosmarinus officinalis*). Antioxidant properties as well as light barrier properties of gelatin-based edible films containing oregano or rosemary aqueous extracts have been assessed by Gómez-Estaca et al. (2009). The essential oil polyphenols–protein interaction was found to be more extensive when tuna-skin gelatin was employed. However, this did not clearly affect the antioxidant properties of the films, although it could affect diffusion of phenolic compounds in the essential oil from film to food. The light barrier properties were improved by the addition of oregano or rosemary extracts, irrespective of the type of gelatin employed. The shelf life of cold-smoked sardine (*Sardina pilchardus*) was improved also by gelatin-based films using, singly or in combination, high pressure (300 MPa/20°C/15 min) and films enriched by adding an extract of oregano (*Origanum vulgare*) or rosemary, (*Rosmarinus officinalis*) or by adding chitosan (Gomez-Estaca et al., 2007). Gelatin seems to be a good way for encapsulating both antimicrobial or antioxidant volatile essential oils. Seaweeds extracts such as alginates and carrageenans have been extensively studied by Hambleton et al (2008, 2009a,b, 2010, 2011) and Fabra et al (2008, 2009). These authors displayed that both carrageenans and alginates are able to be used as films or coatings having encapsulated volatile compounds such as aroma or essential oils. The adding of lipids such as acetylated monoglycerides, beeswax, and oleic acid used alone or as mixture allows reducing moisture transfer but affects their ability to retain and release volatile compounds. Lipid could have positive or negative influences, depending on the polarity of the volatile compounds. Zein-based mono- and multilayer films were loaded with spelt bran and thymol (35% w/w) to obtain edible polymeric materials. Various composite systems were developed to control thymol release (Mastromatteo et al., 2009). Madene *et al* (2006) described the process for encapsulation of sensitive volatile compounds. Encapsulation can be employed to retain aroma in a food product during storage, protect the flavor from undesirable interactions with food, minimize flavor/flavor interactions, guard against light-induced reactions and/or oxidation, to increase flavors shelf-life, and to allow a controlled release. Incorporation of small amounts of flavors into foods can greatly influence finished product quality, cost, and consumer satisfaction. The food industry is continuously developing ingredients, processing methods, and packaging materials to improve flavor preservation and delivery. The stability of the matrices is an important condition to preserve the properties of the flavor materials. Many factors such as the kind of wall material, the ratio of the core material to wall material, the encapsulation method, and storage conditions affect the antioxidative stability of the encapsulated flavor. According to the technological process used, the matrices of encapsulation will present various shapes (films, spheres, particles irregular), various structures (porous or compact), and various physical structures (vitreous or crystalline dehydrated solid, rubbery matrix) that will influence the diffusion of flavors properties or external substances (oxygen, solvent) and the stability of the product during its storage.

11.5.2 Retention and release of volatile compounds in edible films and coatings

Food matrix ingredients, among them food proteins, have little flavor of their own but are known to bind and trap aroma compounds. In function, the nature and strength of the binding, and release of aroma compounds in the gas phase will be more or less decreased. The mechanism of flavor binding was dependent on the role of the protein structure as well as the type of flavor compound (aldehyde, alcohol, ketone, and ester) involved in the binding process (Heng et al., 2004). The most extensively studied proteins are milk proteins, known for their emulsifying properties. For instance, the affinity for β-lactoglobulin increases with hydrophobic chain length or overall hydrophobicity of flavor compounds, except for terpenes, main constituents of essential oils. However, it was not possible to find a simple explanation for the binding strength for aroma compounds from different chemical classes. Among the other food proteins, β-lactalbumin, caseins, bovine serum albumin, and soy proteins are studied to a lesser extent for their binding properties toward flavor compounds. β-Lactalbumin was found to bind ketones and aldehydes but with a poor flavor binding capacity compared to other whey proteins. That explains why whey protein films are often desired for flavor or essential oil encapsulation. Moreover, whey proteins have both film-forming properties and emulsifying capacity.

Carbohydrates can have a measurable influence on the release and perception of flavors. Carbohydrates change the volatility of compounds relative to water, but the effect depends on the interaction between the particular volatile molecule and the particular carbohydrate. As a general rule, carbohydrates, especially polysaccharides, decrease the volatility of compounds relative to water by a small to moderate amount, as a result of molecular interactions. However, some carbohydrates, especially the mono- and disaccharides, exhibit a salting-out effect, causing an increase in volatility relative to water (Godshall, 1997).

Lipids are rather homogeneous, hydrophobic, and nonpolar materials, existing in aqueous mediums in the form of distinct regions. In these systems, flavor compounds are distributed between the lipid and the aqueous phases, following the physical laws of partition (Solms et al., 1973). Lipids are carriers and release modulators of aroma, but they are also flavor precursors. Lipid oxidation as well as lipolysis generates numerous short-chain compounds to which we are extremely sensitive because of their low levels of olfactory detection. The global effect of lipids on aroma compound release is to decrease their volatility. Generally, aroma compounds are hydrophobic and show greater affinity for the lipidic phase than for the aqueous or vapor phase. The influence of the physicochemical properties of both the flavor compound and the fat content has been demonstrated by the determination of air–oil partition coefficients. Solid fat content and crystallinity of fat tend to improve the controlled release by diffusion fall. However, temperature induces the fat melting or solid fat content that is unfavorable to flavor retention.

A food product with active molecules on the surface and a food product coated with an edible film with active volatile molecules are compared in Figure 11.3. In food products, surface contamination and oxidation are the most probable and need to be prevented (Han, 2002).

In both systems, with and without edible film, the food layers that do not contain active agents (flavoring, antimicrobial, and antioxidant) have initially very large volume compared to the volume of thin films, coatings, or food surfaces where active volatile molecule was dispersed. Because of almost an infinite volume of food layer without active molecules, migration of active molecules from surface into food will be favored. If the agent is sprayed onto the surface of food, the initial surface concentration will be very high and start to decrease due to dissolution and diffusion of the agent toward the center of the food (Marcuzzo et al., 2011). Therefore, solubility (or partition coefficient) and diffusion coefficient (diffusivity) of the agent in a food are very important characteristics to maintain the surface concentration needed to hinder microorganism growth, oxidation or flavor loss oxidation, or flavor loss during the expected shelf life. Otherwise, active molecule concentration in the surface layer will be reduced or depleted. The release rate must be controlled to prevent the early depletion of active molecules due to fast migration. Application of edible coatings or films

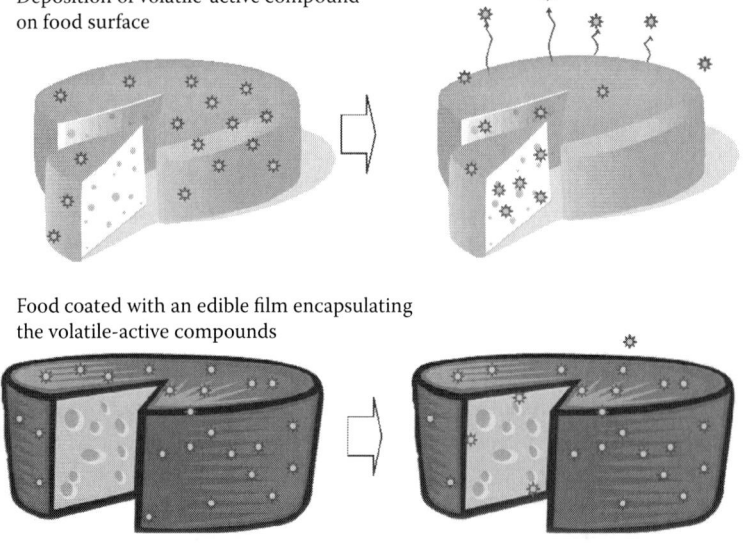

Figure 11.3 Retention and controlled release of aroma compounds added at the food surface or entrapped in edible coatings/films.

with active molecules could allow a very low diffusion rate of active molecules into the food to maintain its efficiency on the surface. Moreover, the effectiveness of the surface function could allow a lower active molecule concentration in the food. If the active molecule dispersed in the edible film or coating is an aroma compound, the factors that influence flavor release have to be considered. The release rate of the volatile agent from the packaging system is highly dependent on the volatility, which relates to the chemical interaction between the volatile agent and packaging materials. The absorption rate of headspace volatiles into a food surface is related to the composition of the food, as the ingredients undergo chemical interactions with the gaseous agents. Because most volatile agents are generally lipophilic, the lipid content of the food is an important factor of headspace concentration. The use of volatile antimicrobial agents has many advantages. This system can be used effectively for highly porous foods, powdered, shredded, irregularly shaped, and particulate foods, such as ground beef, shredded cheese, small fruits, and mixed vegetables (Han, 2002).

The ability of edible films to entrap or to retain flavor compounds during film storage and above all during film processing is of importance. If flavors are lost during drying of hydrocolloid-based film-forming liquid (solution, emulsion, or suspension), the resultant aroma will become unbalanced. The more volatile or the less interactive volatile compounds will be preferentially lost during the film/coating process, inducing a loss of the fresh aroma notes. Using emulsions or suspension with hydrocolloids interacting strongly with the aroma compounds tends to reduce the aroma loss during drying. When edible films are enriched with essential oils, the drying temperatures usually employed to form the edible coating are high enough to volatilize a high percentage of the aromatic components. Sanchez-Gonzalez (2010a) showed that D-limonene (main component of the bergamot essential oil) loss during drying ranged from 39% to 99% when added from 0.5% to 3% (w/w) in chitosan films. Moreover, Monedero et al. (2010) found that losses during drying are tremendously greater than during film storage. Soy protein–beeswax–oleic acid emulsified films, dried at moderate temperature (20°C, 45% relative humidity, RH), lost up to 98% of encapsulated n-hexanal, but this could be reduced by using a higher amount of beeswax. Beeswax particles decreased by half the aroma diffusivity in the protein-based matrix. However, the greater the loss during drying, the faster is the release to the vapor phase. Most of the remaining n-hexanal in the film after drying was lost after 35 days storage in an open ventilated chamber (Figure 11.4). Similar trends were observed by Tunc and Duman (2011) when the montmorillonite (nanoclays) content in chitosan films increased. Solid particles such as beeswax or montmorillonite probably reduce the diffusivity of the volatile compound because of increasing "tortuosity" and then delay of the aroma loss by the films. A significant increase of thymol release rate with the increase of the bran concentration was observed by Mastromatteo et al. (2009). So, in this case, incorporation of nanoparticles (spelt bran) had the opposite effect and was unfavorable for volatile compound retention.

The advantage of substituting essential oils or aroma compounds for their corresponding food-grade oleoresins could lie in the introduction of other nonvolatile components positively affecting food quality (Ponce et al., 2008).

Flavor retention by the film matrix during processing depends as much on the temperature as on the film or coating composition. On one hand, temperature increases

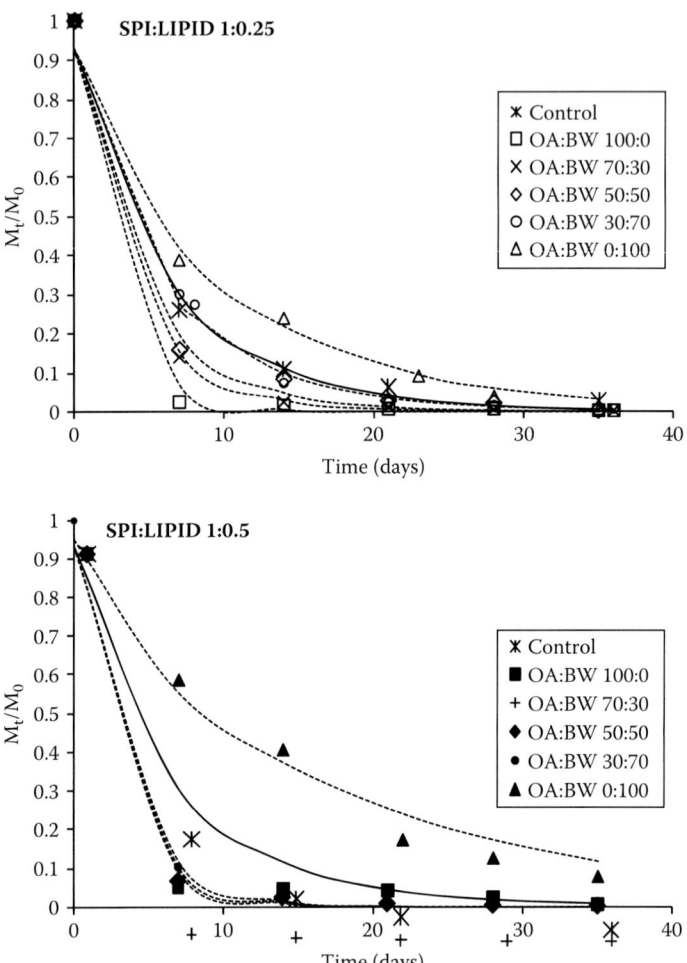

Figure 11.4 N-hexanal release kinetic for the control film and films containing soy protein isolate (SPI) combined with oleic acid (OA) and beeswax (BW) in SPI:LIPID ratio 1:0.25 and 1:0.5 (experimental data: symbols, and fitted model: lines). (From Monedero M., Hambleton A., Talens P., Debeaufort F., Chiralt A., and Voilley A., *Journal of Food Engineering*, 100, 128–133, 2010. With permission.)

the aroma volatility according to Henry's laws (Buttery et al., 1971) and the diffusivity. On the other hand, many film constituents allow reduced volatility and diffusivity. Polyols used as plasticizers in film formulation are very good for aroma support and are able to significantly reduce flavor release. Acacia gum provides the same advantages for flavor retention and could also strengthen film mechanical properties. Recent studies showed that the addition of nanoparticles such as nanoclays is significant to improve the retention efficiency of chitosan-based edible films (Tunc and Duman, 2011). Usually, the mass partition coefficient K_{mass} (ratio between aroma

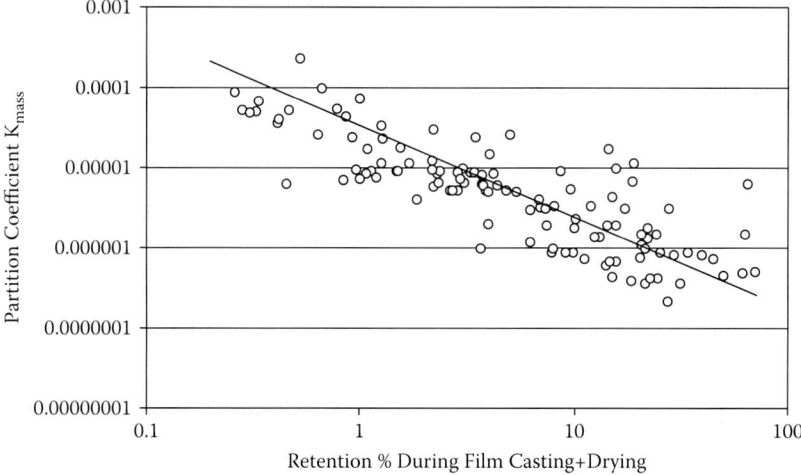

Figure 11.5 Relationship between carvacrol retention during film processing and air/film partition coefficient during film storage. (See Kurek M., and Debeaufort F., Doctoral school intermediate PhD report, University of Burgundy, 2010.)

concentration in air and aroma concentration in film) is inversely proportional to the retention capacity during film processing.

As displayed in Figure 11.5, the partition coefficient (during film storage) can be related to carvacrol retention during film processing (casting and drying) by chitosan-based films (more than 200 recipes in a 25 to 80°C drying temperature range) at various humidities (0 to 98% RH).

The release of various aroma compounds (ethyl esters, methylketones, and alcohols), from either carrageenan-based, or carrageenan-acetylated monoglycerides emulsion-based, or acetylated-monoglycerides-based films strongly differs (Marcuzzo et al., 2010). In lipid films, the aroma compound release is more affected by factors related to diffusivity, whereas in carrageenan emulsified films, the affinity between volatile compounds and polymer preponderantly influences sorption phenomena and thus the release. Carrageenan films resulted in possible encapsulating matrixes: they display better performances for retention of more polar aroma compounds than pure lipid or emulsified films. Carrageenan films were able to retain volatile compounds during film processing and released them gradually with time.

The surrounding medium of encapsulated aroma such as polysaccharides, proteins, lipids, and salts could play an important role in release in liquid media and then on aroma compound retention by the film matrix. Different behaviors have been observed in the presence of salt or sucrose molecules of aroma compound volatility in food products, observing that some of them presented a "salting out" effect (favored release by volatilization), others an opposite "salting in" effect, and for some others no modification (Lubbers et al., 1998; Van Ruth et al., 2002). Similar effects could be observed in the release of the aroma compounds encapsulated in films. While the "salting out" effect should accelerate the release, the "salting in" could decrease the rate of the release.

Table 11.1 Percentage of *n*-hexanal and D-limonene retained in the film at the equilibrium of release kinetics in liquid media

Type of film	Aqueous medium	Retention of n-hexanal (%)	Retention of D-limonene (%)
Carrageenans without lipid	Water	9.3 (1.6)	6.2 (1.7)
	0.9% NaCl	57 (3)	79 (2)
Carrageenans with lipid	Water	11 (1)	7 (3)
	0.9% NaCl	44 (2)^e	100

Note: Mean value (standard deviation).

The effect of aqueous media (containing 0.9% of NaCl) and temperature (25 and 37°C) on the release of encapsulated aroma compounds (*n*-hexanal and D-limonene) in iota-carrageenan-based films with and without lipid have been studied by Fabra et al. (2011). D-limonene was released quickly from water at higher temperatures. However, no effect of temperature was observed on *n*-hexanal release in water. Only between 40% and 56% of the hexanal was released, depending on film composition, while D-limonene was completely retained in the fat material (Table 11.1). Salt presence in the liquid release media favored significantly and tremendously the aroma retention by carrageenan-based films. Retention was 4 to 15 times higher according to aroma compound hydrophobicity and fat particle presence in the film matrix.

In a similar way, Sanchez-Gonzalez et al. (2011) studied chitosan films enriched with different concentrations of bergamot oil and the migration of D-limonene (the major oil component) into five liquid simulated foods (water, 10% ethanol, 50% ethanol, 95% ethanol, and isooctane). D-Limonene migration (release) was significant in 95% ethanol, whereas in the other food simulants, its release remained less impressive. Composite films remained intact with isooctane CH-BO, and no release of limonene was observed. Polarity of simulant and migrant seems to be a key factor to explain these results. However, in the reports of both Sanchez-Gonzalez et al. (2010c) and Fabra et al. (2011), chitosan- or carrageenan-based films immersed in liquid media probably swelled and some of their soluble components (plasticizers) probably migrated into the contacting liquid. Because film integrity was not maintained, the release phenomenon could not be attributed only to the partition (solubility) and diffusion coefficient of the aroma compounds. Other parameters must also be taken into account to better explain and describe the release phenomenon of aroma compounds in contact to liquid or gel foods, such as swelling, partial dissolution of film component, counterdiffusion of other solute from food to film, and osmotic gradients.

Volatile compound retention by edible films is complex, and many factors could affect it and interact. Figure 11.6 is a simplified summary view of the main parameters influencing the flavor release from edible films.

11.5.3 Barrier performances of edible films and coatings to aroma compounds

D-Limonene is the flavor compound most studied. It is considered as the typical aroma probe of citrus juice and soda beverage, and many data are available in the literature. Figure 11.7 gives a comparison between D-limonene permeabilities of

Encapsulation of flavors, nutraceuticals, and antibacterials 349

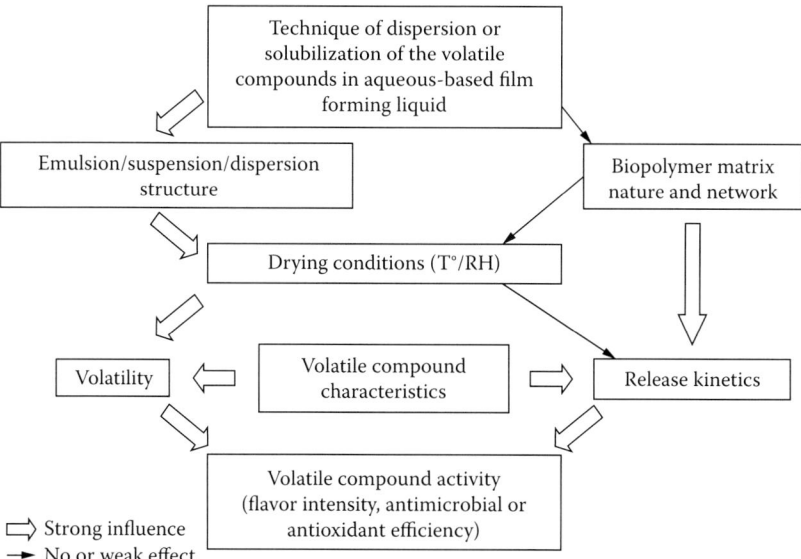

Figure 11.6 Overall mechanism of flavor release in air from edible films: main influencing factors.

edible and plastic films. Polysaccharide-based films are of the same order as heat sealable polymers such as polyolefins, while protein-based films and coatings have barrier properties as good as the best plastics. The permeability of wheat gluten and glycerol or whey protein and glycerol films is about 10^{-17}g.m^{-1}.s^{-1}.Pa^{-1} and from 10^{-14} to 10^{-16}g m^{-1} s^{-1} Pa^{-1} in the case of polysaccharides and other protein-based films (Miller et al., 1998; Quezada-Gallo, 1999; Hambleton et al., 2010).

Although D-limonene is probably the flavor compound most studied in the field of polymer permeability, it is not often used in the flavoring of solid food products. So, Quezada-Gallo et al. (1999a) and Debeaufort and Voilley (1995) studied the mass transfers through edible films of several other molecules, commonly found in cheese, fruits, and dairy products. Table 11.2 displays the permeability of some flavor compounds through edible films. It seems that permeability is always much lower through protein-based than from carbohydrate films, from 100 to 100,000 times lower. These results allow considering their use as a protecting layer against scalping and permeation through classically used plastic films. Edible barriers are able to retain food flavors within the food matrix due to very low permeabilities. However, because of a lack of aroma permeability data, no trend or general rules could be given about the barrier efficiency of edible barriers. Interactions between film constituents and aroma compounds have a great impact on the permeability mechanism. Very few papers focused on the aroma transfers related to interactions, sorption, and diffusion phenomena. It seems that chemical affinities between flavor compounds and film components have more influence on the permeability value

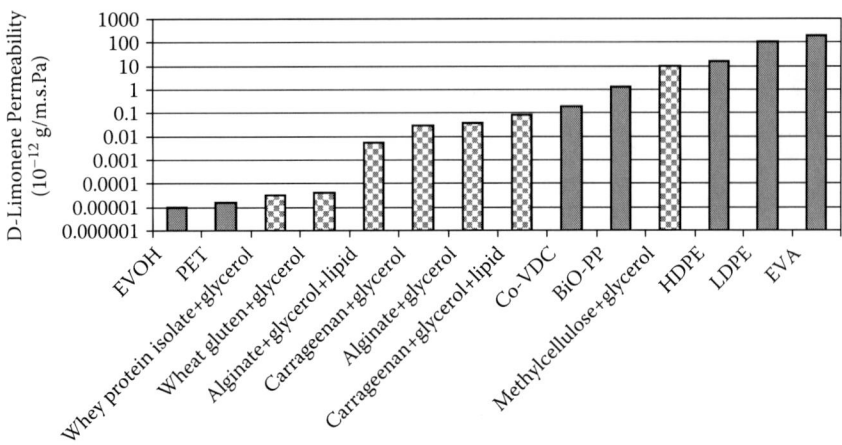

Figure 11.7 D-Limonene permeability (10^{-12} g m^{-1} s^{-1} Pa^{-1}) of some edible and plastic films at 25°C and 0% relative humidity (RH) (thickness ranging from 60 to 130 µm). (From Becker, K., Koszinowski, J., and Piringer, O., Parfurmerie und Kosmetik, 68, 268–278, 1987; Franz, R., Packaging Technology Science, 6, 91, 1993; Debeaufort, F., PhD Dissertation, ENS.BANA, Université de Bourgogne, Dijon, France, 1994; Debeaufort, F., and Voilley, A., Journal of Agricultural and Food Chemistry, 42, 2871–2875, 1994; Debeaufort, F., and Voilley, A., Cellulose, 2, 1–10, 1995; Kobayashi, M., Kanno, T., Hanada, K., and Osanai, S.I., Journal of Food Science, 60, 205–209, 1995; Paik, J.S., and Writer, M.S., Journal of Agriculture and Food Chemistry, 43, 175–178, 1995; Miller, K.S., and Krochta, J.M., Trends in Food Science and Technology, 8, 228–237, 1997; Miller, K.S., Upadhyaya, S.K., and Krochta, J.M., Journal of Food Science, 63, 244–247, 1998; Quezada-Gallo, J.A., Debeaufort, F., and Voilley, A., Journal of Agriculture and Food Chemistry, 47: 108–113, 1999; Quezada-Gallo, A., Debeaufort, F., and Voilley, A., New Developments in the Chemistry of Packaging Materials, ACS Books, Dallas, TX, 1999; Quezada-Gallo, J.A., Ph.D. Dissertation, Université de Dijon, France, 1999; Hambleton, A., Debeaufort, F., Beney, L., Karbowiak, T., and Voilley, A., Biomacromolecules, 9(3), 1058–1063, 1999; Hambleton, A., Debeaufort, F., Bonnotte, A., and Voilley, A., Food Hydrocolloids, 23(8), 2116–2124, 2009; Hambleton, A., Fabra, M.J., Debeaufort, F., Brun-Dury, C., and Voilley, A., Journal of Food Engineering, 93, 80–88, 2009. With permission) (EVOH, ethylene vinyl alcohol; PET, polyethylene terephthalate; Co-VDC, polyvinylidene chloride copolymer; Bio-PP, bioriented polypropylene; HDPE and LDPE: high- and low-density polyethylene; EVA, ethylene vinyl acetate.)

than the classical sorption-diffusion models. Aroma induces significant changes in physical-chemical and structural properties of polysaccharide or protein-based edible films when entrapped or when transferred as described below.

11.5.4 Protection of flavor compounds against chemical degradation by edible films and coatings

The most common problem occurring during flavor storage is deterioration due to oxidation. Any flavoring containing citrus oils or oils based on aldehydes is susceptible

Table 11.2 Permeability (P) to aroma compounds of several edible films at 25°C[a]

Edible film composition	Aroma compounds	$P (10^{-12}\ g\ m^{-1}\ s^{-1}\ Pa^{-1})$
Methylcellulose+glycerol	1-Octen-3-ol	122
	2-Pentanone	19
	2-Heptanone	39
	2-Octanone	338
	2-Nonanone	420
	Ethyl acetate	128
	Ethyl butyrate	119
	Ethyl isobutyrate	106
	Ethyl hexanoate	668
	D-Limonene	10
Iota-Carrageenan+glycerol	Ethyl acetate	0.046
	Ethyl butyrate	1.25
	Ethyl hexanoate	<0.014
	2-Hexanone	0.196
	1-Hexanol	<0.0011
	Cis-3-hexenol	<0.012
	n-Hexanal	0.0189
	D-Limonene	0.0291
Iota-Carrageenan+glycerol+acetylated monoglycerides+beeswax (emulsion)	Ethyl acetate	0.022
	Ethyl butyrate	0.33
	Ethyl hexanoate	<0.014
	2-Hexanone	0.111
	1-Hexanol	<0.0011
	Cis-3-hexenol	<0.017
	n-Hexanal	<0.00005
	D-Limonene	0.0803
		(Continued)

Table 11.2 Permeability (P) to aroma compounds of several edible films at 25°C[a] (Continued)

Edible film composition	Aroma compounds	$P(10^{-12}\ g\ m^{-1}\ s^{-1}\ Pa^{-1})$
Iota-Carrageenan+glycerol + oleic acid + beeswax (emulsion)	Ethyl acetate	0.83
	Ethyl butyrate	0.75
	Ethyl hexanoate	194
	2-Hexanone	<0.0011
	1-Hexanol	1.12
	Cis-3-hexenol	<0.0012
Sodium alginate+glycerol	n-Hexanal	0.0134
	D-Limonene	0.00034
Sodium alginate+glycerol+acetylated monoglycerides+beeswax (emulsion)	Ethyl butyrate	0.033
	Ethyl hexanoate	134.2
	2-Hexanone	0.094
	n-Hexanal	<0.00005
	D-Limonene	0.000053
Wheat gluten+glycerol	1-Octen-3-ol	4.6
	2-Pentanone	0.12
	2-Heptanone	0.50
	2-Octanone	<0.005
	Ethyl acetate	0.059
	Ethyl butyrate	0.670
	Ethyl isobutyrate	0.04
	Ethyl hexanoate	<0.005
	D-Limonene	0.00004
Sodium caseinate+glycerol	Ethyl acetate	0.006
	Ethyl butyrate	0.19
	Ethyl hexanoate	<0.0004
Sodium caseinate+glycerol+oleic acid (emulsion)	Ethyl acetate	36.1

	Ethyl butyrate	1732.7
	Ethyl hexanoate	1632.1
	2-Hexanone	1.85
	1-Hexanol	3116
	Cis-3-hexenol	404
Sodium caseinate+glycerol+oleic acid+beeswax (emulsion)	Ethyl acetate	0.73
	Ethyl butyrate	134.7
	Ethyl hexanoate	<0.0004
	2-Hexanone	18.7
	1-Hexanol	238.8
	Cis-3-hexenol	639.1
Sodium caseinate+glycerol+beeswax (emulsion)	Ethyl acetate	0.0061
	Ethyl butyrate	0.08
	Ethyl hexanoate	<0.00045
	2-Hexanone	<0.00003
	1-Hexanol	<0.005
	Cis-3-hexenol	<0.0004
Whey protein isolate+glycerol	D-Limonene	0.0003

[a] Adapted from Debeaufort, F., PhD Dissertation, ENS.BANA, Université de Bourgogne, Dijon, France, 1994; Debeaufort, F., and Voilley, A., *Journal of Agricultural and Food Chemistry*, 42, 2871–2875, 1994; Debeaufort, F., and Voilley, A., *Cellulose*, 2, 1–10, 1995; Kobayashi, M., Kanno, T., Hanada, K., and Osanai, S.I., *Journal of Food Science*, 60, 205–209, 1995; Miller, K.S. and Krochta, J.M., *Trends in Food Science and Technology*, 8, 228–237, 1997; Miller, K.S., Upadhyaya, S.K., and Krochta, J.M., *Journal of Food Science*, 63, 244–247, 1998; Quezada-Gallo, A., Debeaufort, F., and Voilley, A., *New Developments in the Chemistry of Packaging Materials*, ACS Books, Dallas, TX, 1999; Quezada-Gallo, J.A., Debeaufort, F., and Voilley, A., *Journal of Agriculture and Food Chemistry*, 47, 108–113, 1999; Quezada-Gallo, J.A., PhD Dissertation, Université de Dijon, France, 1999; Hambleton, A., Debeaufort, F., Beney, L., Karbowiak, T., and Voilley, A., *Biomacromolecules*, 9(3), 1058–1063, 2008; Hambleton, A., Debeaufort, F., Bonnotte, A., and Voilley, A., *Food Hydrocolloids*, 23(8), 2116–2124, 2009; Hambleton, A., Fabra, M.J., Debeaufort, F., Brun-Dury, C., and Voilley, A., *Journal of Food Engineering*, 93, 80–88, 2009; Fabra, M.J., Hambleton, A., Talens, P., Debeaufort, F., Chiralt, A., and Voilley, A., *Carbohydrate Polymers*, 76, 325–332, 2009; Fabra, M.J., Hambleton, A., Talens, P., Debeaufort F., Chiralt, A., and Voilley, A., *Biomacromolecules*, 9(5), 1406–1410, 2008.

to oxidative reactions and development of off-flavors. Thus, an important function of an edible film is protection of the flavoring from oxygen.

Choosing an encapsulating agent with adequate barrier properties (to oxygen) will lead to more stable dry flavors. Because D-limonene is sensitive to oxidative degradation, several studies were focused on the possibility of encapsulating this aroma compound. Different ingredients were chosen as wall materials, such as starch, maltodextrin, and gum arabic (Wyler and Solms, 1982; Anandaraman and Reineccius, 1986; Bertolini et al., 2001). Generally, lemon oils are added to food in the form of water-in-oil emulsions. Djordevic et al. (2008) studied the possibility of stabilizing oil in water emulsions with whey proteins instead of gum arabic, to inhibit D-limonene degradation. Formation of the limonene oxidation products limonene oxide and carvone were less in the whey protein isolate (WPI)- than gum Arabic (GA)-stabilized emulsions. This was in agreement with Kim and Morr (1996), who found that limonene oxide, carvone, and carveol formation in microencapsulated orange oil was less in emulsions stabilized with WPI and soy protein isolate than emulsions stabilized with GA. Among wall materials that can be applied to preserve aroma compounds, κ-carrageenans and wheat gluten were selected for their useful gas barrier properties. Hambleton et al. (2008) showed that carrageenans-based edible film prevents the oxidation of n-hexanal to hexanoic acid in oxidative conditions. Marcuzzo et al. (2011) also showed that both carrageenan and wheat gluten–based edible films seem to prevent D-limonene oxidation. Wheat gluten–based film protected D-limonene from degradative reactions, and the increase in carvone release was probably due to oxidation in the headspace once D-limonene was released and not within the matrix as was confirmed by Fourier transform infrared (FTIR) analysis of edible film containing the aroma compounds.

The oxygen permeability is then a key factor for an improved efficiency of film and coating matrices to protect aroma compounds encapsulated or initially present in coated food products. Table 11.3 gives the oxygen permeability of the more studied edible films and coatings. Protein-based edible films are the more efficient barriers against oxygen transfer and are more suitable for flavor protection. Protein also interacts to a greater extent with flavor compounds. But the oxygen protection strongly varies with the moisture level. Increasing water activity or relative humidity tends to increase from 10 to 10^5 times the oxygen permeability. For example, oxygen permeability of collagen film rises from 6.6 10^{-19} to 14.68 10^{-15} g m^{-1} s^{-1} Pa^{-1} when water activity increased from 0 to 0.93, but this behavior is also observed for water-sensitive plastic films such as EVOH or nylon. Plasticization of food macromolecules by water increases molecular mobility that favors both oxygen diffusivity and aroma release and oxidation.

11.5.5 Volatile compound effect on structural and physical-chemical properties of edible films and coatings

Incorporating volatile active compounds may affect significantly the structural organization of the film-forming substance network and, consequently, its physical-chemical properties.

Table 11.3 Oxygen permeability of edible films and coatings compared to common plastic films (10^{-15} g m^{-1} s^{-1} Pa^{-1})[a]

Film	Oxygen permeability	T (°C)	a_w
Low-density polyethylene	16.050	23	0
Polyesters	0.192	23	0
Ethylene vinyl alcohol	0.003	23	0
Polyvinylidene chloride	0.0066	23	0
Methylcellulose+glycerol	8.352	30	0
Carrageenan+glycerol	0.72	25	0
Carrageenan+glycerol + GBS	0.86	25	0
Alginate+glycerol	9.4	25	0
Alginate+glycerol + GBS	2.7	25	0
Chitosan+glycerol	0.009	25	0
Starch alginate	0.087	23	0
Beeswax	7.68	25	0
Carnauba wax	1.296	25	0
Collagen	0.00066	25	0
Zein+glycerol	0.56	38	0
Wheat gluten+glycerol	0.016	25	0
Soy protein isolate+glycerol	0.1	25	0
Casein+glycerol	0.0115	25	0
Casein+sorbitol	0.0132	25	0
Peanut protein + glycerol	0.034	23	0
Fish myofibrillar protein+glycerol	0.031	25	0
Low-density polyethylene	30.85	23	0.5
Polyesters	0.192	23	1
Ethylene vinyl alcohol	0.096	23	0.95
Polyvinylidene chloride	0.084	23	0.95
Pectin+glycerol	21.44	25	0.96
Starch+glycerol	17.36	25	1
Hydroxypropylmethylcellulose	4.88	25	0.50
Chitosan+glycerol	7.552	25	0.93
Collagen	0.48	25	0.63
Collagen	14.68	25	0.93
Wheat gluten+glycerol	20.64	25	0.95
Whey protein+glycerol	33	23	0.34
Whey protein+glycerol	435	23	0.56
Fish myofibrillar protein+glycerol	27.93	25	0.93

[a] From Gennadios, A., *Protein-Based Edible Films and Coatings*. CRC Press, Boca Raton, FL, 2002; Embuscado, M.E., and Huber, K.C., *Edible Films and Coatings for Food Applications*, Springer Science, New York, 2009. With permission.

11.5.5.1 Impact on film appearance and transparency

Incorporation of aroma compounds or essential oils always decreases the transparency and gloss of edible films made of carbohydrate or protein, because of emulsion structure formation during drying. Solubility of most aroma compounds and essential oils is much lower than the amount added to film forming suspensions. All the authors who measured optical properties of films containing essential oils or flavor compounds observed this behavior, regardless of the nature of the matrix (chitosan, HPMC, methylcellulose, soy protein isolate, whey protein isolate, carrageenans, sodium alginate) or the volatile compounds (tea tree oil, bergamot, cloves, ginger, onion, carvacrol, thymol, etc.) (Pranoto et al., 2005a; Hambleton et al., 2009a, 2009b, 2010; Atarés et al., 2010; Fabra et al., 2008, 2009, 2011; Sanchez Gonzalez et al., 2009, 2010a, 2010b; Tunc and Duman, 2011). Changes in appearance could generally be related to the microstructure of the film containing the volatile compound.

11.5.5.2 Changes in film microstructure induced by volatile compound encapsulation

Hambleton et al. (2009a) showed that incorporation of 3% of n-hexanal in alginate matrices induced a more homogeneous structure as observed by an environmental scanning electron microscope (ESEM). To the contrary, n-hexanal provoked a more heterogeneous distribution of the emulsified fat (beeswax+acetylated monoglycerides) in the film cross section. This was attributed to the competition between the emulsifier (glycerol monostearate) and n-hexanal, which is amphipolar. The same behavior was also observed for carrageenan-based films and h-hexanal encapsulated (Hambleton et al., 2009b). Sanchez-Gonzalez et al. (2009) observed a more heterogeneous structure with increasing tea tree oil content in hydroxypropyl methylcellulose (HPMC) films. The essential oil was at a concentration higher than the solubility limit of the volatile in the film-forming suspension, which resulted in an emulsion of tea tree oil droplets in the HPMC network. While a continuous structure was observed for the HPMC film, the presence of tea tree oil caused discontinuities associated with the formation of two phases in the matrix: lipid droplets embedded in a continuous polymer network. Lipid droplets, whose number increased with the tea tree oil concentration, were homogenously distributed across the film. This reveals that very little creaming occurred during the film drying, probably due to the highly viscous effect of HPMC.

11.5.5.3 Influence of volatile compound encapsulation on film mechanical properties

The oil droplets or discontinuities usually induce a loss of mechanical properties, such as a decrease of the tensile strength and Young (elastic) modulus as given in Table 11.4.

The addition of the tea tree oil in the 0.5% to 3% concentration range caused a significant decrease in the elastic modulus and in tensile strength at break of HPMC films, although with no significant effect on deformation at break (Sanchez-Gonzalez

Table 11.4 Influence of volatile compound incorporation on mechanical properties of edible films[a]

Film matrix	Volatile compounds	Elongation (%)	Tensile strength (MPa)	Elastic modulus (MPa)
Soy protein isolate	/	4	11	412
	Cinnamon oil 1%	7.5	14.1	354
	Ginger oil 1%	3	8	340
Chitosan	/	22	113	2182
	Bergamot oil 3%	1.7	22	682
	Tea tree oil 2%	8	54	653
Alginate	/	4.1	66.1	/
	Garlic oil 0.4%	2.7	38.7	/
Sodium alginate	/	4.7	55	3280
	n-hexanal 1%	1.5	28	2247
Sodium alginate+GBS (emulsion)	/	2.1	31	2320
	n-hexanal 1%	1.9	20	1605
Carrageenan	/	1.2	8.8	95
	n-hexanal 1%	1.6	15	1259
Carrageenan+GBS (emulsion)	/	2.6	10	927
	n-hexanal 1%	2.4	10	751
HPMC	/	0.1	59	1697
	Tea tree oil 40%	0.11	42	956

[a] From Atarés, L., De Jesús, C., Talens, P., and Chiralt, A., *Journal of Food Engineering*, 99, 384–391, 2010; Sanchez-Gonzalez et al., 2010b; Sanchez-Gonzalez, L., Vargas, M., Gonzalez-Martinez, C., Chiralt, A., and Chafer, M., *Food Hydrocolloids*, 23, 2102–2109, 2009; Pranoto et al., 2005b; Hambleton et al., *Food Chemistry*, 2011.

et al., 2009). As elongation did not increase, the loss of mechanical properties cannot be attributed to a plasticization of HPMC by tea tree oil. This coincides with the results reported by other authors when adding essential oil to a chitosan matrix (Pranoto et al., 2005a, 2005b; Zivanovic et al., 2005) and is in agreement with the effect of the structural discontinuities provoked by the incorporation of the oil on the mechanical behavior. These discontinuities reduced the film's resistance to fracture. Bergamot oil (0.5%) added to chitosan films reduced the tensile strength and elongation two and three times, respectively (Sanchez-Gonzalez et al., 2010). Cinnamon oil seemed to have some plasticizing effect on soy protein isolate–based film, making them more extensible as the oil content increased (Atarés et al., 2010). On the contrary, cinnamon oil may have caused some degree of rearrangement in the protein network, thus strengthening it and increasing the film resistance to elongation. This effect was not observed when ginger oil was added. Films with ginger oil were less resistant and less elastic than those with cinnamon oil ($p < 0.01$). The discontinuities in the protein matrix may imply a decrease in the deformability of the films with ginger oil, because these reach the break point at lower deformation. This tendency could be explained by the fact that lipids are unable to form a cohesive and continuous matrix. A very different behavior was observed by Hambleton et al. (2011) for carrageenan films incorporating n-hexanal. In fact, the addition of incorporated n-hexanal tends to increase the elastic modulus and tensile strength. This is probably due to the stabilizing effect of the aroma compound on the film matrix. n-Hexanal interacts with κ-carrageenan's lateral chains and plays a stabilizing role in the interface due to its amphipolar character leading to a much more homogeneous structure that increases the film's stiffness. Nevertheless, n-hexanal does not affect the film's capacity to stretch as the elongation percentages are not significantly different from the same film without n-hexanal. The addition of incorporated n-hexanal has the same effect as the presence of fat material; it reduces the elastic modulus and tensile strength, contrary to ι-carrageenan-based film. Incorporated n-hexanal weakly interacts with the sodium alginate, because this type of film has a well-organized structure in an egg-box model, stabilized by divalent ions to form stronger gels and thus stronger films. But n-hexanal interacts with the other components in the film like glycerol, which being a polyol has a great affinity for flavors of this type. These interactions lead to a reduction of the stiffness and to the film's resistance to elongation. The presence of both n-hexanal and fat material has a significant effect, reducing elastic modulus and tensile strength more than in other types of films, probably because the aroma compound interacts primarily with the fat material when added to the film. Film plasticization can occur during aroma transfer, and it depends on the aroma concentration as observed by Quezada-Gallo et al. (1999a, 1999b) for permeation of 2-heptanone and 2-pentanone through methylcellulose-based films. The transfer rate of 2-heptanone strongly increased for an aroma gradient higher than 10 µg/mL. Mechanical film properties changed upon exposure to flavor concentrations higher than 10 µg/mL. 2-Heptanone and 2-pentanone increased film elongation, suggesting that polymer plasticization occurred. The ketone group of the flavor compound interacts with hydroxyl groups of methylcellulose. 2-Heptanone forms weak hydrogen bonds with methylcellulose. This likely widens the spaces among the polymer chains, resulting in swelling and plasticization of the film network which decreases the mechanical properties and enhances the transfer of volatiles and water vapor.

11.5.5.4 Impact of flavor compound encapsulation on the barrier properties of film

Both microstructure and mechanical properties are affected by flavor compound or essential oil encapsulation including permeability of films to other volatile compounds such as water vapor, oxygen, or other aroma compounds. This suggests that the choice of matrix for encapsulation of a volatile compound on the basis of its oxygen barrier performance, for instance, could not be counted on because of permeability changes due to encapsulation. When 1% n-hexanal was encapsulated, oxygen permeability of carrageenan-glycerol and carrageenan-glycerol-lipid edible films increased 15% and 100%, respectively (Hambleton et al., 2008). Incorporation of 1% n-hexanal in a sodium alginate film induced a doubling in oxygen permeability and a tenfold increase when the sodium alginate film contained lipid emulsions

(Hambleton et al., 2009a). On the contrary, Rojas-Grau et al. (2007) did not show any change in the oxygen permeability of alginate-apple puree film when oregano, carvacrol, lemongrass oil, citral, cinnamon oil, or cinnamaldehyde were added in a range of 0.1% to 0.5%. In these cases, the low content of encapsulated volatile compounds did not disturb the network structure.

The effect of RH on flavor transfer rates is probably as important as the effect of temperature. Moisture increases the mass transfer rates of gases and vapors through hydrophilic biopolymer films. Miller et al. (1998) observed substantial increases of D-limonene permeability through whey protein films when RH increased, increasing 2- to 20-fold when RH varied from 40% to 80%. Quezada-Gallo (1999) observed similar behavior for the permeability of 2-pentanone, 2-heptanone, and ethyl esters through both methylcellulose and wheat gluten films. This behavior was attributed to plasticization of the biopolymer network by water. But, if the effect of water on the permeability of volatile compounds is well known and seems obvious, the effect of aroma compounds or essential oils on the water vapor permeability is quite a bit more complicated as displayed in Table 11.5.

Table 11.5 Influence of volatile compound incorporation on water vapor permeability of edible films[a]

Film matrix	Volatile compounds	WVP (10^{-10} g/m s Pa)	T (°C)	ΔRH (%)
Soy protein isolate	/	1.38	25	33–53
	Cinnamon oil 1%	1.52	25	33–53
	Ginger oil 1%	1.89	25	33–53
Chitosan	/	12.4	25	100–54
	Bergamot oil 3%	6.5	25	100–54
	Tea tree oil 2%	7.4	25	100–54
Alginate	/	2.35	25	0–50
	Garlic oil 0.4%	72.64	25	0–50
Sodium alginate	/	2.13	25	30–84
	n-hexanal 1%	2.96	25	30–84
Sodium alginate+GBS (emulsion)	/	1.68	25	30–84
	n-hexanal 1%	1.21	25	30–84
Carrageenan	/	23.5	25	30–84
	n-hexanal 1%	22.2	25	30–84
Carrageenan+GBS (emulsion)	/	29	25	30–84
	n-hexanal 1%	25.3	25	30–84
HPMC	/	8.2	25	100–54
	Tea tree oil 40%	5.5	25	100–54

[a] From Atarés, L., De Jesús, C., Talens, P., Chiralt, A., *Journal of Food Engineering*, 99, 384–391, 2010; Sanchez-Gonzalez, L., Vargas, M., Gonzalez-Martinez, C., Chiralt, A., and Chafer, M., *Food Hydrocolloids*, 23, 2102–2109, 2009; Sanchez-Gonzalez, L., 2010b; Pranoto et al., 2005a; Hambleton et al., *Food Chemistry*, 2011. With permission.

Water vapor permeability (WVP) of soy protein isolate or alginate based films without lipid emulsions tends to increase when essential oils or aroma compounds were encapsulated, whereas that of sodium alginate plus lipid, carrageenan-based, chitosan or HPMC decreased. Atarés et al. (2010) showed that the addition of small proportions of ginger and cinnamon essential oils resulted in a reduction in the water vapor barrier properties of soy protein isolate films. This could be due to the interactions of oil components with some protein tails that could promote a decrease in the hydrophobic character of the protein matrix. Because the amount of oil incorporated is very low, the lipid discontinuities seem not to be relevant to increasing the tortuosity factor for transfer of water molecules, responsible for the reduction of water vapor permeability, although a further increase in cinnamon oil content resulted in WVP decrease. The effectiveness of cinnamon oil, as compared to ginger oil, in reducing WVP at a fixed protein-to-oil ratio, suggests that the former remains partially integrated in the protein network of the dry films. The difficulties in integrating the essential oils or aromas in a hydrophilic network may be due to matrix disruptions and creation of void spaces at the protein–essential oil interface. Therefore, it cannot be assumed that the WVP of edible films is reduced simply by adding a hydrophobic component such as an aroma or essential oil to the formulation, but the impact of lipid addition on the microstructure of emulsified film is a determining factor in water barrier efficiency. In the case of HPMC, chitosan, or carrageenan, the previous explanation for increasing the hydrophobicity does not fit, because water vapor permeability increased with essential oil content. The WVP values showed a significant decrease in line with the increase in tea tree oil concentration, following a linear trend reaching a maximum WVP reduction of about 40% with an incorporation of 2% of the essential oil (Sanchez-Gonzalez et al., 2010a, 2010b). This behavior is expected as an increase in the hydrophobic compound fraction usually leads to an improvement in the water barrier properties of films, as was previously reported for essential oil addition in chitosan films (Zivanovic et al., 2005).

Incorporation of n-hexanal in an ι-carrageenan film induced a twofold decrease in permeability of ethyl acetate, ethyl butyrate, and 2-hexanone and a sixfold decrease for the D-limonene. The encapsulated n-hexanal interacted with CH_2OH or sulfated groups of ι-carrageenan lateral chains inducing a lower permeability as displayed by Hambleton et al. (2008, 2010). When the film contained a lipid dispersed in the carrageenan matrix, the behavior of the aroma compound permeability varied, decreasing from 30% (2-hexanone), twofold (ethyl acetate), 100-fold (D-limonene), but decreasing by 15-fold for ethyl butyrate. The higher the log K, the greater the hydrophobicity of the aroma compound, thereby limiting the ethyl butyrate solubility in the hydrophilic ι-carrageenan matrix. On the contrary, it promotes interactions with the fat. Ethyl butyrate is then more retained in fat globules. The films with fat are also less uniform, fat globule particle size increases, and there are less open spaces in the film structure to facilitate the aroma diffusivity. Therefore, it limits the aroma compound transfer. Alginate-lipid emulsion films exhibit a significant increase in ethyl butyrate (×2000), ethyl hexanoate (×10,000) and D-limonene permeability (×300), whereas permeability drop fourfold for 2-hexanone. Permeability of aroma compounds is much more complex than that of gases or water vapor. It depends on the solubility of the volatile compound in the edible film, on their vapor pressure, volatility, hydrophobicity (often expressed as the partition coefficient between water and octanol), solubility (governed by the chemical nature of both film and volatile compound), molecular mobility of the matrix, volatile diffusion coefficient, and many external parameters such as temperature, pressure, moisture levels, and so forth. Therefore, previous observations when aroma or essential oils are added in edible films cannot easily predict what would happen in other situations or conditions because the physical-chemical explanations have not always been reported.

11.6 Edible films and coatings for nonvolatile molecule encapsulation (peptides, polyunsaturated fatty acids, antioxidants, and antibacterials)

Over the last few years, consumer demand for foods of natural origin, high quality, elevated safety, longer shelf life, fresh taste, and appearance has been strongly increasing. The minimally processed and easy-to-eat foods are also requested. Currently, there is an escalating tendency to employ environmentally friendly packaging materials with the intention of substituting nondegradable materials, thus reducing the environmental impact resulting from waste accumulation. To address the environmental issues, and concurrently extend the shelf life and food quality while reducing packaging waste has catalyzed the exploration of new bio-based packaging materials such as edible and biodegradable films. One of the approaches is to use renewable biopolymers such as polysaccharides, proteins, gums, lipids, and their derivatives, from animal and plant origin as described previously, not only to form capsules but also to form films and coatings able to encapsulate particular biomolecules. Such biodegradable/edible packaging/coatings not only ensure food safety but at the same time are good source of nutrition.

11.6.1 Nutraceuticals

In parallel with natural foods and wrapping materials, consumer demand is more and more focused on healthy foods to provide more than the basic needs. Encapsulation of nutraceuticals is at present a very hot topic. A great number of research articles report on matrix and encapsulation processes. Milk proteins are largely used for encapsulation and controlled delivery using the micelle structures or any other nanostructure (Semo et al., 2007; Livney, 2010). These proteins are often conjugated with polysaccharides to improve the encapsulation and controlled release properties. Maltodextrins and other modified starches are also largely used as encapsulation matrices due to their low cost and ease of use. Cyclodextrins showed high potentiality for encapsulation but are limited in use by their industrial cost.

Several nutraceutical molecules can be incorporated in edible coatings such as vitamins, peptides, polyunsaturated fatty acids, or antioxidants to increase the food nutritional value. The main problem in incorporating nutraceuticals in food is related to stability during storage. These reactive molecules rapidly lose their activity due to oxidation or other chemical reactions. Edible films and coatings are more and more used to protect these active biomolecules from contact with foods. While incorporated in coatings or encapsulated, their bioactive effect is preserved, and nanoencapsulation could even increase the molecule's bioavailability. All above cited techniques are commonly used for nutraceutical encapsulation with good efficiency. The main research topics are now the controlled release of the nutrient, and also, and this is the more complex problem, there are concerns about the bioavailability and the focused release of the active compound (Chen et al., 2006; Kosaraju et al., 2006; Gonnet et al., 2010; Nair et al., 2010). As an example, release of polyunsaturated fatty acids (PUFA) in the brain to limit Alzheimer's disease is a very hot research topic. Polyphenols are used as active compounds to reduce oxidative stress. Fang and Bandhari (2010) reviewed research on the application of polyphenols. The unpleasant taste of most phenolic compounds can be completely masked by encapsulation. The technologies of encapsulation of polyphenols are commonly spray-drying, coacervation, liposome entrapment, inclusion complexion, cocrystallization, nanoencapsulation, freeze-drying, yeast encapsulation, and use of emulsions. In parallel to the development of preservation techniques, advanced research is being developed on flavonoid functionalization to improve their nutraceutical use. Common research on simultaneous functionalization and encapsulation should be developed to accelerate commercialization.

11.6.2 Antibacterials

Considering antibacterial molecules, postprocess contamination caused by product mishandling and faulty packaging is responsible for about two-thirds of all microbiologically related recalls in the United States, with most of these recalls originated from contamination of ready-to-eat food products. Antimicrobial agents are components that hinder growth of microorganisms, sometimes called food preservatives. According to the definition used by the Commission of the European Communities, preservatives are substances that extend the shelf life of foodstuffs by protecting them against deterioration caused by microorganisms (EU Directive 95/2/EC). Similar rules are applied in the United States, where the U.S. Food and Drug Administration (FDA) defines preservatives as any chemical that when added to food tends to prevent or retard deterioration. Antimicrobials are used in food to control natural spoilage and to prevent or control growth of microorganisms, including pathogenic microorganisms (da Silva Malheiros et al., 2010a, 2010b; Drulis-Kawa and Dorotkiewicz, 2010).

Natural antimicrobials can be defined as substances produced by living organisms in their fight with other organisms for space and their competition for nutrients. The main sources of these compounds are plants (secondary metabolites in essential oils and phytoalexins), microorganisms (bacteriocins and organic acids), and animals (lysozyme from eggs, lactoferrins from milk). Across the various sources, the same types of active compounds can be encountered (e.g., enzymes, peptides, and organic acids). Reducing the need for antibiotics, controlling microbial contamination in food, improving shelf-life extension technologies to eliminate undesirable pathogens, delaying microbial spoilage, decreasing the development of antibiotic resistance by pathogenic microorganisms, or strengthening immune cells in humans are some of the benefits (Tajkarimi et al., 2010). Most approved food antimicrobials have limited application due to pH or food component interactions. They are amphiphilic and can solubilize or be bound by lipids or hydrophobic proteins in foods, making them less available to inhibit microorganisms in the food product.

The term *bacteriocin* is mostly used to describe the small, heat-stable cationic peptides synthesized by Gram-positive bacteria, namely lactic acid bacteria (LAB), which display a wider spectrum of inhibition (Cotter et al., 2005). The bacteriocins produced by LAB offer several desirable properties that make them suitable for food preservation. They are generally recognized as safe (GRAS) substances, not active and nontoxic on eukaryotic cells, become inactivated by digestive proteases, and have little influence on the gut microbiota. Bacteriocins are usually pH and heat-tolerant, and they have a relatively broad antimicrobial spectrum against many food-borne pathogenic and spoilage bacteria.

Nisin has been increasingly used as a bio-preservative for direct incorporation in food as well as in active/edible films. Nisin effectively inhibits Gram-positive bacteria and outgrowth spores of *Bacillus* and *Clostridium*. If nisin is efficient against particular microorganisms, its activity rapidly decreases as it hydrolyses in the food product and bacterial inactivation stops. All studies showed a restart of bacterial growth a few days after nisin incorporation. Encapsulation and controlled release is an efficient way to avoid this resumption of bacterial growth, because active nisin is slowly delivered into the food or onto the food surface. In a recent study for edible films (Sebti et al., 2007) using HPMC/chitosan and incorporating the pure nisin, the author evaluated the effect of nisin on the physical characteristics of films.

11.6.3 Edible materials for nutraceuticals or antibacterial molecule encapsulation

The matrix used for nutraceuticals or antimicrobial molecules is of prime importance to allow good preservation or controlled release of these active compounds.

11.6.3.1 HPMC

Cellulose-based materials are being widely used as they offer advantages like edibility, biocompatibility, barrier properties, and aesthetic appearance as well as being nontoxic, nonpolluting, and having low cost (Vasconez et al., 2009). Hydroxypropyl methylcellulose (HPMC) edible films are attractive for food applications because HPMC is a readily available nonionic edible plant derivative shown to form transparent, odorless, tasteless, oil-resistant, water-soluble films with very efficient oxygen, carbon dioxide, aroma, and lipid barriers, but with moderate resistance to water vapor transport. HPMC is used in the food industry as an emulsifier, film former, protective colloid, stabilizer, suspending agent, or thickener. HPMC is approved for food uses by the FDA (21 CFR 172.874) and the EU (EU, 1995); its safety in food use has been affirmed by the Joint Food and Agriculture Organization (FAO)/World Health Organization (WHO) Expert Committee on Food Additives (JECFA) (Burdock, 2007). The tensile strength of HPMC films is high, and flexibility is neither too high nor too fragile, which make them suitable for edible coating purposes (Imran et al., 2010b).

11.6.3.2 PLA

As a GRAS and biodegradable material, and also because of its biosorbability and biocompatible properties in the human body, polyacetic acid (PLA) and its copolymers (especially polyglycolic acid) attracted the pharmaceutical and medical scientist researchers as a carrier for releasing various drugs and agents like bupivacaine and many others. In food domains, little research has been done on the suitability of PLA as an active packaging polymer. PLA is a new corn-derived polymer and needs time to be an accepted and effective active packaging material in the market.

Van Aardt et al. (2007) studied the release of antioxidants from loaded poly (lactide-co-glycolide) (PLGA) (50:50) films with 2% α-tocopherol, and a combination of 1% butylated hydroxytoluene (BHT) and 1% butylated hydroxyanisole (BHA) into water, oil (food stimulant: Miglyol 812), and milk products at 4 and 25°C in the presence and absence of light. They concluded that in a water medium, PLGA (50:50) showed hydrolytic degradation of the polymer, and release of BHT into the water. In Miglyol 812, no degradation or antioxidant release took place, even after 8 weeks at 25°C. Milkfat was stabilized to some extent when light exposed dry whole milk and dry buttermilk was exposed to antioxidant loaded PLGA (50:50). They also suggested potential use of degradable polymers as a unique active packaging option for sustained delivery of antioxidants, which could be of benefit to the dairy industry by limiting the oxidation of high-fat dairy products, such as ice cream mixes.

11.6.4 Plasticizers

Plasticizers impressively affect the physical properties of biopolymer films (Zhang and Han, 2008). The plasticizer helps to decrease inherent brittleness of films by reducing intermolecular forces, increasing the mobility of polymer chains, decreasing the glass transition temperature of these materials, and improving their flexibility (Zhang and Han, 2008; Galdeano et al., 2009). Thus, it is important to study the effect of commonly used polyol glycerol on the homogenous dispersion of Nisaplin® (nisin, salt, and milk solid) for the formation of composite active films of improved quality. However, plasticizers generally cause increased water permeability, so they must be added at a certain level to obtain a film with desired flexibility, thickness, and transparency without significant decrease of mechanical strength and barrier properties to mass transfer (Möller et al., 2004; Jongjareonrak et al., 2006; Brindle and Krochta, 2008).

11.6.5 Molecular diffusion from film and coating to food surface

Migration from capsules included in films or coatings can be represented as shown in Figure 11.8.

Diffusion and erosion lead to slow release of the antibacterials, allowing food stabilization for a longer period than by adding the antibacterials directly into the food. Quantitative measurement of the rate at which a diffusion process occurs is usually expressed in terms of diffusivity (also called the *diffusion coefficient*), expressed in $m^2\ s^{-1}$. The classical theory used to model the diffusion process is based on Fick's laws (Crank, 1975; Stannet, 1978). Diffusion in a homogeneous media is based

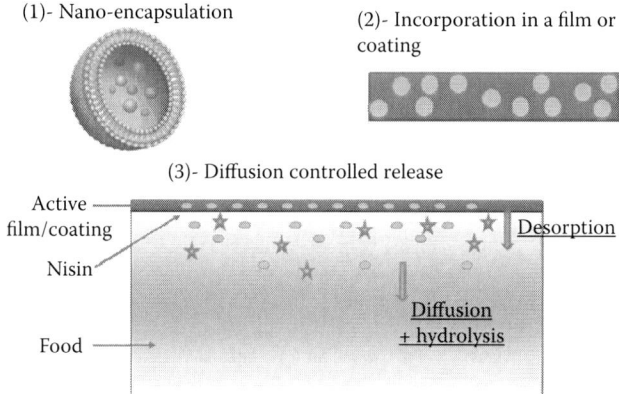

Figure 11.8 Nanoencapsulation, film/coating inclusion, and controlled release into food.

upon the assumption that the rate of transfer, R, of a migrant passing perpendicularly through the unit area of a section is proportional to the concentration gradient between the two sides of the packaging:

$$R = \frac{dM}{A\,dt} = -D\frac{dC}{dx}$$

where D is the diffusion coefficient (m² s⁻¹) ; A is the film area (m²).

In general, D is a function of the local diffusant concentration, C (g m⁻³), t is time (s), and x is the thickness of the film or coating (m). The amount of package components that may migrate from a packaging material into liquid or solid food depends on the chemical and physical properties of food and polymer. Various factors like migrant concentration, molecular weight, solubility, diffusivity, partition coefficient between polymer and food, time, temperature, polymer and food composition, and structures (density, crystallinity, chain branching) are the main controlling factors in migration.

Legally, polymers for packaging are regulated through global or specific migration levels. Global migration measures the total amount of all compounds migrating into food simulants independently of migrant composition. Specific migration concerns a given migrant. Several studies have measured global and specific migration from packaging materials to foods (Baner et al., 1991; Jamshidian et al., 2010). In the case of active films, the global migration limit does not apply, because active compound migration is required. Regulations are now more precise about the possible activity of the packaging and coating.

11.6.6 Chemical structure of migrant and matrix state

The chemical structure of an encapsulated molecule is an important parameter that can influence the partition coefficient and then the controlled release into a food. Alcohols and short-chained esters had higher partition coefficients in the oil/polymer system, than in the water/polymer system. Several studies have attempted to model the relationship between the encapsulated molecule, the composition of the food, and the partition coefficient (Arab Tehrany et al., 2004). It is also known that matrix crystallinity and glass transition of the matrix are key factors for an efficient controlled release of an active compound. A controlled transition from glassy to rubbery state (temperature, water activity) leads to the best system for good food preservation. A lot of work still has to be done to allow perfect control of an active compound release.

11.7 Conclusion

A hot topic in the functional food and pharmaceutical industry is the efficient encapsulation of high value-added ingredients, such as polyunsaturated fatty acids, flavors, vitamins, and health-promoting ingredients, in relation to improved functionalities. In addition, because many of the most popular nutritional ingredients on the market today have unpleasant sensory characteristics, keeping the objectionable flavors out of products can sometimes be as important as keeping the enjoyable flavors in. Numerous developments have been made in the field of encapsulated food flavors. This is because of several favorable properties of the encapsulated form of flavors: ease in handling and mixing; stability against air, light, and evaporation; masking of undesirable tastes and aromas; and delivery of ingredients at the desired stage and at a specifically targeted release sites. Advances in the development of new wall materials and microencapsulation methods have paved the way for value-added ingredients of higher quality, consistency, and enhanced performance and improved prices. Each encapsulation process, generally developed to solve a particular problem encountered by product development, presents advantages and disadvantages. The relationships among problems, capabilities, and encapsulation methods were presented. Microencapsulation by spray-drying is the most economical and flexible way that the food industry can encapsulate ingredients. Thus, this technology is now becoming available to satisfy the increasingly specialized needs of the market. In addition, the fluid-bed process is a promising encapsulation technique for large-scale production of flavor powders to be applied in the food industry. The choice of an appropriate technique of encapsulation depends on the properties of the compounds, the degree of stability required during storage and processing, the properties of the food components, the specific release properties required, the maximum obtainable molecule load in the powder, and the production cost.

References

Anandaraman, S., Reineccius, G.A. (1986). Stability of encapsulated orange peel oil. *Food Technology*, 11: 88–93.

Arab Tehrany, E., Desobry, S. 2004. Partition coefficient in food/packaging systems. *Food Additives and Contaminants,* 21(12): 1186–1202.

Atarés, L., De Jesús, C., Talens, P., Chiralt, A. 2010. Characterization of SPI-based edible films incorporated with cinnamon or ginger essential oils. *Journal of Food Engineering,* 99: 384–391.

Baner, A.L. 1991. Prediction of solute partition coefficients between polyolefins and alcohols using the regular solution theory and group contribution methods. *Industrial and Engineering Chemistry Research*, 30(7): 1506–1515.

Becker, K., Koszinowski, J., Piringer, O. 1987 Permeation von Riech- und Aromastoffen Durch Polyolefine, *Parfurmerie und Kosmetik*, 68: 268–278.

Ben Arfa, A., Chrakabandhu, Y., Preziosi-Belloy, L., Chalier, P., Gontard, N. 2007. Coating papers with soy protein isolates as inclusion matrix of carvacrol. *Food Research International*. 40: 22–32.

Bertolini, A.C., Siani, A.C., Grosso, C.R.F. 2001. Stability of monoterpenes encapsulated in gum arabic by spray-drying. *Journal of Agricultural and Food Chemistry*, 49: 780–785.

Bouwmeester, H., Dekkers, S., Noordam, M.Y., Hagens, W.I., Bulder, A.S., de Heer, C., ten Voorde, S., Wijnhoven, S., Marvin, H., Sips, A. 2009. Review of health safety aspects of nanotechnologies in food production. *Regulatory Toxicology and Pharmacology*, 53(1): 52–62.

Brazel, C.S. (1999). Microencapsulation: offering solution for the food industry. *Cereal Foods World*, 44, 388–393.

Brindle, L.P., Krochta, J.M. 2008. Physical properties of whey protein hydroxypropyl methylcellulose blend edible films. *Journal of Food Science*, 73(9): 446–454.

Burdock, G.A. 2007. Safety assessment of hydroxypropyl methylcellulose as a food ingredient. *Food and Chemical Toxicology*, 45(12): 2341–2351.

Burt, S., 2004. Essential oils: their antibacterial properties and potential applications in foods—a review. *International Journal of Food Microbiology*. 94: 223–253.

Buttery, R.G., Bomben, J.L., Guadagni, D.G., Ling, L.C. 1971. Volatilities of organic flavor compounds in foods. *Journal of Agricultural and Food Chemistry*, 19(6): 1045–1048.

Chen, L., Remondetto, G.E., Subirade, M. 2006. Food protein-based materials as nutraceutical delivery systems *Trends in Food Science and Technology*, 17(5): 272–283.

Cotter, P.D., Hill, C., Ross, P.R. 2005. Bacteriocins: developing innate immunity for food. *Nature Reviews Microbiology*, 3(10): 777–788.

Crank, J. 1975. *Mathematics of Diffusion* (2nd ed.). Clavedon Press, Oxford, England.

Da Silva Malheiros, P., Joner Daroit, D., Pesce da Silveira, N., Brandelli, A. 2010a. Effect of nanovesicle-encapsulated nisin on growth of *Listeria monocytogenes* in milk. *Food Microbiology*, 27(1): 175–178.

Da Silva Malheiros, P., Joner Daroit, D., Brandelli, A. 2010b. Food applications of liposome-encapsulated antimicrobial peptides *Trends in Food Science and Technology*, 21(6): 284–292.

Debeaufort, F. 1994. Étude des Transferts de Matière au Travers de Films d'Emballages: Perméation de l'Eau et de Substances d'Arôme en Relation avec les Propriétés Physico-chimiques des Films Comestibles. PhD Dissertation, ENS.BANA, Université de Bourgogne, Dijon, France.

Debeaufort, F., Voilley, A. 1994. Aroma compound and water vapor permeability of edible films and polymeric packagings, *Journal of Agricultural and Food Chemistry*, 42: 2871–2875.

Debeaufort, F., Voilley, A. 1995. Methylcellulose-based edible films and coatings: 1. Effect of plasticizer content on water and 1-octen-3-ol sorption and transport, *Cellulose*, 2: 1–10.

Debeaufort, F., Tesson, N., Voilley, A. 1995. Aroma compounds and water vapour permeability of edible films and polymeric packagings. In *Food and packaging materials—Chemical interactions*, Ackermann, P., Jägerstad, M., and Ohlsson, T. (eds.), The Royal Society of Chemistry, Cambridge, UK, pp. 169–175.

Debeaufort, F., Quezada-Gallo, J.A., Voilley, A. 2002. Edible films and coatings as aroma barrier. In *Protein-based edible films and coatings,* Gennadios, A. (eds.), CRC Press, Boca Raton, FL, Chap 24, pp. 579–600.

Del Nobile, M.A., Conte, A., Incoronato, A.L., Panza, O. 2008. Antimicrobial efficacy and release kinetics of thymol from zein films. *Journal of Food Engineering*, 89: 57–63.

Djordevic, D., Cercaci, L., Alamed, J., McClements, D.J., Decker, E.A. (2008). Chemical and physical stability of protein- and gum arabic- stabilized oil-in-water emulsions containing limonene. *Journal of Food Science*, 73: C167–C172.

Drulis-Kawa, Z., Dorotkiewicz-Jach, A. 2010. Liposomes as delivery systems for antibiotics. *International Journal of Pharmaceutics*, 387(1–2): 187–198.

Embuscado, M.E., Huber, K.C. 2009. *Edible films and coatings for food applications*, Springer Science, New York. pp. 403

Fabra, M.J., Hambleton, A., Talens, P., Debeaufort, F., Chiralt, A., Voilley, A. 2009. Influence of interactions on the water and aroma permeabilities of iota-carrageenan-oleic acid-beeswax edible films used for flavour encapsulation. *Carbohydrate Polymers*, 76: 325–332.

Fabra, M.J., Hambleton, A., Talens, P., Debeaufort F., Chiralt, A., Voilley, A. 2008. Aroma barrier properties of sodium caseinate-based edible films. *Biomacromolecules*, 9(5): 9, 1406–1410.

Fabra, M.J., Chambin, O., Assifaoui, A., Debeaufort, F. 2011. Influence of temperature and salt concentration on the release in liquid media of aroma compounds encapsulated in edible films. Submitted to *Journal of Controlled Release*.

Fang, Z., Bhandari, B. 2010. Encapsulation of polyphenols—a review. *Trends in Food Science and Technology*, 21(10): 510–523.

Fayoux, S., Seuvre, A.M., Voilley, A. 1997a. Aroma transfers in and through plastic packagings: orange juice and d-limonene. A review. Part 1: orange juice aroma sorption. *Packaging Technology Science*, 10: 69–82.

Fayoux, S., Seuvre, A.M., Voilley, A. 1997b. Aroma transfers in and through plastic packagings: orange juice and d-limonene. A review. Part 2: overall sorption mechanism and parameter—a literature survey. *Packaging Technology Science*, 10: 145–160.

Fenaroli, G. (Ed.). 1995. *Fenaroli's handbook of flavor ingredients*. CRC Press, Boca Raton, FL.

Franz, R. 1993. Permeation of volatile organic compounds across polymer films. Part I: Development of a sensitive test method suitable for high barrier packaging films at very low permeant vapour pressures, *Packaging Technology Science*, 6: 91.

Friedman, M., Henika, P.R., Levin, C.E., Mandrell, R.E. 2004. Antibacterial activities of plant essential oils and their components against *Escherichia coli* O157:H7 and *Salmonella enterica* in apple juice. *Journal of Agricultural and Food Chemistry*, 52: 6042–6048.

Galdeano, M.C., Mali, S., Grossmann, M.V.E., Yamashita, F., Garcia, M.A. 2009. Effects of plasticizers on the properties of oat starch films. *Materials Science and Engineering C*, 29(2): 532–538.

Gennadios, A. 2002. *Protein-based edible films and coatings*. CRC Press, Boca Raton, FL, pp 639.

Godshall, M.A. 1997. How carbohydrate influence food flavor. *Food Technology*, 51(1): 63–67.

Gómez-Estaca, J., Montero, P., Giménez, B., Gómez-Guillén, M.C. 2007. Effect of functional edible films and high pressure processing on microbial and oxidative spoilage in cold-smoked sardine (*Sardina pilchardus*). *Food Chemistry*, 105(2): 511–520.

Gómez-Estaca, J., Montero, P., Fernández-Martín, F., Alemán, A., Gómez-Guillén, M.C. 2009. Physical and chemical properties of tuna-skin and bovine-hide gelatin films with added aqueous oregano and rosemary extracts. *Food Hydrocolloids*, 23(5): 1334–1341.

Gómez-Estaca, J., López de Lacey, A., López-Caballero, M.E., Gómez-Guillén, M.C., Montero, P. 2010. Biodegradable gelatin–chitosan films incorporated with essential oils as antimicrobial agents for fish preservation *Food Microbiology*, 27(7): 889–896.

Gonnet, M., Lethuaut, L., Boury, F. 2010. New trends in encapsulation of liposoluble vitamins. *Journal of Controlled Release*, 146(3): 276–290.

Green, B.K., Scheicher, L. 1955. Pressure Sensitive Record Materials. US Patent no. 2, 217, 507, Ncr C.

Guilbert, S., Gontard, N. 2005. Agro-polymers for edible and biodegradable films: review of agricultural polymeric materials, physical and mechanical characteristics. In *Innovations in food packaging*, Han, J.H. (ed.), Oxford, U.K.: Elsevier Academic Press. pp. 263–276.

Hambleton, A., Debeaufort, F., Beney, L., Karbowiak, T., Voilley, A. 2008. Protection of active aroma compound against moisture and oxygen by encapsulation in biopolymeric emulsion-based edible films. *Biomacromolecules*, 9(3): 1058–1063.

Hambleton, A., Debeaufort, F., Bonnotte, A., Voilley, A. 2009b. Influence of alginate emulsion-based films structure on its barrier properties and on its protection of microencapsulated aroma compound. *Food Hydrocolloids*, 23(8): 2116–2124.

Hambleton, A., Fabra, M.J., Debeaufort, F., Brun-Dury, C., Voilley, A. 2009a. Interface and aroma barrier properties of iota-carrageenan emulsion-based films used for encapsulation of active food compounds. *Journal of Food Engineering*, 93: 80–88.

Hambleton, A., Voilley, A., Debeaufort, F. 2010. Transport parameters for aroma compounds through i-carrageenan and sodium alginate-based edible films. *Food Hydrocolloids*, 10.1016/j.foodhyd.2010.10.010.

Hambleton, A., Perpiñan-Saiz, N., Fabra, M.J., Voilley, A., Debeaufort, F. 2011. The Schroeder paradox or how the state of water affects the moisture transfers through edible films. *Food Chemistry*. In press.

Han, J. 2002. Protein-based edible films and coatings carrying antimicrobial agents. In *Protein-based films and coatings*. Gennadios, A. (ed.), CRC Press, Boca Raton, FL. pp. 485–500.

Heng, L., Van Koningsveld, G.A., Gruppen, H., Van Boekel, M., Vincken, J.P., Roozen, J.P., Voragen, A.G. 2004. Protein-flavour interactions in relation to development of novel protein foods. *Trends Food Science and Technology*, 15(3): 217–224.

Imran, M., El-Fahmy, S., Revol-Junelles, A.M., Desobry, S. 2010a. Cellulose derivative based active coatings: Effects of nisin and plasticizer on physico-chemical and antimicrobial properties of hydroxypropyl methylcellulose films. *Carbohydrate Polymers*, 81: 219–225.

Imran, M., Revol-Junelles, A.-M., Martyn, A., Tehrany, E.A., Jacquot, M., Linder, M., Desobry, S. 2010b. Active food packaging evolution: Transformation from micro- to nanotechnology. *Critical Reviews in Food Science and Nutrition*, 50(9): 799–821.

Jamshidian, M., Arab Tehrany, E., Imran, M., Jacquot, M., Desobry, S. 2010. PLA: production, application and controlled release. *Comprehensive Reviews: Food Science and Food Safety*, 9(5): 552–571.

Jongjareonrak, A., Benjakul, S., Visessanguan, W., Tanaka, M. 2006. Effects of plasticizers on the properties of edible films from skin gelatin of bigeye snapper and brownstripe red snapper. *European Food Research and Technology*, 222(3–4): 229–235.

Kim, Y.D., Moor, C.V. 1996. Microencapsulation properties of gum arabic and several food proteins: spray-dried orange oil emulsion particles. *Journal of Agricultural Food Chemistry*, 44(5): 1314–1320.

Kobayashi, M., Kanno, T., Hanada, K., Osanai, S.I. 1995. Permeability and diffusivity of d-limonene vapor in polymeric sealant films, *Journal of Food Science*, 60: 205–209.

Kosaraju, S.L., D'ath, L., Lawrence, A. 2006. Preparation and characterisation of chitosan microspheres for antioxidant delivery. *Carbohydrate Polymers*, 64(2): 163–167.

Kurek, M., Debeaufort, F. 2010, Development of an antimicrobial coating for packaging films: physico-chemical and microbiological approaches. Doctoral school intermediate PhD report, University of Burgundy.

Laohakunjit, N., Kerdchoechuen, O. 2007. Aroma enrichment and the change during storage of non-aromatic milled rice coated with extracted natural flavour. *Food Chemistry*, 101: 339–344.

Livney, Y.D. 2010. Milk proteins as vehicles for bioactives. *Current Opinion in Colloid and Interface Science*, 15(1–2): 73–83.

Lubbers, S., Landy, P., Voilley, A. 1998. Retention and release of aroma compounds in foods containing proteins. *Food Technology*, 52(5): 68–74, 208–214.

Madene, A., Jacquot, M., Scher, J., Desobry, S. 2006. Flavour encapsulation and controlled release—a review. *International Journal of Food Science and Technology*, 41(1): 1–21.

Marcuzzo, E., Sensidoni, A., Debeaufort, F., Voilley, A. 2010. Encapsulation of aroma compounds in biopolymeric emulsion based edible films to control flavour release. *Carbohydrate Polymers*, 80(3): 984–988.

Marcuzzo, E., Debeaufort, F., Hambleton, A., Sensidoni, A., Tat, L., Beney, L., Voilley, A. 2011. Encapsulation of aroma compounds in biopolymeric emulsion emulsion-based edible films to prevent oxidation. *Food Research International*, accepted 2011.

Martin-Belloso, O., Rojas-Grau, M.A., Soliva-Fortuny, R. 2009. Delivery of flavour and active ingredients using edible films and coatings. In *Edible films and coatings for food applications*, Embuscado, M.E., Huber, K.C. (eds.), Springer Science, New York, pp. 295–313.

Mastromatteo, M., Barbuzzi, G., Conte, A., Del Nobile, M.A. 2009. Controlled release of thymol from zein based film. *Innovative Food Science and Emerging Technologies*, 10(2): 222–227.

McHugh, T.H., Huxsoll, C.C., Krochta, J.M. 1996. Permeability properties of fruit puree edible films. *Journal of Food Science*, 61: 88–91.

Miller, K.S., and Krochta, J.M. 1997. Oxygen and aroma barrier properties of edible films: a review. *Trends in Food Science and Technology*, 8: 228–237.

Miller, K.S., Upadhyaya, S.K., Krochta, J.M. 1998. Permeability of d-limonene in whey protein films. *Journal of Food Science*, 63: 244–247.

Möller, H., Grelier, S., Pardon, P., Coma, V. 2004. Antimicrobial and physicochemical properties of chitosan–HPMC-based films. *Journal of Agricultural and Food Chemistry*, 52(21): 6585–6591.

Monedero, M., Hambleton, A., Talens, P., Debeaufort, F., Chiralt, A., Voilley, A. 2010. Study of the retention and release of n-hexanal from soy protein isolate-lipid composite films. *Journal of Food Engineering*, 100: 128–133.

Nair, H.B., Sung, B., Yadav, V.R., Kannappan, R., Chaturvedi, M.M., Aggarwal, B.B. 2010. Delivery of antiinflammatory nutraceuticals by nanoparticles for the prevention and treatment of cancer. *Biochemical Pharmacology*, 80(12): 1833–1843.

Paik, J.S., Writer, M.S. 1995. Prediction of flavor sorption using the Flory-Huggins equation, *Journal of Agriculture and Food Chemistry*, 43:175–178.

Ponce, A.G., Roura, S.I., del Valle, C.E., Moreira, M.R. 2008. Antimicrobial and antioxidant activities of edible coatings enriched with natural plant extracts: *In vitro* and *in vivo* studies. *Postharvest Biology and Technology*, 49: 294–300.

Pranoto, Y., Salokhe, V.M., Rakshit, S.K. 2005a. Physical and antibacterial properties of alginate-based edible film incorporated with garlic oil. *Food Research International*, 38: 267–272.

Pranoto, Y., Rakshit, S.K., Salokhe, V.M. 2005b. Enhancing antimicrobial activity of chitosan films by incorporating garlic oil, potassium sorbate and nisin. *LWT—Food Science and Technology*, 38(8): 859–865.

Quezada-Gallo, J.A. 1999. Influence de la Structure et de la Composition de Réseaux Macromoléculaires sur les Transferts de Molécules Volatiles (eau et Arômes). Application aux Emballages Comestibles et Plastiques. Ph.D. Dissertation, Université de Dijon, France.

Quezada-Gallo, A., Debeaufort, F., Voilley, A. 1999b. Mechanism of aroma transport through edible and plastic packagings. In *New developments in the chemistry of packaging materials*. Rish, S. (ed.), ACS Books, Dallas, TX, pp. 125–140.

Quezada-Gallo, J.A., Debeaufort, F., Voilley, A. 1999a. Interactions between aroma and edible films. 1. Permeability of methylcellulose and polyethylene films to methyl ketones, *Journal of Agriculture and Food Chemistry*, 47: 108–113.

Raybaudi-Massilia, R., Mosqueda-Melgar, J., Martín-Belloso, O. 2006. Antimicrobial activity of essential oils on *Salmonella Enteritidis*, *Escherichia coli*, and *Listeria innocua* in fruit juices. *Journal of Food Protection,* 69: 1579–1586.

Reineccius, G. 2009. Edible films and coatings for flavour encapsulation. In *Edible films and coatings for food applications*, Embuscado, M.E., Huber, K.C. (eds.), Springer Science, New York, pp. 269–294.

Rojas-Grau, M.A., Avena-Bustillos, R., Friedman, M., Henika, P., Martın-Belloso, O., McHugh, T., 2006. Mechanical, barrier and antimicrobial properties of apple puree edible films containing plant essential oils. Journal of Agricultural and Food Chemistry. 54: 9262–9267.

Rojas-Grau, M., Raybaudi-Massilia, R.M., Soliva-Fortuny, R.S., Avena-Bustillos, R.J., McHugh, T.H., Martın-Belloso, O. 2007. Apple puree-alginate edible coating as carrier of antimicrobial agents to prolong shelf-life of fresh-cut apples. *Postharvest Biology and Technology*, 45: 254–264.

Rupasinghe, H.P., Boulter-Bitzer, J., Ahn, T., Odumeru, J. 2006. Vanillin inhibits pathogenic and spoilage microorganisms *in vitro* and aerobic microbial growth in fresh-cut apples. *Food Research International*, 39: 575–580.

Sánchez-González, L., Vargas, M., Gonzalez-Martinez, C., Chiralt, A., Chafer, M. 2009. Characterization of edible films based on hydroxypropylmethylcellulose and tea tree essential oil. *Food Hydrocolloids,* 23: 2102–2109.

Sánchez-González, L. 2010c. Caracterizacion y aplicacion de recubrimientos antimicrobianos a base de polisacaridos y aceites esenciales. PhD Dissertation, Universidad Politecnica de Valencia, pp 309.

Sánchez-González, L., Chafer, M., Chiralt, A., Gonzalez-Martinez, C. 2010a. Physical properties of edible chitosan films containing bergamot essential oil and their inhibitory action on *Penicillium italicum*. *Carbohydrate Polymers,* 82: 277–283.

Sánchez-González, L., Gonzalez-Martinez, C., Chiralt, A., Chafer, M. 2010b. Physical and antimicrobial properties of chitosan–tea tree essential oil composite films. *Journal of Food Engineering*, 98: 443–452.

Sánchez-González, L., Cháfer, M., González-Martínez, C., Chiralt, A., Desobry, S. 2011. Study of the release of limonene present in chitosan films enriched with bergamot oil in food simulants *Journal of Food Engineering*, 105(1), 138–143.

Sebti, I., Chollet, E., Degraeve, P., Noel, C., Peyrol, E. 2007. Water sensitivity, antimicrobial, and physicochemical analyses of edible films based on HPMC and/or chitosan. *Journal of Agriculture and Food Chemistry,* 55(3): 693–699.

Semo, E., Kesselman, E., Danino, D., Livney, Y.D. 2007. Casein micelle as a natural nano-capsular vehicle for nutraceuticals. *Food Hydrocolloids*, 21(5–6): 936–942.

Shimoni, E. 2009. Nanotechnology for foods: delivery systems. *Global Issues in Food Science and Technology*, 411–424.

Solms, J., Osman-Ismail, F., Beyler, M. 1973. The interaction of volatiles with food components. *Canadian Institute of Food Science and Technology Journal*, 6: A10–A16.

Stannet. 1978. The transport of gases in synthetic polymeric membranes—an historic perspective. *Journal of Membrane Science*, 3: 97–115.

Tajkarimi, M.M., Ibrahim, S.A., Cliver, D.O. 2010. Antimicrobial herb and spice compounds in food. *Food Control*, 21(9): 1199–1218.

Tunc, S., Duman, O. 2011. Preparation of active antimicrobial methyl cellulose/carvacrol/montmorillonite nanocomposite film and investigation of carvacrol release. *LWT—Food Science and Technology*, 44: 465–472.

Van Aardt, M., Duncan, S.E., Marcy, J.E., Long, T.E., O'Keefe, S.F., Sims, S.R. 2007. Release of antioxidants from poly(lactide-co-glycolide) films into dry milk products and food simulating liquids. *International Journal of Food Science and Technology*, 42(11): 1327–1337.

Vasconez, M.B., Flores, S.K., Campos, C.A., Alvarado, J., Gerschenson, L.N. 2009. Antimicrobial activity and physical properties of chitosan-tapioca starch based edible films and coatings. *Food Research International*, 42(7): 762–769.

van Ruth, S.M., King, C., Giannouli, P. 2002. Influence of lipid fraction, emulsifier fraction, and mean particle diameter of oil-in-water emulsions on the release of 20 aroma compounds. *Journal of Agricultural and Food Chemistry*. 50(8): 2365–2371.

Wyler, L., Solms, J. 1982. Starch flavour complexes III. Stability of dried starch-flavor complexes and other dried flavour preparations. *Lebensmittel-Wissenschaft und Technologie*, 15(2): 93–97.

Zhang, Y., Han, J.H. 2008. Sorption isotherm and plasticization effect of moisture and plasticizers in pea starch film. *Journal of Food Science*, 73(7): 313–324.

Zivanovic, S., Chi, S., Draughon, A.F. 2005. Antimicrobial activity of chitosan films enriched with essential oils. *Journal of Food Science*, 70(1): 1145–1151.

12 Overview of pharmaceutical coatings

Anthony Palmieri III

Contents

12.1	Introduction	373
12.2	Sugar coatings	375
12.3	Modern coatings	375
12.4	Coating application	377
12.5	Interaction of polymeric films with drugs and excipients	378
12.6	Suggested readings	380
References		380

12.1 Introduction

Pharmaceutical coatings serve a number of functions, probably the most important of which is to control the release of active pharmaceutical ingredient. Some coatings are employed to mask objectionable taste or to protect the ingredient inside from the environment, either air or water. Another function is to separate two physically or chemically incompatible ingredients packaged together in one tablet or capsule. Coatings are most commonly applied to tablets, the most common pharmaceutical dosage form now in use. Tablets are formed by compression of a powdered mixture of the active ingredients and binders and sometimes other *excipients*, the pharmaceutical term for any ingredient other than the active ingredient. Pharmaceutical coatings are often quite different from food coatings in composition, function, mathematical modeling, method of application, and technical sophistication (Table 12.1).

Coating of pills is an ancient activity. Rhazes, around the year 900, used mucilage of psyllium seed to mask the taste of drugs. Arguably the first coating of a dose form was that of applying a coat of silver or gold to the pill. The earliest reference is attributed to an Arab physician, Avicenna, around the year 1000. He was most likely the first to silver, or gild, pills so that they reputedly had a stronger medicinal effect. In Europe, and to a lesser degree in the United States, pills were covered with silver or gold to make them look nicer, and to mask the taste (Kremer and Urdang, 1976).

The earliest coated dose forms were prepared extemporaneously by druggists, usually for patients of wealth. Many druggists had small equipment to coat the pills. The first successful commercial application of a coating for a dose form appears to have been in 1866 when William Warner, a Philadelphia druggist, began manufacturing sugar-coated tablets. By the late 1890s, many dose forms were available, including a patented enteric coating that contained cellulose nitrate and cellulose acetate. By 1940, stearic acid-phthalate esters were used (Kremer and Urdang, 1976).

Table 12.1 A comparison of typical food and pharmaceutical coatings

	Food coatings	Pharmaceutical coatings
Amount used	Thickness of a fruit coating is 2 μm, and this makes up 0.02% of fruit weight.	Tablet film coatings are much thicker, often about 50 μm. Sugar coatings make up about 40% of pill weight.
Complexity	Typically one uniform layer.	Multiple layers are common.
What is inside and outside the coating	Usually food inside and air outside.	*Before ingestion*: active ingredient inside and air outside. *After ingestion*: active ingredients inside, stomach and intestines outside.
Main functions	Reduce gas (O_2 or water vapor) movement into or out of the food.	*Before ingestion*: mask and impart color, reduce gas movement into the active ingredient. *After ingestion*: mask and impart flavor, reduce gas movement; control release rate of active ingredient.
Important modeling parameters	Permeance of gas diffusion through coatings; coating thickness, and gas effusion through holes in the coating.	*Before ingestion*: same as for food coatings, plus diffusion of active ingredient within the tablet. *After ingestion*: diffusion of active ingredient through wetted (and possibly gelled) coating, including leaking through holes in the coating, possibly driven by osmotic flow.
Permitted ingredients	Generally recognized as safe (GRAS) food additives plus relatively few other ingredients.	Wide range of ingredients, including many synthetic polymers.
Coating technology	Processing to apply coatings must be cheap, gentle, and fast to avoid product damage. Slow development of new technology.	Freedom to use extensive handling, higher temperatures, more ingredients, and higher-cost techniques have led to ever more sophisticated and highly engineered coatings.

Early coatings made use of shellac and pectin, both natural products. Shellac has largely been replaced with other film-formers, and pectin by other gelling agents.

12.2 Sugar coatings

In this process, the tablet is first sealed with a thin coating of hydrophobic material (e.g., polymer, shellac, or zein). Coating with sugar involves uniform deposition of the coating material as well as rapid controlled drying. Both processes occur simultaneously in the coating pan with techniques similar to those used for confectionaries. The pan, once a solid piece open at one end, is now enclosed and perforated along the side panels for increased ventilation and drying power (Kitt, 1988; Rowell, 1950). Sugar syrup is delivered by a pumping system to spray guns mounted inside the pan. Successful coating depends on drying to produce a uniform tablet. This depends on pumping rate, droplet size, pan speed, and drying rate. Tumbling action of the product inside the pan is important to ensure an evenly coated and dried product. Care must be taken to avoid unnecessary or excessive tumbling, or the coating will fracture and break away from the product, dust will develop, and uniformity will suffer. Polishing the now-rounded surface is achieved by glazing the panned tablets with waxes, oils, or shellac dissolved in acceptable solvents (Isganitis, 1988).

Sugar coatings have many advantages and are widely used. The ingredients are cheap and have no regulatory issues. The equipment is simple and inexpensive, and unlike film coatings, the sugar coating process allows reworking of defective product. The final product is well suited to humid conditions and has wide consumer acceptance.

Sugar coatings do have some disadvantages, however. The final product is somewhat variable because the process is more art than science. This variability makes it difficult to control dissolution rates and release rates. More importantly, sugar coatings begin to dissolve immediately upon ingestion, in contrast to film coatings, which can be designed to control the delayed release of active ingredient. For these reasons, modern coatings tend to replace sugar coatings in new dosage forms.

12.3 Modern coatings

Films make up many modern pharmaceutical coatings. The great majority of these are based on cellulose derivatives, or acrylic polymers and copolymers.

The cellulose derivatives include ethyl cellulose, methylcellulose, hydroxypropyl methyl cellulose, hydroxyethyl cellulose, and the like. These have been employed for decades to coat pharmaceutical dose forms. Either granules or the entire dosage form may be coated. All are extremely inexpensive and well studied.

Ethyl cellulose is widely used for delayed or sustained release of the drug by control of diffusion through the coating layer. Hydroxyethyl cellulose acts by increasing the viscosity of the microenvironment. Hydroxypropyl cellulose acts as a gel-forming agent. Hydroxypropyl methyl cellulose, used for delayed release, is arguably the most common cellulose derivative employed as a coating material for pharmaceutical dosage forms. Different blends of these cellulose materials are used to obtain the desired sustained release profile. Phthalate derivatives of cellulose are added for protect against stomach acidity and allow for release of the payload in the higher pH of the intestinal tract.

Chitosan has been used for a long time in cosmetics and is increasingly used as a coating agent. It has also been used as a coating for gene delivery. It is produced from partial deacetylation of chitin. Almost all of its functions depend on the appropriate chain length.

Carbomers are a group of high molecular weight synthetic derivatives of acrylic acid. These are modified to control release of active ingredient by cross-linkage with allyl sucrose or allyl esters of pentaerytol, in order to modify the rheology of the area immediately around the dose form.

Some excipients delay release by causing a viscous layer to form upon contact with water or gastric juices. These compounds are called gelling agents, hydrogels, thickening agents, or viscosity increasing agents. Examples include carbomers, polyethylene glycols, sorbitol, pectin, and pectin derivatives. Pectin, mostly from citrus, has been widely replaced by the other gelling agents.

Plasticizers are commonly used in coating formulations. These impart mobility to the polymeric chain and affect many properties of the film, including flexibility, mechanical properties, and cracking. They also affect permeability and active ingredient release rates. Plasticizers regularly used in pharmaceutical coatings include dibutyl sebacate, diethyl phthalate, triethyl citrate, tributyl citrate, acetyl tributyl citrate, and polyethylene glycol. Polyethylene glycol compounds, which also function as a hydrophilic polishing substance, are used in conjunction with hydrogels to control release. The higher is molecular weight of the polyethylene glycol, the slower is the release of active pharmaceutical ingredient.

Opacifers are excipients added to mask the color of the active ingredient and also to impart desired color, much as they do in ordinary house paint. These also change physical properties of the coating, for example, they reduce the stickiness caused by use of plasticizers. Coatings are colored not only for aesthetics but also to distinguish one capsule or tablet from another. This is especially useful for highly potent compounds.

Gelatin is often used in the pharmaceutical industry and is best known as the material to prepare oral capsules. It is a term used for a mixture of purified protein fractions made from partial acid hydrolysis or by partial alkaline hydrolysis. Gelatin may be obtained from a variety of animal sources, such as horses, cows, or pigs. Fish gelatin is also available but not used very often. As well as providing form for oral capsules, gelatin may be used as a sustained releasing agent for controlled release products.

Recently developed techniques for controlled release of active ingredients make simultaneous use of diffusion and osmotic pressure through coatings that have opening hole sizes controlled by nanotechnology or selective leaching of ingredients (Loh et al., 2010; Marucci et al., 2010). Such sophisticated technology will probably never be used for food coatings because of cost and also because such a high degree of control is not needed for foods.

As mentioned in Table 12.1, one very big difference between pharmaceutical and food coatings is the number of permitted ingredients, which vary from country to country. Active ingredients are discussed in The Food Chemical Codex, the pharmacopoeias of Europe, China, India, Japan, and Great Britain, and in Title 21 of the U.S. Food and Drug Administration (FDA), Code of Federal Regulations (CFR).

In addition to these printed lists, the FDA has made available on the Internet their *Inactive Ingredient Search for Approved Drug Products*. The Web site is www.accessdata.fda.gov/scripts/cder/iig/index.cfm. This Web site seems of much value, especially because it is unique. No similar list is available for the rest of the world.

12.4 Coating application

Several types of coating equipment may be used to film coat pellets or granules with coatings. These include fluid-bed equipment, perforated coating pans, and conventional coating pans. Application by coating pans was discussed in Section 12.2; thus, in this section we will discuss fluid-bed coating and compare the performances of three major fluid-bed processes.

Fluid-bed processes are more efficient at removing water due to high air throughput and are characterized by short processing times. The setups in fluid-bed family include top spray (granulation or conventional mode), bottom spray (Wurster), and tangential spray (rotary granulator) (Figure 12.1).

In the top-spray system, the coating material is sprayed downward on to the fluid bed such that as the solid or porous particles move to the coating region, they become encapsulated. Increased encapsulation efficiency and the prevention of cluster formation are achieved by opposing flows of the coating materials and the particles. Dripping of the coated particles depends on the formulation of the coating material. Top-spray fluid-bed coaters produce higher yields of encapsulated particles than either bottom or tangential sprays.

The bottom spray is also known as "Wurster's coater" in recognition of its development by Wurster (1953). This technique uses a coating chamber that has a cylindrical nozzle and a perforated bottom plate. The cylindrical nozzle is used for spraying the coating material. As the particles move upward through the perforated bottom plate and pass the nozzle area, they are encapsulated by the coating material. The coating material adheres to the particle surface by evaporation of the solvent or cooling of the encapsulated particle. This process is continued until the desired thickness and weight are obtained. Although it is a time-consuming process, the multilayer coating procedure helps in reducing particle defects.

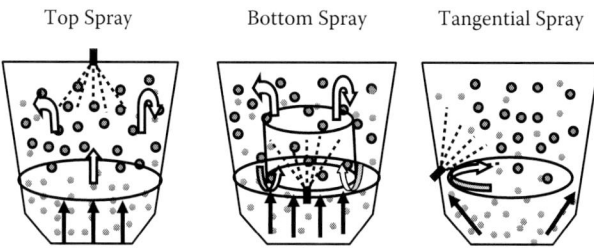

Figure 12.1 A fluid-bed coater. From left: top spray, bottom spray, and tangential spray.

The tangential spray consists of a rotating disc at the bottom of the coating chamber, with the same diameter as the chamber. During the process, the disc is raised to create a gap between the edge of the chamber and the disc. The tangential nozzle is placed above the rotating disc through which the coating material is released. The particles move through the gap into the spraying zone and are encapsulated. As they travel a minimum distance, there is a higher yield of encapsulated particles.

12.5 Interaction of polymeric films with drugs and excipients

For pharmaceutical coating development, controlled release of drug ingredient after ingestion is very important. Figures 12.2, 12.3, and 12.4 show examples of how release may be altered over a considerable range by selection of polymer, plasticizer, or curing time.

Figure 12.2 shows how polymer type affects the release of drug substances. The release of warfarin in pH 7.4 phosphate buffer from tablets cast by Eudragit S is much faster than with those cast by Eudragit RL, and Eudragit E is the slowest one (Figure 12.2) (Lin et al., 1994).

Figure 12.3 shows how various plasticizers affect drug release from ethylcellulose latex coating. When coated with Citroflex-2 as plasticizers, phenylpropanolamine hydrochloride (PPA HCl), a model drug released two to three times faster than Citroflex A-4 and dibutyl sebacate (DBS) as plasticizer within 6 hours (Figure 12.3) (Carlin et al., 2008).

Figure 12.4 shows how curing and storage time influence drug release. During storage of coated pellets, physical aging or enthalpy relaxation occurs which influences the drug-release rate. Generally, the longer is the storage time, the faster the pellets release active ingredients. Curing after coating and storage condition can

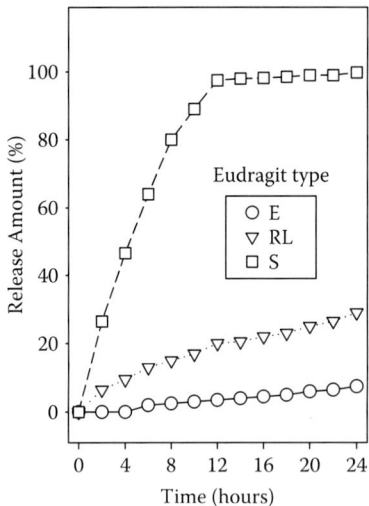

Figure 12.2 Release behavior of warfarin from tablets cast by different Eudragit® resin films. (Modified from Lin, S.Y., Cheng, C.L., and Perng, R.I., *Eur. J. Pharm. Sci.*, 1, 313, 1994.)

Overview of pharmaceutical coatings 379

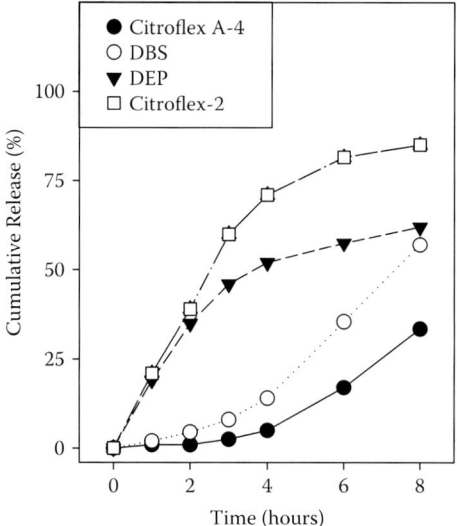

Figure 12.3 Effect of four different plasticizers (30%) on drug release from ethylcellulose latex. (DBS, dibutyl sebacate; DEP, dithyl phthalate.) (Modified from Carlin, B., Li, J.-X., and Felton, L.A., *Aqueous Polymeric Coatings for Pharmaceutical Dosage Forms*, 3rd ed., Informa Healthcare USA, New York, 2008, p. 26.)

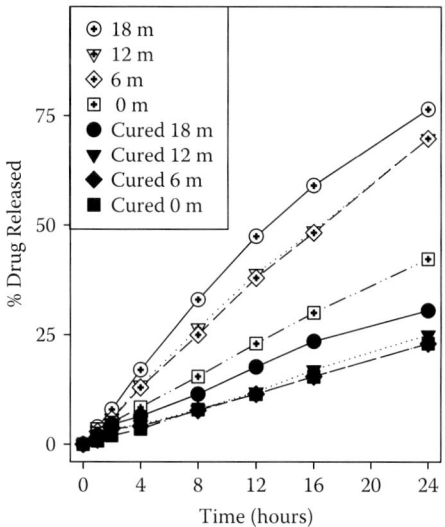

Figure 12.4 Influence of curing (60°C for 1 hour [solid marks] or 0 hour [open marks]) and storage time (0 to 18 months, m) on the theophylline release from Aquacoat®/ATBC-coated pellets (high-dose pellets) in 0.1N HCl. (ATBC, acetyl tributyl citrate.) (Modified from Wesseling, M., and Bodmeier, R., *Pharm. Dev. Tech.*, 6, 325, 2001.)

affect the aging speed. Curing is a process for film-forming after completion of the coating process. Figure 12.4 shows that as extended storage time, drug release rate remarkably increased, and curing decreased the release rate (Wesseling and Bodmeier, 2001).

12.6 Suggested readings

While this discussion is intended for nonpharmaceutical formulators, the reader is directed to a number of well-recognized textbooks and other sources for more information.

Handbook of Pharmaceutical Excipients, Sixth Edition, 2009, Published by the Pharmaceutical Press and the American Pharmacist Association *(Rowe et al., 2009)*. A necessary and premier reference for excipients and formulations. The science and art of pharmaceutical coatings have been the subject of many textbooks.

History of Pharmacy, Fourth Edition, 1976, JB Lippincott Co. (Kremer and Urdang, 1976). A very useful text to learn of the history of coated dosage forms.

Novel Drug Delivery Systems, Volume 14, 1982, Marcel Decker (Chien, 1982). An excellent discussion on basic systems as well as theoretical considerations.

Martin's Physical Pharmacy and Pharmaceutical Sciences, Fifth Edition, Lippincott, Williams and Wilkins (Sinko, 2006). The premier text for physical chemical considerations.

Modern Pharmaceutics, Third Edition, 1996, Marcel Decker (Banker and Rhodes, 1996). A good resource for formulators.

Remington's Pharmaceutical Sciences, 18th Edition, 1996, Mack Publishing Co. (Gennaro *1990*). An excellent starting point.

References

Banker, G.S., and C.T. Rhodes. 1996. *Modern Pharmaceutics, 3rd ed.* Marcel Dekker, New York.
Carlin, B., J.-X. Li, and L.A. Felton. 2008. Pseudolatex dispersions for controlled drug delivery. In: McGinity, J.W., and L.A. Felton (Eds.), *Aqueous Polymeric Coatings for Pharmaceutical Dosage Forms. 3rd ed.* Informa Healthcare USA, New York, pp. 1–46.
Chien, Y.W. 1982. *Novel Drug Delivery Systems*. Marcel Dekker, New York.
Gennaro, A.G. 1990. *Remington's Pharmaceutical Sciences. 18th ed.* Mack, Easton, PA.
Isganitis, D.K. 1988. *Polishing: A Review of Processes and Techniques. The Manufacturing Confectioner* October: 75–78.
Kitt, J.S. 1988. *Panning Problems—Causes and Remedies.* The Manufacturing Confectioner October: 57–62.
Kremer, E., and G. Urdang. 1976. *The History of Pharmacy.* Lippincott, Philadelphia, PA.
Lin, S.Y., C.L. Cheng, and R.I. Perng. 1994. Solid state interactions studies of drug-polymers (II): Warfarin-Eudragit® E, RL or S resins. *Eur. J. Pharm. Sci.* 1: 313–322.

Loh, X.M., P. Peh, S. Liao, C. Sng, and J. Li. 2010. Controlled drug release from biodegradable thermoresponsive physical hydrogel nanofibers. *J. Controlled Release*. 143: 175–182.

Marucci, M., G. Ragnarsson, B. Nilsson, and A. Axelsson. 2010. Osmotic pumping release from ethyl-hydroxypropyl-cellulose-coated pellets: A new mechanistic model. *J. Controlled Release* 142: 53–60.

Rowe, R.C., P.J. Sheskey, and M.E. Quinn. 2009. *Handbook of Pharmaceutical Excipients, 6th ed.* Pharmaceutical Press, London.

Rowell, T.H. 1950. The Art of Coating Tablets. Industry publication of F.J. Stokes Co., Chicago, IL, 7–33.

Sinko, P.J. (Ed.). 2006. *Martin's Physical Pharmacy and Pharmaceutical Sciences, 5th ed.* Lippincott, Williams and Wilkins, Philadelphia, PA.

Wesseling, M., and R. Bodmeier. 2001. Influence of plasticization time, curing conditions, storage time, and core properties on the drug release from aquacoat-coated pellets. *Pharm. Dev. Tech.* 6: 325–331.

Wurster, D.E. 1953. Method of applying coating onto edible tablets or the like. US patent 2 648 609.

13 Regulatory aspects of coatings

Guiwen A. Cheng and Elizabeth A. Baldwin

Contents

13.1	Introduction	383
13.2	Major international and national food safety administration	385
	13.2.1 Food and Agriculture Organization of the United Nations (FAO) and World Health Organization (WHO)	385
	13.2.2 European Union	387
	13.2.3 United States	387
	13.2.4 Japan	388
	13.2.5 Canada	389
	13.2.6 Australia and New Zealand	389
13.3	Substances in food and food products	389
	13.3.1 Food and food ingredients	389
	13.3.2 Food additives	393
	13.3.3 Processing aids	393
	13.3.4 Food contaminants	393
	13.3.5 Nanotechnology	400
	13.3.6 Substance classification variations	401
13.4	Qualification of substances as food additives	401
	13.4.1 Food safety	401
	13.4.2 Technical function need	402
13.5	Major food additive standards	402
	13.5.1 FAO/WHO's Codex Alimentarius	402
	13.5.2 EU's directives on food additives	402
	13.5.3 U.S. FDA's Code of Federal Regulation	405
	13.5.4 Japan's Specifications and Standards for Food Additives	406
	13.5.5 Canada's Food and Drug Regulations	406
	13.5.6 Australia and New Zealand Food Standards Code	406
13.6	Substances permitted for fruit and vegetable coatings	407
	13.6.1 Coating ingredients	407
	13.6.2 Organic ingredients	407
13.7	Conclusions	413
References		413

13.1 Introduction

Substances used to prepare coatings for food and pharmaceutical products are considered food additives even though the coating may or may not eventually be consumed (as in fruits where the peel is not eaten). Therefore, the ingredients used for coatings are generally subjected to the same level of regulation as other food additives. The functions of coatings, as they are used on finished products, are numerous, even though uses are not as broad as those listed in the general food additives section of most regulations.

Coatings can be used for preserving product shelf life, enhancing appearance (color and gloss) (Bai et al., 2002; Hagenmaier and Grohmann, 1999), improving handling through reduction of stickiness, as carriers of useful ingredients such as antibrowning agents, and so on (Baldwin, 1994; Cuppett, 1994). These effects are the results of the individual functions and interactions of carefully selected ingredients based on their chemical and physical properties. Coating ingredients can be found in many food additive substance groups with broad designated technical functions but could also be intentionally separated as an independent group for regulation, for example, U.S. Food and Drug Administration (FDA), Code of Federal Regulations (CFR) title 21 part 175.

To understand regulation of food and pharmaceutical coatings ingredients, one must understand the development of food regulations. Food safety regulation is vital in modern societies as food supply chains have dramatically changed over the last century, becoming more global. Pursuit of food product innovation, convenience, and economical success in food processing has led us to an era of abundant foods with numerous artificial ingredients, massive commercial preparations, and heavy chemical inputs for cultivation, manufacturing, processing, and packaging. As a result, consumers' health and safety, as affected by consumption of these food items, has become a subject of increasing concern.

An attempt to regulate food identity and standards can be traced back to as early as the 1800s. The Codex *Alimentarius Austriacus* was a collection of standards and product descriptions used in the Austro-Hungarian Empire, and this may be one of the first known food regulation documents. During the colonial era in the United States, regulation of food was by the state and local governments. The first federal food protection law was enacted by Congress in 1883 to prevent the importation of adulterated tea. Soon after in 1896, the oleo-margarine statute was passed due to pressure from the dairy industry concerning adulterated butter and fats colored to look like butter (Fortin, 2009). During the 1900s, many more industrialized countries in Europe and North America started to establish independently strong regulatory systems on food safety due to the rapid spread of food commercialization and trade. This was followed by rapid progress in food science and technology and an increasing awareness of food safety by the public. In 1883, Dr. Harvey Wiley, chief chemist of the U.S. Bureau of Chemistry (part of the U.S. Department of Agriculture, USDA), campaigned for a national food and drug law. He galvanized public awareness, leading to advances in food safety and the signing of the Pure Food and Drug Act and the Meat Inspection Act, both in 1906, which began the modern era of U.S. food regulation (Fortin, 2009). This legislation was later followed by the Food, Drug, and Cosmetic Act (FDCA) of 1938, which required premarketing approval and proof of the safety of drugs; the Food Additives Amendment to the FDCA in 1958; and the Delaney clause, which forbade use of substances in food found to be carcinogenic in laboratory animals, including color additives. Then came the Low Acid Food Processing Regulations in 1973 after outbreaks of botulism in canned foods, and the Federal

Anti-Tampering Act in 1983 for packaged consumer products. In the 1980s, there was growing interest in nutrition and health leading to the Nutritional Labeling and Education Act, requiring nutritional labeling, in 1990 (Fortin, 2009). Federal responsibility for regulation of food in the United States has mostly been delegated to FDA and USDA; however, the Environmental Protection Agency (EPA) and the Alcohol and Tobacco Tax and Trade Bureau (TTB) also share regulation for drinking water, pesticide residues, and use of alcohol, and the National Marine Fisheries Service (NMFS) oversees fish and seafood products and has a Seafood Inspection Program that inspects and certifies fishing vessels, seafood processing plants, and retail facilities.

True food safety and globalization of food supplies, however, cannot be achieved without the help of universal food standards (Huggett et al., 1998). The desire to harmonize food standards and protect the health of consumers while ensuring fair food trade practices led to the creation of the Association of Food and Drug Officials (AFDO), Food and Agriculture Organization of the United Nations (FAO), and the World Health Organization (WHO) in the late 1940s. The FAO has the responsibility of covering nutrition and associated international food standards, while the WHO is organized to cover human health and, in particular, mandate and establish food standards. Since inception, the two international organizations have taken on the responsibility of developing international food regulatory standards. Despite variation in historical diet traditions, regional technology, commercial development, and commencement and progression of regionalized food regulation efforts, international efforts to improve food safety have progressed significantly in recent years. The concept of risk assessment, risk management, and risk communication in food risk analysis as well as independent scientific work on risk assessment are widely accepted now in the international community. The adoption of this concept no doubt will further facilitate the progress of international food safety progress. Nevertheless, the above organizations provide opportunities for government officials to exchange information, with more formal cooperative arrangements put into memoranda of understandings (MOUs) that are approved on the U.S. side by the U.S. Secretary of State (Fortin, 2009). A list of FDA international agreements is found at www.fda.gov/oia/. The FDA Modernization Act of 1997 required the FDA to proceed to accept mutual recognition agreements to reduce the burden of regulation and harmonize regulatory requirements to facilitate trade (FDCA Sec 803) (Fortin, 2009).

13.2 Major international and national food safety administration

13.2.1 Food and Agriculture Organization of the United Nations (FAO) and World Health Organization (WHO)

Prior to 1953, the governing body of the WHO, the World Health Assembly, proposed that FAO and WHO should conduct relevant studies on chemicals used in food due to their widening use, which presented a new issue for public health safety. One such study identified the use of food additives (including ingredients of coatings) as a critical factor. As a result, these two organizations convened the first joint FAO/WHO Conference on Food Additives held in Geneva in 1955. This marked the initiation of an effort to track and regulate food additives on a global basis. The Conference recommended to the Directors-General of FAO and WHO that one or more expert committees should be convened to address the technical and administrative aspects of chemical additives and their safety in food. This recommendation provided the basis for the first meeting of the Joint FAO/WHO Committee on Food Additives in Rome, in December 1956. One of the expert committees is the Joint FAO/WHO Expert Committee on Food Additives (JECFA).

In 1960, at the first FAO European Regional Conference, there was a desire to convert the regional food standards, proposed as European Codex Alimentarius (Latin for "food code" or "food book") in 1954 into an international standard. It invited the FAO Director-General to submit proposals for a joint FAO/WHO program on food standards to the FAO Conference. In 1961, with the support of WHO, the United Nations Economic Commission for Europe (UNECE), the Organization for Economic Co-operation and Development (OECD), and the Council of the Codex Alimentarius Europaeus, at a FAO conference established the Codex Alimentarius and resolved to create an international food standards program. The work of the Council of the Codex Alimentarius Europaeus was then resolved to be taken over by FAO and WHO in pursuit of an international food code book. In 1962, the Joint FAO/WHO Food Standards Conference requested that the newly created Codex Alimentarius Commission implement a joint FAO/WHO food standards program and create the Codex Alimentarius. In 1963, the World Health Assembly, WHO's highest governing body, approved establishment of the Joint FAO/WHO Food Standards Program and adopted the Statutes of the Codex Alimentarius Commission. The Codex Alimentarius Commission membership included 174 member countries and represented 99% of the world population as of 2008.

Both FAO and WHO have complementary functions in selecting experts to serve on the Committee. The FAO is responsible for selecting members with chemical expertise for the development of specifications for the identity and purity of food additives. The WHO is responsible for selecting members for the toxicological evaluations of additives in order to establish Acceptable Daily Intakes (ADI). Both FAO and WHO invite experts, called rosters, who are responsible for assessing risk exposure.

The substances submitted by member states for consideration are first subject to priority determination by the Codex Committee on Food Additives (CCFA). The scientific assessment process is then conducted by JECFA, as the expert body responsible for establishing specifications of identity and purity and performing the risk assessment, to provide advice mainly on the allocation of ADI for food additives. To do this, they follow the risk analysis principles established by CCFA. The JECFA also adopts the specifications for the food additives proposed by the FAO. The CCFA's major task is to decide the permitted uses of food additives in all foods, using the JECFA's allocated ADI to establish appropriate limits on an additive's use in food. The CCFA then makes recommendations on the specifications for food additives and flavorings (starting in 2006) to the Codex Alimentarius Commission to be included as Codex Specifications.

13.2.2 European Union

The European Union's (EU) food additive regulation is under Regulation (EC) 1331/2008 (EU, 2008a). This regulation set out the general principles and requirements of food law. The current regulations were the result of proposals by the European Commission in 2006 to develop a new single regulation for all food additives and regulation on food flavoring and food enzymes not used as food additives. The regulation on food additives (EU, 2008b) includes provision on additives currently in different directives, including the Sweeteners Directive (European Parliament and council Directive 94/35/EC) and the Colors Directive (European Parliament and Council Directive 94/36/EC). The regulations for food additives (Regulation EC 1333/2008) apply to all food additives but not substances when they are used for the purpose of imparting flavor or taste, for nutritional purposes, or as processing aids.

Regulation (EC) 1331/2008 also established the European Food Safety Authority (EFSA) to take over the food safety responsibility of the European Commission and laid down procedures for food safety issues. EFSA is responsible for safety evaluations of new food additives, review of existing food additives (based on significant new scientific information and changing conditions), and systematic reevaluation of all authorized food additives in the EU. EFSA has its scientific committees and a panel on food additives, flavorings, processing aids, food enzymes, and materials in contact with food which is responsible for safe use of food additives, flavorings, processing aids, materials in contact with food, and other deliberately added substances to food, which would include food coatings. These regulations on the common authorization procedure for additives, flavorings, and food enzymes as processing aids became effective in 2010.

13.2.3 United States

In the United States, one of the early major food laws was the Food and Drug Act (known simply as the Wiley Act) enacted in 1906 (U.S. Statutes at Large, 1906). The law prohibited the addition of any ingredients that would substitute for the food, conceal damage, pose a health hazard, or constitute a "filthy or decomposed substance." The Food, Drug, and Cosmetic Act, passed in 1938, mandated premarketing approval, proof of safety of drugs, authorized standards of identity, quality, and safe tolerances for unavoidable substances in food, as well as factory inspections (Fortin, 2009). Over the years, numerous amendments and other acts have been passed on food safety.

In the area of food additives, FDA (www.fda.gov) of the Department of Health and Human Services is responsible for domestically produced and imported products except meat, poultry, and frozen, dried, and liquid eggs, which are under the authority of USDA's Food Safety and Inspection Service (FSIS, www.fsis.usda.gov). FDA, established in 1927, as the Food, Drug, and Insecticide Administration, was changed to the current name in 1930.

The Division of Petition Review (DPR), Office of Food Additive Safety of FDA has the primary responsibility of overseeing the premarket review of food additives that have a technical effect in food. A food additive means "any substance the intended use of which results or may reasonably be expected to result, directly or indirectly, in its becoming a component or otherwise affecting the characteristics of any food (including any substance intended for use in producing, manufacturing, packaging, processing, preparing, treating, packaging, transporting or holding food," if such a substance is not generally recognized as safe (GRAS) (FDCA Sec.201[321] (s), FDA 21 CFR Part 170). A substance is listed as GRAS by FDA when used in accordance with FDA's good manufacturing practices (GMP) and contains no residues of heavy metals or other contaminants in excess of tolerances set by FDA; and the substance is essential for the handling of organically produced agricultural products.

In addition, the DPR is responsible for managing the safety review of new color additives to be used in food, cosmetics, drugs, and medical devices.

The Food Additives Amendment exempted two groups of substances from the food additive regulation process: all substances that FDA or the USDA determined were safe for use in specific foods prior to the 1958 amendment were designated as prior-sanctioned substances, and GRAS substances whose use is generally recognized by experts as safe based on the extensive history or use in food before 1958, or based on published scientific evidence (Fortin, 2009).

The Color Additive Amendment mandates food colors used in foods (including edible coatings), drugs, or cosmetics to be approved by FDA. However, unlike food additives, colors used before this legislation were not allowed continued use without testing (Fortin, 2009).

Indirect additives are substances that become part of a food in trace amounts due to its packaging, storage, or other handling (not usually listed on the ingredient label) (i.e., materials in contact with food, which in some cases, applies to coatings) (FDA, 21 CFR Parts 175, 176, 177, and 178).

GMP regulations limit the use of a food or color additive in foods, where manufacturers use only the amount of an additive necessary to achieve the desired effect (Fortin, 2009).

For example, in the United States, FDA 21CFR Part 172 contains a list of "Food additives Permitted for Direct Addition to Food for Human Consumption," Subpart C "Coatings, Films, and Related Substances"; Part 175 "Indirect Food Additives: Adhesives and Components of Coatings"; Part 184 "Direct Food Substances Affirmed as GRAS"; and Part 184 Subpart B contains a "List of Specific Substances Affirmed as GRAS" all contain possible ingredients for use in coatings.

13.2.4 Japan

In Japan, the Food Sanitation Law is the regulation for establishment of standards/ specifications for food, additives, apparatus, and food containers/packages, inspection of compliance, hygiene management of the manufacture and sale of food, and business licenses. While the Law for the Control of Foods and Things Relating to Foods (Law No. 15) was released in 1900, it was not until 1947 that the comprehensive law concerning food safety and food additives was established. In 2003, the Food Safety Basic Law was enacted to enhance the management of risk assessment.

The Food Safety Commission under the Ministry of Health, Labour, and Welfare is responsible for risk assessment and recommendation of ADI, while the Pharmaceutical Affairs and Food Sanitation Council is responsible for the specification on identity and purity.

13.2.5 Canada

In Canada, the Canadian Environmental Protection Act (CEPA) 1999 (EC, 1999) provides guidelines on food safety. It is based on the Food and Drug Act created in the late 1980s and is a strong law that requires every new chemical substance, made in Canada or imported from other countries since 1994, to be assessed against specific criteria. It covers a range of activities that can affect human health and the environment, and acts to address any pollution issues not covered by other federal laws. Health Canada is responsible for the risk assessment of food additives under the CEPA 1999. The Bureau of Chemical Safety is responsible for policy, standard setting, risk assessment, and research and evaluation activities with respect to chemicals in foods.

13.2.6 Australia and New Zealand

Australia and New Zealand have their own national food safety programs: Australian Food Standards Code and New Zealand Food Regulations 1984. However the two nations decided to work together and develop an integrated food regulatory system. Food Standards Australia New Zealand (FSANZ) was established by the Food Standards Australia New Zealand Act 1991 (ANZ, 1991) as an independent statutory agency for the two countries. The Ministerial Council formulates policy guidelines and notifies the FSANZ, who follows these guidelines when developing or reviewing food regulatory measures.

13.3 Substances in food and food products

Food or finished food products can generally be considered to include the naturally common food materials or food ingredients and other substances whose presence may be intentional and unintentional. Based on their intended purpose, all materials can be classified into four categories: food ingredients, food additives, food contaminants, and processing aids. Ingredients of food coatings can be found in all four categories.

13.3.1 Food and food ingredients

Food ingredients are usually materials considered as foods in themselves. They are derived directly from raw agricultural commodities, although processing is usually involved. An example related to coatings would be fruit-derived wraps (Ravishankar et al., 2009). Their acceptance as food is the result of consumption over many centuries. Some improvements made to the raw agricultural commodities may, however, lead to concerns as to their identity as the original safe food items. One example is the genetically modified insect-resistant goods due to biomolecular breeding technologies.

Currently there are substantial differences in categorization of food substances among regulations. In the Codex system, they are categorized into 16 groups. Each food category is associated with a number that reflects the hierarchical structure of the naming system (Table 13.1).

Under the EU's regulation (EC 178/2002, EU, 2002), food is "any substance or product, whether processed, partially processed or unprocessed, intended to be, or

Table 13.1 Categories under Food and Agriculture Organization of the United Nations (FAO)/World Health Organization (WHO), United States and European Union standards

Codex Alimentarius, FAO/WHO[a]

1	Dairy products and analogues, excluding products of food category 02.0
2	Fats and oils, and fat emulsions
3	Edible ices, including sherbet and sorbet
4	Fruits and vegetables (including mushrooms and fungi, roots and tubers, pulses and legumes, and aloe vera), seaweeds, and nuts and seeds
5	Confectionery
6	Cereals and cereal products, derived from cereal grains, from roots and tubers, pulses and legumes, excluding bakery wares of food category 07.0
7	Bakery wares
8	Meat and meat products, including poultry and game
9	Fish and fish products, including mollusks, crustaceans, and echinoderms
10	Eggs and egg products
11	Sweeteners, including honey
12	Salts, spices, soups, sauces, salads, protein products (including soybean protein products) and fermented soybean products
13	Foodstuffs intended for particular nutritional uses
14	Beverages, excluding dairy products
15	Ready-to-eat savouries
16	Composite foods—foods that could not be placed in categories 01–15.

Food Classification System, EU[b]

1	Cereals and cereal products (Bread and rolls, bakery products, rice, cereals and products, flour, and pasta)
2	Meat, meat products, and dishes (Red meat, offal, poultry, canned meat and meat products, and meat dishes)
3	Fish seafood and dishes
4	Milk and milk products (Milk, cheese, and milk products)

5   Eggs
6   Added lipids (Animal origin and vegetable origin)
7   Potatoes and other starchy roots
8   Pulses
9   Nuts
10 Vegetables (Fresh and processed vegetables)
11 Fruits (Fresh and processed fruits)
12 Fruit and vegetable juices
13 Sugar and sugar products
14 Nonalcoholic beverages (Stimulants, mineral water, and soft drinks)
15 Alcoholic beverages (Wine, beer, and spirits)

General Food Category, United States[c]

1   Baked goods and baking mixes
2   Beverages, alcoholic
3   Beverages and beverage bases, nonalcoholic
4   Breakfast cereals
5   Cheeses
6   Chewing gum
7   Coffee and tea
8   Condiments and relishes
9   Confections and frostings
10 Dairy product analogs
11 Egg products
12 Fats and oils
13 Fish products
23 Grain products and pastas
24 Gravies and sauces
25 Hard candy and cough drops
26 Herbs, seeds, spices, seasonings, blends, extracts, and flavorings
27 Jams and jellies, home prepared
28 Jams and jellies, commercial
29 Meat products
30 Milk, whole and skim
31 Milk products
32 Nuts and nut products
33 Plant protein products
34 Poultry products
35 Processed fruits and fruit juices

(*Continued*)

Table 13.1 Categories under Food and Agriculture Organization of the United Nations (FAO)/World Health Organization (WHO), United States and European Union standards (Continued)

14	Fresh eggs	36	Processed vegetables and vegetable juices
15	Fresh fish	37	Snack foods
16	Fresh fruits and fruit juices	38	Soft candy
17	Fresh meats	39	Soups, home prepared
18	Fresh poultry	40	Soups and soup mixes
19	Fresh vegetables, tomatoes, and potatoes	41	Sugar, white, granulated
20	Frozen dairy desserts and mixes	42	Sugar substitutes
21	Fruit and water ices	43	Sweet sauces, toppings, and syrups
22	Gelatins, puddings, and fillings		

^a Codex STAN 192-1995 (version 2010) (Summarized from www.codexalimentarius.net/web/standard_list.jsp.)
^b The DAFNE Food Classification System (Summarized from http://ec.europa.eu/health/ph_projects/2002/monitoring/dafne_code_en.pdf.)
^c CFR 21, 170.3 General Food Category (Summarized from http://edocket.access.gpo.gov/cfr_2004/aprqtr/pdf/21cfr170.3.pdf.)

reasonably expected to be ingested by humans." It includes drink, chewing gum, coatings, and any substance, including water, intentionally incorporated into the food during its manufacture, preparation, or treatment. Animal feed, live animals (unless they are prepared for placing on the market for human consumption), plants prior to harvesting, residues, and contaminants are not considered as food. There are a total of 15 food groups under the EU regulation (Table 13.1).

The U.S. FDCA (Chapter IV: Food, Sec. 401) states that definitions and standards for food shall promulgate regulations fixing and establishing for any food, under its common or usual name so far as practical: a reasonable definition and standard of identity, a reasonable standard of quality, or reasonable standards of fill of container. As a result, there are a total of 43 food groups under the U.S. regulation (Table 13.1).

For the Canadian regulatory system, food is classified as standardized and unstandardized foods. Fruits and vegetables fall in the unstandardized food category.

13.3.2 Food additives

A general definition of food additives is that they are substances added into food products intentionally to fulfill certain specified functions such as preservation and glazing as in case of food coatings. More specific definitions by different regulatory bodies are given in Table 13.2. They are usually used in small quantities and remain with the food but may or may not be consumed together with the edible portion of the food items. For example, food additives and coatings used for fresh fruits remain on the fruit peel, which is not consumed as part of the fresh fruit as in case of oranges but is consumed in case of apples. Therefore, there are coating ingredients, like resins, that are allowed on oranges but not on apples (Baldwin, 2004). There are so many food additives currently in use. These are the largest group of food components for processed food products, including coatings. This is especially true of food coatings used as carriers of fungicides (Baldwin, 2004), color (also used on oranges), flavorings, antioxidants (nuts) (Baldwin and Wood, 2006), acidulants, firming agents, and preservatives (cut apple) (Baldwin et al., 1996).

13.3.3 Processing aids

Processing aids are substances used as part of a manufacturing process and intentionally added into food (Table 13.3). Processing aids are normally absent in the food before the end of the process as a result of some type of removal processing. There may, however, be residues in the food, and because of this, processing aids are generally nontoxic or of low toxicity. As a result, processing aid residue is normally of only limited concern when good processing practices and controls are adopted. Some coatings are considered to be processing aids if these fulfill a processing function and are mostly removed in processing (Baldwin, 2004).

13.3.4 Food contaminants

Food contaminants are substances that find their way into food by their natural environmental presence and by accident or as residues from use of chemical substances during production of food items (Table 13.4).

Table 13.2 Definitions of "food additives" by major food standards

FAO/WHO Codex Alimenterius[a]

Any substance not normally consumed as a food by itself and not normally used as a typical ingredient of the food, whether or not it has nutritive value, the intentional addition of which to food for a technological (including organoleptic) purpose in the manufacture, processing, preparation, treatment, packing, packaging, transport, or holding of such food results, or may be reasonably expected to result (directly or indirectly), in it or its by-products becoming a component of or otherwise affecting the characteristics of such foods. The term does not include contaminants or substances added to food for maintaining or improving nutritional qualities.

EU Directives[b]

Any substance not normally consumed as a food in itself and not normally used as a characteristic ingredient of food, whether or not it has nutritive value, the intentional addition of which to food for a technological purpose in the manufacture, processing, or preparation. The following are not considered to be food additives:

(i) monosaccharides, disaccharides, or oligosaccharides and foods containing these substances used for their sweetening properties;

(ii) foods, whether dried or in concentrated form, including flavorings incorporated during the manufacturing of compound foods, because of their aromatic, sapid, or nutritive properties together with a secondary coloring effect;

(iii) substances used in covering or coating materials, which do not form part of foods and are not intended to be consumed together with those foods;

(iv) products containing pectin and derived from dried apple pomace or peel of citrus fruits or quinces, or from a mixture of them, by the action of dilute acid followed by partial neutralization with sodium or potassium salts (liquid pectin);

(v) chewing gum bases;

(vi) white or yellow dextrin, roasted or dextrinated starch, starch modified by acid or alkali treatment, bleached starch, physically modified starch and starch treated by amylolytic enzymes;

(vii) ammonium chloride;

(viii) blood plasma, edible gelatin, protein hydrolysates and their salts, milk protein, and gluten;

(ix) amino acids and their salts other than glutamic acid, glycine, cysteine, and cystine and their salts having no technological function;

(x) caseinates and casein;

(xi) inulin;

U.S. Code of Federal Regulations[c]

Includes all substances not exempted by section 201(s) of the Federal Food, Drug, and Cosmetic Act, the intended use of which results or may reasonably be expected to result, directly or indirectly, either in their becoming a component of food or otherwise affecting the characteristics of food. A material used in the production of containers and packages is subject to the definition if it may reasonably be expected to become a component, or to affect the characteristics, directly or indirectly, of food packed in the container.

[a] Codex General Standard For Food Additives, Codex STAN 192-1995; Codex Alimentarius Commission Procedural Manual (Summarized from www.codexalimentarius.net/web/standard_list.jsp.)

[b] Regulation (EC) No 1333/2008 on food additives; Directive 89/107/EEC (Summarized from http://eur-lex.europa.eu/LexUriServ/LexUriServ.do?uri=OJ:L:20 08:354:0016:0033:en:PDF.)

[c] Federal Food, Drug, and Cosmetic Act, 201(s) and 21 CFR 170.3(e)(1) (Summarized from http://edocket.access.gpo.gov/cfr_2004/aprqtr/pdf/21cfr170.3.pdf.)

Table 13.3 Definitions of "processing aids" by major food standards

FAO/WHO Codex Alimentarius[a]

Any substance or material, not including apparatus or utensils, and not consumed as a food ingredient by itself, intentionally used in the processing of raw materials, foods, or its ingredients, to fulfill a certain technological purpose during treatment or processing and which may result in the nonintentional but unavoidable presence of residues or derivatives in the final product.

EU Regulation[b]

Any substance which: (i) is not consumed as a food by itself; (ii) is intentionally used in the processing of raw materials, foods or their ingredients, to fulfill a certain technological purpose during treatment or processing; and (iii) may result in the unintentional but technically unavoidable presence in the final product of residues of the substance or its derivatives provided they do not present any health risk and do not have any technological effect on the final product.

U.S. Code of Federal Regulations[c]

Incidental additives that are present in a food at insignificant levels and do not have any technical or functional effect in that food. For the purposes of this paragraph (a)(3), incidental additives are:

(i) Substances that have no technical or functional effect but are present in a food by reason of having been incorporated into the food as an ingredient of another food, in which the substance did have a functional or technical effect.

(ii) Processing aids, which are as follows:

(a) Substances that are added to a food during the processing of such food but are removed in some manner from the food before it is packaged in its finished form.

(b) Substances that are added to a food during processing, are converted into constituents normally present in the food, and do not significantly increase the amount of the constituents naturally found in the food.

(c) Substances that are added to a food for their technical or functional effect in the processing but are present in the finished food at insignificant levels and do not have any technical or functional effect in that food.

(iii) Substances migrating to food from equipment or packaging or otherwise affecting food that are not food additives as defined in section 201(s) of the act; or if they are food additives as so defined, they are used in conformity with regulations established pursuant to section 409 of the act.

---

[a] Codex General Standard For Food Additives, Codex STAN 192-1995; Codex Alimentarius Commission Procedural Manual (Summarized from www.codexalimentarius.net/web/standard_list.jsp.)

[b] Regulation (EC) No. 1333/2008 on food additives; Directive 89/107/EEC (Summarized from http://eur-lex.europa.eu/LexUriServ/LexUriServ.do?uri=OJ:L:2008:354:0016:0033:en:PDF.)

[c] Federal Food, Drug, and Cosmetic Act, 201(s) and 21 CFR 170.3(e)(1) (Summarized from http://edocket.access.gpo.gov/cfr_2004/aprqtr/pdf/21cfr170.3.pdf.)

Table 13.4 Definitions of "contaminant" by major food standards

FAO/WHO Codex Alimentarius[a]

Any substance not intentionally added to food, which is present in such food as a result of the production (including operations carried out in crop husbandry, animal husbandry, and veterinary medicine), manufacture, processing, preparation, treatment, packing, packaging, transport, or holding of such food or as a result of environmental contamination.

The term does not include insect fragments, rodent hairs, and other extraneous matter.

EU Regulation[b]

Any substance not intentionally added to food which is present in such food as a result of the production (including operations carried out in crop husbandry, animal husbandry, and veterinary medicine), manufacture, processing, preparation, treatment, packing, packaging, transport, or holding of such food, or as a result of environmental contamination. Extraneous matter, such as, for example, insect fragments, animal hair, etc., is not covered by this definition.

U.S. Code of Federal Regulations[c]

Incidental additives that are present in a food at insignificant levels and do not have any technical or functional effect in that food. For the purposes of this paragraph (a)(3), incidental additives are:

(i) Substances that have no technical or functional effect but are present in a food by reason of having been incorporated into the food as an ingredient of another food, in which the substance did have a functional or technical effect.

(ii) Processing aids, which are as follows:

(a) Substances that are added to a food during the processing of such food but are removed in some manner from the food before it is packaged in its finished form.

(b) Substances that are added to a food during processing, are converted into constituents normally present in the food, and do not significantly increase the amount of the constituents naturally found in the food.

(c) Substances that are added to a food for their technical or functional effect in the processing but are present in the finished food at insignificant levels and do not have any technical or functional effect in that food.

(iii) Substances migrating to food from equipment or packaging or otherwise affecting food that are not food additives as defined in section 201(s) of the act; or if they are food additives as so defined, they are used in conformity with regulations established pursuant to section 409 of the act.

a Codex General Standard For Food Additives, Codex STAN 192-1995; Codex Alimentarius Commission Procedural Manual (Summarized from www.codexalimentarius.net/web/standard_list.jsp.)
b Council Regulation (EEC) No. 315/93 (Summarized from http://eur-lex.europa.eu/LexUriServ/LexUriServ.do?uri=CONSLEG:1993R 0315:20090807:EN:PDF.)
c Federal Food, Drug, and Cosmetic Act, 201(s) and 21 CFR 170.3(e)(1) (Summarized from: http://edocket.access.gpo.gov/cfr_2004/aprqtr/pdf/21cfr170.3.pdf.)

Environmental pollutants (e.g., lead), microbiological toxins (fungal mycotoxins, e.g., aflatoxins), agricultural chemicals (pesticides), and veterinary drugs (antibiotics) are considered as food contaminants by major regulations. Certain contaminants are treated differently among the various regulations. Insect and animal filth are considered as "quality" issues by EU regulations. However, in the United States, their presence is considered an indication of improper sanitation and elevated health risks and the food product sale is restricted.

EU regulation also restricts the sale of food from genetically engineered crops. On the contrary, U.S. regulations consider them safe. This could affect coating materials coming from, for example, corn protein (zein) if the corn was genetically engineered.

13.3.5 Nanotechnology

There is a need for global polices on food nanotechnology, which is a new technology that uses nanoscale materials and which has ramifications for use in food and coatings. Nanoscale technologies can have applications related to food security, food-borne pathogen detection, food packaging, food processing, and food ingredient technologies (Bugusu et al., 2009). FDA has established a Nanotechnology Task Force to determine regulatory approaches for nanoscale materials. The resulting report (FDA, 2007) made recommendations on scientific and policy issues related to safety of nanoscale food ingredients and concluded that FDA authorities will be expected to regulate such products. A case-by-case approach may be necessary initially, because the technology is still evolving. FDA held a public forum in 2008 to elicit comments and collect data to facilitate implementation of task force recommendations (Bugusu et al., 2009). Similarly, Health Canada requested the Council of Canadian Academics (CCA) to review existing nanomaterials and appointed an expert panel. Their report concluded that knowledge on nanoparticles was limiting, although there was no evidence that current applications posed a risk that could not be addressed through available risk management strategies (CCA, 2008). The EU has established specialized agencies to investigate the risk of nanotechnology to consumers. The European Food Safety Authority (EFSA) identified certain risk assessment uncertainties due to limited information on toxicokinetics and toxicology and recommended a case-by-case evaluation (EFSA, 2009). However, potential use of nanotechnology for food will be regulated by the existing framework including EU Food Law (EC 178/2002, EU, 2002) or by specific approval processes as in the "novel food" regulation. Current legislation for food additives may be sufficient to cover nanomaterials (Bugusu et al., 2009). Recent food additive regulation (EC 1333/2008, EU, 2008b) indicates that a significant change in particle size, such as nanotechnology, would require reevaluation of approved food additives. Another EU regulatory program, Registration, Evaluation, and Authorization of Chemicals (REACH), asserts that new chemicals should not be allowed on the market without adequate safety data. For REACH, not only chemical composition but also chemical structure and physical state including dimension of particles are subject to regulation (Bugusu et al., 2009). Food Standards Australia New

Zealand (FSANZ) has no standards that specifically regulate nanomaterials; thus, the regulatory framework in FSANZ applies to nanomaterials in food as it does for conventional food additives. In Japan, the Ministry of Economic Trade, and Industries works to standardize testing methods for safety and evaluation of nanoparticles, and the Ministry of Health, Labor, and Welfare (JMHLW) evaluates health impacts (Chau et al., 2007). In China, the Chinese Academy of Sciences (CAS) and Ministry of Education cofounded the National Center for Nanoscience and Technology (NCNST), which has several divisions doing basic and applied research in nanoscience. The Commission on Nanotechnology Standardization, affiliated with NCNST, is responsible for developing national standards, including safety requirements for nanomaterials, and governs/guides assessment and authorization of nanoproducts to improve product quality and reduce health risks (NCNST, 2009).

13.3.6 Substance classification variations

The definitions of the food additive, contaminant, and processing aid are similar among the major regulatory authorities (Tables 13.2–13.4). The substances covered under each of the three categories, however, differ. First, the inclusion of substances in any particular classification are various. For example, the regulation of food additives in the United States does not include colors, because coloring additives were first regulated and kept separated from other food additives when control of the latter were developed. In contrast, the current EU food regulation, color agents are included. On the other hand, flavoring additives are food additives in U.S. regulation but not in EU regulation. Second, the same substance may be classified differently among regulation authorities. For example, postharvest pesticides are classified as food additives by the Japanese but not by U.S. or European food regulations.

13.4 Qualification of substances as food additives

13.4.1 Food safety

The first principle on whether or not a substance can be used as a food additives is whether or not it presents a hazard to the health of the consumer at the level of use proposed, based on the scientific evidence available. Due to independent development of regulation by various authorities, different traditional usage, historical legislation practices, and conclusions on risk assessments, the acceptance of a chemical or substance as food additive may vary around the world. However, permission to use a chemical or substance in food or food processing is generally based on the food safety assessment and technical necessity. For example, polyethylene is allowed for use in citrus coatings in the United States but not in Japan.

Food safety assessment is a risk, and so the maximum permitted level (MPL) is an important indicator of a substance's tolerance in foods. The primary objective is to ensure that the intake of additives from all its uses does not exceed its Acceptable Daily Intake (ADI). For many additives, the use level is not regulated and is required to be in compliance with GMPs.

13.4.2 Technical function need

The use of a food additive in food must fulfill a technical function besides being safe to human health under the specified condition and food type. The use is justified when the substance serves one or more of the purposes of preserving the nutritional quality of the food, providing necessary ingredients or constituents for dietary needs, enhancing the keeping quality or stability of a food, improving its organoleptic properties and providing aids in manufacture, processing, preparation, treatment, packing, transport, or storage of food. Coatings fulfill several of these criteria, including carrying useful ingredients (antioxidants, for nuts or fresh-cut apples) (Baldwin et al., 1996; Baldwin and Wood, 2006), preserving quality (inhibiting water loss, delaying ripening of fresh fruits and vegetables) (Baldwin, 1994), and providing an aid in sorting, packing, or storing of fresh or fresh-cut fruits and vegetables or nuts (Bai et al., 2006).

Some substances and chemicals could have more than one function declared, and their usage must follow the actual function in the finished products. Their usage will be subjected to all relevant regulation if they fulfill multiple functions. Many of the same technical function classes are used among the major regulations, but there are some differences, as seen in Table 13.5.

13.5 Major food additive standards

13.5.1 FAO/WHO's Codex Alimentarius

The specification of food additives by the JECFA was first implemented in 1992 as the Compendium of Food Additive Specifications (FNP 52). The Compendium was updated in 2005 to include all the additions and revisions after 1992 (JECFA Monographs 1, 2005 and 2006). The Codex Alimentarius is published in Arabic, Chinese, English, French, and Spanish. However, not all texts are available in all languages. Currently, an ongoing effort is underway that will lead to the adoption of a Codex General Standard for Food Additives. It is not clear when the next update will occur.

The listing of food additives is by no means a complete work and is continually updated. The lack of reference to a particular additive or to a particular use of an additive in a food in the General Standard, as currently drafted, does not imply that the additive is unsafe or unsuitable for use in food as it is noted in the Codex General Standard for Food Additives (Codex STAN 192-1995).

To help provide an agreed international numerical system for identifying food additives in label ingredient lists, CCFA released a list of class names and assigned identification numbers, creating the International Numbering System (INS), for most common additives to allow easy recognition (CAC/GL 36-1989). The INS was developed adopting the Codex General Standard for Labeling, and therefore, it does not include flavoring substances, dietetic and nutritive additives as well as chewing gum bases. The INS is an open list subject to the inclusion and removal of additives on an ongoing basis.

13.5.2 EU's directives on food additives

The guidelines on food additive regulation are provided by the four initial directives: the Frame Directive (Council Directive 89/107/EEC), the Sweeteners Directive

Table 13.5 Similarity and differences of technical function classification among some major food additive regulations[a]

FAO/WHO[b]	EU[c]	U.S.[d]
Functional Classes	Food Additive Category	Physical or technical function effects
Acidity regulator	Acidity regulators	pH control agents
Acid	Acids	
Anticaking agent	Anti-caking agents	Anticaking agents and free-flow agents
Antifoaming agent	Anti-foaming agents	Surface-active agents
Antioxidant	Anti-oxidants	Antioxidants
Bleaching agent		
Bulking agent	Bulking agents	
Carbonating agent		
Carrier		Solvents and vehicles
Preservative	Preservatives	Antimicrobial agents
Color	Colors	Colors and coloring adjuncts
Color retention agent		
Emulsifier	Emulsifiers	
Emulsifying salt	Emulsifiers and emulsifier salts	
Emulsifying salts		
Firming agent	Firming agents	Firming agents
Flavor enhancer	Flavor enhancers	Flavor enhancers
		Flavoring agents and adjuvants
Flour treatment agent	Flour treatment agents	Dough strengtheners
Gelling agent	Gelling agents (include lubricants)	
Glazing agent	Glazing agents	Surface-finishing agents Texturizers
Humectant	Humectants	Humectants
		Fumigants

(Continued)

Table 13.5 Similarity and differences of technical function classification among some major food additive regulations[a] (Continued)

FAO/WHO[b]	EU[c]	U.S.[d]
Propellant	Propellent gas and packaging gas	Propellants, aerating agents and gases
Package gas		
Raising agent	Raising agents	Leavening agents
Sweeteners	Sweeteners	Nutritive sweeteners
		Non-nutritive sweeteners
Stabilizer	Stabilizers (also comprises foam stabilizers)	Stabilizers and thickeners
Thickener	Thickeners	
Sequestrant	Sequestrants	Sequestrants
	Enzymes (Only those used as additives)	Enzymes
	Modified starches	
		Curing and pickling agents
		Drying agents
		Formulation aids
		Lubricants and release agents
		Nutrient supplements
		Oxidizing and reducing agents
		Processing aids
		Synergists

[a] The function comparisons among regulations are for illustration only. Functions listed on the same row are not necessarily the same. Please see relevant documents for detailed definitions.
[b] Codex CAC/GL 36-1989 (Summarized from www.codexalimentarius.net/download/standards/7/CXG_036e.pdf).
[c] Directive 89/107/EEC (Summarized from http://eur-lex.europa.eu/LexUriServ/LexUriServ.do?uri=OJ:L:2008:354:0016:0033:en:PDF.)
[d] CFR 21, 170.3 (Summarized from http://edocket.access.gpo.gov/cfr_2004/aprqtr/pdf/21cfr170.3.pdf.)

(European Parliament and Council Directive 94/35/EC), the Colors Directive (European Parliament and Council Directive 94/36/EC), and the Food Additives Other than Colors and Sweeteners Directive (European Parliament and Council Directive No 95/2/EC). Flavoring additives are regulated by The Flavor Directive (88/388/EEC).

In 2006 the European Commission published a proposal for a new regulation on food additives together with proposals for regulations on food flavorings and on food enzymes. If the proposal is adopted, the provisions on additives in the different existing Directives will be brought together in one Regulation.

Under the European regulation, food additives must be explicitly authorized before they can be used in foods. Therefore, food additives acceptable outside of the EU may not be permitted. Use of unauthorized additives can be requested by submitting a formal application to the European Commission on the substance, including scientific data concerning safety. If the application is accepted, the Commission will formally ask EFSA to issue an opinion on the safety of the substance for its intended uses. Once a request for an opinion has been accepted by EFSA, it is included in the Register of requested opinions where its status is can be monitored, including date of reception and anticipated timing of finalization and adoption.

13.5.3 U.S. FDA's Code of Federal Regulation

The food additive standards are published first in the *Federal Register* and collected in the FDA/21CFR. The relevant section is the 21CFR Part 170.3 (e)(1) in which substances that may be expected to be used in food or as components of food. Also 21CFR, Part 172 (subparts A "General Provisions," B "Food preservatives," C "Coatings, Films, and Related Substances," D "Special Dietary and Nutritional Additives," E "Anticaking Agents," F "Flavoring Agents and Related Substances," G "Gums, Chewing Gum Bases, and Related Substances," H "Other Specific Usage Additives," and I "Multipurpose Additives").

Although the CFR is the ultimate source for food additive regulation, the Food Additives Status List and Color Additive Status List summary is a quick reference guide on the use limitations for food additives and selected pesticide chemicals from 40 CFR 180, for which EPA has set tolerances in food. Another helpful reference on food additives is the FDA, Everything Added to Food in the United States (FDA/EAFUS) by the FDA, Center for Food Safety and Applied Nutrition (FDA/CFSAN, 2010). This list organizes additives into one alphabetized list and is updated annually.

According to the definition, under Sec. 201(s) of the FDCA are the substances in the category of GRAS. These substances are not (necessarily) considered food additives; however, some food additives are considered GRAS and are on the GRAS list. Therefore, most GRAS substances have no quantitative restrictions as to use, although their use must conform to good manufacturing practices. Some GRAS substances, such as sodium benzoate, do have a quantitative limit for use in foods. Such substances are GRAS in foods but limited in standardized foods where the standard provides for its use. Prior to being compiled in the CFR, new regulations and revisions are daily published in the *Federal Register*.

13.5.4 Japan's Specifications and Standards for Food Additives

Japan's Specifications and Standards for Food Additives (JMHLW, 2000) is the official reference for food additive standards in Japan. At this time the latest edition is the seventh, which made an effort to provide new information on specifications and standards on both synthetic and natural substances and to harmonize with international standards.

Approved Food Additives are currently listed in the List of Designated Additives (JMHLW, 2010), List of Plant or Animal Sources of Natural Flavoring Agents (JMHLW, 1996a), and List of Existing Food Additives (JMHLW, 1996b), as well as substances that are generally provided for the eating or drinking of foods, and which are used as additives and substances appearing in foods, plus those generally considered as food. All additives and foods used will comply with established specifications.

Other substances that are not parts of the above lists could potentially be permitted. The substance should be considered safe within a certain level by JECFA, widely used in foreign countries, including the United States and EU, and be recognized internationally. Furthermore, the substances should be considered as safe and necessary as a food additive by the Japan Ministry of Health, Labor, and Welfare (JMHLW). The Pharmaceutical Affairs and Food Sanitation Council will review the draft list for candidates for pharmaceuticals and authorize it for review of safety, quality, and necessity.

13.5.5 Canada's Food and Drug Regulations

For Canada, the food additive standards are listed in the Food and Drug Regulations C.R.C., c. 870. The permitted food additives and their allowable areas of use and maximum levels of use appear in the Tables of Part B Division 16 (Canada, 2010). New regulations as well as amendments to existing regulations are published in the Canada Gazette (http://canadagazette.gc.ca/index-e.html) on a weekly basis. For substances not listed in the Food and Drug Regulations, Division 16 Tables, the Letter of Opinion from Health Canada's Bureau of Chemical Safety (BCS), which coordinates the assessment of food additive submissions, and is issued to the substance manufacturers, will confirm the regulatory status of the product in terms of the Food and Drug Regulations.

13.5.6 Australia and New Zealand Food Standards Code

The joined food standard regulation of Australia and New Zealand is the Australia New Zealand Food Standards Code (ANZFSC). The food additives permitted to be used in food are listed in the Schedules of the Standard 1.3.1 (ANZ, 2010).

Unless expressly permitted in this Standard, food additives should not be added to food. The food additives standard lists the food additives that may be used in all foods together into one generic standard that covers all foods. New regulations as well as amendments to existing regulations are published in the Gazette Notices which are periodically updated.

13.6 Substances permitted for fruit and vegetable coatings

13.6.1 Coating ingredients

Regulations for coatings of fresh fruit and vegetables in FAO/WHO, EU, United States, and Australia/New Zealand are shown in Tables 13.6, 13.7, 13.8, and 13.9, respectively.

Like other food additives, the ingredients used to manufacture coatings applied to harvest fruits and vegetables are regulated to ensure that the application does not pose a health risk beyond the allowable levels. In principle, the use of additives as coating ingredients must be stated in the intended uses of the food additives.

Additives are generally listed by their level of health risk and technical function following the type of foods. In the United States, food additives classified as GRAS can be used in all foods but may be subjected to limitation of maximum use levels or restricted technical function. Similarly, the Codex Alimenterius contains a list of Additives Permitted for Use in Food in General, Unless Otherwise Specified, in Accordance with GMP (Codex STAN 192-1995).

Due to differences in culinary customs and coating practice development, some additives approved for fruit and vegetable coatings in a country may be not acceptable in another country. One example is that oxidized polyethylene is not accepted in Japan but is accepted in the United States and EU (Baldwin, 2004), and morpholine is accepted in fruit coatings in the United States, but not in the EU (Hagenmaier, 2004) as shown in Table 13.7 and Table 13.8. With coated fruits and vegetables as the finished food items, coatings are considered as a food additive or composite food additives. The food additives used should be either GRAS or permitted for use in coatings for fruits and vegetables in general, or for specifically identified commodities.

In Canada, components of fruit and vegetable coatings are not regulated as food additives (with the exceptions of mineral oil, paraffin wax, and petrolatum). Fresh fruits and vegetables are not included in the Food and Drug Regulation of Canada as food items. Therefore, fruits and vegetables belong to the food category of "unstandardized foods." All food additives that can be used on unstandardized foods are, therefore, permitted to be used on fruits and vegetables unless mentioned otherwise.

13.6.2 Organic ingredients

The National Organic Program (NOP) of the USDA develops, implements, and administers national production handling and labeling standards for organic agricultural products in the United States. The National List of Allowed and Prohibited Substances (USDA, 2002) gives information on allowed and prohibited substances and ingredients. An organic coating or coating for organic produce must be formulated from ingredients on this list.

In Canada, the Organic Production Systems Permitted Substances Lists provides the food additives that can be used (CAN/CGSB-32.311-2006). The standards developed by the Canadian General Standards Board (CGSB) are accredited by the Standards Council of Canada and used as the national standards. All the updates are published in the annual document CGSB Catalogue.

Table 13.6 Food additives listed for surface-treated fresh fruit and vegetables in Codex Alimenterius[a]

INS[b]	Additive	Max level (mg/kg)	Comments
901	Beeswax	GMP[c]	Surface-treated fresh fruit and vegetables[b]
902	Candelilla wax	GMP	Surface-treated fresh fruit and vegetables
903	Carnauba wax	400	Surface-treated fresh fruit and vegetables
904	Shellac	GMP	Fresh fruit and nuts
905ci	Microcrystalline wax	50	Surface-treated fresh fruit and vegetables
101i,ii	Riboflavins	300	Surface-treated fresh fruit and vegetables
110	Sunset yellow FCF	330	Surface-treated fresh vegetables
120	Carmines	500	
172i-iii	Iron oxides	1000	Fresh fruit
220–225, 227, 228, 539	Sulfites	20	Surface-treated fresh fruit. As residual SO_2
231, 232	Ortho-phenylphenols	12	For use on citrus fruits only
338, 339i-iii, 340i-iii, 341i-iii, 342i-ii, 343i-iii, 450i-iii,v-vii, 451 1–11, 452i-v, 542	phosphates	1760	Vegetables and nuts. As phosphorus
445	Glycerol ester of wood rosin	110	Surface-treated fresh fruit and vegetables
474	Sucroglycerides	GMP	Surface-treated fresh fruit
1521	Polyethylene glycol	GMP	Surface-treated fresh fruit
1201	Polyvinylpyrrolidone	GMP	Surface-treated fresh fruit

[a] Codex STAN 192-1995, 2009 Ver. Food Category No. 04.1.1.2 and 04.2.1.2. Vegetables include mushrooms and fungi, roots and tubers, pulses and legumes, and aloe vera (www.codexalimentarius.net/web/standard_list.jsp).
[b] INS: International Numbering System for Food Additives (www.cfs.gov.hk/english/whatsnew/whatsnew_fstr/whatsnew_fstr_13_ins.html).
[c] GMP: No maximum level defined, but they must be used in accordance with Good Manufacturing Practice at a level that is sufficient to achieve the intended purpose.
[d] Surface-treated fresh vegetables include mushrooms and fungi, roots and tubers, pulses and legumes, and aloe vera.

Table 13.7 Food additives listed for surface-treated fresh fruit and vegetables under Regulation of European Union[a]

INS[b]	Additive	Max level (mg/kg)	Comments
Annex III			
220–228	Sulfur dioxide and sulfite	10 (as SO_2)	Fresh lychees (measured on edible parts) Table grapes
230	Biphenyl, diphenyl	70	Surface treatment of citrus fruits
231	Orthophenyl phenol, 2-hydroxybiphenyl	12	
232	Sodium orthophenyl phenol	12	
Annex IV			
322	Lecithins	GMP[c]	Glazing agents for fruit
432	Polysorbate 20		
433	Polysorbate 80		
434	Polysorbate 40		
435	Polysorbate 60		
436	Polysorbate 65		
445	Glycerol esters of wood rosins	50	Surface treatment of citrus fruits
451	Triphosphates (pentapotassium)	4000	Glazing for meat and vegetable products
470a	Sodium, potassium, and calcium salts of fatty acids	GMP	Glazing agents for fruit
471	Mono- and diglycerides of fatty acids		
473	Sucrose esters of fatty acids	GMP	Fresh fruits, surface treatment
474	Sucroglycerides		
491	Sorbitan monostearate	GMP	Glazing agents for fruit

(*Continued*)

Table 13.7 Food additives listed for surface-treated fresh fruit and vegetables under Regulation of European Union[a] (Continued)

INS[b]	Additive	Max level (mg/kg)	Comments
492	Sorbitan tristearate		
493	Sorbitan monolaurate		
494	Sorbitan monooleate		
495	Sorbitan monopalmitate		
570	Fatty acids		
900	Dimethylpolysiloxane		
901	Beeswax, white and yellow	GMP	Glazing agents only for nuts
902	Candelilla wax	GMP	Surface-only treatment of citrus fruits, melons, apples, pears, peaches, and pineapples
903	Carnauba wax	200	
904	Shellac	GMP	
905	Microcrystalline wax	GMP	Surface-only treatment of fresh melons, papayas, mangoes, avocadoes
912	Montan acid esters	GMP	Surface-only treatment of fresh citrus, melons, mangoes, papayas, avocadoes, pineapples
914	Oxidized polyethylene wax		

[a] European Parliament and Council Directive No. 95/2/EC (http://eur-lex.europa.eu/LexUriServ/LexUriServ.do?uri=CONSLEG:1995L0002:20060815:EN:PDF).
[b] INS: International Numbering System for Food Additives (www.cfs.gov.hk/english/whatsnew/whatsnew_fstr/whatsnew_fstr_13_ins.html).
[c] GMP: No maximum level defined, but they must be used in accordance with Good Manufacturing Practice at a level that is sufficient to achieve the intended purpose.

Table 13.8 Food additives listed for surface-treated fresh fruit and vegetables in Code of Federal Regulation (United States)[a]

Additives[b]	Section	Use limits
Fatty acids	172.210	Coatings on fresh citrus fruit
Oleic acid derived from tall oil fatty acids		
Partially hydrogenated rosin		
Pentaerythritol ester of maleic anhydride-modified wood rosin		
Polyethylene glycol		
Polyhydric alcohol diesters of oxidatively refined (Gersthofen process) montan wax acids		
Sodium lauryl sulfate		
Wood rosin		
Vinyl chloride-vinylidene chloride copolymer repeat		
Polyvinylpyrrolidone		
Potassium persulfate		
Propylene glycol alginate		
Sodium decylbenzenesulfonate		
Calcium salt of partially dimerized rosin		
Petroleum naphtha		
Sperm oil		
Coumarone-indene resin	172.215	Grapefruit, lemons, limes, oranges, tangelos, and tangerines

(Continued)

Table 13.8 Food additives listed for surface-treated fresh fruit and vegetables in Code of Federal Regulation (United States)[a] (Continued)

Additives[b]	Section	Use limits
Morpholine	172.235	As the salt(s) of one or more of the fatty acids meeting the requirements of 172.860, as a component of protective coatings applied to fresh fruits and vegetables.
Petroleum naphtha	172.25	As a solvent in protective coatings on fresh citrus fruit in compliance with 172.210.
Oxidized polyethylene	172.26	As protective coating/component of protective coatings for fresh avocados, bananas, beets, coconuts, eggplant, garlic, grapefruit, lemons, limes, mangos, muskmelons, onions, oranges, papaya, peas (in pods), pineapple, plantain, pumpkin, rutabaga, squash (acorn), sweet potatoes, tangerines, turnips, watermelon, Brazil nuts, chestnuts, filberts, hazelnuts, pecans, and walnuts (all nuts in shells)
Synthetic paraffin and succinic derivatives	172.275	Protective coating/component of protective coatings for fresh grapefruit, lemons, limes, mangos, muskmelons, oranges, sweet potatoes, and tangerines
Terpene resin	172.28	Grapefruit, lemons, limes, oranges, tangelos, and tangerines

[a] USA CFR Title 21, Part 172 - Food Additives Permitted For Direct Addition To Food For Human Consumption, Subpart C--Coatings, Films, and Related Substances (www.accessdata.fda.gov/scripts/cdrh/cfdocs/cfcfr/CFRSearch.cfm?CFRPart=172&showFR=1&subpartNode=21:3.0.1.1.3.3).
[b] Additives that can be used for fresh fruits and vegetables are also listed elsewhere beside Subpart C-Coatings, Films, and Related Substances.

Table 13.9 Food additives listed for surface-treated fresh fruit and vegetables (Australia and New Zealand)[a]

INS	Additive	Max level (mg/kg)	Comments
342	Ammonium phosphates	GMP	Surface-treated fruits and vegetables
473	Sucrose esters of fatty acids	100	
901	Beeswax, white and yellow	GMP	
903	Carnauba wax	GMP	
904	Shellac	GMP	
914	Oxidized polyethylene	250	Citrus
1520	Propylene glycol	30000	Citrus

[a]*Source:* Australia New Zealand Food Standards Code, Standard 1.3.1, Food Additives 4.1.2. Surface-treated fruit and vegetables (www.comlaw.gov.au/Details/F2010C00718).

13.7 Conclusions

Therefore, substances used to prepare coatings for food and pharmaceutical products are considered food additives and are regulated as such subject to their function in coatings. Coatings are generally used somewhat in the context of "packaging" food and drugs in that they impart containment and protection, thereby extending the shelf life and quality of the product. Coatings are also used as carriers of functional ingredients that are additives, such as antioxidants, antimicrobials, coloring, flavoring, and so forth. For that reason, coating ingredients are regulated and can be found in substance groups with many other designated technical functions unless they are intentionally separated as an independent group for regulation. Regulations can differ for different products, such as fruits or cheeses, depending on whether the coating is consumed, and regulations differ for food and pharmaceutical coatings. Regulations also obviously differ among different regions of the world, but efforts are underway to unify regulations.

References

ANZ—Australia and New Zealand. 1991. Food Standards Australia New Zealand Act 1991. Available at: www.comlaw.gov.au/comlaw/Legislation/ActCompilation1.nsf/0/34FDA538E7B40ACFCA256F71004DA6FE/$file/FoodStandANZ91.pdf. Accessed January 18, 2011.

ANZ—Australia and New Zealand. 2010. Australia New Zealand Food Standards Code (ANZC). Standard 1.3.1. Food Additives. Available at www.comlaw.gov.au/Details/F2010C00718. Accessed April 27, 2011.

Bai, J., E.A. Baldwin, and R.H. Hagenmaier. 2002. Alternatives to shellac coatings provide comparable gloss, internal gas modification, and quality for 'Delicious' apple fruit. *HortScience.* 37:559–563.

Bai, J., E.A. Mielke, P.M. Chen, R.A. Spotts, M. Serdani, J.D. Hansen, and L.G. Neven. 2006. Effect of high-pressure hot-water washing treatment on fruit quality, insects, and disease in apples and pears: Part I. System description and the effect on fruit quality of 'D'Anjou' pears. *Postharvest Biology and Technology.* 40:207–215.

Baldwin, E.A. 1994. Edible coatings for fresh fruits and vegetables: Past, present, and future, pp. 25–64. In: J.M. Krochta, E.A. Baldwin, and M.O. Nisperos-Carriedo (eds.). *Edible coating and films to improve food quality.* Technomic, Lancaster, PA.

Baldwin, E.A. 2004. Edible coatings, pp. 301–314. In: S. Ben-Yehoshua (ed.). *Environmentally friendly technologies for agricultural produce quality.* CRC Press/Taylor and Francis, Boca Raton, FL.

Baldwin, E.A., M.O. Nisperos, X. Chen, and R.D. Hagenmaier. 1996. Improving storage life of cut apple and potato with edible coating. *Postharvest Biology and Technology.* 9:151–163.

Baldwin, E.A., and B. Wood. 2006. Use of edible coating to preserve pecans at room temperature. *HortScience.* 41:188–192.

Bugusu, B., C. Meija, B. Magnuson, and S. Tafazoli. 2009. Global regulatory policies on food nanotechnology. *Food Technology.* 63:24–28.

CAC—Codex Alimentarius Commission. 1989. CAC/GL 36-1989. Codex Class Names and the International Numbering Systems for Food Additives. Available at: www.codexalimentarius.net/download/standards/7/CXG_036e.pdf. Accessed December 3, 2010.

Canada. 2010. Food and Drug Regulations C.R.C., c. 870. Available at: http://laws-lois.justice.gc.ca/PDF/Regulation/C/C.R.C.,_c._870.pdf. Accessed January 19, 2011.

CCA—Council of Canadian Academies. 2008. Report in Focus: Small is Different: A Science Perspective on the Regulatory Challenges. Available at: www.nanolawreport.com/JulyCanadaReport.pdf. Accessed December 3, 2010.

Chau, C.-F., S.-H. Wu, and G.-C. Yen. 2007. The development of regulations for food nanotechnology. *Trends in Food Science and Technology.* 18:269–280.

Cuppett, S.L. 1994. Edible coatings as carriers of food additives, fungicides, and natural antagonists, pp. 121–137. In: J.M. Krochta, E.A. Baldwin, and M.O. Nisperos-Carriedo (eds.). *Edible coating and films to improve food quality.* Technomic, Lancaster, PA.

EFSA—European Food Safety Authority. 2009. Scientific Opinion: The Potential Risks Arising from Nanoscience and Nanotechnologies on Food and Feed Safety. Available at: www.efsa.europa.eu/en/scdocs/doc/s958.pdf. Accessed December 3, 2010.

EC—Environment Canada. 1999. Canadian Environmental Protection Act 1999 (CEPA 1999). Available at: http://laws-lois.justice.gc.ca/PDF/C-15.31.pdf. Accessed May 9, 2011.

EU—European Union. 2002. Regulation (EC) No 178/2002. Available at: http://eur-lex.europa.eu/pri/en/oj/dat/2002/l_031/l_03120020201en00010024.pdf. Accessed December 3, 2010.

EU—European Union. 2008a. Regulation (EC) 1331/2008. Available at: http://eur-lex.europa.eu/LexUriServ/LexUriServ.do?uri=OJ:L:2008:354:0001:0006:EN:PDF. Accessed January 18, 2011.

EU—European Union. 2008b. Regulation (EC) No 1333/2008. Available at: http://eur-lex.europa.eu/LexUriServ/LexUriServ.do?uri=OJ:L:2008:354:0016:0033:en:PDF. Accessed December 3, 2010.

FDA—Food and Drug Administration. 2007. Nanotechnology Task Force Report 2007. Available at: www.fda.gov/ScienceResearch/SpecialTopics/Nanotechnology/NanotechnologyTaskForceReport2007/default.htm. Accessed December 21, 2010.

FDA, CFSAN—Food and Drug Administration, Center for Food Safety and Applied Nutrition. 2010. Everything Added to Food in the United States (EAFUS). Available at: www.accessdata.fda.gov/scripts/fcn/fcnNavigation.cfm?rpt=eafusListing. Accessed December 3, 2010.

Fortin, N.D. 2009. *Food regulation: Law, science, policy, and practice*. John Wiley and Sons, Hoboken, NJ.

Hagenmaier, R.D. 2004. Fruit coatings containing ammonia instead of morpholine. *Proc. Fla. State Hort. Soc.* 117:396–402.

Hagenmaier, R.D., and K. Grohmann. 1999. Polyvinyl acetate as a high-gloss edible coating. *Journal of Food Science*. 64:1064–1067.

Huggett, A., B.J. Petersen, R. Walker, C.E. Fisher, S.H.W. Notermans, F.M. Rombouts, P. Abbott, M. Debackere, S.C. Hathaway, E.F.F. Hecker, A.G.A. Knaap, P.M. Kuznesof, I. Meyland, G. Moy, J.-F. Narbonne, J. Paakkanen, M.R. Smith, D. Tennant, P. Wagstaffe, J. Wargo, and G. Wurtzen. 1998. Towards internationally acceptable standards for food additives and contaminants based on the use of risk analysis. *Environmental Toxicology and Pharmacology*. 5:227–236.

JMHLW—Japan Ministry of Health, Labour and Welfare. 1996a. List of Plant or Animal Sources of Natural Flavoring Agents. Available at: www.ffcr.or.jp/zaidan/FFCRHOME.nsf/pages/list-nat.flavors. Accessed January 18, 2011.

JMHLW—Japan Ministry of Health, Labour and Welfare. 1996b. List of Existing Food Additives. Available at: www.ffcr.or.jp/zaidan/FFCRHOME.nsf/pages/list-exst.add. Accessed January 18, 2011.

JMHLW—Japan Ministry of Health, Labour and Welfare. 2000. Japan's Specifications and Standards for Food Additives (7th ed). Available at: www.ffcr.or.jp/zaidan/FFCRHOME.nsf/pages/spec.stand.fa. Accessed December 3, 2010.

JMHLW—Japan Ministry of Health, Labour and Welfare. 2010. List of Designated Additives. Available at: www.ffcr.or.jp/zaidan/FFCRHOME.nsf/pages/list-desin.add-x. Accessed January 18, 2011.

NCNST—National Center for NanoScience and Technology of China. 2009. Nanotoxicological Study of Manufactured Nanomaterials: China Nanosafety Research Project. Available at: http://english.nanoctr.cas.cn/rh/rps/200907/t20090703_19578.html. Accessed December 3, 2010.

Ravishankar, S., L. Shu, C.W. Olsen, T.H. McHugh, and M. Friedman. 2009. Edible apple film wraps containing plant antimicrobials inactivate foodborne pathogens on meat and poultry products. *Journal of Food Science*. 74:M440–M445.

USDA—U.S. Department of Agriculture. 2002. The National List of Allowed and Prohibited Substances. Available at: www.ams.usda.gov/AMSv1.0/getfile?dDocName=STELPRDC5068682&acct=nopgeninfo. Accessed January 19, 2011.

U.S. Statutes at Large. 1906. Pure Food and Drug Act of 1906. Available at: www.usd116.org/ProfDev/AHTC/lessons/DPLhealth/DPLfooddrugact.pdf. Accessed December 3, 2010.

Index

A

AA, *see* Ascorbic acid (AA)
Acacia gums
   flavor retention, 346
   fundamentals, 104
   polysaccharide coatings, 104
   properties, 114–115
Acacia senegal, 114
Acacia seyal, 114
acid-converted starches, 109
Actinidia arguta, 220
Actinidia deliciosa, 220
additives, role of
   antibrowning agents, 173–175
   antimicrobial agents, 170–173
   antioxidants, 173–175
   colorants, 175
   emulsifiers, 167–170
   flavors, 175
   fundamentals, 157–158
   nanoparticles, 176–177
   nutrients, 175
   plasticizers, 158–167
   polysaccharide films and coatings, 165–167
   protein films and coatings, 159–165
agar, 114
albedo, 216
albumins, 29
alcohol, 5
alginates
   development, 187
   emulsifiers, 170
   incorporation in films, 342
   minimally processed fruits and vegetables, 264
   plasticizers, polysaccharide films/coatings, 166
   polysaccharide coatings, 126–127
   properties, 111–113
   response to coatings, 224
   water vapor permeability, 360
alginic acid, 111
Alimentarius Austriacus, 104, 384, 402
Aloe vera
   development of, 187
   lipids, 270
   response to coatings, 218
Alternaria sp., 218
Alternaria alternata, 220
Alzheimer's disease, 362
Amasya apples, 311
American Society for Testing and Materials, 142
ammonia, 5
amphiphilic character, proteins, 168
amylopectin
   modfied starches, 109
   raw starches, 108–109
   starch and starch derivatives, 267
amylose
   plasticizers, polysaccharide films/coatings, 166
   raw starches, 108–109
   starch and starch derivatives, 267
anaerobic conditions, 216
analytical errors, 144–145
Anastrepha suspensa, 201
anionic cellulose ethers, 107–108, 166
Anjou pears
   Nutri-Save, 308
   polysaccharide coatings, 206
   Pro-long, 309

418 *Index*

response to coatings, 214–215
superficial scald, 199
Ankara pears, 311
Anogeissus latifolia, 116
antibacterials, 362–363, *see also specific type*
antibrowning agents, *see also* Browning
   components, edible coatings, 5
   role of additives, 173–175
   uses, edible coatings, 4
antifoam agents, 5
antifungal properties, 171, *see also* Fungicides
antimicrobial activity and agents
   collagen and gelatin coatings, 49
   components, edible coatings, 5
   role of additives, 170–173
   uses, edible coatings, 4
   whey protein films, 39–40
   zein protein films, 24
antioxidants
   edible coatings, 259
   gellan gum, 265
   nutritional and flavor changes, 247
   role of additives, 173–175
   uses, edible coatings, 4
appearance
   future directions, 98
   volatile molecules, 356
apples
   alginates, 264
   antimicrobial agents, 171
   antioxidants and antibrowing agents, 173, 175
   carrageenans, 265
   casein protein coatings, 44
   cellulose and cellulose derivatives, 266, 267
   coating optimization, 272
   composite and bilayer coatings, 271
   composites and bilayer coatings, 213
   decreased nutritional quality, 197
   discoloration and gloss reduction, 195
   emulsion coatings, 98
   ethanol concentrations, 3
   ethylene production, 245
   extended shelf life, 194
   films *vs.* emulsion coatings, 259
   firmness, loss of, 194
   flavor loss, 198
   gellan gum, 265
   incorporation into films, 341
   internal gases, 3
   lipids, 202, 270
   minimally processed foods, 260
   NatureSeal, 306–307
   nitrogen and mass transfer, 151
   Nutri-Save, 308
   pectins and derivatives, 111
   permeance equations, 148, 150
   polysaccharide coatings, 122, 206, 212
   postprocessing practices, 256
   Pro-long, 309
   properties, successful application, 7
   resins, 202
   respiration, oxygen/carbon dioxde effect, 191
   response to coatings, 213–215, 221
   Semperfresh, 310–311
   shellac coating, 96
   soy protein, 270
   spray application, 93
   superficial scald, 198–199
   surface characteristics, 327
   sweating, 91
   volatile molecules, films, 340
   water loss, 193
   waxes, 87, 96
   whey protein, 40, 268
   zein coatings, 24
apple wraps
   lipids, 270
   polysaccharide coatings, 122
applications
   pharmaceutical coatings, 377–378
   properties for successful, 6
applications, commercial coatings
   confectionaries, 321–322
   dip application, 322–323
   drip application, 323
   electrostatic coating, 325–326
   fluidized-bed coating, 324
   foam application, 323–324
   fresh and fresh-cut fruits and vegetables, 320–322
   functional sustances, incorporation, 328–329
   fundamentals, 319–320, 329
   gas barrier properties, 326–327
   keys to successful application, 326–329

method of application, 322–326
moisture barrier properties, 326–327
nuts, 322
panning application, 324–325
processed foods, 321
sensory attributes, 327–328
spraying application, 325
surface characteristics, 327
applications, polysaccharide coatings
cereal-based products, 125–126
fish, 124–125
fruits, 120, 122–124
fundamentals, 120
meats, 124–125
nuts, 125–126
other applications, 126–127
poultry, 124–125
seafood, 124–125
vegetables, 120, 122–124
applications, to processed foods
cereal coatings, 297–298
confectionary coatings, 300–303
fresh fruits and vegetables, 305–312
fundamentals, 291
meat films and coatings, 291–297
NatureSeal, 305–307
nut coatings, 298–300
Nutri-Save, 307–308
pouches, 303–305
Pro-long, 308–309
raisin coatings, 298–300
Semperfresh, 309–312
shellac, 312
strips, 303–305
waxes, 312
applicators
fundamentals, 92–95
future directions, 98
apricots, 197, 218
arabic gum, 300, 302
areas of concern, edible coatings, 7, 8
aroma compounds, 350–354
arrowroot jelly, 303
artichoke, 251
Ascophyllum nodosu, 112
Ascophyllum vinelandii, 112
ascorbic acid (AA)
antioxidants and antibrowing agents, 173
future trends, 276
nutritional and flavor changes, 247–248

asparagus
coatings, 187–188
cultivar selection, 253
discoloration and gloss reduction, 195
disease and physical injury, 201
ethylene production, 245
postharvest hardening, 195
surface microbial flora, 252
Astragalus spp., 115
Atlantic cod, 125
atmosphere modification, 2
Australia, 389, 406
Autumn Seedless grapes, 218
Avicel, 105
avocados
coatings, minimally processed foods, 260
lipids, 270
NatureSeal, 307
respiration, oxygen/carbon dioxde effect, 191
response to coatings, 221–222

B

Bacillus sp., 363
Bacillus cereus, 172
Bacillus natto, 28
bacon simulation, 303
bacteriocin, 363
baked products
antimicrobial agents, 171
applications, films/coating, 104
polysaccharide coatings, 104, 122, 126
seed gums, 117
bananas
carrageenans, 265
coatings, minimally processed foods, 260
disease and physical injury, 200
film development, 187
lipids, 270
polysaccharide coatings, 122, 123, 206, 212
Pro-long, 308
respiration, stress response, 246
response to coatings, 220–221
Semperfresh, 310
ultraviolet light, 257

Barlett pears
    NatureSeal, 306
    Nutri-Save, 308
    response to coatings, 214
    Semperfresh, 311
barrier properties measurement
    analytical errors, 144–145
    coating permeability, 145
    equipment, 144
    methods, 142–143
barriers
    moisture and gas, 326–327
    volatile molecules, 348–350, 358–361
beam sliders, 94
beans, fresh, 206
BeeCoat, 222
beef
    antimicrobial agents, 172
    polysaccharide coatings, 125
    wheat gluten protein coatings, 33
beeswax
    casein films, 42
    cellulose and cellulose derivatives, 266
    coating optimization, 272
    components, edible coatings, 5
    composites and bilayer coatings, 212
    emulsifiers, 168
    fundamentals, 84
    historical developments, 87
    response to coatings, 219
    wheat gluten protein films, 32
    whey protein, 35–38, 269
bell peppers, *see also* Peppers
    cultivar selection, 253
    mineral oil, 79
benefits, edible coatings, 7, 8
bergamot oil
    film mechanical properties, 357
    incorporation into films, 341
    retention and release, 348
berries, response to coatings, 219, *see also specific type of berry*
beverages
    Acacia gums, 115
    antimicrobial agents, 171
    seed gums, 117
Bifidobacterium lactis, 122, 276
bilayer coatings
    minimally processed fruits and vegetables, 271–272
    properties, 212–213
    zein protein films, 19
biopolymer materials, 164–165
biscuits, 126
blades, cutting, 254
Blanquilla pears, 256
bleaching, 109
blueberries, 191
Blue goose plums, 307
bologna
    polysaccharide coatings, 125
    whey protein coatings, 40
Bosc pears, 214–215
Botrytis cinerea, 218, 219, 224
bottom spray systems, 377
Braeburn apples
    ethanol concentrations, 3
    flavor loss, 198
    internal gases, 3
    polysaccharide coatings, 206
    response to coatings, 213–214
    water loss, 193
breadfruit, 199, 223
breads, 53, 297, *see also* Baked products
Breath Strips, 295
Brevipalpus chilensis, 201
British gums, 109
broccoli
    cultivar selection, 253
    enzyme activity, 247
    heat treatments, 257
    surface microbial flora, 252
    volatile molecules, films, 340
Brochothrix thermosphacta, 293
brown algae, 112
brownies, 298
browning, *see also* Antibrowning agents
    litchi fruit, 222–223
    minimally processed fruits and vegetables, 249–250
brown seaweed, 264
brushes
    application, historical developments, 87
    material, 91
    spiral cut, 93
    straight-cut, 93
    tumble-trim, 93
    wear, 90

buffalo meat, 296
bulk laxatives, 117
buns, 297, *see also* Baked products
Burlat cherries, 218

C

cabbage
    nutritional and flavor changes, 248
    surface microbial flora, 252
Caesalpinia spinosa, 117
cake decorations, 301
cakes, 115, *see also* Frosting, cake
calcium salts, 27–28
Campylocater jejuni, 172
Canada, 389, 406
candelilla wax
    components, edible coatings, 5
    flavor loss, 198
    fundamentals, 84
    historical developments, 87, 187
    lipids, 270
    response to coatings, 213–214
    whey protein films, 35–36
candies
    applications, films/coating, 301
    beeswax, 84
    candelilla wax, 84
    historical developments, 2
    zein coatings, 24
cantaloupes
    emulsion coatings, 97
    ethylene production, 246
    modified atmosphere packaging, 258
    nutritional and flavor changes, 248, 249
    respiration, stress response, 246
    response to coatings, 220
    soy protein, 270
capsules
    carbohydrates, 336
    coacervation, 338
    controlled release, 339
    diffusion, controlled release, 339
    extrusion, 338
    freeze-drying method, 337–338
    matrices, 335–337
    matrix degradation, controlled release, 339
    melting, controlled release, 339
    methods, 337–339
    proteins, 337
    spray-cooling/chilling method, 338
    spray-drying method, 337
    swelling, controlled release, 339
carambolas
    dip application, 92
    drip application, 96
carbohydrates
    capsules, 336
    uses, edible coatings, 3
carbomers, 376
carbon dioxide
    casein films, 43
    coatings, fruits and vegetables, 203–206
    effect on respiration, 188, 191
    gelatin films, 48
    zein protein films, 19
carboxymethylcellulose (CMC)
    anionic cellulose ethers, 107
    flavor loss, 198
    fundamentals, 105, 266
    nonionic cellulose ethers, 106
carnuba wax
    cellulose and cellulose derivatives, 266
    components, edible coatings, 5
    discoloration and gloss reduction, 197
    fundamentals, 80
    historical developments, 87, 186–187
    lipids and resins, 202
    permeance equations, 148
    response to coatings, 213–214, 222
    whey protein, 269
    whey protein films, 36
    zein protein films, 16
carob bean gum, 117, *see also* Locust bean gum
carotenoid degeneration, 248
carrageenans
    antioxidants and antibrowing agents, 173
    fundamentals, 104
    incorporation in films, 342
    minimally processed fruits and vegetables, 265
    properties, 113–114
    seed gums, 117
carrots
    antioxidants and antibrowing agents, 173
    casein protein coatings, 44

422 Index

cellulose and cellulose derivatives, 267
chitosan coating, 263
coatings, minimally processed foods, 260–261
composites and bilayer coatings, 213
cultivar selection, 253
enzyme activity, 247
ethylene production, 245
NatureSeal, 305
nutrients, 175
nutritional and flavor changes, 248
polysaccharide coatings, 123, 124, 212
respiration, stress response, 246
response to coatings, 225
soy protein, 270
volatile molecules, films, 340
whey protein, 268
xanthan gum, 266
carvacrol
   antimicrobial agents, 171, 172
   barrier properties, 359
   meat film/coatings, 296
CarWax
   casein films, 42
   whey protein films, 37–38
casein-based films and coatings
   applications, films/coating, 300
   coatings, 44
   formation of, 41
   functional properties, 41–44
   fundamentals, 40–41
   mechanical strength, 161–162
   plasticizers, 161–162
caseins
   emulsifiers, 170
   minimally processed fruits and vegetables, 268–269
casings
   collagen and gelatin protein films, 46, 291–292
   historical developments, 2
cassava and cassava roots, 92, 224
casting method, 30, 45
cauliflower
   postharvest hardening, 195
   surface microbial flora, 252
celery
   casein protein coatings, 44
   coatings, minimally processed foods, 261

nutritional and flavor changes, 248
respiration, stress response, 246
cellulose and cellulose derivatives, *see also specific derivative*
   anionic cellulose ethers, 107–108
   emulsifiers, 167
   fundamentals, 105
   microcrystalline cellulose, 105–106
   microfibrillated cellulose, 106
   minimally processed fruits and vegetables, 266–267
   nonionic cellulose ethers, 106–107
   plant cell walls, 194
   plasticizers, polysaccharide films/coatings, 166
   properties, 105–108
cellulose ethers, 170
Ceratonia siliqua, 117
cereal-based products
   applications, films/coating, 298, 302
   coating applications, 297–298
   collagen and gelatin coatings, 49
   edible coatings, 3–4
   egg white protein coatings, 53
   polysaccharide coatings, 125–126
   properties, successful application, 6–7
CertiCoat, 302
Certified, 301
Certiseal, 302
CFR, *see* Code of Federal Regulations (CFR)
chalking, 91, *see also* Whiting
cheese, *see also* Dairy products
   applications, films/coating, 303–304
   barrier performances, 349
   wheat gluten protein coatings, 33
cheese-seasoned crackers, 302
cheese wheels, wax coating not eaten, 6
chemical treatments, postprocessing, 255–256
cherimoyas, 201
cherries
   composites and bilayer coatings, 212
   decreased nutritional quality, 197
   extended shelf life, 194
   Nutri-Save, 308
   polysaccharide coatings, 206
   response to coatings, 218
   Semperfresh, 312

cherry tomatoes, 33
chewing gum
   antimicrobial agents, 171
   beeswax, 84
   candelilla wax, 84
chicken, *see also* Poultry
   applications, films/coating, 296
   meat film/coatings, 297
   polysaccharide coatings, 124
   zein coatings, 24
chilling injury
   fundamentals, 199
   postprocessing practices, 255
   response to coatings, 215
China, Semperfresh, 310
Chinese pears, 199
Chinese water chestnuts, *see* Water chestnuts
chitosan/chintin
   clove oil, 342
   disease and physical injury, 200
   fundamentals, 376
   minimally processed fruits and vegetables, 263–264
   nutrients, 175
   polysaccharide coatings, 212
   properties, 119–120
   response to coatings, 217–218, 221, 223–224
Chloraseptic Strips, 295
chlorinated starch, 109
chlorinated water, washing with, 254
chocolate
   applications, films/coating, 293, 300–303
   commercial coatings, 321–322
   uses, edible coatings, 3
Chondrus crispus, 265
cinnamaldehyde
   alginates, 264
   antimicrobial agents, 172
   barrier properties, 359
   meat film/coatings, 296
cinnamon oil
   alginates, 264
   antimicrobial agents, 171
   barrier properties, 359
   film mechanical properties, 357
citral oil, 264, 359
Citruseal, 200, 220
citrus fruit
   casein protein coatings, 44
   chilling injury, 199
   dehydration and shrinkage, 193
   dip application, 91
   discoloration and gloss reduction, 195
   emulsion coatings, 97–98
   lipids and resins, 202
   nitrogen and mass transfer, 151
   pectins and derivatives, 111
   permeance equations, 148, 150
   pitting, 199
   polysaccharide coatings, 123, 212
   response to coatings, 215–217
   solvent wax, 87
   spray application, 93
   sweating, 91
   wax, historical developments, 87
Cladosorium sp., 171
clear emulsions, 167
climacteric fruit, 191
Clostridium sp., 363
Clostridium perfringens, 172
clove oil, 171
Clupea harengus, 125
CMC, *see* Carboxymethylcellulose (CMC)
coacervation, 338
coating permeability, 145
coatings, *see also* Films; Wraps
   casein-based films and coatings, 44
   collagen-based films and coatings, 48–49
   egg white protein films and coatings, 53
   food *vs.* pharmaceutical, 374
   fundamentals, 1
   gelatin-based films and coatings, 48–49
   keratin-based films and coatings, 54
   myofibrillar protein-based films and coatings, 51
   soy-based films and coatings, 29
   wheat gluten protein-based films and coatings, 33
   whey protein-based films and coatings, 40
   zein-based films and coatings, 24
coatings, fresh fruits and vegetables
   apples, 213–215
   bananas, 220–221

berries, 219
bilayer coatings, 212–213
carbon dioxide, 188, 191
citrus fruit, 215–217
climacteric fruit, 191
composites, 212–213
dehydration, 193–194
discoloration, 195–197
disease, 199–201
factors affecting, 191–192
flavor loss, 197–198
fruit vegetables, 223–224
fundamentals, 186–188, 225–226
gloss reduction, 195–197
grapes, 218–219
hardening, 194–195
kiwifruit, 219–220
leafy vegetables, 225
lipids, 202
mangoes, 221–223
mealy texture, 194–195
melons, 220
nonclimacteric fruit, 191
nutritional quality, decreased, 197
oxygen, 188, 191
pears, 213–215
physical injury, 199–201
physiological disorders, 198–199
polysaccharides, 206, 212
postharvest deterioration, effect of coatings, 193–202
postharvest physiology, 188–193
product type influence, 193
properties, 202–213
proteins, 212
quarantine treatments, 201–202
relative humidity, 192–193
resins, 202
respiration, 188, 191–192
responses to coatings, 213–225
root crops, 224–225
shrinkage, 193–194
softening, 194–195
stone fruits, 217–218
strawberries, 219
temperature, 192–193
tomatoes, 223–224
transpiration, 192–193
tropical fruits, 221–223
tuber crops, 224–225

coatings, minimally processed fruits and vegetables
alginates, 264
bilayer coatings, 271–272
carrageenan, 265
casein, 268–269
cellulose and derivatives, 266–267
chitosan, 263–264
coating optimization, 272
composite coatings, 271–272
films *vs.* emulsion coatings, 259, 263
fundamentals, 258–259
gellan, 265
lipids, 270–271
pectin, 267
polysaccharides, 263–268
proteins, 268–270
soy protein, 269–270
starch and derivatives, 267–268
whey protein, 268–269
xanthan gum, 265–266
coatings, responses to
apples, 213–215
bananas, 220–221
berries, 219
citrus fruit, 215–217
fruit vegetables, 223–224
grapes, 218–219
kiwifruit, 219–220
leafy vegetables, 225
mangoes, 221–223
melons, 220
pears, 213–215
root crops, 224–225
stone fruits, 217–218
strawberries, 219
tomatoes, 223–224
tropical fruits, 221–223
tuber crops, 224–225
Cochlospernum genus, 116
coconuts, 222
Code of Federal Regulations (CFR), 168, 377, 405
Codex *Alimentarius Austriacus,* 104, 384, 402
Coffi, 292, 295
cold-smoke sardines, 49, 342
collagen-based films and coatings
coatings, 48–49
formation of, 46–47

functional properties, 47
fundamentals, 46
historical developments, 2
meat film/coatings, 296
plasticizers, 162
colorants, 175
coloration, 7
color changes, 249–250
color development, 214
Color Index, 195
colostomy bags, 117
Comice pears, 214
commercial coatings
   brands available, 207–211
   confectionaries, 321–322
   dip application, 322–323
   drip application, 323
   electrostatic coating, 325–326
   fluidized-bed coating, 324
   foam application, 323–324
   fresh and fresh-cut fruits and vegetables, 320–322
   functional sustances, incorporation, 328–329
   fundamentals, 319–320, 329
   gas barrier properties, 326–327
   keys to successful application, 326–329
   method of application, 322–326
   moisture barrier properties, 326–327
   nuts, 322
   panning application, 324–325
   processed foods, 321
   sensory attributes, 327–328
   spraying application, 325
   surface characteristics, 327
   types, 207–211
components, edible coatings, 4–6
composite coatings
   emulsifiers, 167, 168–169
   minimally processed fruits and vegetables, 271–272
composites, 212–213
compression molding/extrusion, 164
concern, areas of, 7, 8
Concord grapes, 219
confectionary products
   applications, films/coating, 297
   beeswax, 84
   coating applications, 300–303

commercial coatings, 321–322
polysaccharide coatings, 104, 122
zein coatings, 24
Congress involvement, 158
controlled release, 24, 339
copal, 85
Copernica cerifera, 80
corn wet-milling industry, 15, *see also* Zein
costs, future directions, 98
cottonseed-based films and coatings
   formation of, 44–45
   functional properties, 45–46
   fundamentals, 44
coumarone indene resin, 85, 87
cowpea past, 24
crabsticks, 113
crackers
   applications, films/coating, 302
   polysaccharide coatings, 126
crenshaw melons, 246
Crimson Seedless grapes, 218, 219
Cripps Pink apples, 193
cross-linking
   anionic cellulose ethers, 107–108
   casein films, 41, 43
   chitosan/chintin, 120
   collagen and gelatin protein films, 47
   cottonseed-protein films, 45–46
   edible coatings, 259
   egg white protein films, 52
   gelatin films, 48
   keratin protein films, 54
   meat film/coatings, 296
   modfied starches, 110
   peanut protein, 55
   pea protein, 56
   polysaccharide coatings, 126
   soybean films and coatings, 26, 27–28
   wheat gluten protein films, 32
   whey protein films, 38–39
cross-over phenomenon, 161–162
Cryptosporidium sp., 252
Crystalac/Crystalac Z2, 294, 301
cucumber pickle brine protein, 58
cucumbers
   chilling injury, 199
   cultivar selection, 253
   disease and physical injury, 200

lipids and resins, 202
mineral oil, 79
respiration, oxygen/carbon dioxde effect, 191
response to coatings, 223, 225
Cucurbita pepo, 311
Cupressus sempervirens, 342
cuticle waxes, 3
cutting blades, 254
cutting shape, 254
Cyamopsis tetragonolobus, 117
cyclodextrins, 361
Cyclospora sp., 252
cypress, 342

D

dairy products, *see also specific type*
　barrier performances, 349
　karaya gum, 117
　polysaccharide coatings, 104
　seed gums, 117
damar, 85
d'Anjou pears
　Nutri-Save, 308
　polysaccharide coatings, 206
　Pro-long, 309
　response to coatings, 214–215
　superficial scald, 199
DE, *see* Degree of esterification (DE)
Decco Lustr, 222
deciduous fruits, 200
defuzzing, 217
degreening
　Pro-long, 309
　response to coatings, 215
degree of esterification (DE), 110, 267
dehydration, 193–194
Delicious apples
　ethanol concentrations, 3
　internal gases, 3
　mass loss and shrinkage, 194
　Nutri-Save, 308
　Pro-long, 309
　response to coatings, 214
dental impression materials, 127
dentures, 117
design, 96–98
desserts, 104, 117
dewaxed shellac, 5

dextrins
　modfied starches, 109
　starch and starch derivatives, 268
diffusion
　controlled release capsules, 339
　to food surfaces, 364–365
Dioscorea sp., 267
dip application
　commercial coatings, 322–323
　historical developments, 87
　lipids, waxes, and resins, 91–92
discoloration
　determination, coating properties, 7
　postharvest deterioration, 195–197
disease, 199–201
D-limonene
　aroma, oxidative degradion, 354
　barrier performances, 348–349
　barrier properties, 359
　loss during drying, 345
　retention and release, 348
　water vapor permeability, 360
dressings, 117
drip application
　commercial coatings, 323
　lipids, waxes, and resins, 95–96
　vs. spray application, 95
drug delivery tablets, 127
drug release, 49, 126
dry mixes, 115
dual-fluid nozzles, 93
Durkex 500, 295
Durvillea Antarctica, 112

E

edible coatings
　adverse effects, 328
　areas of concern, 7, 8
　benefits of, 7, 8
　components, 4–6
　defined, 1
　determination of properties, 7
　historical developments, 2
　important properties, 6–7
　minimally processed foods, 260–263
　studies, fruits and vegetables, 189–190
　uses for, 2–4
edible ink and ink-jet printing, 303
Edonia maxima, 112

Index

effusion
  gas-exchange properties, 138–139
  permeance comparison, 147–148
egg-box model, 111, 112
eggplants
  coatings, minimally processed foods, 261
  mineral oil, 79
  response to coatings, 223
  soy protein, 270
eggs, 166
egg white protein films and coatings
  coatings, 53
  functional properties, 52–53
  fundamentals, 51–52
electrophoresis, 56
electrostatic coating application, 325–326
elemi, 85
elliptical chain applicators, 94
EMC, *see* Ethylmethylcellulose (EMC)
emulsifiers, 167–170
emulsion formulations
  pressure-spray application, 95
  technology, fruit coatings, 88–90
emulsion *vs.* film coatings, 259, 263
encapsulation
  antibacterials, 362–363
  appearance of film, 356
  barrier performances, 348–350
  barrier properties, 358–361
  capsules, 335–337
  carbohydrates, 336
  coacervation, 338
  controlled release, 339
  diffusion, controlled release, 339
  diffusion to food surfaces, 364–365
  encapsulation influence, 356–361
  extrusion, 338
  flavor compounds, 340–343, 358–361
  freeze-drying method, 337–338
  fundamentals, 334–335, 366
  hydroxypropyl methylcellulose, 363
  matrices, 335–337, 340–343
  matrix degradation, controlled release, 339
  melting, controlled release, 339
  methods, 337–339
  microstructure changes, 356
  migrant/matrix state chemical structure, 365–366

  nonvolatile molecules, 361–366
  nutraceuticals, 361–364
  physical-chemical properties, effect on, 354, 356–361
  plasticizers, 364
  polyacetic acid, 364
  protection, flavor compounds, 350
  proteins, 337
  retention and release, 343–348
  spray-cooling/chilling method, 338
  spray-drying method, 337
  structural properties, effect on, 354, 356–361
  swelling, controlled release, 339
  transparency of film, 356
  uses, edible coatings, 4
  volatile molecules, 340–361
endive, 245, 246
Enterobacter aerogenes, 306
Enterobacteriaceae, 125
Enterobacter sp., 251
environmental scanning electron microscope (ESEM), 356
enzyme activity, 246–247
equipment, 144
errors, 144–145
errosion, 364–365
Erwina sp., 251
Escherichia coli
  antimicrobial agents, 171, 172
  clove oil, 342
  disease and physical injury, 201
  essential oils, 125
  NatureSeal, 306
  soybean protein coatings, 29
  surface microbial flora, 252
  whey protein coatings, 40
  ZnO nanoparticles, 187
ESEM, *see* Environmental scanning electron microscope (ESEM)
essential oils, *see also* Volatile molecules, films; *specific type*
  alginates, 264
  antifungal properties, 171
  antimicrobial agents, 171–173
  incorporation into films, 341
  uses, edible coatings, 4
  whey protein films, 39–40

esterification, *see* Degree of
    esterification (DE)
ethanol concentrations, 3
ethyl cellulose, 375
ethylene production, 192, 245–246
ethylmethylcellulose (EMC),
    105, 106
Euphorbia antisphylitica, 84
Euphorbia cerifera, 84
European Communities, 362
European Economic Community, 310
European Union
    food additive standards, 402, 405
    hydroxypropyl methylcellulose, 363
    polysaccharide coatings, 104
    regulatory aspects, coatings, 387
excipients, 373
exopolysaccharides, 118–119
external plasticizers, 164
extruded products, 126
extrusion processes
    capsules, 338
    uses, edible coatings, 4
    wheat gluten protein films, 30
exudate gums
    Acacia, 114–115
    ghatti, 116
    karaya, 116–117
    properties, 114–117
    tragacanth, 115–116

F

FAO, *see* Food and Agriculture
    Organization of the United
    Nations (FAO)
FAO/WHO Codex *Alimentarius
    Austriacus,* 104, 384, 402
fatty acids
    components, edible coatings, 5
    emulsifiers, 168
    nutritional and flavor changes, 248
    response to coatings, 219
    whey protein films, 35–36
FDA, *see* Food and Drug Administration
    (FDA)
fennel, 342
ferulic acid, 27
Fick's laws
    determination, coating properties, 7

diffusion, 364
homogeneous barriers, 139
fillings, 53, 113
films, *see also* Coatings; Wraps
    fundamentals, 1
    volatile compound impact, 356–358
    vs. emulsion coatings, 259, 263
firming agents, 4
fish oil, 24
fish products, *see also* Seafood
    collagen and gelatin coatings, 49
    collagen films, 47
    essential oils, 342
    myofibrillar and sarcoplasmatic protein
        films, 51
    myofibrillar protein films, 49–50
    polysaccharide coatings, 120,
        124–125
flavedo, 216
flavor changes, 247–249
flavor compounds
    barrier property impact, 358–361
    involvement, 340–343
flavor loss, 197–198
flavored film strips, 295
flavors, role of additives, 175
Flavr Savr tomatoes, 253
flesh translucency, 251
fluid-bed processes, 377
fluidized-bed coating application, 324
foam application
    commercial coatings, 323–324
    lipids, waxes, and resins, 92
Foeniculum vulgare, 342
food additive standards
    Australia and New Zealand Food
        Standards Code, 406
    Canada's Food and Drug
        Regulations, 406
    EU's directives, 402, 405
    FAO/WHO Codex Alimentarius, 402
    Japan's Specifications and
        Standards, 406
    US FDA's Code of Federal
        Regulation, 405
food additives, regulatory aspects, 393,
    401–402
Food and Agriculture Organization of the
    United Nations (FAO)
    fundamentals, 385–386

hydroxypropyl methylcellulose, 363
polysaccharide coatings, 104
Food and Drug Administration (FDA)
  antibacterials, 362
  polysaccharide coatings, 104
  Semperfresh, 310
Food and Drug Regulations, 406
food and food ingredients
  coatings, 389–393
  contaminants, 393, 400
  safety, 401
food-borne disease outbreaks, 252
Food Chemical Codex, 376
food grade, defined, 6
food labeling laws, 5–6
Foods Glaze Sheets, 293, 295
Food Standards Code, 406
formation
  casein-based films and coatings, 41
  collagen-based films and coatings, 46–47
  cottonseed-based films and coatings, 44–45
  gelatin-based films and coatings, 46–47
  keratin-based films and coatings, 54
  myofibrillar protein-based films and coatings, 49–50
  soy-based films and coatings, 25
  wheat gluten protein-based films and coatings, 30
  whey protein-based films and coatings, 34–35
Fourier transform infrared (FTIR) analysis, 354
Fourier transform infrared spectroscopy, 56
Fourier transform Raman spectroscopy, 24
fracturing, 91, *see also* Whiting
freeze-drying method, 337–338
fresh-cut fruits and vegetables, *see also* Minimally processed fruits and vegetables
  commercial coatings, 320–322
  polysaccharide coatings, 104, 120, 122–1244
fresh fruits and vegetables, *see also* Fruits; Vegetables
  apples, 213–215
  bananas, 220–221
  berries, 219
  bilayer coatings, 212–213
  carbon dioxide, 188, 191
  citrus fruit, 215–217
  climacteric fruit, 191
  coating applications, 305–312
  commercial coatings, 320–322
  composites, 212–213
  dehydration, 193–194
  discoloration, 195–197
  disease, 199–201
  factors affecting, 191–192
  flavor loss, 197–198
  fruit vegetables, 223–224
  fundamentals, 186–188, 225–226
  gloss reduction, 195–197
  grapes, 218–219
  hardening, 194–195
  kiwifruit, 219–220
  leafy vegetables, 225
  lipids, 202
  mangoes, 221–223
  mealy texture, 194–195
  melons, 220
  NatureSeal, 305–307
  nonclimacteric fruit, 191
  Nutri-Save, 307–308
  nutritional quality, decreased, 197
  oxygen, 188, 191
  pears, 213–215
  physical injury, 199–201
  physiological disorders, 198–199
  polysaccharides, 206, 212
  postharvest deterioration, effect of coatings, 193–202
  postharvest physiology, 188–193
  product type influence, 193
  Pro-long, 308–309
  properties, 202–213
  proteins, 212
  quarantine treatments, 201–202
  relative humidity, 192–193
  resins, 202
  respiration, 188, 191–192
  responses to coatings, 213–225
  root crops, 224–225
  Semperfresh, 309–312
  shellac, 312
  shrinkage, 193–194
  softening, 194–195
  stone fruits, 217–218
  strawberries, 219

temperature, 192–193
tomatoes, 223–224
transpiration, 192–193
tropical fruits, 221–223
tuber crops, 224–225
waxes, 312
fried foods and frying oil, 124, 297
frosting, cake
  applications, films/coating, 302
  beeswax, 84
  uses, edible coatings, 3
frozen desserts, 117
fruit fillings, 113
fruit flies, 201
fruit vegetables, 223–224
fruits, *see also specific type;* Fresh fruits and vegetables, coatings for
  barrier performances, 349
  coatings, as test, 148, 150
  edible coatings, 4
  mass transfer, 150–153
  nitrogen, 150–153
  polysaccharide coatings, 120, 122–124
Fuji apples
  essential oils, 341
  flavor loss, 198
  NatureSeal, 306
  polysaccharide coatings, 122
  response to coatings, 213–214
  surface characteristics, 327
functional properties
  casein-based films and coatings, 41–44
  collagen-based films and coatings, 47
  cottonseed-based films and coatings, 45–46
  egg white protein films and coatings, 52–53
  gelatin-based films and coatings, 47–48
  keratin-based films and coatings, 54
  myofibrillar protein-based films and coatings, 50–51
  sarcoplasmic protein-based films and coatings, 50–51
  soy-based films and coatings, 25–29
  wheat gluten protein-based films and coatings, 30–33
  whey protein-based films and coatings, 35–40
  zein-based films and coatings, 16, 24
functional sustances, incorporation, 328–329
fungicides, *see also* Antifungal properties
  disease and physical injury, 200–201
  postharvest compatibility, 97
Furcellaria lumbricalis, 113
furcelleran, 104, 113
Fusarium semitectum, 220
future developments
  lipids, waxes, and resins, 98
  minimally processed fruits and vegetables, 273–276
  polysaccharide coatings, 127

G

Gadus morhua, 125
Gala apples
  polysaccharide coatings, 122
  propylene glycol effects, 6
  zein effects, 6
galactomannans, 117
garlic oil, 39–40
garnishes, 53
gas barrier properties
  collagen films, 47
  gelatin films, 47
  keys, successful application, 326–327
  wheat gluten protein films, 33
gas exchange
  analytical errors, 144–145
  barrier properties measurement, 142–145
  coating permeability, 145
  effusion, 138–139
  effusion/permeance comparison, 147–148
  equations, 139–153
  equipment, 144
  fruit coatings, as test, 148, 150
  fundamentals, 137, 139
  homogeneous barriers, 139–140
  layered coatings and films, 140–141
  mass transfer, 139, 150–153
  methods, 142–143
  moisture gradients, 141

nitrogen, fresh fruit, 150–153
permeability values, 146–147
permeance, 137–138
properties, successful application, 6
units for permeability/permeance, 141–142
gastric reflux, 127
GDL, *see* Glucono-δ-lactone (GDL)
gelatin-based films and coatings
    coatings, 48–49
    formation of, 46–47
    functional properties, 47–48
    fundamentals, 46, 376
    historical developments, 2
    meat film/coatings, 296
    plasticizers, 162
Gelcote, 293
gellan gum
    fundamentals, 104
    minimally processed fruits and vegetables, 265
    properties, 118–119
German standards, 142
ghatti gum, 116
gingerbread, 298
ginger oil, 357
gliadins, 29, 33
global migration, 364–365
globulins, 24, 29
gloss
    desirability, 91
    determination, coating properties, 7
    historical developments, 186–187
    permeance equations, 148
    postharvest deterioration, 195–197
    proteins, 212
    units comparison, 196
    whey protein coatings and films, 37, 40
glucono-δ-lactone (GDL), 27–28
glutenins, 29, 33
glycerol monostearate, 170
Golden Delicious apples
    coating optimization, 272
    composites and bilayer coatings, 213
    NatureSeal, 306
    Nutri-Save, 308
    polysaccharide coatings, 122
    Semperfresh, 310
    superficial scald, 199

Goldrush apples, 306
Graham's experiments, 138–139
Granny Smith apples
    ethanol concentrations, 3
    flavor loss, 198
    internal gases, 3
    NatureSeal, 306
    response to coatings, 213–215
    Semperfresh, 310
    superficial scald, 198–199
    water loss, 193
grapefruits, *see also* Citrus fruit
    chilling injury, 199
    coatings, minimally processed foods, 261
    dehydration and shrinkage, 193
    polysaccharide coatings, 123
    quarantine treatments, 201
    respiration, oxygen/carbon dioxde effect, 191
    response to coatings, 217
    waxes, 312
grape pomace extract, 172
grapes
    antimicrobial agents, 171
    collagen and gelatin coatings, 49
    response to coatings, 218–219
grape seed extract
    antimicrobial agents, 172
    meat film/coatings, 292–293
GRAS (generally recognized as safe), defined, 6
gray mold, 219
grease-resistant paper, 29
green apples, 191
green peppers, *see also* Peppers
    decreased nutritional quality, 197
    discoloration and gloss reduction, 195
green rot, 218
green tomatoes (mature)
    lipids and resins, 202
    Pro-long, 309
guar gum, 117
guavas
    disease and physical injury, 200
    response to coatings, 221, 222
    ultraviolet light, 257
gum balls, 301

H

HACS, *see* High-amylose corn starch (HACS)
Haden mangoes, 312
ham, 125
HAMFF, *see* Hard anhydrous milkfat fraction (HAMFF)
Hami melons, 197
hard anhydrous milkfat fraction (HAMFF), 36
hardening, 194–195
heat treatment, 256–257
hemicelluloses, 194
Henry's laws
    determination, coating properties, 7
    flavor retention, 346
    homogeneous barriers, 140
herb-of-the-cross essential oil, 342
herring, 125
high-amylose corn starch (HACS), 19
high-pressure spray applicators, 93
historical developments
    coatings, 186
    edible coatings, 2
    technology, fruit coatings, 87
homogeneous barriers, 139–140
honeydew melons
    antimicrobial agents, 171
    chilling injury, 199
    disease and physical injury, 200
    respiration, stress response, 246
honey pineapples, 257
horseradish peroxidaase enzyme, 27
hot dogs, 40
hot fog process, 87
HPMC, *see* Hydroxypropyl methylcellulose (HPMC)
humidity, *see also* Relative humidity
    flavor retention, 346
    surface microbial flora, 251
hydrogen bonding, 31
hydrogen peroxide, 254
hydrophobic coatings, 2
hydroxypropylcellulose (HPC)
    fundamentals, 105, 266
    nonionic cellulose ethers, 106–107
hydroxypropyl methylcellulose (HPMC)
    fundamentals, 105, 266, 375
    nonionic cellulose ethers, 106–107
nonvolatile molecules, encapsulation, 363
hypobaric treatments, 218

I

iceberg lettuce
    nutritional and flavor changes, 248
    surface microbial flora, 251–252
    water loss, 193
ice cream and ice cream cones, 4, 171
Ida Red apples, 310
Inactive Ingredient Search for Approved Drug Products, 377
ingredients, 407
ink and ink-jet printing, 303
insect pheromones, 40
internal gases, 3, 191
internal plasticizers, 164
Irish fish, collagen films, 47
irrigation drippers, 93

J

jams, 113
Japan
    food additive standards, 406
    regulatory aspects, coatings, 388
    water vapor standards, 143
jasmine rice, 126, 340
jelly bean coatings, 301
jelly products
    applications, films/coating, 303
    pectins and derivatives, 113
jerky products, 120
jicama, 247
Jonathan apples, 198
Jonfresh, 311

K

kafarin, 57, *see also* Sorghum protein coating
karaya gum
    fundamentals, 104
    plasticizers, polysaccharide films/coatings, 166
    properties, 116–117

Kensington Pride mangoes, 312
keratin-based films and coatings
  coatings, 54
  formation of, 54
  functional properties, 54
  fundamentals, 53
Kerria lacca, 312, *see also Laccifer lacca*
keys, successful application
  functional sustances, incorporation, 328–329
  fundamentals, 326
  gas barrier properties, 326–327
  moisture barrier properties, 326–327
  sensory attributes, 327–328
  surface characteristics, 327
King Edward potatoes, 309
kinnow citrus
  casein protein coatings, 44
  Semperfresh, 311
kiwifruit
  antimicrobial agents, 171
  ethylene production, 245
  nutritional and flavor changes, 248
  respiration, oxygen/carbon dioxde effect, 191
  response to coatings, 219–220
  Semperfresh, 311
kuzukiri, 303

L

Laccifer lacca, 85, *see also Kerria lacca*
lactic acid bacteria
  antibacterials, 363
  surface microbial flora, 251
Lactobacilus sp., 251
Lactobacilus acidophilus, 342
lactoferrin, 39–40
lactoferrin hydrolysate, 39–40
lactoperoxidate systems (LPOS), 39–40
Laminaria digitata, 112
Laminaria hyberborea, 112
Laminaria japonica, 112
larding, 2
Lavandula angustifolia, 342
lavender, 342
law sprinkler applicators, 94
laxatives, 117
layered coatings and films, 140–141

leafy vegetables
  enzyme activity, 247
  polysaccharide coatings, 212
  respiration, stress response, 246
  response to coatings, 225
  water loss, 193
lemongrass oil
  alginates, 264
  antimicrobial agents, 171
  barrier properties, 359
  incorporation in films, 341
lemon oils, 354
lemons
  dehydration and shrinkage, 193
  emulsion formulations, 89
  ethylene production, 246
Lessonia nigrescens, 112
lettuce
  alginates, 264
  coatings, minimally processed foods, 261
  enzyme activity, 247
  ethylene production, 245
  nutritional and flavor changes, 248
  polysaccharide coatings, 123, 212
  respiration, stress response, 246
  surface microbial flora, 251–252
  water loss, 193
Leuconostoc mesenteroides, 251
Licabee, 302
licorice, 301
lightly-processed fruits and vegetables, 4, *see also* Minimally processed fruits and vegetables
lime, 201
limited availability films and coatings
  cucumber pickle brine protein, 58
  grain sorghum protein, 57
  lupin protein, 57
  peanut protein, 54–55
  pea protein, 56
  pistachio protein, 56–57
  rice protein, 55
  winged bean protein, 57–58
lipids
  components, edible coatings, 5
  dehydration and shrinkage, 194
  minimally processed fruits and vegetables, 270–271

properties, 202
response to coatings, 219
whey protein films, 35–36
lipids, waxes, and resins
application to fruit, 90–96
design, 96–98
dip application, 91–92
drip application, 95–96
emulsion formulations, 88–90
foam application, 92
function, 96–98
fundamentals, 79–80
future developments, 98
historical developments, 87
lipids, 80
resin formulations, 90
resins, 84–87
solvent wax, 87–88
spray application, 92–95
technology, fruit coatings, 87–90
waxes, 80, 82, 84
liquid whey, byproduct, 269
Listeria innocua, 341–342
Listeria monocytogenes
antimicrobial agents, 171, 172
disease and physical injury, 201
essential oils, 341
NatureSeal, 307
soybean protein coatings, 29
surface microbial flora, 252
whey protein coatings, 40
zein coatings, 24
Listerine PocketPak Mint Breath Strips, 295
litchi fruit
chitosan coating, 263
coatings, minimally processed foods, 261
discoloration and gloss reduction, 195
polysaccharide coatings, 212
response to coatings, 222–223
locust bean gum
agar, 114
carrageenans, 114
composites and bilayer coatings, 212
fundamentals, 104
seed gums, 117
xanthan gum, 118
low-calorie foods, 105

LPOS, *see* Lactoperoxidate systems (LPOS)
lupin/lupine protein, 57
lychees, 96

M

Macrocystis pyrifera, 112
maintenance, future directions, 98
maize starches, 108
maltodextrins, 361
mamey sapote fruit, 222
mandarins
dehydration and shrinkage, 193
pitting, 199
postprocessing practices, 256
Semperfresh, 311
mangoes
chilling injury, 199
chitosan coating, 263
coatings, minimally processed foods, 261
cultivar selection, 253
discoloration and gloss reduction, 195
ethylene production, 245
film development, 187
films *vs.* emulsion coatings, 259
heat treatments, 257
modified atmosphere packaging, 258
nutritional and flavor changes, 248, 249
polysaccharide coatings, 122, 123, 206, 212
postprocessing practices, 255
Pro-long, 309
response to coatings, 221–223
ripening, 194
Semperfresh, 312
starch and starch derivatives, 268
ultraviolet light, 257
volatile molecules, films, 340
zein coatings, 24
Mantrocel, 294
MAP, *see* Modified atmosphere packaging (MAP)
mashed potato balls, 24
mass transfer
gas-exchange properties, 139
permeance equations, 150–153

matrices
    capsules, 335–337
    degradation, controlled release capsules, 339
    volatile molecules, 340–343
mayonnaises, 117
MC, see Methylcellulose (MC)
MCC, see Microcrystalline cellulose (MCC)
McIntosh apples
    Nutri-Save, 308
    Pro-long, 309
    Semperfresh, 310
mealy texture, 194–195
meatballs, 124
meat films and coatings
    application, 291–297
    edible coatings, 4
meat products
    collagen and gelatin coatings, 48–49
    karaya gum, 117
    polysaccharide coatings, 104, 120, 124–125
    properties, successful application, 6
    seed gums, 117
mechanical properties, comparison, 22–23
melons
    alginates, 264
    antimicrobial agents, 171
    chilling injury, 199
    coatings, minimally processed foods, 261
    decreased nutritional quality, 197
    disease and physical injury, 200
    gellan gum, 265
    nutritional and flavor changes, 248
    processing practices, 254
    respiration, oxygen/carbon dioxde effect, 191
    respiration, stress response, 246
    response to coatings, 220, 223, 225
    surface microbial flora, 252
melting, controlled release, 339
metering pumps, 93, 95
methylcellulose (MC)
    antioxidants and antibrowing agents, 173
    fundamentals, 105, 266
    nonionic cellulose ethers, 106–107

MFC, see Microfibrillated cellulose (MFC)
microbial fermentation gums
    exopolysaccharides, 118–119
    gellan gum, 118–119
    properties, 118–119
    xanthan gum, 118
microbial flora, surface, 251–252
microcrystalline cellulose (MCC), 105–106
microemulsion coatings, 167
microencapsulation
    Acacia gums, 115
    collagen and gelatin coatings, 49
    future trends, 274
    keratin protein films, 54
    raw starches, 109
microfibrillated cellulose (MFC), 106
microstructure changes, 356
migrant/matrix state chemical structure, 365–366
migration, 364–365
milk proteins, 170, see Caseins; Whey protein
milky emulsions, 167
mineral oil
    components, edible coatings, 5
    response to coatings, 224
    uses, edible coatings, 3
minimally processed fruits and vegetables, see also Fresh fruits and vegetables
    alginates, 264
    bilayer coatings, 271–272
    browning, 249–250
    carrageenan, 265
    casein, 268–269
    cellulose and derivatives, 266–267
    chemical treatments, 255–256
    chitosan, 263–264
    coating optimization, 272
    coatings, 258–276
    color changes, 249–250
    composite coatings, 271–272
    enzyme activity, 246–247
    ethylene production, 245–246
    films vs. emulsion coatings, 259, 263
    flavor changes, 247–249
    fundamentals, 244
    future developments, 273–276
    gellan, 265

lipids, 270–271
nutritional changes, 247–249
pectin, 267
physical treatments, 256–258
physiology, 244–252
polysaccharides, 263–268
postprocessing treatments, 255–258
processing practices, 254
proteins, 268–270
raw product quality, 253
respiration increase, 246
sanitation treatments prior processing, 253–254
shelf life, increasing, 252–258
softening, 250–251
soy protein, 269–270
starch and derivatives, 267–268
surface microbial flora, 251–252
temperature, 255
texture, 250–251
whey protein, 268–269
xanthan gum, 265–266
Mint Breath Strips, 295
modern coatings, 375–377
modified atmosphere packaging (MAP), 256–257
modified starches, 109–110
moisture barrier properties
casein films, 42
keys, successful application, 326–327
zein protein films, 16
moisture gradients, 141
morpholine, 5
murta extracts, 48
mushrooms
chitosan coating, 263
coatings, minimally processed foods, 261
polysaccharide coatings, 122
respiration, oxygen/carbon dioxde effect, 191
muskmelons
respiration, stress response, 246
response to coatings, 225
myofibrillar protein-based films and coatings
coatings, 51
formation of, 49–50
functional properties, 50–51
fundamentals, 49

N

nanoemulsion coatings, 167
nanoencapsulation, 274–276
nanoparticles, *see also* Zinc oxide (ZnO) nanoparticles
disease and physical injury, 201
flavor retention, 346
fundamentals, 335
future trends, 274–276
role of additives, 176–177
nanotechnology, 400–401
natural cuticle waxes, 3
NatureSeal, 294, 305–307
Navel oranges, 216
near-infrared Fourier transform Raman spectroscopy, 15
nectarines
discoloration and gloss reduction, 195
response to coatings, 217
nettings, 292
NewGem Foods, 303
NewGem Foods Glaze Sheets, 293, 295
New Zealand, 47, 389
Nisaplin, 364
nisin
antibacterials, 363
temperature, 33
wheat gluten protein films, 33
nitrogen, fresh fruit, 150–153
nonclimacteric fruit, 191
nonionic cellulose ethers, 106–107
nonvolatile molecules, encapsulation
antibacterials, 362–363
diffusion to food surfaces, 364–365
fundamentals, 361
hydroxypropyl methylcellulose, 363
migrant/matrix state chemical structure, 365–366
nutraceuticals, 361–364
plasticizers, 364
polyacetic acid, 364
nori sheets, 303, 340
nozzles, 92–95
nutraceuticals, 361–364
nutrients, 175
Nutrisave, 123

Nutri-Save
    fresh fruits and vegetables, 307–308
    polysaccharide coatings, 212
    response to coatings, 215, 223
nutritional changes, 247–249
nutritional quality, decreased, 197
nuts
    antioxidants and antibrowing
        agents, 173
    coating applications, 298–300
    commercial coatings, 322
    polysaccharide coatings, 122,
        125–126
    properties, successful application, 6
    zein coatings, 24

O

Ohm's laws, 141
oil burner nozzle, 92–93
oiled wrappers, 87
oilseed milk, 44
onion rings, 113
onions
    coatings, minimally processed
        foods, 261
    respiration, stress response, 246
    water loss, 193
OP, see Xxygen permeability (OP)
opacifers, 376
OPP-Coex-film, 225
optimization of coating, 272
oranges, see also Citrus fruit
    dehydration and shrinkage, 193
    discoloration and gloss reduction,
        195, 197
    disease and physical injury, 200
    flavor loss, 198
    Pro-long, 309
    response to coatings, 216–217
    waxes, 312
oregano oil
    antimicrobial agents, 171, 172
    barrier properties, 359
    collagen and gelatin coatings, 49
    incorporation in films, 341, 342
    whey protein films, 39–40
organic acids, 254–255
organic ingredients, 188, 407
Origami Foods, 305, 340

Origami Wraps, 294, 303
Origanum vulgare, 342
overwraps, boneless meats, 292
oxidized polyethelene
    fundamentals, 80, 82
    lipids and resins, 202
oxygen and oxygen barriers
    collagen films, 47
    effect on respiration, 188, 191
    soybean films and coatings, 29
    uses, edible coatings, 2
    whey protein films, 39
oxygen permeability (OP)
    casein films, 43–44
    coatings, fruits and vegetables,
        203–205
    edible *vs.* plastic films, 355
    fundamentals, 14
    gelatin films, 48
    myofibrillar and sarcoplasmatic
        protein films, 50
    peanut protein, 55
    plasticizers, 163
    rice protein, 55
    whey protein films, 35–39
    zein protein films, 19, 20
ozone, 256–257

P

PAA, see Peroxyacetic acid (PAA)
Pacific Rose apples, 193
packaging films, 22–23
Packham's Triumph pears, 214
palmarosa oil, 264
pandan leaf extract, 126, 340
Pandanus amaryllifolius, 126, 340
panning application, 324–325
papayas
    alginates, 264
    antioxidants and antibrowing
        agents, 175
    chilling injury, 199
    coatings, minimally processed
        foods, 261
    ethylene production, 245
    gellan gum, 265
    nutritional and flavor changes, 248
    permeance equations, 150
    polysaccharide coatings, 122

processing practices, 254
response to coatings, 221, 222
paraffin
  casein films, 42
  components, edible coatings, 5
  dip application, 92
  fundamentals, 84
  historical developments, 87
parsnips, 248
pastrami, 125
pastries, 53
peaches
  antimicrobial agents, 171
  discoloration and gloss reduction, 195
  emulsion coatings, 97
  heat treatments, 257
  nutritional and flavor changes, 248
  polysaccharide coatings, 122, 212
  volatile molecules, films, 340
peanut protein, 54–55
peanuts
  applications, films/coating, 299
  polysaccharide coatings, 125–126
  whey protein coatings, 40
pea protein, 56, 167
pears
  alginates, 264
  antimicrobial agents, 171
  antioxidants and antibrowing
    agents, 174
  cellulose and cellulose derivatives, 266
  coatings, minimally processed
    foods, 261
  ethylene production, 245
  extended shelf life, 194
  gellan gum, 265
  Jonfresh, 311
  NatureSeal, 306
  Nutri-Save, 308
  pectin, 267
  polysaccharide coatings, 206
  postprocessing practices, 256
  Pro-long, 309
  respiration, stress response, 246
  response to coatings, 213–215, 221
  ripening, 194
  Semperfresh, 311
  spray application, 93
  superficial scald, 199
  volatile molecules, films, 340

pea starch, 292–293
pectic polysaccharides, 194
pectin and pectin derivatives, *see also*
    specific derivative
  emulsifiers, 167
  fundamentals, 104
  minimally processed fruits and
    vegetables, 267
  properties, 110–111
Pedilanthus aphyllus, 84
Pedilanthus parvonis, 84
Penicillium expansum, 224, 306
Penicillium italicum, 341
pentasaccharides, 118
pepino dulce, 194
peppers, *see also* Bell peppers; Green
    peppers; Sweet peppers
  cultivar selection, 253
  decreased nutritional quality, 197
  dip application, 91
  discoloration and gloss
    reduction, 195
  response to coatings, 223
permeability/permeance
  aroma compounds, 351–353
  casein films, 43–44
  coatings, fruits and vegetables,
    203–205
  determination, coating properties, 7
  lipids and resins, 202
  units for, 141–142
permeability values, 146–147
permeance
  effusion comparison, 147–148
  fundamentals, 137–138
permeance equations
  analytical errors, 144–145
  barrier properties measurement,
    142–145
  coating permeability, 145
  effusion/permeance comparison,
    147–148
  equipment, 144
  fruit coatings, as test, 148, 150
  fundamentals, 139
  homogeneous barriers, 139–140
  layered coatings and films, 140–141
  mass transfer, fresh fruit, 150–153
  methods, 142–143
  moisture gradients, 141

nitrogen, fresh fruit, 150–153
permeability values, 146–147
units for permeability/permeance, 141–142
peroxyacetic acid (PAA), 254
Persea americana, 307
persimmons
   coatings, minimally processed foods, 261
   nutritional and flavor changes, 248
   soy protein, 270
   whey protein, 268
PFSP, *see* Proteinaceious fibrous material (PFSP)
Phaeophyceae class, 264
pharmaceutical coatings and applications
   applications, 377–378
   fundamentals, 373–375
   modern coatings, 375–377
   polymeric film interaction, 378–380
   polysaccharide coatings, 126
   sugar coatings, 375
   suggested readings, 380
pH levels
   Acacia gums, 115
   anionic cellulose ethers, 108
   casein films, 41
   casein films and coatings, 40
   collagen and gelatin proteins, 46
   cottonseed-protein films, 45
   cucumber pickle brine protein, 58
   egg white protein films, 52
   ghatti gum, 116
   lupin protein, 57
   modfied starches, 110
   myofibrillar and sarcoplasmatic protein films, 51
   myofibrillar protein films, 50
   NatureSeal, 307
   Nutri-Save, 307
   peanut protein, 55
   pectins and derivatives, 113
   proteins, 268
   rice protein, 55
   soybean films and coatings, 26, 28
   surface microbial flora, 252
   tragacanth gums, 115
   wheat gluten protein films, 32
   whey protein, 268
   winged bean protein, 58

Photobacterium phosphoreum, 342
physical-chemical properties, effect on, 354, 356–361
physical injury, 199–201
physical treatments, 256–258
physiological disorders, 198–199
pigmented rice brans, 171
pimento olive fillings, 113
pine, essential oil, 342
pineapple oranges, 309
pineapples
   applications, films/coating, 303
   dip application, 92
   drip application, 96
   nutritional and flavor changes, 248
   postprocessing practices, 255, 256
   response to coatings, 221, 222
   ultraviolet light, 257
Pinus sylvestris, 342
pistachio protein, 56–57
pitting, citrus, 199
pizza bases, 53
PLA, *see* Polyacetic acid (PLA)
plantain, 308
plant extracts, 4, *see also* Essential oils
plasticizers
   carrageenans, 114
   casein films, 43
   collagen and gelatin protein films, 47
   components, edible coatings, 5
   cottonseed-protein films, 45
   egg white protein films, 53
   fundamentals, 158–159, 376
   gelatin films, 48
   myofibrillar protein films, 50
   nonvolatile molecules, encapsulation, 364
   peanut protein, 55
   pea protein, 56
   pistachio protein, 57
   polysaccharide films and coatings, 165–167
   protein films and coatings, 159–165
   proteins, 268
   rice protein, 55
   sorghum protein, 57
   soybean films and coatings, 27
   wheat gluten protein films, 31
   whey protein films, 36–37

plums, 307
PocketPak Mint Breath Strips, 295
Polish sausage, 47
polyacetic acid (PLA), 364
polyethelene wax
    flavor loss, 198
    fundamentals, 80
    historical developments, 187
    lipids and resins, 202
polymeric film interaction, 378–380
polymers, 4
polysaccharide coatings
    Acacia gums, 114–115
    agar, 114
    alginates, 111–113
    anionic cellulose ethers, 107–108
    applications, 120–127
    carrageenans, 113–114
    cellulose and derivatives, 105–108
    cereal-based products, 125–126
    chitosan, 119–120
    components, 4–5
    exopolysaccharides, 118–119
    exudate gums, 114–117
    fish, 124–125
    fruits, 120, 122–124
    fundamentals, 104–105
    future developments, 127
    gellan gum, 118–119
    gum ghatti, 116
    gum karaya, 116–117
    gum tragacanth, 115–116
    main applications, 121
    meats, 124–125
    microbial fermentation gums, 118–119
    microcrystalline cellulose, 105–106
    microfibrillated cellulose, 106
    modified starches, 109–110
    nonionic cellulose ethers, 106–107
    nuts, 125–126
    other applications, 126–127
    pectins and derivatives, 110–111
    poultry, 124–125
    properties, 105–120
    raw starches, 108–109
    seafood, 124–125
    seaweed extracts, 111–114
    seed gums, 117
    starches and derivatives, 108–110
    vegetables, 120, 122–124
    xanthan gum, 118
polysaccharide coatings, minimally processed fruits and vegetables
    alginates, 264
    carrageenan, 265
    cellulose and derivatives, 266–267
    chitosan, 263–264
    fundamentals, 263
    gellan, 265
    pectin, 267
    starch and derivatives, 267–268
    xanthan gum, 265–266
polysaccharides
    properties, 206, 212
polyunsaturated fatty acids (PUFA), 362
pork chops, 125, *see also* Sausage products
postharvest deterioration, effect of coatings
    dehydration, 193–194
    discoloration, 195–197
    disease, 199–201
    flavor loss, 197–198
    gloss reduction, 195–197
    hardening, 194–195
    mealy texture, 194–195
    nutritional quality, decreased, 197
    physical injury, 199–201
    physiological disorders, 198–199
    quarantine treatments, 201–202
    shrinkage, 193–194
    softening, 194–195
postharvest physiology
    carbon dioxide, 188, 191
    climacteric fruit, 191
    factors affecting respiration, 191–192
    nonclimacteric fruit, 191
    oxygen, 188, 191
    product type influence, 193
    relative humidity, 192–193
    respiration, 188, 191–192
    temperature, 192–193
    transpiration, 192–193
postprocessing treatments
    chemical treatments, 255–256
    physical treatments, 256–258
    temperature, 255
potatoes
    antioxidants and antibrowing agents, 173
    coatings, minimally processed foods, 262

collagen and gelatin coatings, 49
cultivar selection, 253
discoloration and gloss reduction, 197
enzyme activity, 247
ethylene production, 245
nutritional and flavor changes, 248
Pro-long, 309
respiration, oxygen/carbon dioxde effect, 191
response to coatings, 225
soy protein, 270
whey protein coatings, 40
pouches, 304–305
poultry, see also Chicken
  meat film/coatings, 293
  polysaccharide coatings, 124–125
  zein coatings, 24
pretzels, 297
prickly pear cactus coating, 219
PrimaFresh 50-V, 220
probiotics, future trends, 276
processed foods
  commercial coatings, 321
processing aids, regulatory aspects, 393
processing practices, 254
product type, influence, 193
programmable spray systems, 94–95
Pro-long
  fresh fruits and vegetables, 308–309
  fundamentals, 294
  polysaccharide coatings, 206
  response to coatings, 217, 221–223
properties
  bilayer coatings, 212–213
  cellulose and derivatives, 105–108
  chitosan, 119–120
  composites, 212–213
  determination, 7
  exudate gums, 114–117
  important, edible coatings, 6–7
  lipids, 202
  microbial fermentation gums, 118–119
  pectins and derivatives, 110–111
  polysaccharides, 206, 212
  proteins, 212
  resins, 202
  seaweed extracts, 111–114
  seed gums, 117
  starches and derivatives, 108–110

propylene glycol, 6
protection, flavor compounds, 350
proteinaceious fibrous material (PFSP), 28
protein-based films and coatings
  casein-based, 40–44
  collagen proteins, 46–49
  cottonseed proteins, 44–46
  cucumber pickle brine protein, 58
  egg white proteins, 51–53
  fundamentals, 14–15, 58
  gelatin proteins, 46–49
  grain sorghum protein, 57
  keratin type, 53–54
  limited availability type, 54–58
  lupin protein, 57
  mechanical properties, comparison, 22–23
  myofibrillar proteins, 49–51
  peanut protein, 54–55
  pea protein, 56
  pistachio protein, 56–57
  rice protein, 55
  sarcoplasmic proteins, 49–51
  soy-based, 24–29
  wheat gluten protein type, 29–33
  whey proteins, 34–40
  winged bean protein, 57–58
  zein type, 15–24
proteins
  capsules, 337
  coatings, 2
  properties, 212
proteins, coatings for minimally processed fruits and vegetables
  casein, 268–269
  fundamentals, 268
  soy protein, 269–270
  whey protein, 268–269
Prunus salicina, 307
Pseudomona putida, 29
Pseudomonas sp., 251
Pseudomonas aeruginosa, 306
Pseudomonas elodea, 265
Pseudomonas fluorescens, 342
Pseudomonoas elodea, 118
puddings, 115
PUFA, see Polyunsaturated fatty acids (PUFA)

pullulan
  development of, 187
  strips and pouches, 304
  volatile molecules, films for, 340
pumpkins, 223
puncture-strength, 45–46
Pusa Early Dwarf tomatoes, 311

Q

qualification of substances, 401–402
quarantine treatments, 201–202

R

radiation, 256–257
radishes
  extended shelf life, 194, 195
  nutritional and flavor changes, 248
  processing practices, 254
  respiration, stress response, 246
  response to coatings, 225
raisins
  coating applications, 298–300
  edible coatings, 3
  egg white protein coatings, 53
random-coil structure, 40
raspberries, 124
raw product quality, 253
raw starches, 108–109
ready-to-eat meals, 104, 171
recyclable coated paper, 24
red algae, 303
red cabbage, 248
Red Chief apples, 199
Red Globe grapes, 218
red seaweeds, 113–114, 265
reducing agents, 256
refrigeration, 248–249
regulatory aspects, coatings
  Australia, 389, 406
  Canada, 389, 406
  European Union, 387, 402, 405
  FAO/WHO Codex Alimentarius, 402
  food additives, 393, 401–406
  Food and Agriculture Organization of the United Nations, 385–386
  food and food ingredients, 389–393
  food contaminants, 393, 400
  food safety, 401

  fundamentals, 1, 383–385, 413
  ingredients, 407
  Japan, 388, 406
  nanotechnology, 400–401
  New Zealand, 389, 406
  organic ingredients, 407
  processing aids, 393
  qualification of substances, 401–402
  substance classification variations, 401
  substances, fruit and vegetable coatings, 407
  substances in food/food products, 389–401
  technical function need, 402
  United States, 387–388, 405
  World Health Organization, 385–386
relative humidity, *see also* Humidity
  effect on transpiration, 192–193
  lipids and resins, 202
  moisture and gas barriers, 327
  plasticizers, 161, 163–164
  uses, edible coatings, 2
resins, *see also* Lipids, waxes, and resins
  components, edible coatings, 4–5
  dehydration and shrinkage, 194
  pressure-spray application, 95
  properties, 202
respiration
  carbon dioxide, effect on, 188, 191
  climacteric fruit, 191
  factors affecting, 191–192
  increase, 246
  nonclimacteric fruit, 191
  oxygen, effect on, 188, 191
  uses, edible coatings, 2
responses to coatings
  apples, 213–215
  bananas, 220–221
  berries, 219
  citrus fruit, 215–217
  fruit vegetables, 223–224
  grapes, 218–219
  kiwifruit, 219–220
  leafy vegetables, 225
  mangoes, 221–223
  melons, 220
  pears, 213–215
  root crops, 224–225
  stone fruits, 217–218

strawberries, 219
 tomatoes, 223–224
 tropical fruits, 221–223
 tuber crops, 224–225
restructured food, 113, 293
retention and release, 343–348
rewetting, 91, *see also* Whiting
Rhizopus sp., 171, 218
Rhodophyta sp., 113
rice brans, 171
rice bran wax, 5, 87
rice-gelatin coating, 49
rice protein, 55
role of additives
 antibrowning agents, 173–175
 antimicrobial agents, 170–173
 antioxidants, 173–175
 colorants, 175
 emulsifiers, 167–170
 flavors, 175
 fundamentals, 157–158
 nanoparticles, 176–177
 nutrients, 175
 plasticizers, 158–167
 polysaccharide films and coatings, 165–167
 protein films and coatings, 159–165
root crops
 dip application, 91–92
 paraffin, 84
 response to coatings, 224–225
rosemary oil
 collagen and gelatin coatings, 49
 incorporation in films, 341, 342
 whey protein films, 39–40
Rosmarinus officinalis, 342
rutabagas
 dip application, 91
 lipids and resins, 202
 paraffin, 84
 response to coatings, 224

S

Saccharomyces cerevisiae, 306
sachets, edible, 304
salad dressings, 117
salad mixtures, 246
sales literature, 5–6
Salmonella sp., 252

Salmonella enterica
 antimicrobial agents, 172
 NatureSeal, 306
 whey protein coatings, 40
Salmonella enteritidis, 201, 264
Salmonella gaminara, 29, 171
Salmonella Montevideo, 124, 171
Salmonella typhimurium, 125, 307
salting in/out effect, 347
sanitation treatments, prior processing, 253–254
sarcoplasmic protein-based films and coatings, 49–51
sardines, 49, 342
Sargassum sp., 112
sauces, 53, 117
sausage products, *see also* Pork chops
 casing, historical developments, 2
 collagen and gelatin coatings, 48–49
 polysaccharide coatings, 120
SDS, *see* Sodium dodecyl sulfate (SDS)
seafood, 124–125, *see also* Fish products
seaweed extracts
 agar, 114
 alginates, 111–113
 carrageenans, 113–114
 incorporation in films, 342
 properties, 111–114
seed gums, 117
Semperfresh
 fresh fruits and vegetables, 309–312
 fundamentals, 294
 NatureSeal, 306
 polysaccharide coatings, 123, 206
 response to coatings, 217–218, 220, 223
 superficial scald, 198
sensory attributes, 327–328
Serratia liquefaciens, 125
service, future directions, 98
Shamouti orange fruit, 148, 150
sharon fruits, 33
shelf life, increasing
 chemical treatments, 255–256
 fundamentals, 252
 physical treatments, 256–258
 postprocessing treatments, 255–258
 processing practices, 254
 raw product quality, 253

sanitation treatments prior processing, 253–254
temperature, 255
shellac and shellac coatings
    applications, films/coating, 293
    components, edible coatings, 5
    discoloration and gloss reduction, 197
    fresh fruits and vegetables, 312
    fundamentals, 85, 312
    historical developments, 2
    lipids and resins, 202
    response to coatings, 213
    waxes, 5, 90–91
shell eggs, 24
Shewanella putrefaciens, 342
Shiga toxin, 172
Shigella flexneri, 307
shipping wax, 80, 89, 98
shrinkage, 193–194
shriveling, 6
slab-wax process, 87
snow wax, 92
sodium dodecyl sulfate (SDS), 28, 54
softening
    minimally processed fruits and vegetables, 250–251
    postharvest deterioration, 194–195
solvent wax, *see also* Waxes
    fundamentals, 80
    historical developments, 186
    technology, fruit coatings, 87–88
sorghum protein coating
    collagen and gelatin coatings, 49
    limited availability films and coatings, 57
    response to coatings, 219
soups, 115
soy-based films and coatings
    coatings, 29
    emulsifiers, 167
    formation of, 25
    functional properties, 25–29
    fundamentals, 24
    plasticizers, 164
    water vapor permeability, 360
    zein protein films, 19
soy protein, 269–270
soy protein concentrate, *see* Soy-based films and coatings
soy protein isolate, *see* Soy-based films and coatings
Spartan apples, 308
Specifications and Standards, Japan, 406
Sphingomonas elodea, 265
Sphingomonas paucimobilis, 118–119
spinach, 193
spinners, drip applicators, 95
spiral-cut brushes, 93
spray application
    historical developments, 87
    lipids, waxes, and resins, 92–95
spray-cooling/chilling method, 338
spray-drying method
    capsules, 337
    uses, edible coatings, 4
spraying application, 325
spray nozzles, 94
sprinkles, 301
squash
    cultivar selection, 253
    essential oils, 341–342
    ethylene production, 245
    mineral oil, 79
Sta-Fresh MP, 223
Staphylococcus aureus, 172
starch and starch derivatives, *see also specific derivative*
    emulsifiers, 167
    minimally processed fruits and vegetables, 267–268
    modified starches, 109–110
    properties, 108–110
    raw starches, 108–109
Starking Delicious apples, 310
stationary spray nozzles, 94
steam pockets, 84
Stella cherries, 308
Sterculia genus, 116
stone fruits
    paraffin-based emulsions, 97
    response to coatings, 217–218
storage wax, 80, 89
straight-cut brushes, 93
strawberries
    antimicrobial agents, 171
    antioxidants and antibrowing agents, 174
    chitosan coating, 263

coatings, minimally processed
  foods, 262
 ethylene production, 245
 nutrients, 175
 nutritional and flavor changes, 248
 polysaccharide coatings, 124, 206, 212
 respiration, oxygen/carbon dioxde
  effect, 191
 respiration, stress response, 246
 response to coatings, 219
Streptococcus faecalis, 171
strips, 303–305
structural properties, effect on, 354,
  356–361
substance classification variations, 401
substances in food/food products, 389–401
sucrose-plasticized whey protein films, 37
sucrose stearate, 170
sugarcane, 84
sugar coatings
 applications, films/coating, 297–298, 302
 pharmaceutical coatings, 373, 375
sulfur-containing proteins, 27
summer sausage, 40
summer squash, 223
superficial scald, 198–199
surface characteristics, 327
surface microbial flora, 251–252
surface wettability, 327
surfactants, 5
surimi, 49–50
sweating, 91, 160
sweet corn, cooked, 24
sweet peppers, 223, *see also* Peppers
sweet potatoes
 discoloration and gloss reduction, 197
 enzyme activity, 247
 nutritional and flavor changes, 248
sweets, 171
swelling, controlled release, 339
swing arm applicators, 94

T

TAL Pro-long, *see* Pro-long
tamarind seed polysaccharide, 117
tangential spray system, 378
tangerines
 flavor loss, 198
 polyethelene wax, 187

response to coatings, 215–216
 waxes, 312
tara gum, 117, 118
taste aroma quality, 249
teas, 171
tea tree oil, 341, 356
Technical Association of the Pulp and
  Paper Industry (TAPPI), 143
technical function need, 402
technology, fruit coatings
 emulsion formulations, 88–90
 historical developments, 87
 resin formulations, 90
 solvent wax, 87–88
temperatures
 agar, 114
 casein films, 42
 chilling injury, 199
 cottonseed-protein films, 45
 cucumber pickle brine protein, 58
 effect on transpiration, 192–193
 ethylene production, 246
 flavor retention, 345–346
 gelatin films, 48
 modfied starches, 110
 moisture and gas barriers, 327
 peanut protein, 55
 pea protein, 56
 plasticizers, 160, 164
 postprocessing treatments, 255
 respiration, 191
 response to coatings, 216, 222
 soybean films and coatings, 26, 28
 superficial scald, 199
 surface microbial flora, 251
 wheat gluten protein films, 31–33
terpene citral, 171
tetracycline, 172
texture, 250–251
TG, *see* Transglutaminase (TG)
thermoplastic processing, plasticizers, 165
Thompson Seedless grapes, 218
thyme oil, 171, 342
Thymus vulgaris, 342
tilapia fish, 49, 51
tissue adhesion, 49
tomatoes
 antimicrobial agents, 171, 172
 chilling injury, 199
 dip application, 91

discoloration and gloss reduction, 195
drip application, 95
ethylene production, 245
firmness, loss of, 194
lipids and resins, 202
mineral oil, 79
nutritional and flavor changes, 248
polysaccharide coatings, 123, 124, 206, 212
respiration, stress response, 246
response to coatings, 223–224, 225
Semperfresh, 311
volatile molecules, films, 340
water loss, 193
wheat gluten protein coatings, 33
tomato films, 296
toppings, 53
top-spray system, 377
tortuosity, 345
tragacanth gums
  polysaccharide coatings, 104
  properties, 115–116
transglutaminase (TG)
  egg white protein films, 53
  gelatin films, 48
  soybean films and coatings, 27
  whey protein films, 38–39
translucency, 251
transparency of film, 356
transpiration
  fundamentals, 192
  product type influence, 193
  relative humidity, effect on, 192–193
  temperature, effect on, 192–193
traveling nozzle system, 93
Tree of Life, 80
tropical fruits, 221–223
tuber crops, 224–225
tumble-trim brushes, 93
tuna-fish gelatin, 48
turmeric starch, 187, 213
turnips
  paraffin, 84
  response to coatings, 224, 225

U

ultraviolet (UV) light and irradiation
  casein films, 41, 43
  egg white protein films, 52–53

postprocessing practices, 256–257
soybean films and coatings, 26–27
whey protein films, 39
United States, 387–388, 405, *see also* Code of Federal Regulations (CFR)
units for permeability/permeance, 141–142
uses, edible coatings, 2–4

V

Valencia oranges
  nitrogen and mass transfer, 150–151
  response to coatings, 217
  waxes, 312
Van cherries, 308
vancomycin, 172
vanillin, 306, 341
vegetable oils, 5
vegetables, 120, 122–124, *see also specific type;* Fresh fruits and vegetables, coatings for
Verbena officinalis, 342
Vibrio cholerae, 307
volatile molecules, films, *see also* Essential oils
  appearance of film, 356
  barrier performances, 348–350
  barrier properties, 358–361
  encapsulation influence, 356–361
  flavor compounds, 358–361
  flavor compounds involved, 340–343
  fundamentals, 340
  matrices involved, 340–343
  microstructure changes, 356
  physical-chemical properties, effect on, 354, 356–361
  protection, flavor compounds, 350
  retention and release, 343–348
  structural properties, effect on, 354, 356–361
  transparency of film, 356

W

washing, 254
water barrier
  gelatin films, 47
  soybean films and coatings, 28

water chestnuts
  chitosan coating, 263
  coatings, minimally processed foods, 262
  polysaccharide coatings, 122
watermelon
  nutritional and flavor changes, 248
  respiration, oxygen/carbon dioxde effect, 191
water-to-wax method, 89
water vapor, 163
water vapor permeability (WVP)
  barrier properties, 360
  casein films, 41–42
  cellulose and cellulose derivatives, 266
  coatings, fruits and vegetables, 203–206
  edible coatings, 259
  egg white protein films, 53
  fundamentals, 14, 17–18
  myofibrillar and sarcoplasmatic protein films, 51
  peanut protein, 55
  pea protein, 56
  pistachio protein, 57
  polysaccharide coatings, 206
  rice protein, 55
  sorghum protein, 57
  soybean protein, 27–28
  wheat gluten protein films, 32–33
  whey protein films, 35–39
  zein protein films, 16
water vapor transfer rate, 161
water wax, 80, 89
waxes, *see also* Lipids, waxes, and resins
  components, edible coatings, 4–5
  dehydration and shrinkage, 194
  fresh fruits and vegetables, 312
  historical developments, 87
  uses, edible coatings, 3
weight loss control, 97
wettability, surface, 327
wheat gluten protein-based films and coatings
  applications, films/coating, 299
  coatings, 33
  formation of, 30
  functional properties, 30–33
  fundamentals, 29–30
  response to coatings, 219
  zein protein films, 19
whey protein
  applications, films/coating, 299
  emulsifiers, 170
  minimally processed fruits and vegetables, 268–269
whey protein films/coatings, 37
  antioxidants and antibrowing agents, 173
  coatings, 40
  emulsifiers, 168
  formation of, 34–35
  functional properties, 35–40
  fundamentals, 34
  plasticizers, 162–164
white cabbage, 248
white dextrins, 109
whitening/whiting
  discoloration and gloss reduction, 197
  lipids and resins, 202
  response to coatings, 213
  sweating effect, 91
white wax, 84
WHO, *see* World Health Organization (WHO)
wig-wag nozzles, 94
winged bean protein, 57–58
winter squash, 245
wood rosins, 85, 186
wool, 53
World Health Organization (WHO)
  hydroxypropyl methylcellulose, 363
  polysaccharide coatings, 104
  regulatory aspects, coatings, 385–386
  Semperfresh, 310
wound dressings, 127
wraps, 1, *see also* Coatings; Films
Wurster's coater, 377
WVP, *see* Water vapor permeability (WVP)

X

xanthan gum
  antioxidants and antibrowing agents, 173
  fundamentals, 104
  minimally processed fruits and vegetables, 265–266
  properties, 118

seed gums, 117
tragacanth gums, 116
Xanthomonas genus, 118
Xanthomonas campestris, 118, 265

Y

yam starch, 267
yellow dextrins, 109
yuba films, 296
yucca
    paraffin, 84
    response to coatings, 224

Z

zein-based films/coatings
    applications, films/coating, 299, 302, 305
    coatings, 24
    discoloration and gloss reduction, 197
    film formation, 15–16
    functional properties, 16, 24
    fundamentals, 15
    Gala apples, effects on, 6
    plasticizers, 159–161
    response to coatings, 224
zinc oxide (ZnO) nanoparticles, *see also* Nanoparticles
    disease and physical injury, 201
    Escherichia coli, 187
    future trends, 275
zucchini
    casein protein coatings, 44
    nutritional and flavor changes, 248
    respiration, stress response, 246
    Semperfresh, 311